Practice Masters Levels A, B, and C

HOLT, RINEHART AND WINSTON

A Harcourt Classroom Education Company

Austin • New York • Orlando • Atlanta • San Francisco • Boston • Dallas • Toronto • London

To the Student

Practice Masters Levels A, B, and C consist of three levels of exercises graded by level of difficulty. There is a full page for each of the three levels of exercises for each lesson in the *Pupil's Edition.* Level A exercises are the least difficult, designed for students to practice the lesson objectives through the use of examples. Level B exercises are middle-range exercises that students can handle with the use of current examples together with some prior knowledge. Level C exercises are the most challenging exercises, relevant to the lesson and appropriate for the student who has mastered the lesson.

Printed in the United States of America

ISBN 0-03-064829-7

1 2 3 4 5 6 7 066 05 04 03 02 01 00

Table of Contents

NAME ______________________ CLASS __________ DATE __________

Practice Masters Level A

1.1 Using Differences to Identify Patterns

Find the next three terms of each sequence by using constant differences.

1. 4, 9, 14, 19, 24, ...

   Constant difference: ______

   Next 3 terms: ______________________

2. 17, 35, 53, 71, 89, ...

   Constant difference: ______

   Next 3 terms: ______________________

3. 9, 28, 47, 66, 85, ...

   Constant difference: ______

   Next 3 terms: ______________________

4. 27, 59, 91, 123, 155, ...

   Constant difference: ______

   Next 3 terms: ______________________

5. 0.5, 1.1, 1.7, 2.3, 2.9, ...

   Constant difference: ______

   Next 3 terms: ______________________

6. 8.9, 9.8, 10.7, 11.6, 12.5, ...

   Constant difference: ______

   Next 3 terms: ______________________

7. 79, 76, 73, 70, 67, ...

   Constant difference: ______

   Next 3 terms: ______________________

8. 97, 89, 81, 73, 65, ...

   Constant difference: ______

   Next 3 terms: ______________________

9. 114, 98, 82, 66, 50, ...

   Constant difference: ______

   Next 3 terms: ______________________

10. 161, 140, 119, 98, 77, ...

   Constant difference: ______

   Next 3 terms: ______________________

11. 6.5, 5.8, 5.1, 4.4, 3.7, ...

   Constant difference: ______

   Next 3 terms: ______________________

12. 17.0, 14.7, 12.4, 10.1, 7.8, ...

   Constant difference: ______

   Next 3 terms: ______________________

Find the next three terms of each sequence.

13. 5, 6, 8, 11, 15, ... ______________________
14. 11, 13, 16, 20, 25, ... ______________________
15. 14, 22, 31, 41, 52, ... ______________________
16. 11, 21, 41, 71, 111, ... ______________________
17. 56, 55, 53, 50, 46, ... ______________________
18. 87, 85, 81, 75, 67, ... ______________________
19. 100, 95, 88, 79, 68, 55, ... ______________________
20. 302, 292, 272, 242, 202, ... ______________________
21. 8, 18, 27, 19, 26, ... ______________________
22. 14, 29, 33, 46, 58, 69, ... ______________________
23. 5, 25, 43, 59, 73, ... ______________________
24. 2, 102, 192, 272, 342, ... ______________________
25. 92, 84, 77, 71, 66, ... ______________________
26. 900, 790, 690, 600, 520, ... ______________________

NAME ______________________ CLASS __________ DATE __________

Practice Masters Level B

1.1 Using Differences to Identify Patterns

Find the next three terms of each sequence.

1. 3, 5, 12, 24, 41, ... ______________

2. 200, 199, 193, 182, 166, ... ______________

3. 1, 4, 2, 5, 3, 6, ... ______________

4. 40, 38, 43, 41, 46, 44, ... ______________

Use constant differences to fill in the missing values of each table.

5.

Term Number	1	2	3	4	5	6	7	8
Term	9	10	12	15		24		

6.

Term Number	1	2	3	4	5	6	7	8
Term	54	46	39		28	24		

7.

Term Number	1	2	3	4	5	6	7	8
Term		7	17	32	55			

8.

Term Number	1	2	3	4	5	6	7	8
Term	8	9	6	7			2	

9. If the first differences of a sequence are a constant 3 and the second term is 8, find the first 5 terms of the sequence. ______________

10. If the first differences of a sequence are a constant −7 and the third term is 15, find the first 5 terms of the sequence. ______________

11. The second and third terms of a sequence are 79 and 92. If the first differences are constant, then find the first 5 terms of the sequence. ______________

12. The third and fourth terms of a sequence are 101 and 72. If the first differences are constant, then find the first 5 terms of the sequence. ______________

13. If the second differences of a sequence are a constant 2, the first of the first differences is 8, and the first term is 11, find the first 5 terms of the sequence. ______________

14. If the second differences of a sequence are a constant 4, the first of the first differences is 7, and the second term is 40, find the first 5 terms of the sequence. ______________

NAME ______________________ CLASS ______________ DATE ______________

Practice Masters Level C

1.1 Using Differences to Identify Patterns

Use constant differences to fill in the missing values of each table.

1.

Term Number	1	2	3	4	5	6	7	8
Term		74	71		75	82		86

2.

Term Number	1	2	3	4	6	8	10	12
Term	5	7	10		25			

3.

Term Number	1	2	3	4	10	15	20	25
Term	18	30		54				306

4.

Term Number	1	2	3	4	10	15	20	100
Term	802	794	786	778				

5. If the second differences of a sequence are a constant 5, the first of the first differences is 8, and the third term is 56, find the first 5 terms of the sequence. ______________

6. If the second differences of a sequence are a constant -2, the first of the first differences is 4, and the second term is 60, find the first 5 terms of the sequence. ______________

7. If the second differences of a sequence are a constant -3, the first of the first differences is -6, and the third term is 127, find the first 5 terms of the sequence. ______________

Find each sum. Think of a geometric dot pattern, but do not draw a sketch.

8. $1 + 2 + ... + 50$ ______________

9. $1 + 2 + ... + 100$ ______________

10. Suppose that there are 12 computers in a small office. If each computer is to be linked to each of the other computers by exactly one cable, how many cables are necessary? ______________

11. If there are 9 people in a room, how many handshakes will be exchanged if each person shakes hands exactly once with every other person? ______________

12. There are 16 people playing a two-person game. If each player wants to play every other player exactly once, then how many games must be played? ______________

Practice Masters Level A

1.2 Variables, Expressions, and Equations

Given the values of the variable, complete each table to find the values of each expression.

1.

x	1	2	3	4	5	6
$7x$						

2.

z	1	2	3	4	5	6
$8z$						

3.

r	1	2	3	4	5	6
$r + 9$						

4.

q	5	6	7	8	9	10
$q - 3$						

5.

x	1	2	3	4	5	6
$4x + 3$						

6.

p	1	2	3	4	5	6
$5p - 2$						

7.

w	1	2	4	8	10	12
$3w + 6$						

8.

g	1	2	4	8	16	32
$9g - 9$						

Use guess-and-check to solve the following equations:

9. $q + 8 = 14$ ______

10. $q - 8 = 14$ ______

11. $x + 7 = 42$ ______

12. $x - 13 = 42$ ______

13. $9k = 72$ ______

14. $5j = 105$ ______

15. $4k = 108$ ______

16. $12j = 84$ ______

For Exercises 17–20, write an equation and solve by guess-and-check.

17. Notebooks cost $2 each. How many notebooks can you buy with $28? ______

18. If tickets for a concert cost $13 each, how many can you buy with $91? ______

19. If candles cost $3 each, how many can you buy with $48? ______

20. How many $12 toolkits can be purchased for $108? ______

NAME ______________________ CLASS ____________ DATE ____________

Practice Masters Level B

1.2 Variables, Expressions, and Equations

Given the values of the variable, complete each table to find the values of each expression.

1.

m	2	5	9	18	81	142
$2m + 1$						

2.

n	3	8	14	30	100	201
$13n - 3$						

If small binders cost \$3 and large binders cost \$5, find the cost of each of the following:

3. 4 small binders ____________
4. 5 large binders ____________
5. 2 small binders and 1 large binder ____________
6. 3 small binders and 6 large binders ____________
7. s small binders ____________
8. s small binders and l large binders ____________

Use guess-and-check to solve the following equations:

9. $3x + 2 = 17$ ____________
10. $8y + 3 = 43$ ____________
11. $4v + 2 = 10$ ____________
12. $6b + 5 = 59$ ____________
13. $7h + 15 = 57$ ____________
14. $11r + 18 = 51$ ____________
15. $4f + 7 = 51$ ____________
16. $10h + 4 = 144$ ____________
17. $9r - 2 = 70$ ____________
18. $8r - 7 = 25$ ____________
19. $5x - 2 = 33$ ____________
20. $9u - 9 = 72$ ____________
21. $11w - 18 = 59$ ____________
22. $12q - 15 = 45$ ____________
23. $13d + 20 = 163$ ____________
24. $11y + 22 = 154$ ____________

For Exercises 25–29, write an equation and solve by guess-and-check.

25. If keys cost \$0.95 to make, how many keys can be made for \$4.75? ____________
26. If keys cost \$2 to make, how many keys can be for \$27? ____________
27. A video costs \$6. How many videos can be purchased with \$133? ____________
28. If earplugs cost \$1.10 each, how many can be bought for \$10? ____________
29. A large candy bar is sold at a fundraiser for \$1.50.
How many of these bars must be sold in order to raise \$70? ____________

NAME ______________________ CLASS ____________ DATE ____________

Practice Masters Level C

1.2 Variables, Expressions, and Equations

If widgets cost $0.42 each and globs cost $0.51 each, find the cost of each:

1. 2 widgets and 1 glob ____________
2. 3 widgets and 4 globs ____________
3. 18 widgets and 8 globs ____________
4. 29 widgets and 30 globs ____________
5. m widgets ____________
6. n globs ____________
7. 3 widgets and n globs ____________
8. w widgets and g globs ____________

Write an equation that describes each linear pattern.

9.

x	1	2	3	4	5	6
y	4	8	12	16	20	24

____

10.

x	1	2	3	4	5	6
y	11	22	33	44	55	66

____

11.

x	1	2	3	4	5	6
y	8	9	10	11	12	13

____

12.

x	1	2	3	4	5	6
y	3	5	7	9	11	13

____

13.

x	1	2	3	4	5	8
y	2	7	12	17	22	37

____

Find the 20th term of each sequence.

14. 6, 8, 10, 12, 14, ... ____________
15. 9, 14, 19, 24, 29, ... ____________
16. 22, 29, 36, 43, 50, ... ____________
17. 116, 256, 396, 536, 676, ... ____________
18. 98, 97, 96, 95, 94, ... ____________
19. 104, 101, 98, 95, 92, ... ____________
20. 1000, 960, 920, 880, 840, ... ____________
21. 800, 761, 722, 683, 644, ... ____________

NAME ______________________ CLASS ______________ DATE ______________

Practice Masters Level A

1.3 The Algebraic Order of Operations

Evaluate each expression.

1. $7(12) + 2$ ______________
2. $7 + 12 \cdot 2$ ______________
3. $(7 + 12) \cdot 2$ ______________
4. $12 - 8 \div 4$ ______________
5. $36 \div 4 - 2$ ______________
6. $3 \cdot 19 + 5 \cdot 7$ ______________
7. $3(19 + 5) \div 6$ ______________
8. $7 \cdot 4 - 10 \div 2$ ______________
9. $(7 \cdot 4 - 10) \div 2$ ______________
10. $8 \div 4 \cdot 2$ ______________
11. $4 \cdot 9^2$ ______________
12. $(4 \cdot 9)^2$ ______________
13. $6 + 5^2 - 5$ ______________
14. $(6 + 5)^2 - 5$ ______________
15. $78 \div 13 + 13 - 1^2$ ______________
16. $78 \div (13 + 13) - 1^2$ ______________

Place inclusion symbols to make each equation true.

17. $5 + 6 \cdot 7 = 77$ ______________
18. $3 \cdot 11 - 1 = 30$ ______________
19. $14 \div 4 + 3 - 1 = 1$ ______________
20. $3 + 6 - 5 \cdot 7 = 10$ ______________
21. $3 + 5 \cdot 6 \div 2 = 24$ ______________

Given $a = 4$, $b = 6$, *and* $c = 5$, evaluate each expression.

22. $a + b - c$ ______________
23. $a \cdot b - c$ ______________
24. $a \cdot (b - c)$ ______________
25. $a \cdot b + a \cdot c$ ______________
26. $a \cdot (b + a) \cdot c$ ______________
27. $a \cdot b - a \cdot c$ ______________
28. $a \cdot (b - a) \cdot c$ ______________
29. $(a + a + a) \div b$ ______________
30. $a \div (b - c)$ ______________
31. $b^2 - a$ ______________
32. $a \cdot c^2$ ______________
33. $(a \cdot c)^2$ ______________

NAME ______________________ CLASS __________ DATE __________

Practice Masters Level B

1.3 The Algebraic Order of Operations

Evaluate each expression.

1. $50 \div 5^2 + 25$ ______
2. $(50 \div 5)^2 + 25$ ______
3. $40 \div 4 \cdot 2 - 3$ ______
4. $6(4 + 5 - 7) + 10$ ______
5. $3 \cdot 5^2 - 15 \cdot 2$ ______
6. $2 \cdot 3 + [16 - 4 \div (2 - 1)]$ ______
7. $7 \cdot 8 - [20 - 10 \div (2 + 3)]$ ______
8. $2 \cdot 4^2 + [(4 + 2 \cdot 3) - 3]$ ______
9. $(7 - 4) \cdot (4 + 2) \div (2 + 8 - 1)$ ______
10. $(6 - 3)^2 \cdot 2 \div (6^2 \div 2)$ ______

Place inclusion symbols to make each equation true.

11. $3 + 6^2 \div 2 = 21$ ______
12. $3 + 6^2 - 1 = 80$ ______
13. $2 + 3 \cdot 4 - 4 \cdot 4 = 64$ ______
14. $4 + 6 \cdot 5 - 3^2 = 40$ ______

Given $r = 2$, $s = 7$, and $t = 9$, evaluate each expression.

15. $2s^2 - t$ ______
16. $3t^2 - s - r$ ______
17. $t^2 - (s - r)$ ______
18. $t^2 - 2r^2 + s$ ______
19. $5(t - s)^2 + r$ ______
20. $3(s + t) \div r^2$ ______

Evaluate each expression. Round to the nearest thousandth if necessary.

21. $\dfrac{52 \cdot 51}{20 - 3}$ ______
22. $\dfrac{9 \cdot 17}{89 - 86}$ ______
23. $\dfrac{8 \cdot 22}{96 - 92} + 12$ ______
24. $\dfrac{82 - 37}{82 - 73} + \dfrac{37 + 47}{37 - 25}$ ______
25. $\dfrac{30 + 80}{15 + 17}$ ______
26. $\dfrac{41 + 76}{31 \cdot 3}$ ______
27. $\dfrac{17 \cdot 43}{12 \cdot 14}$ ______
28. $\dfrac{35 \cdot 28}{16 - 7} + 2$ ______

NAME ______________________ CLASS __________ DATE __________

Practice Masters Level C

1.3 The Algebraic Order of Operations

Evaluate each expression.

1. $3\{4[5(2)6]\} + 3$ ______________
2. $7\{4[6(3) - 5]\} - 2$ ______________
3. $3\{3[3 + 3(3)]\} + 3^2$ ______________
4. $18 + 5\{6[7 - (3 - 1)] + 3\}$ ______________
5. $1 + [2 \cdot (3 + 4)]^2$ ______________
6. $(14 - [(9 + 16) \div 5^2]\} - 13$ ______________

Given $m = 8$, $n = 10$, and $p = 6.5$, evaluate each expression.

7. $\dfrac{n + m}{n - m}$ ______________
8. $\dfrac{n + 2m}{4p}$ ______________
9. $5m + 6n^2 - 2p$ ______________
10. $5(m + 6n) - p$ ______________
11. $[m + n \cdot (p + 1)] \cdot m$ ______________
12. $n + \{[n + m(n - m)] \cdot m\}$ ______________

Evaluate each expression. Round to the nearest thousandth if necessary.

13. $\dfrac{19 + 6^2}{73 - 68}$ ______________
14. $\dfrac{(19 + 6)^2}{73 - 68}$ ______________
15. $\dfrac{18 \cdot 27}{2^2 + 2} + (14 + 13)^2$ ______________
16. $\dfrac{37 \cdot 29}{41 \cdot 83} + (14 + 3)19$ ______________
17. $\dfrac{18 \cdot 12}{17 \cdot 13} + \dfrac{4 \cdot 21}{2 \cdot 19}$ ______________
18. $\dfrac{12 \cdot 16}{51 - 27} - \dfrac{31 \cdot 53}{40 \cdot 41}$ ______________

19. The area of a trapezoid, A, can be found using the formula, $A=0.5(b_1+b_2)h$, where b_1 and b_2 are the lengths of the bases and h is the height. Find A.

    a. $b_1 = 3$ in., $b_2 = 5$ in., and $h = 7$ in. ______________

    b. $b_1 = 5$ in., $b_2 = 9$ in., and $h = 2.5$ in. ______________

    c. $b_1 = 3.5$ in., $b_2 = 6.5$ in., and $h = 12$ in. ______________

20. Place inclusion symbols to make the following equation true:
    $4 \cdot 3 - 1 \div 5 \cdot 6 - 4 \div 5 = 4$ ______________

NAME ______________________ CLASS ____________ DATE ____________

Practice Masters Level A

1.4 Graphing With Coordinates

Determine the quadrant and coordinates of each point.

1. A ____________ 2. B ____________

3. C ____________ 4. D ____________

5. E ____________ 6. F ____________

7. G ____________ 8. H ____________

9. P ____________ 10. Q ____________

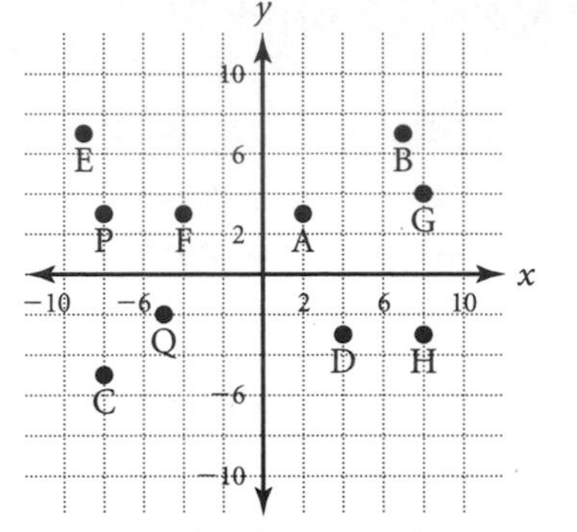

Graph each list of ordered pairs. State whether they lie on a straight line.

11. $(-2, -4), (-1, -1), (0, 2)$ 12. $(-3, 3), (-1, 1), (1, -1)$ 13. $(-3, -2), (0, 0), (4, 4)$

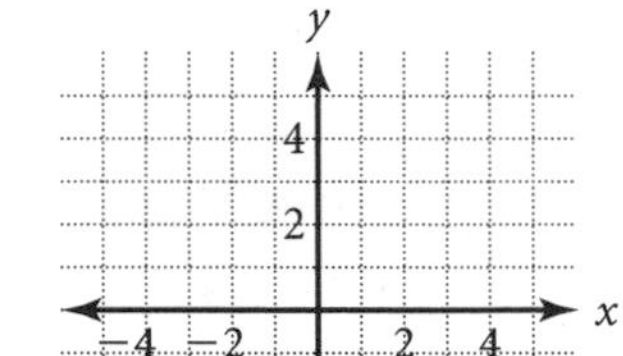

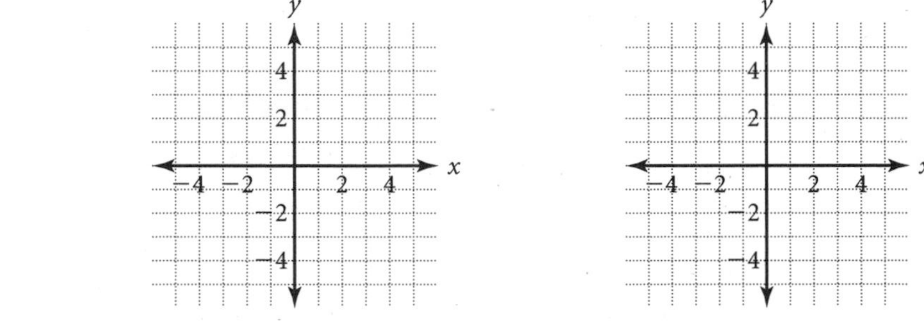

____________ ____________ ____________

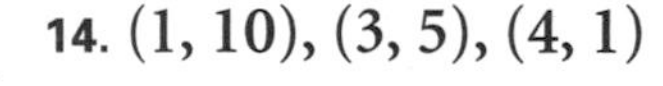

14. $(1, 10), (3, 5), (4, 1)$ 15. $(-6, -8), (-3, -4), (0, 0)$ 16. $(-10, 10), (-8, 5), (-6, 0)$

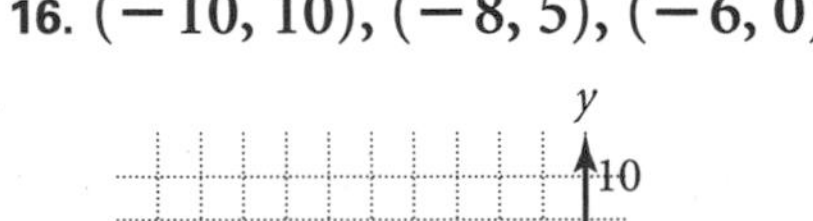

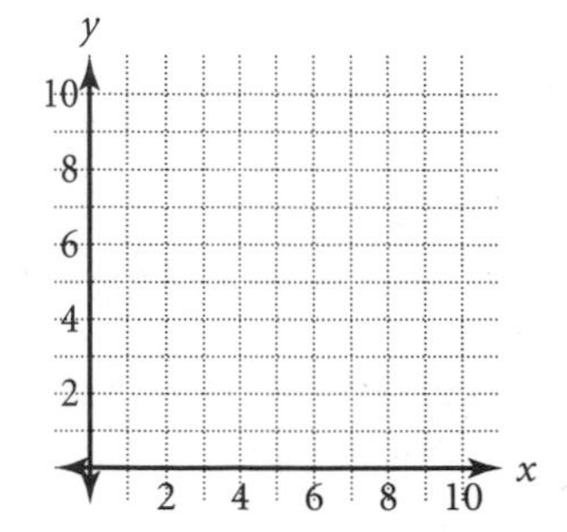

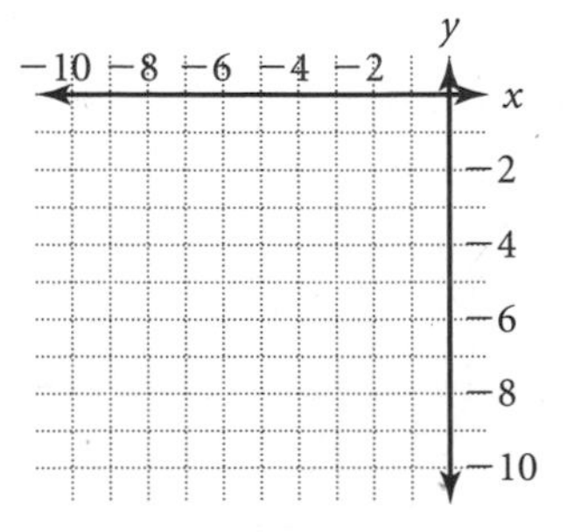

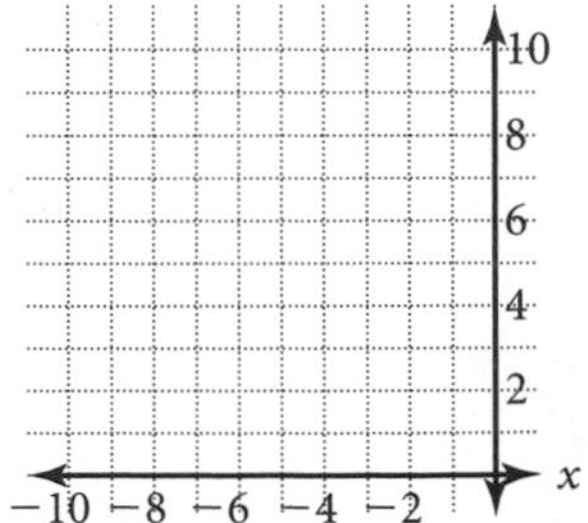

____________ ____________ ____________

Practice Masters Level B

1.4 Graphing With Coordinates

Make a table for each equation, and find the values for *y* by substituting 1, 2, 3, 4, and 5 for *x*. Graph the ordered pairs, and connect the points to make a line.

1. $y = x$

x	1	2	3	4	5
y					

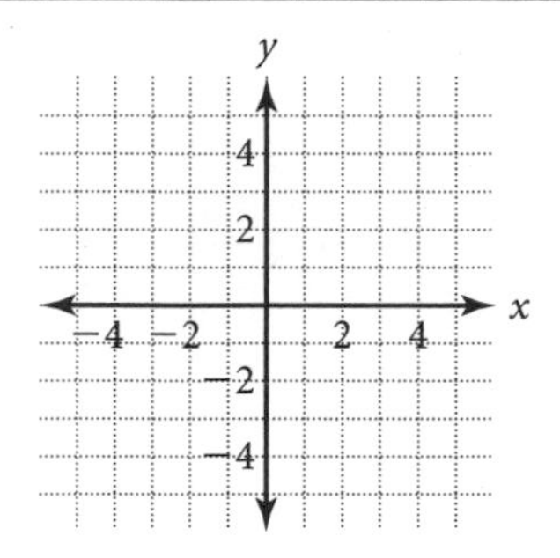

2. $y = x + 2$

x	1	2	3	4	5
y					

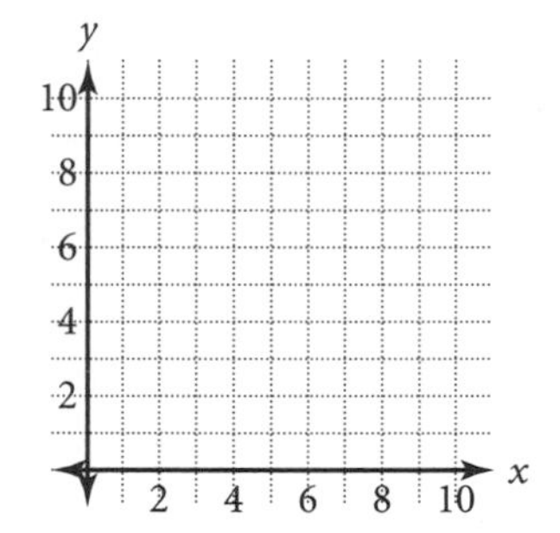

3. $y = x + 4$

x	1	2	3	4	5
y					

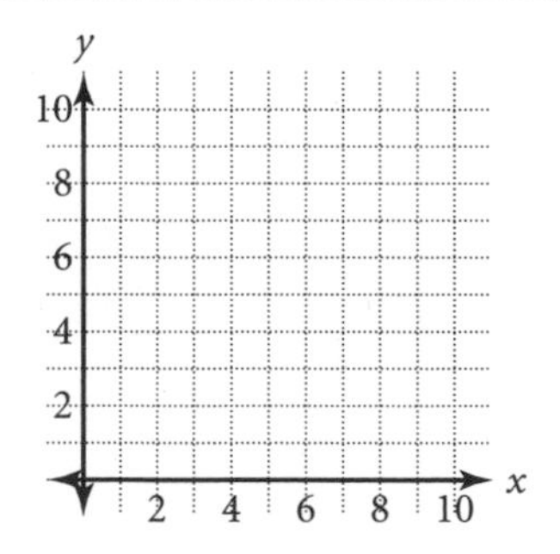

4. $y = x - 1$

x	1	2	3	4	5
y					

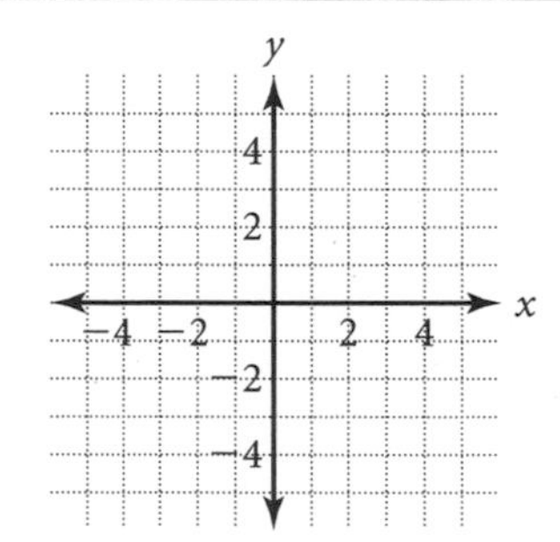

5. $y = 2x - 1$

x	1	2	3	4	5
y					

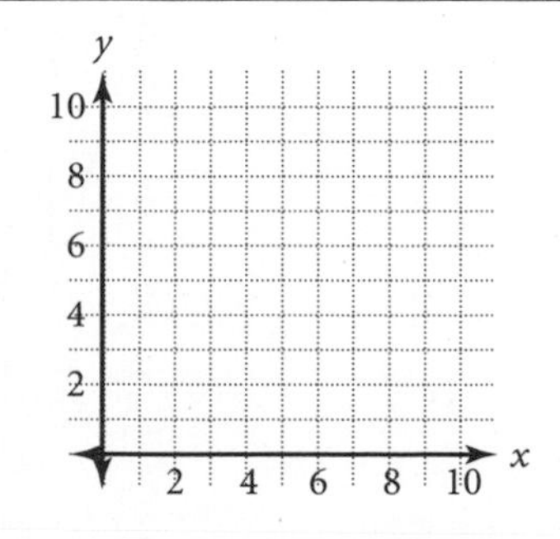

6. $y = -x$

x	1	2	3	4	5
y					

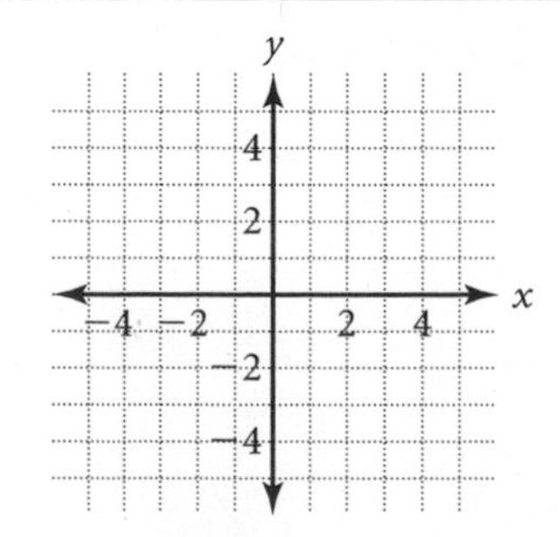

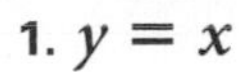

NAME ______________________ CLASS ____________ DATE ____________

Practice Masters Level C

1.4 Graphing With Coordinates

Make a table for each equation, and find the values for y by substituting the given values for x. Graph the ordered pairs, and connect the points to make a line.

1. $y = -(2x)$

x	1	2	3	4	5
y					

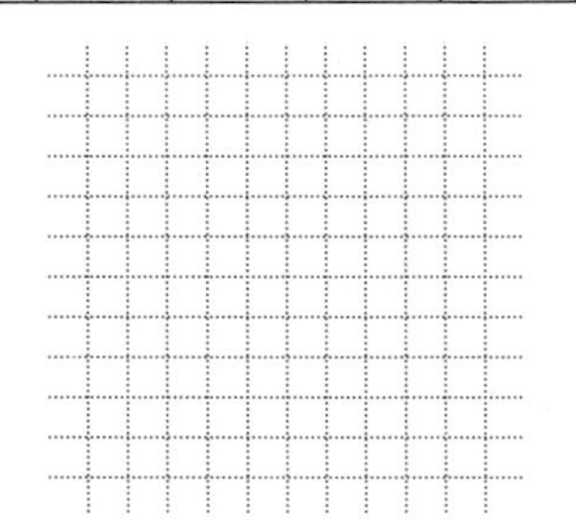

2. $y = -(x + 1)$

x	1	2	3	4	5
y					

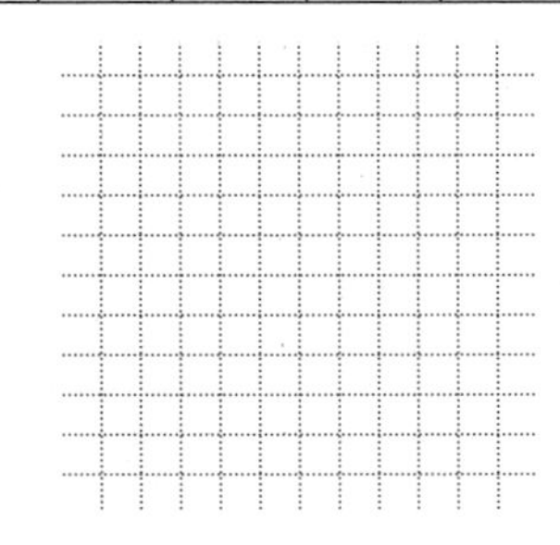

3. $y = \frac{1}{2}x + 5$

x	0	2	4	6	8
y					

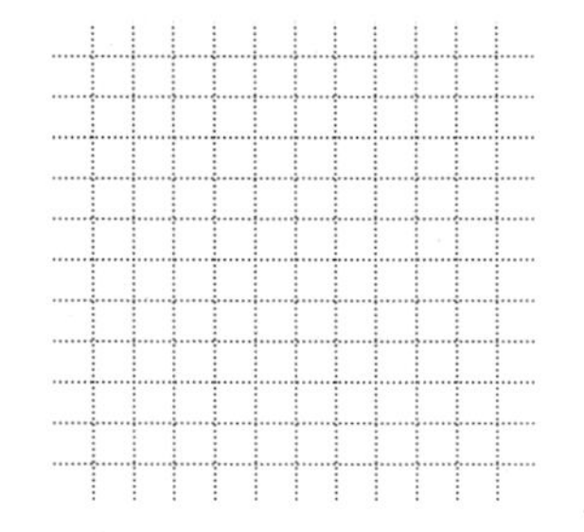

4. $y = 2.5x - 2.5$

x	1	3	5	7	9
y					

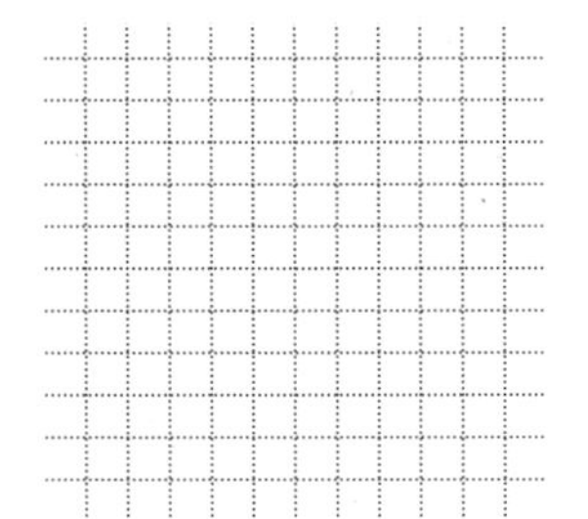

5. Suppose that you are planning to place a single order with an internet bookstore that charges shipping fees of $1 per book and $3 per order. Make a set of ordered pairs from the given information, and plot them as points on a graph. Do the points lie on a straight line?

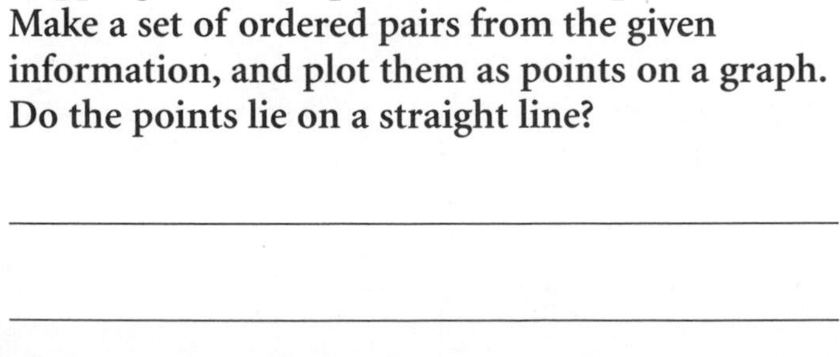

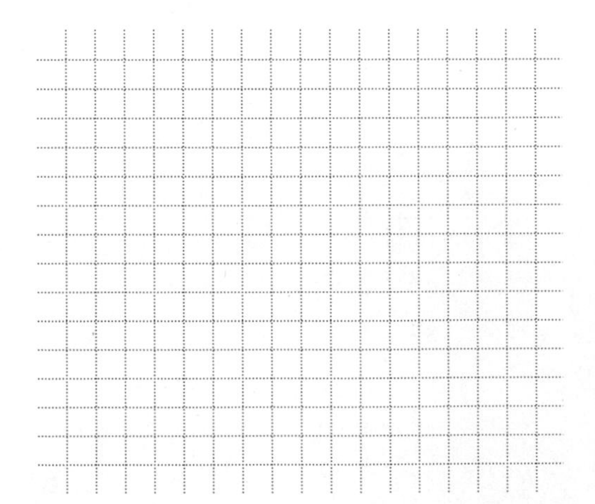

Practice Masters Level A

1.5 Representing Linear Patterns

Find the first differences for each data set, and write an equation to represent the data pattern.

1.

x	0	1	2	3	4	5	6
y	0	6	12	18	24	30	36

____

2.

x	0	1	2	3	4	5	6
y	5	12	19	26	33	40	47

____

3.

x	0	1	2	3	4	5	6
y	19	30	41	52	63	74	85

____

Make a table of values for each equation, using 1, 2, 3, 4, and 5 as values for *x*. Draw a graph for each equation by plotting points from your data set.

4. $y = 20x$

x	1	2	3	4	5
y					

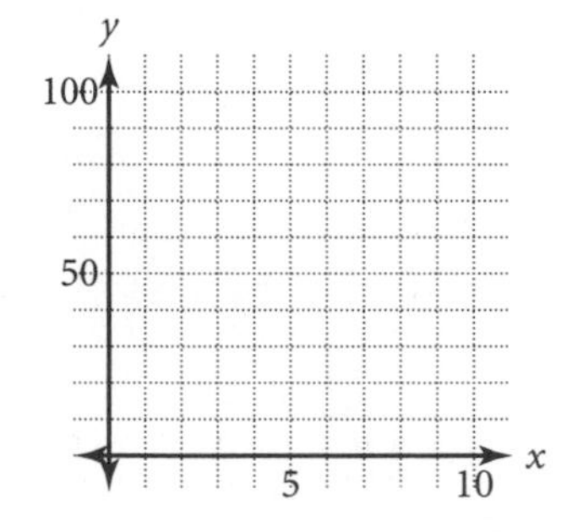

5. $y = 10x + 10$

x	1	2	3	4	5
y					

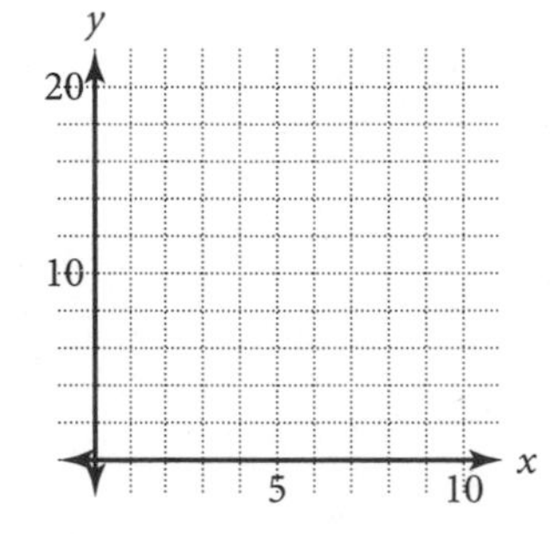

6. $y = 6x + 13$

x	1	2	3	4	5
y					

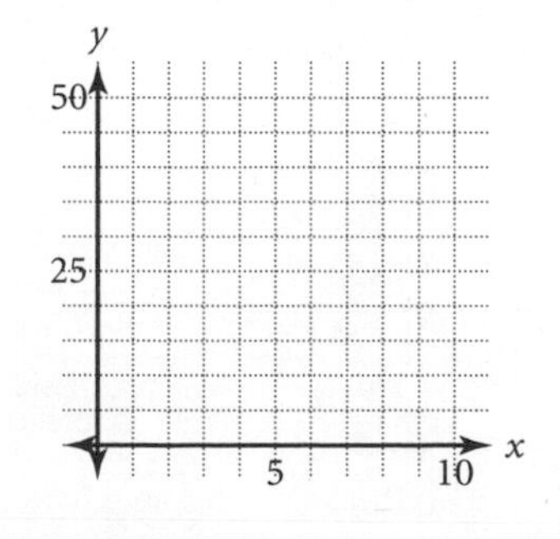

7. $y = 12x - 9$

x	1	2	3	4	5
y					

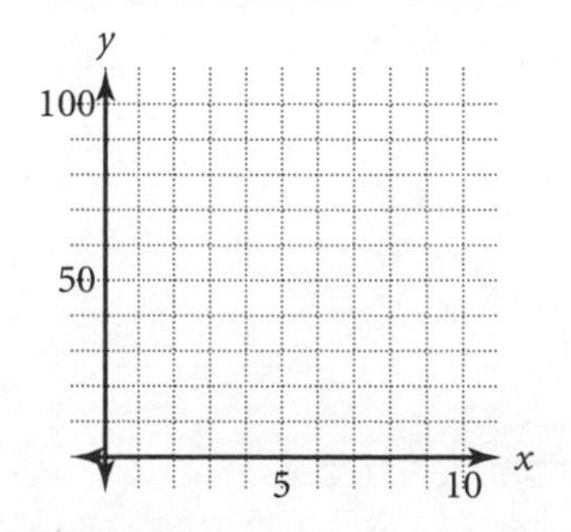

Practice Masters Level B

1.5 Representing Linear Patterns

Find the first differences for each data set, and write an equation to represent the data pattern.

1.

x	0	1	2	3	4	5	6
y	358	495	632	769	906	1043	1180

____

2.

x	0	1	2	3	4	5	6
y	2	3.5	5	6.5	8	9.5	11

____

3.

x	0	1	2	3	4	5	6
y	$12\frac{3}{4}$	12	$11\frac{1}{4}$	$10\frac{1}{2}$	$9\frac{3}{4}$	9	$8\frac{1}{4}$

____

Make a table of values for each equation, using 1, 2, 3, 4, and 5 as values for *x*. Draw a graph for each equation by plotting points from your data set.

4. $y = 2x - 0.1$

x	1	2	3	4	5
y					

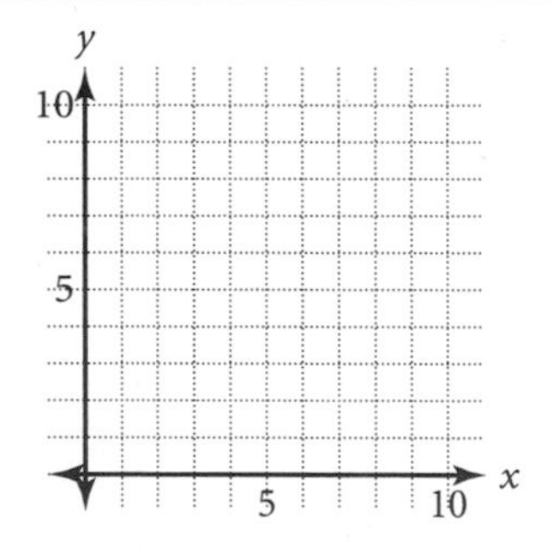

5. $y = 2.5 + 5x$

x	1	2	3	4	5
y					

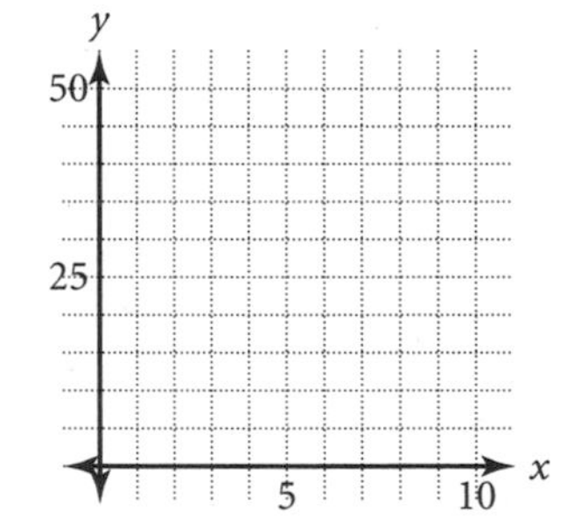

6. $y = 4 + \frac{1}{2}x$

x	1	2	3	4	5
y					

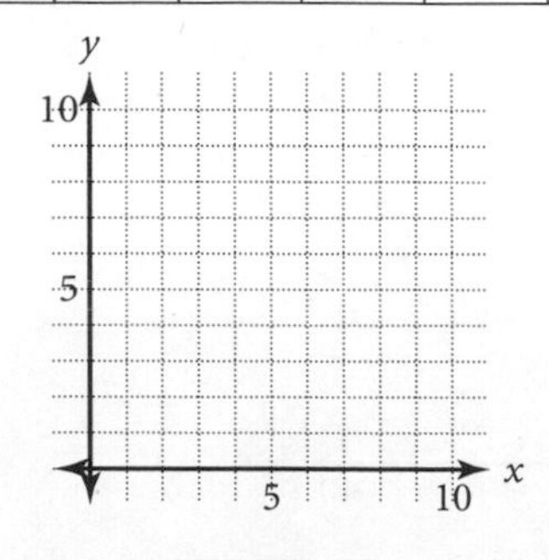

7. $y = 3 - \frac{1}{2}x$

x	1	2	3	4	5
y					

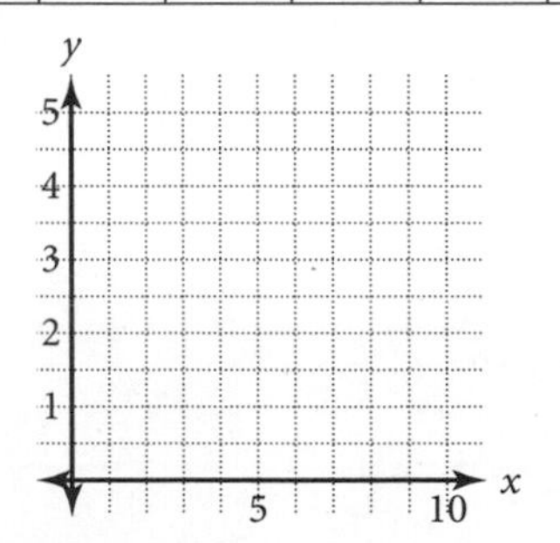

NAME ______________________ CLASS ____________ DATE ____________

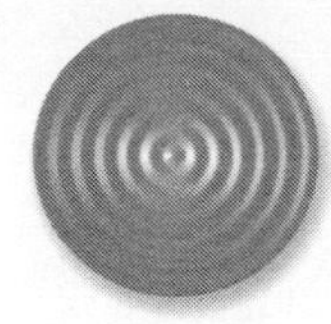

Practice Masters Level C

1.5 Representing Linear Patterns

Find the first differences for each data set, and write an equation to represent the data pattern.

1.

x	0	1	2	3	4	5	6
y	221.45	214.07	206.69	199.31	191.93	184.55	177.17

____

2.

x	0	1	2	3	4	5	6
y	$\frac{1}{8}$	$1\frac{29}{40}$	$3\frac{13}{40}$	$4\frac{37}{40}$	$6\frac{21}{40}$	$8\frac{1}{8}$	$9\frac{29}{40}$

____

____

3. A game show contestant can win $12,000 for answering a question. For each hint he receives from the host, $1500 is deducted from his winnings.

a. Represent the number of hints given by h, and write an equation for the amount of winnings, w. ______________

b. Complete the table that shows the amount of money won given 0 through 5 hints.

h	0	1	2	3	4	5
w						

c. Graph each ordered pair from the table, and use the graph to determine the number of hints for which the contestant will win $0.

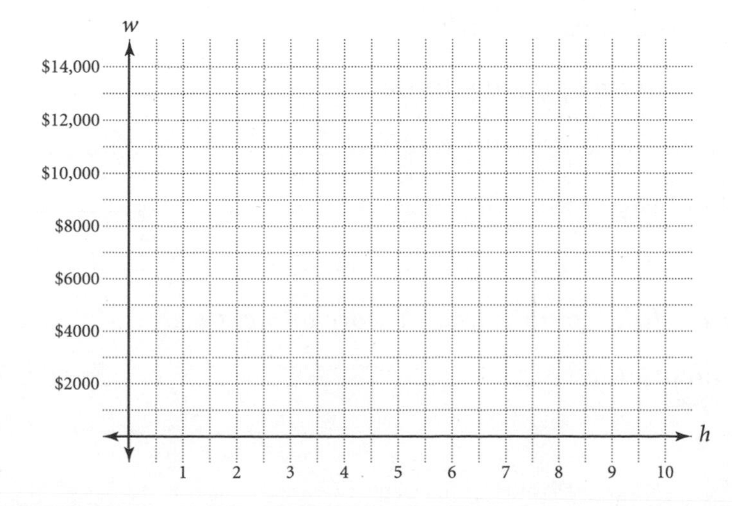

Number of hints: ______________

NAME ______________________ CLASS ____________ DATE ____________

Practice Masters Level A

1.6 Scatter Plots and Lines of Best Fit

Describe the correlation in each as *strong positive, strong negative,* or *little to none.* Then draw a line of best fit if the correlation is *strong positive* or *strong negative.*

1.

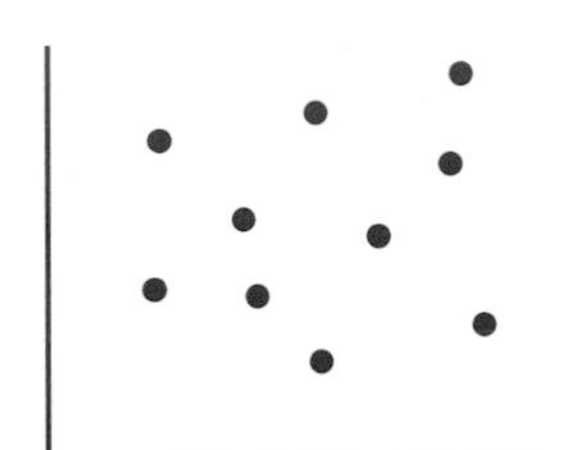

____

2.

____

3.

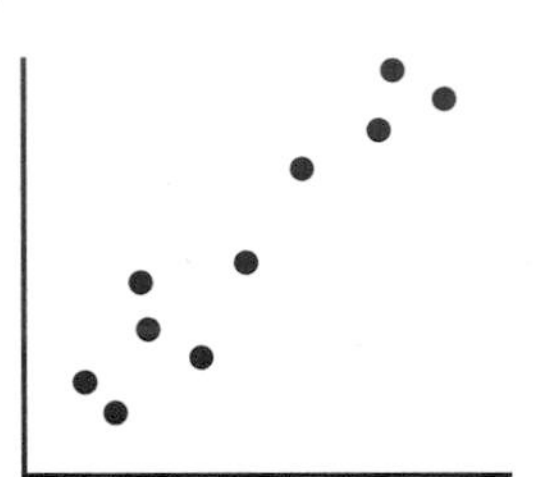

____

Match scatter plots from Exercises 1, 2, and 3 with statements 4, 5, and 6.

4. The variables studied are *gas tank capacity* and *length of time between fill-ups.* ________
5. The variables studied are *cost of video game* and *average score on video game.* ________
6. The variables studied are *number of missing homework assignments* and *grade in class.* ________

Describe the likely correlation between the variables studied in each case as *positive, negative,* or *little to none.*

7. *average driving speed* and *dashboard length* ________
8. *number of times brushing and flossing* and *number of trips to the dentist* ________
9. *number of pages in book* and *length of time required to read book* ________
10. *height* and *shirt size* ________
11. *hair length* and *number of haircuts received in past month* ________
12. *loudness of music* and *frequency of radio station playing music* ________
13. *number of satisfied customers* and *number of complaints received* ________
14. *length of time candle burns* and *height of wax remaining* ________
15. *income* and *fingernail length* ________
16. *number of homework assignments successfully completed* and *score on next test* ________
17. *weight of product* and *serial number on product* ________

Practice Masters Level B

1.6 Scatter Plots and Lines of Best Fit

The table shows the number of hours during two school days that 10 teenagers spent sleeping, surfing the internet, and studying. Use the data for Exercises 1–2.

Student number	Hours sleeping	Hours on internet	Hours studying
1	7	2	2
2	8	2	3
3	2	5	4
4	6	5	0
5	6	3	4
6	9	1	3
7	8	0	0
8	8	3	2
9	4	4	3
10	7	4	1

1. Make a scatter plot that compares the number of hours a student spent sleeping to the number of hours the student spent on the internet. Plot the hours sleeping along the x-axis and the hours of internet usage along the y-axis. Draw a line of best fit if appropriate. Describe the correlation.

____

2. Make a scatter plot that compares the number of hours a student spent on the internet to the number of hours the student spent studying. Plot the hours of internet usage along the x-axis and the hours studying along the y-axis. Draw a line of best fit if appropriate. Describe the correlation.

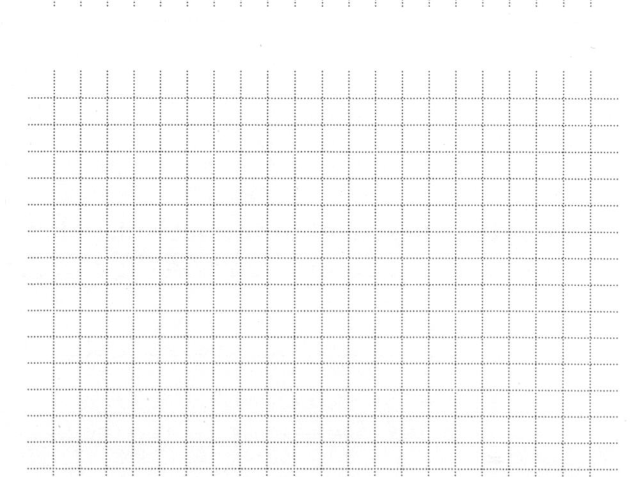

____

NAME ______________________ CLASS __________ DATE __________

Practice Masters Level C

1.6 Scatter Plots and Lines of Best Fit

The table shows the scores of 10 teenagers on their algebra final exams and their semester grade point averages (GPA). Use the data for Exercises 1–3.

Exam score	42	76	91	97	86	58	70	74	76	83
GPA	1.1	1.9	4.0	3.9	3.2	2.2	2.3	2.9	3.0	3.1

1. Make a scatter plot that compares the final exam score of a student to the student's GPA. Plot the exam score along the x-axis and the GPA along the y-axis. Draw a line of best fit if appropriate. Describe the correlation.

   ____

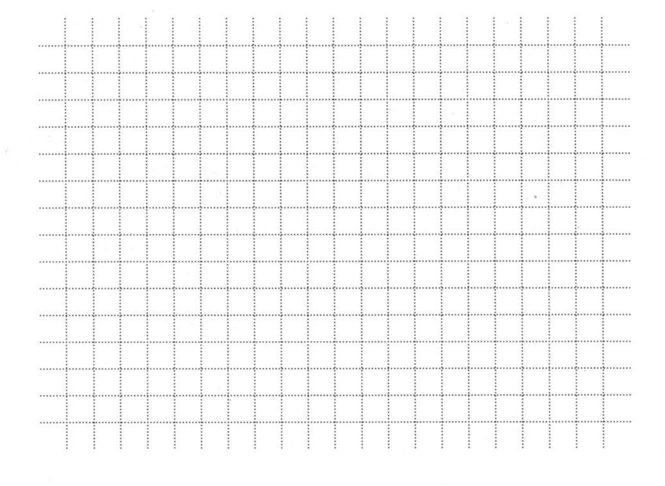

2. Make a scatter plot that compares the final exam score of a student to the student's GPA. This time, plot the GPA along the x-axis and the exam score along the y-axis. Draw a line of best fit if appropriate. Describe the correlation.

   ____

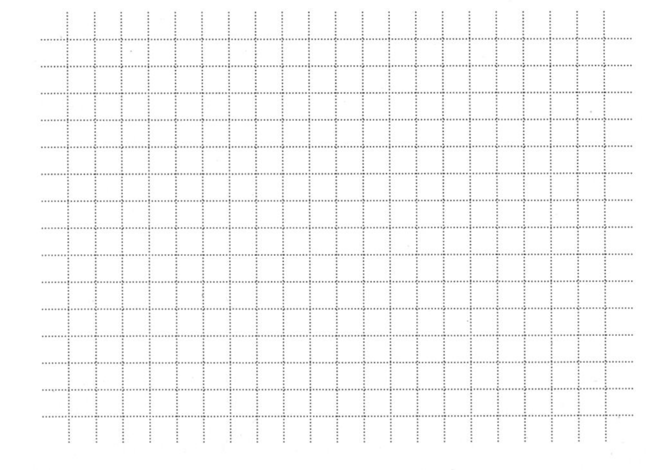

3. Does the placement of the variables along the axes appear to make a difference in the correlation? Explain. ______________________

   ____

   ____

NAME ______________________ CLASS ____________ DATE ____________

Practice Masters Level A

2.1 The Real Numbers and Absolute Value

Insert $<$, $>$, or = to make each statement true.

1. 11 ______ 6
2. 0 ______ 6
3. 0 ______ -6
4. -7 ______ 5
5. 4 ______ -3
6. 3 ______ -4
7. -2 ______ 1
8. -19 ______ 1
9. 0 ______ -28
10. -5 ______ -4
11. -8 ______ -1
12. -18 ______ -4
13. $-2\frac{1}{2}$ ______ -2
14. $-1\frac{3}{4}$ ______ -2
15. $-5\frac{1}{4}$ ______ -5.25
16. $-\frac{1}{3}$ ______ $-\frac{2}{3}$
17. $-\frac{2}{5}$ ______ -0.4
18. $-\frac{1}{2}$ ______ $-\frac{1}{4}$
19. -1 ______ -1.1
20. -2.83 ______ -2.8
21. $-3\frac{3}{8}$ ______ -3.375

Find the opposite of each number.

22. 18 ______
23. -5 ______
24. -9 ______
25. $(8 - 2)$ ______
26. 0 ______
27. -1.4 ______
28. $4\frac{1}{8}$ ______
29. $-3\frac{2}{7}$ ______
30. x ______
31. m ______
32. $-x$ ______
33. $-z$ ______

Find the absolute value of each number.

34. 18 ______
35. -5 ______
36. -9 ______
37. $(8 - 2)$ ______

Simplify each expression.

38. $-(-1)$ ______
39. $-(-6)$ ______
40. $-(8 + 4)$ ______
41. $-(8 - 4)$ ______
42. $-(8 \cdot 4)$ ______
43. $-(8 \div 4)$ ______
44. $-(7)$ ______
45. $-(5 - 5)$ ______
46. $-(7 - 6.5)$ ______
47. $|6|$ ______
48. $|101|$ ______
49. $|-13|$ ______
50. $-|12|$ ______
51. $|18 \div 2|$ ______
52. $-|9 \cdot 5|$ ______

NAME ______________________ CLASS ______________ DATE ______________

Practice Masters Level B

2.1 The Real Numbers and Absolute Value

Insert <, >, or = to make each statement true.

1. -7.1 ______ $-7\frac{1}{5}$
2. -6.2 ______ -6.25
3. $-10\frac{1}{2}$ ______ $-10\frac{1}{4}$
4. -2 ______ $-|-2|$
5. $-(-2)$ ______ -2
6. $-(-2)$ ______ $|-2|$

Simplify each expression.

7. $-|\frac{91}{7}|$ ______________
8. $-|-\frac{91}{7}|$ ______________
9. $-(-14) + 2$ ______________
10. $-|-(-4)|$ ______________
11. $|12 + 12|$ ______________
12. $-|12 + 12|$ ______________
13. $|12| + |-12|$ ______________
14. $|12| - |-12|$ ______________
15. $|12| \cdot |12|$ ______________
16. $|12| \cdot |-12|$ ______________
17. $|6| \div |-3|$ ______________
18. $|-6| \div |-3|$ ______________
19. $-(-0.5) \cdot 8$ ______________
20. $-|3.5 - 1.7|$ ______________
21. $1\frac{3}{4} - |-\frac{1}{2}|$ ______________
22. $-|\frac{5}{6} \div \frac{1}{3}|$ ______________
23. $-(-x)$ ______________
24. $-(-z)$ ______________

Determine whether each statement is true or false.

25. The expression $-x$ is always negative. ______________
26. If $-x$ is positive, then x is negative. ______________
27. There are two real numbers whose absolute values are each 0.9. ______________
28. $\sqrt{13}$ is an irrational number. ______________
29. If x is negative, then $-x$ is to the left of x on the number line. ______________

NAME ______________________ CLASS ____________ DATE ____________

Practice Masters Level C

2.1 The Real Numbers and Absolute Value

Given $a = 6$, $b = 3$, and $c = -2$, evaluate each expression.

1. $-(a + b)$ ____________
2. $-(a - b)$ ____________
3. $-|a + b|$ ____________
4. $-|a - b|$ ____________
5. $-(-a)$ ____________
6. $-(-c)$ ____________
7. $-|-b|$ ____________
8. $-|-c|$ ____________
9. $-|-(-c)|$ ____________
10. $-\left|\frac{a}{b}\right|$ ____________
11. $\left|\frac{a}{b}\right| - (-c)$ ____________
12. $b^2 \cdot |c|$ ____________
13. $2b - |c|$ ____________
14. $\left|\frac{3a}{2b}\right| + |c|$ ____________
15. $(-c)^2 - \frac{a}{b}$ ____________
16. $2b^2 + 3|c| - a$ ____________

Determine whether each statement is true or false.

17. For all values of x, $|x| \geq x$. ____________
18. A real number may be less than its opposite. ____________
19. If x is negative, then $|x| < x$. ____________
20. The number 0.13113111311113... is irrational. ____________
21. A number may be both a whole number and irrational. ____________
22. Determine the final point on a number line if you start at -8, move 5 spaces to the right, and then move $2\frac{1}{2}$ spaces to the left. Write an equation that models this situation. ____________
23. Determine the next 3 terms in the sequence 28, 23, 18, 13, 8, ... ____________

NAME ______________________ CLASS ____________ DATE ____________

Practice Masters Level A

2.2 Adding Real Numbers

Use algebra tiles to find the following sums:

1. $2 + 5$ ________
2. $-2 + (-5)$ ________
3. $-2 + 5$ ________
4. $2 + (-5)$ ________
5. $-5 + 5$ ________
6. $5 + (-5)$ ________
7. $9 + (-3)$ ________
8. $2 + (-7)$ ________
9. $-6 + 5$ ________
10. $-1 + 8$ ________
11. $4 + (-3)$ ________
12. $-5 + (-4)$ ________

Find each sum.

13. $-1 + 5$ ________
14. $3 + (-6)$ ________
15. $-9 + (-5)$ ________
16. $-6 + 2$ ________
17. $-7 + (-7)$ ________
18. $8 + (-4)$ ________
19. $-13 + 41$ ________
20. $-86 + 98$ ________
21. $58 + (-24)$ ________
22. $-75 + (-5)$ ________
23. $-22 + (-19)$ ________
24. $33 + (-62)$ ________
25. $20 + (-58)$ ________
26. $-19 + (-13)$ ________
27. $-37 + 80$ ________
28. $-44 + (-27)$ ________
29. $-91 + 100$ ________
30. $-62 + (-28)$ ________
31. $-103 + 212$ ________
32. $-212 + 103$ ________
33. $-102 + (-212)$ ________
34. $-4.5 + 2.5$ ________
35. $-6.2 + 1.8$ ________
36. $3.6 + (-1.9)$ ________
37. $-2.1 + (-2.2)$ ________
38. $-3.7 + 0.8$ ________
39. $-4.2 + (-0.8)$ ________
40. $3.4 + (-6.3)$ ________
41. $9.2 + (-2.5)$ ________
42. $-3.3 + 8.1$ ________
43. $4.7 + (-5.5)$ ________
44. $-7.3 + (-6.2)$ ________
45. $-8.5 + (-4.3)$ ________
46. $\frac{2}{5} + \left(-\frac{1}{5}\right)$ ________
47. $-\frac{2}{5} + \left(-\frac{1}{5}\right)$ ________
48. $-\frac{2}{5} + \frac{1}{5}$ ________
49. $-\frac{3}{4} + \frac{5}{4}$ ________
50. $\frac{3}{4} + \left(-\frac{5}{4}\right)$ ________
51. $-\frac{3}{4} + \left(-\frac{5}{4}\right)$ ________
52. $-1\frac{1}{2} + \left(-3\frac{1}{2}\right)$ ________
53. $-1\frac{1}{2} + 3\frac{1}{2}$ ________
54. $1\frac{1}{2} + \left(-3\frac{1}{2}\right)$ ________

NAME ______________________ CLASS ______________ DATE ______________

Practice Masters Level B

2.2 Adding Real Numbers

Use algebra tiles to find the following sums:

1. $-10 + (-11)$ ________
2. $-10 + 11$ ________
3. $10 + (-11)$ ________
4. $-12 + 9$ ________
5. $12 + (-9)$ ________
6. $-12 + 12$ ________

Find each sum.

7. $101 + (-992)$ ________
8. $241 + (-320)$ ________
9. $-331 + 875$ ________
10. $-312 + (-453)$ ________
11. $-822 + 1000$ ________
12. $1312 + (-840)$ ________
13. $-697 + 1250$ ________
14. $1064 + (-1182)$ ________
15. $1584 + (-1448)$ ________
16. $-6.28 + 9.01$ ________
17. $-5.48 + 4$ ________
18. $4.79 + (-8)$ ________
19. $13.24 + (-8.1)$ ________
20. $-15.2 + 9.73$ ________
21. $-20 + 18.49$ ________
22. $-19.1 + 10.006$ ________
23. $-8.35 + 5.243$ ________
24. $14.847 + (-16.08)$ ________
25. $-\frac{1}{4} + \frac{5}{7}$ ________
26. $2\frac{3}{5} + \left(-8\frac{1}{3}\right)$ ________
27. $-4\frac{4}{9} + 12\frac{2}{5}$ ________
28. $\frac{11}{5} + \left(-\frac{8}{3}\right)$ ________
29. $-1\frac{3}{4} + \left(-3\frac{5}{8}\right)$ ________
30. $6\frac{2}{9} + \left(-8\frac{2}{3}\right)$ ________
31. $9\frac{12}{25} + \left(-3\frac{4}{5}\right)$ ________
32. $-10\frac{1}{8} + \left(-11\frac{3}{7}\right)$ ________
33. $13 + \left(-15\frac{4}{5}\right)$ ________

Let $a = 2$, $b = 6$, and $c = -9$. Evaluate each expression.

34. $a + b$ ________
35. $a + (-b)$ ________
36. $-a + b$ ________
37. $-a + (-b)$ ________
38. $b + c$ ________
39. $b + (-c)$ ________
40. $-b + c$ ________
41. $-b + (-c)$ ________
42. $-c + a$ ________
43. $c + a$ ________
44. $-c + (-a)$ ________
45. $c + (-a)$ ________
46. $2a + c$ ________
47. $2a + (-c)$ ________
48. $a + b + c$ ________

NAME ______________________ CLASS ______________ DATE ______________

Practice Masters Level C

2.2 Adding Real Numbers

Use algebra tiles to find the following sums:

1. $6 + 2 + (-1)$ ________
2. $6 + (-2) + (-1)$ ________
3. $-6 + (-2) + (-1)$ ________
4. $-3 + 4 + 5$ ________
5. $-3 + (-4) + 5$ ________
6. $-10 + 11 + (-8)$ ________

Find each sum.

7. $3 + 4 + (-2)$ ________
8. $6 + (-7) + 8$ ________
9. $-5 + 12 + (-7)$ ________
10. $20 + (-18) + (-19)$ ________
11. $-2.4 + 3.3 + 2.8$ ________
12. $6.2 + (-1.7) + 3$ ________
13. $20 + (-13.7) + (-0.5)$ ________
14. $[6 + (-3.8)] + [2 + (-1.8)]$ ________
15. $3\frac{1}{8} + \left(-\frac{1}{3}\right) + \frac{2}{9}$ ________
16. $-8 + 2\frac{2}{5} + (-1)$ ________
17. $-6\frac{4}{5} + 10\frac{2}{3} + \left(-1\frac{2}{7}\right)$ ________
18. $-2\frac{3}{8} + \left(-7\frac{1}{4}\right) + \left(-8\frac{5}{9}\right)$ ________

Let $a = 3.5$, $b = -2.5$, and $c = -4$. Evaluate each expression.

19. $a + b + (-c)$ ________
20. $(-c)(a + b)$ ________
21. $(-b)^2 + c + (-a)$ ________
22. $(-b + a)(-c)$ ________
23. $\dfrac{-b}{-c + (-a)}$ ________
24. $a^2 + (-b)^2 + (-c)$ ________
25. $\dfrac{b + a}{(-c)}$ ________
26. $\dfrac{(-c)^2}{a + (-b)}$ ________
27. Suppose that you have \$35 in a checking account. You deposit \$27 and then write a check for \$70. The bank informs you that your account is overdrawn. Is the bank correct? If so, by how much is your account overdrawn? ________
28. The temperature of a substance in a test tube is 90°F. During an experiment, the temperature falls 103°F and then rises 24°F. What is the final temperature of the substance? ________

NAME ______________________ CLASS ______________ DATE ______________

Practice Masters Level A

2.3 Subtracting Real Numbers

Use algebra tiles to find each difference.

1. $4 - 2$ ________
2. $4 - (-2)$ ________
3. $-4 - 2$ ________
4. $-4 - (-2)$ ________
5. $-3 - 5$ ________
6. $-3 - (-5)$ ________
7. $3 - 5$ ________
8. $3 - (-5)$ ________
9. $-6 - (-6)$ ________
10. $-6 - 6$ ________
11. $6 - (-6)$ ________
12. $0 - 7$ ________

Find each difference.

13. $5 - 9$ ________
14. $0 - 8$ ________
15. $9 - (-2)$ ________
16. $-9 - 1$ ________
17. $-6 - 3$ ________
18. $-2 - (-10)$ ________
19. $-6 - (-2)$ ________
20. $-4 - (-7)$ ________
21. $3 - 12$ ________
22. $6.4 - 9.5$ ________
23. $-2.8 - 3.1$ ________
24. $-4.7 - (-2.4)$ ________

Find each sum or difference.

25. $13 - 17$ ________
26. $-12 - 19$ ________
27. $-11 - (-18)$ ________
28. $-22 - (-13)$ ________
29. $34 - (-61)$ ________
30. $20 - (-32)$ ________
31. $47 + (-26)$ ________
32. $26 - 47$ ________
33. $-26 - 47$ ________
34. $-52 - 49$ ________
35. $-52 + (-49)$ ________
36. $-52 + 49$ ________
37. $312 - 801$ ________
38. $-312 - 801$ ________
39. $-312 + 801$ ________
40. $-622 - 187$ ________
41. $-622 + 187$ ________
42. $622 - (-187)$ ________
43. $6.4 + (-8.2)$ ________
44. $3.8 - 5.5$ ________
45. $-1.4 + (-2.1)$ ________
46. $-1.4 - 2.1$ ________
47. $1.4 - (-2.1)$ ________
48. $-1.4 - (-2.1)$ ________
49. $\frac{1}{5} - \frac{2}{5}$ ________
50. $\frac{4}{7} + \left(-\frac{6}{7}\right)$ ________
51. $-\frac{4}{7} - \frac{6}{7}$ ________
52. $-\frac{1}{8} + \frac{7}{8}$ ________
53. $-\frac{3}{11} - \left(-\frac{2}{11}\right)$ ________
54. $-1\frac{4}{5} - \left(-\frac{3}{5}\right)$ ________

NAME ______________________ CLASS ______________ DATE ______________

Practice Masters Level B

2.3 Subtracting Real Numbers

Use algebra tiles to find each difference.

1. $8 - 12$ ______
2. $8 - (-12)$ ______
3. $-10 - (-13)$ ______
4. $(6 - 2) - 5$ ______
5. $6 - (2 - 5)$ ______
6. $(6 - 2) - (-5)$ ______

Find each sum or difference.

7. $227 - 1000$ ______
8. $-227 - 900$ ______
9. $-1111 - (-999)$ ______
10. $1.17 - 3.48$ ______
11. $-1.17 + 3.48$ ______
12. $-1.17-3.48$ ______
13. $54.8 - 62.18$ ______
14. $-8.73 + 0.8$ ______
15. $-1 - 9.37$ ______
16. $-31.2 - (-40.14)$ ______
17. $-2.401 - 5.9$ ______
18. $1.01 - 2.001$ ______
19. $-\frac{2}{5} - \frac{6}{11}$ ______
20. $-\frac{2}{5} + \frac{6}{11}$ ______
21. $-\frac{1}{8} - \left(-\frac{4}{7}\right)$ ______
22. $2\frac{5}{8} - 2\frac{3}{4}$ ______
23. $-6\frac{5}{9} - \left(-2\frac{7}{9}\right)$ ______
24. $-4\frac{3}{8} + 7\frac{4}{9}$ ______
25. $2 - 5\frac{4}{11}$ ______
26. $-\frac{11}{8} - \left(-\frac{41}{8}\right)$ ______
27. $-11 - 3\frac{5}{7}$ ______

Let $p = 2$, $q = -4$, and $r = -7$. Evaluate each expression.

28. $p - q$ ______
29. $p - (-q)$ ______
30. $-p - q$ ______
31. $-p - (-q)$ ______
32. $q - r$ ______
33. $-q - r$ ______
34. $q - (-r)$ ______
35. $p - r$ ______
36. $-p - (-r)$ ______

Find the distance between each pair of points on a number line.

37. 0, 8 ______
38. 1, 13 ______
39. −4, 5 ______
40. −2, 18 ______
41. −12, −1 ______
42. −18, −5 ______
43. −48, −19 ______
44. −20, 57 ______
45. −103, 219 ______

NAME ______________________ CLASS __________ DATE __________

Practice Masters Level C

2.3 Subtracting Real Numbers

Use algebra tiles to evaluate each expression.

1. $6 - 10 + 2$ ________
2. $4 - (-2) - 9$ ________
3. $-3 - (-4) + (-3)$ ________

Find each sum or difference.

4. $7 - 13 - (-8)$ ________
5. $-12 + 18 - (-15)$ ________
6. $310 - (-158) + 72$ ________
7. $-123 - 456 - (-789)$ ________
8. $6.28 + (-9.04) - 8$ ________
9. $-2.1 + 4.03 - 1.78$ ________
10. $2\frac{4}{5} - \left(-8\frac{1}{3} + \frac{1}{2}\right)$ ________
11. $-6\frac{2}{7} - 1\frac{4}{5} - 2\frac{2}{3}$ ________
12. $12\frac{8}{9} + \left(-2\frac{1}{9} - 4\frac{2}{3}\right)$ ________
13. $-10\frac{3}{7} + \left(2 - 13\frac{1}{2}\right)$ ________

Let $a = \frac{3}{4}$, $b = -1\frac{1}{2}$, and $c = -5\frac{1}{2}$. Evaluate each expression.

14. $a - (-b)$ ________
15. $b - c$ ________
16. $2a + b$ ________
17. $2a - b$ ________
18. $a + b - c$ ________
19. $a - b - c$ ________
20. $a - (b - c)$ ________
21. $b^2 - c^2$ ________

Find the distance between each pair of points on a number line.

22. $-13.4, -1.8$ ________
23. $-1\frac{1}{5}, 3\frac{4}{5}$ ________
24. $-2\frac{4}{7}, 5\frac{4}{5}$ ________
25. Use constant differences to find the next 3 terms in the sequence 21, 11, 2, -6, ... ________
26. If the distance between x and $-5\frac{1}{3}$ on a number line is $\frac{8}{9}$, what are the possible values of x? ________

NAME ______________________ CLASS ____________ DATE ____________

Practice Masters Level A

2.4 Multiplying and Dividing Real Numbers

Find each product or quotient.

1. $-12 \cdot 4$ ________
2. $12 \cdot (-4)$ ________
3. $(-12)(-4)$ ________
4. $12 \div (-4)$ ________
5. $-12 \div (-4)$ ________
6. $-12 \div 4$ ________
7. $6(-1)$ ________
8. $-5(2)$ ________
9. $-19 \cdot 0$ ________
10. $0 \div (-2)$ ________
11. $-13 \cdot (-3)$ ________
12. $17(-5)$ ________
13. $21(-4)$ ________
14. $-24 \div 6$ ________
15. $-33 \div (-11)$ ________
16. $-60 \div (-5)$ ________
17. $-6 \div 0$ ________
18. $-6 \cdot (-15)$ ________
19. $70 \div (-14)$ ________
20. $(-16)(-12)$ ________
21. $-84 \div (-12)$ ________
22. $(0.5)(-6)$ ________
23. $-1.5 \cdot 8$ ________
24. $(-3.7)(-10)$ ________
25. $-8 \div 0.4$ ________
26. $2.1 \div (-0.7)$ ________
27. $-30 \div (-1.5)$ ________
28. $(0)(-1.7)$ ________
29. $-12 \cdot (0.6)$ ________
30. $-12 \div (-0.6)$ ________

Write the reciprocal of each number.

31. $\frac{2}{5}$ ________
32. $-\frac{2}{3}$ ________
33. -7 ________
34. $-\frac{1}{4}$ ________
35. 0 ________
36. $1\frac{1}{2}$ ________

Evaluate.

37. $\frac{3}{4} \cdot \left(-\frac{2}{3}\right)$ ________
38. $-\frac{1}{2}\left(\frac{4}{7}\right)$ ________
39. $-\frac{3}{8} \cdot \left(-\frac{1}{5}\right)$ ________
40. $-\frac{5}{8} \div \frac{1}{4}$ ________
41. $-\frac{3}{10} \div \left(-\frac{1}{2}\right)$ ________
42. $\frac{5}{9} \div \left(-\frac{2}{3}\right)$ ________
43. $-4 \cdot \frac{3}{4}$ ________
44. $\frac{5}{8} \cdot (-16)$ ________
45. $\left(-\frac{6}{7}\right)(-42)$ ________
46. $3 \div \frac{1}{2}$ ________
47. $-2 \div \frac{2}{3}$ ________
48. $-6 \div \frac{3}{8}$ ________

NAME ______________________ CLASS ______________ DATE ______________

Practice Masters Level B

2.4 Multiplying and Dividing Real Numbers

Evaluate.

1. $-884 \div 17$ ________
2. $0 \div \left(-1\frac{4}{7}\right)$ ________
3. $(-1.6)(-2.1)$ ________
4. $1\frac{3}{4} \cdot (-4)$ ________
5. $(2.5)(-3.4)$ ________
6. $1\frac{1}{2} \cdot \left(-1\frac{1}{3}\right)$ ________
7. $(-6.22)(0.5)$ ________
8. $-7\frac{1}{2} \div 1\frac{1}{2}$ ________
9. $-\left(-\frac{2}{3}\right) \div 3$ ________
10. $0 \cdot (-72.1)$ ________
11. $37.5 \div (-0.25)$ ________
12. $(-132) \cdot 47$ ________
13. $-1\frac{1}{5}\left(-4\frac{3}{4}\right)$ ________
14. $-2\frac{5}{9} \cdot 3$ ________
15. $-6.28 \div (-3.14)$ ________
16. $1.72(-3.1)$ ________
17. $2\frac{5}{6} \div \left(-\frac{1}{2}\right)$ ________
18. $-22.8 \div 1.2$ ________
19. $-3\frac{1}{5} \cdot (-4)$ ________
20. $-3834 \div 71$ ________
21. $101(-99)$ ________
22. $-12 \div 2\frac{1}{2}$ ________
23. $-384 \cdot 0$ ________
24. $7\frac{2}{9} \cdot (-9)$ ________
25. $-990 \div (-22)$ ________
26. $-3\frac{2}{7} \cdot 1\frac{1}{3}$ ________
27. $-1331 \div 0$ ________

Let $x = 3\frac{1}{2}$, $y = 4$, and $z = -2$. Evaluate each expression.

28. yz ________
29. $(-y) \cdot z$ ________
30. $(-y)(-z)$ ________
31. xz ________
32. $(-x) \cdot z$ ________
33. $x \cdot (-z)$ ________
34. $-x \cdot y$ ________
35. xy ________
36. $(-x) \cdot (-y)$ ________
37. $y^2 + z$ ________
38. $z^2 + y$ ________
39. $x^2 + z$ ________
40. $x(y + z)$ ________
41. $\frac{x}{y + z}$ ________
42. $x \div (-y + z)$ ________

NAME ______________________ CLASS ____________ DATE ____________

Practice Masters Level C

2.4 Multiplying and Dividing Real Numbers

Evaluate.

1. $\frac{-144}{-16} + (-8)$ ______

2. $-144 \div [-16 + (-8)]$ ______

3. $(-3.75) \cdot 4 + (-8.1)$ ______

4. $-6\frac{1}{4} \div \left(-\frac{1}{2} + 3\right)$ ______

5. $-3 \cdot \left|\frac{1}{2} \div \left(-\frac{1}{3}\right)\right|$ ______

6. $-\left(\frac{8}{9}\right)^2 + \left(-\frac{8}{9}\right)^2$ ______

7. $3\frac{6}{7} - \left(-2\frac{1}{2} \cdot 5\right)$ ______

8. $(1.28)(-3.31) + 7$ ______

Let $a = -0.75$, $b = -3$, and $c = -2$. Evaluate each expression.

9. $(b - c)(-a)$ ______

10. $ac - ab$ ______

11. $8a + 2b - c$ ______

12. $6a + \frac{b}{c}$ ______

13. $a^2 - b^2 + c^2$ ______

14. $2a^2 + cb^2$ ______

15. $2a^2 + (cb)^2$ ______

16. $3b - |5c| + a$ ______

Suppose that r is a positive integer and that s is a negative integer. Classify each expression as *positive, negative, zero, undefined,* or *cannot be determined.*

17. rs ______

18. $(-r) \cdot s$ ______

19. $(-r)(-s)$ ______

20. $0 \cdot r$ ______

21. $r + s$ ______

22. s^2 ______

23. $r \div (-s)$ ______

24. $s \div 0$ ______

25. $r - s$ ______

26. Find the next 3 terms in the sequence 243, −81, 27, −9, ... ______

27. A business has \$3128 in its bank account. It deposits 7 payments averaging \$201 each into the account. How much more must the business deposit into the account in order to pay 9 bills averaging \$612 each and have \$1000 left in the account? ______

Practice Masters Level A

2.5 Properties and Mental Computation

Complete each step, and name the property used.

1. $82 + (76 + 18)$

   $= 82 + (\_\_\_\_\_ + 76)$ Commutative Property of Addition

   $= (82 + \_\_\_\_\_) + 76$ ______________ Property of Addition

   $= \_\_\_\_\_ + 76$

   $= \_\_\_\_\_$

2. $(40 \cdot 7) \cdot 5$

   $= (7 \cdot \_\_\_\_\_) \cdot 5$ ______________ Property of Multiplication

   $= 7 \cdot (\_\_\_\_\_ \cdot 5)$ ______________ Property of Multiplication

   $= 7 \cdot \_\_\_\_\_$

   $= \_\_\_\_\_$

3. $4(80 + 7)$

   $= 4 \cdot \_\_\_\_\_ + \_\_\_\_\_ \cdot 7$ ______________ Property

   $= \_\_\_\_\_ + \_\_\_\_\_$

   $= \_\_\_\_\_$

Use the Associative, Commutative, and Distributive Properties to find each sum or product. Show your work and name the properties used.

4. $20 \cdot (17 \cdot 5)$ ______________

5. $(52 + 37) + 48$ ______________

Use the Distributive Property and mental computation to calculate each product.

6. $5 \cdot 32 = 5(30 + 2)$

   $= 5 \cdot 30 + 5 \cdot 2$

   $= 150 + 10$

   $= \_\_\_\_\_$

7. $9 \cdot 13 = 9(10 + 3)$

   $9 \cdot \_\_\_\_\_ + 9 \cdot \_\_\_\_\_$

   $= \_\_\_\_\_ + \_\_\_\_\_$

   $= \_\_\_\_\_$

NAME ______________________ CLASS ____________ DATE ____________

Practice Masters Level B

2.5 Properties and Mental Computation

Complete each step, and name the property used.

1. $5 \cdot (91 \cdot 2)$

   $= 5 \cdot (\_\_\_\_ \cdot \_\_\_\_)$ ____________________

   $= (5 \cdot \_\_\_\_) \cdot \_\_\_\_$ ____________________

   $= \_\_\_\_ \cdot \_\_\_\_$

   $= \_\_\_\_$

2. $(161 + 81) + 39$

   $= (81 + \_\_\_\_) + \_\_\_\_$ ____________________

   $= 81 + (\_\_\_\_ + \_\_\_\_)$ ____________________

   $= \_\_\_\_ + \_\_\_\_$

   $= \_\_\_\_$

Use the Distributive Property and mental computation to calculate each product.

3. $5 \cdot 595 = 5(600 - 5)$

   $= \_\_\_\_ \cdot \_\_\_\_ - \_\_\_\_ \cdot \_\_\_\_$

   $= \_\_\_\_ - \_\_\_\_$

   $= \_\_\_\_$

4. $4(28.5) = 4(30 - 1.5)$

   $= \_\_\_\_ \cdot \_\_\_\_ - \_\_\_\_ \cdot \_\_\_\_$

   $= \_\_\_\_ - \_\_\_\_$

   $= \_\_\_\_$

Use the Distributive Property to rewrite each expression.

5. $8x + 8y$ ____________________
6. $8p + 28q$ ____________________
7. $6a - 9b$ ____________________
8. $ab + ac$ ____________________
9. $5rs + 25rt$ ____________________
10. $3fg - hg$ ____________________

Name the property illustrated. Be specific.

11. $14 + 2.2 = 2.2 + 14$ ____________________
12. $3(-2 + 4) = 3(-2) + 3(4)$ ____________________
13. $5(8x) = (5 \cdot 8)x$ ____________________
14. $(6 + 19) + 11 = 6 + (19 + 11)$ ____________________
15. $xy = yx$ ____________________
16. $(x - 3)7 = x \cdot 7 - 3 \cdot 7$ ____________________

NAME ______________________ CLASS ____________ DATE ____________

Practice Masters Level C

2.5 Properties and Mental Computation

Use the Distributive Property and mental computation to calculate each product.

1. $6(5.9) =$ ______(______ + ______)

$=$ ______ · ______ + ______ · ______

$=$ ______ + ______

$=$ ______

2. $9\left(7\frac{1}{3}\right) = 9($______ + ______$)$

$=$ ______ · ______ + ______ · ______

$=$ ______ + ______

$=$ ______

Name the property illustrated. Be specific.

3. $-3\frac{1}{2}\left(8 + 7\frac{1}{3}\right) = \left(8 + 7\frac{1}{3}\right)\left(-3\frac{1}{2}\right)$ ______________________

4. $(7 - 4a)xy = 7(xy) - (4a)(xy)$ ______________________

5. $(5a + 9.1b)3 = (9.1b + 5a)3$ ______________________

Prove each statement. Justify each step.

6. If $x = y$, then $xa = ya$.

Statements	Reasons
______	______
______	______
______	______

7. If $m = n$, then $m + r = n + r$.

Statements	Reasons
______	______
______	______
______	______

8. $(x + y) + z = (z + y) + x$

Statements	Reasons
______	______
______	______
______	______
______	______

9. $\frac{1}{x} \cdot (y \cdot x) = y (x \neq 0)$

Statements	Reasons
______	______
______	______
______	______
______	______

NAME ______________________ CLASS ____________ DATE ____________

Practice Masters Level A

2.6 Adding and Subtracting Expressions

Use the Distributive Property to show that the following are true statements:

1. $5x + 3x = 8x$ ______________
2. $6y + 7y = 13y$ ______________
3. $6r - 3r = 3r$ ______________
4. $14z - 5z = 9z$ ______________
5. $-3q + 5q = 2q$ ______________
6. $8a + (-7a) = a$ ______________
7. $2m - 9m = -7m$ ______________
8. $-3p - 11p = -14p$ ______________

Give the opposite of each expression.

9. $2x + 5$ ______________
10. $-3z + 4$ ______________
11. $8m - 7$ ______________
12. $-4b - 6$ ______________
13. $a + b - c$ ______________
14. $2f - 3g + 1$ ______________
15. $-2m + 5n - 9$ ______________
16. $-3x + 4y - z$ ______________

Simplify the following expressions:

17. $8x + 5x$ ______________
18. $9a + 7a$ ______________
19. $12p - 6p$ ______________
20. $15d - 11d$ ______________
21. $-y + 2y$ ______________
22. $-7w + 5w$ ______________
23. $j - 7j$ ______________
24. $-3h - 14h$ ______________
25. $(2t + 3) + (9t + 5)$ ______________
26. $(4n + 7) + (5n - 6)$ ______________
27. $(5b - 4) + (7b - 8)$ ______________
28. $(z + 2) + (-3z - 5)$ ______________
29. $(-2u + 3v) + (4u - 7v)$ ______________
30. $(4q - 3r) + (-2q - 2r)$ ______________
31. $(10l + 5m) - (4l + 2m)$ ______________
32. $(7a + 3b) - (4a + 3b)$ ______________
33. $(8f + 9g) - (10f + 11g)$ ______________
34. $(8c - 3d) - (4c + 7d)$ ______________
35. $(17v - 9w) - (9v - 8w)$ ______________
36. $(-5g + 2h) - (6g + 5h)$ ______________
37. $(-5s + 6t) - (-3s + 5t)$ ______________
38. $(-7r + s) - (-3r - 2s)$ ______________

NAME ______________________ CLASS ____________ DATE ____________

Practice Masters Level B

2.6 Adding and Subtracting Expressions

Use the Distributive Property to show that the following are true statements:

1. $8.5m + 1.3m = 9.8m$ ____________
2. $\frac{11}{9}a - \frac{2}{9}a = a$ ____________
3. $-6x - 6.1x = -12.1x$ ____________
4. $2q + \frac{1}{3}q = \frac{7}{3}q$ ____________

Simplify the following expressions:

5. $(0.5a + 3) + (1.2a + 1)$ ____________
6. $(g + 4) + (-1.9g - 2)$ ____________
7. $(9.1f - 2g) - (10f + g)$ ____________
8. $(-7p + 0.1q) - (2.7p - 0.3q)$ ____________
9. $-(-0.7x + y) + (x - 0.7y)$ ____________
10. $(11.3j - 2.7k) - (-1.8j - 0.5k)$ ____________
11. $2.2w - 3.98w$ ____________
12. $(-1.4m - 0.3n) - (-0.17m)$ ____________
13. $\frac{3}{5}h + 1\frac{1}{5}h$ ____________
14. $-4y - 5\frac{1}{3}y$ ____________
15. $2\frac{1}{3}k + \left(-\frac{2}{3}k\right)$ ____________
16. $-b - \left(-1\frac{3}{7}b\right)$ ____________
17. $\left(2c + 4\frac{1}{2}d\right) - \left(-\frac{1}{2}c + 1\frac{1}{2}d\right)$ ____________
18. $\left(-\frac{3}{4}l + \frac{1}{8}m\right) - \left(-\frac{1}{4}l + \frac{5}{8}m\right)$ ____________

Give an expression in simplified form for the perimeter of each figure.

19.

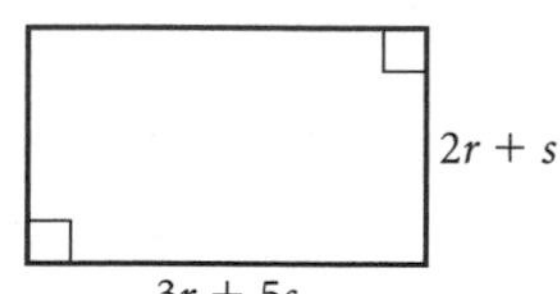

____

20.

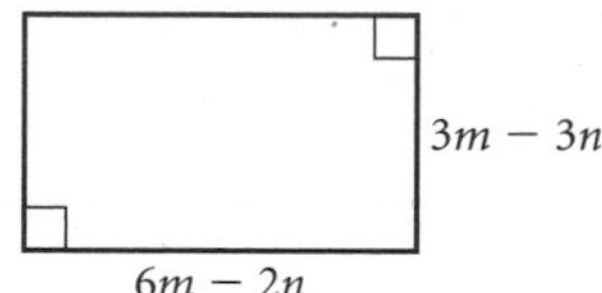

____

21.

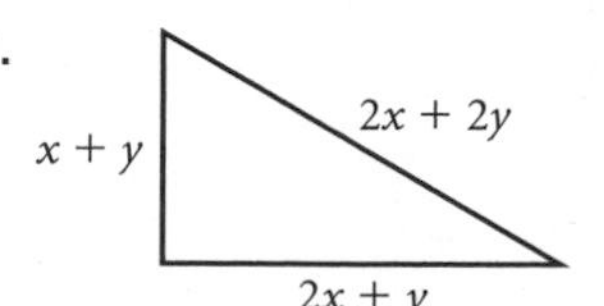

____

22.

____

NAME ______________________ CLASS ____________ DATE ____________

Practice Masters Level C

2.6 Adding and Subtracting Expressions

Use the Distributive Property to show that the following are true statements:

1. $13x = -14.1x + 27.1x$ ______________

2. $-2.05z = z - 3.05z$ ______________

3. $-1\frac{1}{8}a - 2\frac{1}{4}a = -3\frac{3}{8}a$ ______________

4. $-\frac{5}{9}b + 2\frac{1}{3}b = 1\frac{7}{9}b$ ______________

Simplify the following expressions:

5. $2a + 3b - c + (a - b - c) - (2a - 4b + 3c)$ ______________

6. $-8x + y - 2z - (-x - y - z) - (3x - y + z)$ ______________

7. $(3j - 0.8k) - (0.2j + l) - (-0.3k - 1.1l)$ ______________

8. $0.1m + 0.2n - 0.3p + (0.4m - 0.5n + 0.6p) - (0.7m - 0.8n + 0.9p)$ ______________

9. $1\frac{2}{9}r - 3s + 5\frac{1}{4}t - \left(-2\frac{1}{9}r + s - 3\frac{1}{2}t\right) - (-2s - t)$ ______________

10. $\left(\frac{3m}{8} - 4\right) - \left(\frac{m}{4} - 2\right)$ ______________

Perform each operation.

11. Add $7q + 8$ to $3q - 4r + 1$. ______________

12. Subtract $x + 3y$ from $2x + 5y$. ______________

13. Subtract $a - 3b$ from $6a + 7b - 2$. ______________

14. Subtract $-3f - 2g + h$ from $-g + 6h - 1$. ______________

15. By how much does the perimeter of the rectangle exceed the perimeter of the triangle?

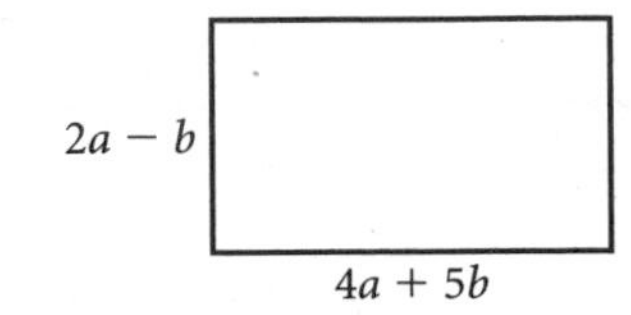

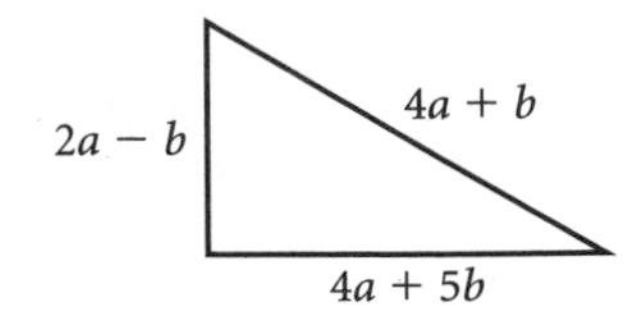

____

NAME ______________________ CLASS __________ DATE __________

Practice Masters Level A

2.7 Multiplying and Dividing Expressions

Simplify the following expressions. Use the Distributive Property if needed.

1. $3 \cdot 4y$ ______

2. $-2 \cdot 8a$ ______

3. $-5 \cdot (-6r)$ ______

4. $6b \cdot 9$ ______

5. $-4q \cdot 7$ ______

6. $-9c \cdot (-2)$ ______

7. $8n \div 2$ ______

8. $12h \div 4$ ______

9. $-10q \div 5$ ______

10. $6f \div (-3)$ ______

11. $-14p \div (-2)$ ______

12. $-5z \div (-5)$ ______

13. $2k \cdot 3k$ ______

14. $-6j \cdot 9j$ ______

15. $8w \cdot (-4w)$ ______

16. $-12u \cdot (-2u)$ ______

17. $h \cdot 7h$ ______

18. $-v \cdot (-v)$ ______

19. $3(2x + 1)$ ______

20. $5(4l + 3)$ ______

21. $6(2d - 8)$ ______

22. $7(7g - 1)$ ______

23. $-3(2m + 5)$ ______

24. $-11(m - 2)$ ______

25. $2b(5b + 2)$ ______

26. $7c(4c + 8)$ ______

27. $4s(2s - 1)$ ______

28. $-3y(2y + 5)$ ______

29. $-6f(10f - 7)$ ______

30. $-g(3g - 1)$ ______

31. $6d \cdot 2 + 5d \cdot 3$ ______

32. $8l \cdot 2 - 5l \cdot 5$ ______

33. $3j + 2(2j + 1)$ ______

34. $h - (4h - 1)$ ______

35. $2t - 4(5t + 9)$ ______

36. $3m - 2(-5m - 8)$ ______

37. $\frac{15 + 10a}{5}$ ______

38. $\frac{22x + 33}{11}$ ______

39. $\frac{14 - 7y}{7}$ ______

40. $\frac{50m - 25}{25}$ ______

NAME ______________________ CLASS ______________ DATE ______________

Practice Masters Level B

2.7 Multiplying and Dividing Expressions

Simplify the following expressions. Use the Distributive Property if needed.

1. $4y^2 - (3 - y^2)$ ______________
2. $6k^2 - 3(4 - k^2)$ ______________
3. $-1.5r \cdot 2$ ______________
4. $(-3.6m)(-2m)$ ______________
5. $\frac{1}{2}p \cdot 12$ ______________
6. $-\frac{3}{5}x \cdot 20$ ______________
7. $-14 \cdot \frac{2}{7}q$ ______________
8. $-26\left(-\frac{5}{13}t\right)$ ______________
9. $2.5(4y - 6)$ ______________
10. $-5.1(2a - 4)$ ______________
11. $-3\frac{1}{3}(9b - 6)$ ______________
12. $-1\frac{1}{4}(-4m - 8)$ ______________
13. $\frac{6s + 3}{-3}$ ______________
14. $\frac{16r - 24}{-8}$ ______________
15. $\frac{-9p - 27}{-3}$ ______________
16. $\frac{52c^2 - 28c}{-2}$ ______________
17. $\frac{75 + 2.5k}{0.5}$ ______________
18. $\frac{8d^2 + 6d + 4}{2}$ ______________

The formula for the area of a rectangle is $A = lw$. Find the area for each by evaluating the formula for the given length and width.

19. $w = 3$ inches; $\ell = 5$ inches

____

20. $w = 1.5$ feet; $\ell = 12$ feet

____

21. $w = 3a$; $\ell = 4a$

____

22. $w = 6.5x$; $\ell = 8x$

____

NAME ______________________ CLASS ____________ DATE ____________

Practice Masters Level C

2.7 Multiplying and Dividing Expressions

Simplify the following expressions. Use the Distributive Property if needed.

1. $2(a + b^2) - 3(-2a - b^2)$ ____________

2. $-1.5(4m + n) - 2(m - n)$ ____________

3. $\dfrac{16p^2 + 8p - 24}{-8}$ ____________

4. $\dfrac{-4.75r^2 + 2.5r - 1}{-0.25}$ ____________

5. $3(6a - 4b + 2c) - 2(1.5a - c) - 3(-b + c)$ ____________

6. $-\dfrac{1}{2}x(2x - 4) - \dfrac{1}{5}(10x^2 + 20y) + \dfrac{2}{3}(-9y + 6)$ ____________

The formula for the volume of a rectangular prism is *V* = *lwh*. Find the volume for each rectangular prism by evaluating the formula for the given length, width, and height.

7.

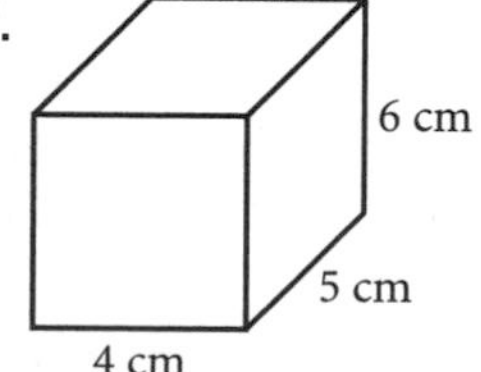

____

8.

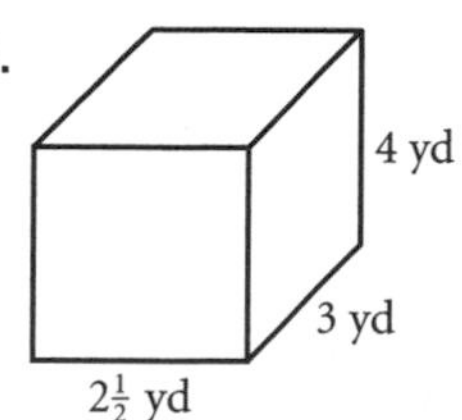

____

9.

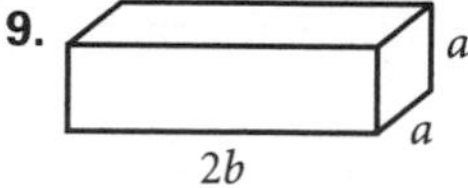

____

10.

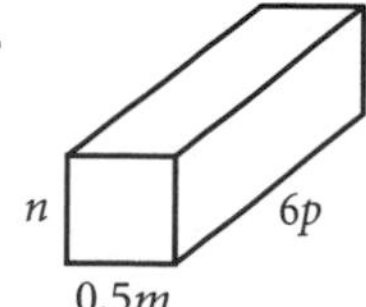

____

11. A student has a part-time job paying \$8 per hour. How much more money will the student make by working $(h + 6)$ hours in a week than by working $(h - 2)$ hours in a week? ____________

12. If the same student in Exercise 11 works $(h + 6)$ hours per week for 2 weeks and then $(h - 2)$ hours per week for 3 weeks, how much money will she earn? ____________

NAME ______________________ CLASS ______________ DATE ______________

Practice Masters Level A

3.1 Solving Equations by Adding and Subtracting

State the property needed to solve each equation.

1. $x + 3 = 7$ ______________________
2. $z - 8 = 6$ ______________________
3. $-28 + z = 14$ ______________________
4. $13 = k - 7$ ______________________
5. $x - 0.5 = 3$ ______________________
6. $m + 12 = -14$ ______________________
7. $n - 6 = -2$ ______________________
8. $-12 = k + 3$ ______________________
9. $-5 = j + 3$ ______________________
10. $17 = z - 7$ ______________________
11. $15 + n = -3$ ______________________
12. $f - 6 = -7$ ______________________
13. $g + 20 = -50$ ______________________
14. $-30 = 7 + h$ ______________________
15. $21 = 0.75 + m$ ______________________

Solve each equation. Check the solution.

16. $y - 6 = 10$ ________
17. $13 = h - 6$ ________
18. $z + 7 = -6$ ________
19. $t + 4 = -2$ ________
20. $-12 = -3 + y$ ________
21. $d - 4 = -2$ ________
22. $m - 7 = -8$ ________
23. $28 + k = -6$ ________
24. $x + 7 = -16$ ________
25. $15 = 17 + x$ ________
26. $-16 = z + 4$ ________
27. $j + 18 = 52$ ________
28. $-7 + y = 14$ ________
29. $25 + h = -23$ ________
30. $-52 = z - 12$ ________

NAME ______________________ CLASS ____________ DATE ____________

Practice Masters Level B

3.1 Solving Equations by Adding and Subtracting

State the property needed to solve each equation.

1. $z - (-2) = -3$ ______________
2. $j + (-3) = 5$ ______________
3. $-5 = m + (-7)$ ______________
4. $7 = j - (-8)$ ______________

Solve each equation. Check the solution.

5. $x + \frac{1}{2} = 1$ ______________
6. $-6.2 + k = -8.5$ ______________
7. $z + \frac{2}{3} = \frac{4}{3}$ ______________
8. $7.1 = -2.5 + h$ ______________
9. $4.3 + x = -2.1$ ______________
10. $a - 6\frac{1}{3} = -5\frac{1}{2}$ ______________
11. $y - \frac{4}{5} = \frac{3}{10}$ ______________
12. $b + 4\frac{1}{2} = 7\frac{2}{3}$ ______________
13. $13.2 = m + 2.1$ ______________
14. $d - \frac{4}{9} = \frac{5}{9}$ ______________
15. $-17.5 = n + 3.2$ ______________
16. $f - 3.7 = 28.5$ ______________
17. $\frac{6}{7} + j = \frac{5}{21}$ ______________
18. $-1.59 + t = -28.8$ ______________
19. $1\frac{1}{5} + p = -2\frac{1}{3}$ ______________
20. $4.62 = -3.5 + q$ ______________
21. $-8.4 + w = -2.7$ ______________
22. $y + \frac{7}{8} = -\frac{9}{10}$ ______________
23. $1\frac{3}{5} = m - 2\frac{1}{4}$ ______________
24. $x - 16.3 = 12.15$ ______________
25. $x - \frac{3}{4} = 2$ ______________
26. $5.1 + z = -7.3$ ______________
27. $b + 3.15 = -4.2$ ______________
28. $-6.2 + k = -8.5$ ______________

NAME ______________________ CLASS ____________ DATE ____________

Practice Masters Level C

3.1 Solving Equations by Adding and Subtracting

Solve each equation. Check the solution.

1. $\frac{1}{5} + x = -2.7$ ____________
2. $6 - k = 12$ ____________
3. $1\frac{2}{3} + h = 3\frac{1}{5}$ ____________
4. $m + \frac{7}{8} = -5.8$ ____________
5. $-16 = 5 - g$ ____________
6. $\frac{3}{4} - f = 0.75$ ____________
7. $p - 0.5 = \frac{6}{5}$ ____________
8. $n - 1.75 = -3.78$ ____________
9. $3.5 = k - \frac{7}{2}$ ____________
10. $15 = 12 - z$ ____________

Write and solve an equation for each situation.

11. Lilly bought a shirt for $18, and a pair of earrings for $7.50. If her sales total was $27.75, how much was the tax? ____________
12. The measures of two interior angles of a triangle are 30° and 68°. What is the measure of the third angle? ____________
13. Caroline is allowed to have 2150 calories each day. If she has already eaten 1375 calories, how many more calories can she eat? ____________
14. The population in a town is 315,000 this year. Five years ago, the population was 295,000. How much has the population grown in five years? ____________
15. A recipe calls for $1\frac{2}{3}$ cups of flour. If $\frac{3}{4}$ cups have already been used, how much more needs to be added? ____________
16. Jasmine has saved $350.28 for her trip to Europe. She needs $652.75 to pay for the entire trip. How much more money does she need? ____________
17. Jeffrey's test score increased by 15 points. If his last test grade was a 62, what was his score on this test? ____________

NAME ______________________ CLASS ______________ DATE ______________

Practice Masters Level A

3.2 Solving Equations by Multiplying and Dividing

State the property needed to solve each equation.

1. $3x = -21$ ______________

2. $-z = 12$ ______________

3. $\frac{m}{5} = 21$ ______________

4. $54 = -7k$ ______________

5. $\frac{1}{2} = \frac{y}{-3}$ ______________

6. $-9 = \frac{k}{7}$ ______________

7. $-12n = 98$ ______________

8. $14r = -112$ ______________

9. $-\frac{b}{7} = 32$ ______________

10. $-17 = 17z$ ______________

Solve each equation and check your solution.

11. $6y = -24$ ______________

12. $1.5 = 0.75h$ ______________

13. $-q = 100$ ______________

14. $-3t = -12$ ______________

15. $\frac{g}{0.5} = -6$ ______________

16. $\frac{d}{6} = -7$ ______________

17. $\frac{k}{4} = -12$ ______________

18. $-150h = 450$ ______________

19. $0.25j = -6.5$ ______________

20. $8 = -\frac{p}{2}$ ______________

21. $-16 = -z$ ______________

22. $\frac{s}{3} = 21$ ______________

23. $14 = -7y$ ______________

24. $-56 = 8f$ ______________

25. $-72 = -12z$ ______________

NAME ______________________ CLASS ____________ DATE ____________

Practice Masters Level B

3.2 Solving Equations by Multiplying and Dividing

State the property needed to solve each equation.

1. $\frac{3x}{5} = 2.1$ ______________________

2. $\frac{1}{3}p = 7$ ______________________

3. $1.5 = -5y$ ______________________

4. $-\frac{2.6}{3}g = -6$ ______________________

Solve each equation and check your solution.

5. $5x = -1$ ______________________

6. $-6.2k = 18.6$ ______________________

7. $\frac{2}{3}z = 4$ ______________________

8. $-5m = \frac{10}{11}$ ______________________

9. $\frac{3n}{4} = 21$ ______________________

10. $6\frac{1}{2}a = -45\frac{1}{2}$ ______________________

11. $\frac{1.2}{0.5}b = 12$ ______________________

12. $\frac{15}{2} = -\frac{5f}{7}$ ______________________

13. $13.2 = 2.2m$ ______________________

14. $-\frac{4}{9}d = \frac{5}{9}$ ______________________

15. $2 = -6n$ ______________________

16. $-10.2 = -5.1g$ ______________________

17. $8 = -\frac{3}{8}v$ ______________________

18. $\frac{6}{7} = 3m$ ______________________

19. $35 = -\frac{7n}{8}$ ______________________

20. $15\frac{9}{10} = 5\frac{3}{10}x$ ______________________

21. $15 = \frac{0.5}{3.5}k$ ______________________

22. $\frac{11p}{3} = -\frac{22}{5}$ ______________________

NAME ______________________ CLASS ____________ DATE ____________

Practice Masters Level C

3.2 Solving Equations by Multiplying and Dividing

Solve each equation.

1. $-1 = \frac{m}{-521}$ ______________

2. $\frac{2n}{5} = 1.5$ ______________

3. $-1\frac{2}{3}h = 3\frac{1}{5}$ ______________

4. $\frac{6}{7}k = \frac{6}{7}$ ______________

5. $-\frac{m}{5} = 0$ ______________

6. $\frac{3}{4}f = -0.75$ ______________

Solve each formula for the indicated variable.

7. $V = lwh$ for h ______________

8. $w_1d_1 = w_2d_2$ for d_2 ______________

9. $w = \frac{m}{r}$ for m ______________

10. $\frac{p}{q} = \frac{m}{n}$ for p ______________

11. $A = \frac{1}{2}h(b_1 + b_2)$ for h ______________

12. $E = mc^2$ for m ______________

Write an equation for each situation. Then solve the equation.

13. The sum of the measures of the interior angles of a regular hexagon is 705°. What is the measure of each angle? ______________

14. A recipe calls for 4 cups of flour when making 8-dozen cookies. How many cups of flour are needed to make 6-dozen cookies? ______________

15. Leslie purchases 9-dozen pencils for $16.20. How much does each pencil cost? ______________

16. Mrs. Howard has $64 to spend on supplies for her classroom. If all she needs to buy are overhead projector light bulbs, which cost $16 each, how many is she able to purchase? ______________

17. If 12 bags of dog food cost $8.50 each, what is the total cost? ______________

18. What is the price of one apple if a dozen costs $3.99? ______________

19. Lydia arrives at work in 20 minutes. If she drives 20 miles to work, how fast is she driving? ______________

NAME ______________________ CLASS ____________ DATE ____________

Practice Masters Level A

3.3 Solving Two-Step Equations

Solve each equation.

1. $6x + 7 = 49$ ____________
2. $5m - 8 = -63$ ____________
3. $-27 = 2x + 3$ ____________
4. $8 = \frac{k}{3} + 2$ ____________
5. $3x - 6 = 15$ ____________
6. $4n + 8 = 16$ ____________
7. $\frac{b}{7} - 8 = 1$ ____________
8. $3f - 12 = 18$ ____________
9. $13y + 1 = 27$ ____________
10. $74 = 3z - 10$ ____________
11. $2p - 8 = 14$ ____________
12. $9 = \frac{g}{5} - 1$ ____________
13. $\frac{h}{4} - 5 = 9$ ____________
14. $27q + 6 = -48$ ____________

Write an equation for each situation. Do not solve.

15. Twice Andy's age plus 3 is 25. How old is Andy? ____________

16. The number of CDs sold last month was 8 more than 3 times the number of CDs sold this month. Last month, 107 CDs were sold. How many were sold this month? ____________

17. The perimeter of triangle *ABC* is 28 inches. Side *AB* is 6 inches long and side *BC* is 8 inches long. Find the length of side *AC*. ____________

18. Seven more than the average of 4 numbers is 52. Find the sum of the 4 numbers. ____________

19. Eleven less than the quotient of a number and 3 is 18. Find the number. ____________

20. The height of a plant is 8 more inches than twice the height of the plant last month. The plant is 32 inches this month. Find the height of the plant last month. ____________

21. Jerry is making $5000 more than 3 times his salary 20 years ago. Jerry made $71,000 now. How much did he make twenty years ago? ____________

22. The sum of 9 and the quotient of a number and 5 is 21. Find the number. ____________

NAME ______________________ CLASS ______________ DATE ______________

Practice Masters Level B

3.3 Solving Two-Step Equations

Solve each equation.

1. $8 - 4d = 12$ ____________
2. $-11 - 10f = -41$ ____________
3. $5.3 = -b + 1.8$ ____________
4. $9 = 7 - \frac{j}{4}$ ____________
5. $12h - 2.5 = -26.5$ ____________
6. $1 - 4n = 65$ ____________
7. $\frac{b}{4} - 3.2 = 12.3$ ____________
8. $12 - f = 22$ ____________
9. $0.5y + 5.2 = -2.8$ ____________
10. $-18 = -4z + 12$ ____________
11. $-p + 10 = -25$ ____________
12. $12 = -\frac{r}{6} + 3$ ____________
13. $8 - \frac{h}{3} = -10$ ____________
14. $0.75q + 2.5 = -6.5$ ____________
15. $15 = -7 - \frac{z}{6}$ ____________
16. $-1.25q - 12.5 = 8.35$ ____________
17. $-\frac{k}{3} - \frac{1}{5} = \frac{4}{5}$ ____________
18. $9 = 2.1 - 0.25m$ ____________
19. $9 - \frac{b}{5} = 12$ ____________
20. $11 - 2q = 5$ ____________

Write and solve an equation for each situation.

21. Alex has some quarters and \$5.15 in nickels. If he has \$7.90 total, how many quarters does he have? ____________

22. The difference of the quotient of a number and -2 from 12, is 15. Find the number. ____________

23. An Internet service offers internet access for \$29.95 a month with a \$15.75 initial charge for the hookup. If Glenda spent a total of \$187.45 on her Internet access, how many months did she pay for? ____________

24. If 11.2 less than the average of 5 numbers is 27.75, what is the sum of the 5 numbers? ____________

NAME ______________________ CLASS ____________ DATE ____________

Practice Masters Level C

3.3 Solving Two-Step Equations

Solve each equation.

1. $\frac{x}{-2} - 2.3 = -1.2$ ____________

2. $2.5 + \frac{m}{-2} = 5.5$ ____________

3. $\frac{n}{3} - 3.5 = 4.75$ ____________

4. $7.1 - \frac{p}{4} = -2.15$ ____________

5. $\frac{q}{-3} + 4\frac{1}{5} = \frac{3}{5}$ ____________

6. $\frac{7}{8} - \frac{r}{4} = -3.5$ ____________

7. $\frac{s}{-3} + 1.5 = 1.5$ ____________

8. $\frac{t}{2} - 6\frac{1}{3} = -8\frac{1}{3}$ ____________

9. $\frac{w}{4} + 5\frac{3}{4} = -8\frac{1}{2}$ ____________

10. $31.5 - x = 20.5$ ____________

11. $-\frac{4}{7} + \frac{y}{3} = 1\frac{3}{7}$ ____________

12. $\frac{z}{5} - 3\frac{3}{10} = -4\frac{1}{2}$ ____________

13. $-a - 19.2 = 7.8$ ____________

14. $2.75 + \frac{b}{-4} = 3.25$ ____________

15. $\frac{c}{4} - \frac{7}{12} = \frac{5}{6}$ ____________

16. $-d - 8.2 = 6.8$ ____________

17. $-12\frac{1}{3} + \frac{f}{-3} = -5\frac{2}{3}$ ____________

18. $\frac{g}{2} + 4\frac{1}{3} = -2\frac{5}{6}$ ____________

19. $h - 86 = (-2.6)(-5)$ ____________

20. $7\frac{3}{4} + \frac{k}{-4} = -1\frac{1}{4}$ ____________

Write and solve an equation for each situation.

21. Kati has some dimes and $15.75 in quarters. If she has $27.95 total, how many dimes does she have? ____________

22. The price of a disposable camera was reduced by half and then another $5.50 was taken off the price. The new selling price is $6.00. What was the original price of the camera? ____________

23. After making 6 craft projects and spending $21.20 on lunch, the Rheam family realized they had spent a total of $52.70. How much did the family pay per craft? ____________

NAME ______________________ CLASS __________ DATE __________

Practice Masters Level A

3.4 Solving Multistep Equations

Solve and check each equation.

1. $6y - 4 = y + 6$ ______
2. $8n + 5 = 6n - 3$ ______
3. $10y - 13 = 3y + 8$ ______
4. $3y + 4 = 5y + 6$ ______
5. $4x - 8 = 10 - 2x$ ______
6. $5x + 4 = 2x - 5$ ______
7. $3 - 12m = 15 - 8m$ ______
8. $6z - 6 = 3z + 9$ ______
9. $8y - 2 = 6 + 4y$ ______
10. $16m - 12 = 12 + 4m$ ______
11. $2h - 8 = 3h - 10$ ______
12. $5y + 8 = -7 - 10y$ ______
13. $3x - 4 = 5x + 6$ ______
14. $6z - 3 = 4z + 5$ ______
15. $2h + 8 = 5h - 7$ ______
16. $12a - 8 = 16 + 4a$ ______
17. $3x - 5 = 2x + 6$ ______
18. $9y - 6 = 12 + 3y$ ______
19. $7h + 2 = 3h - 10$ ______
20. $4 - 4x = 2x - 8$ ______
21. $2a - 5 = 3a - 3$ ______
22. $3 + 2x = 7x - 7$ ______
23. $5y - 14 = 2y + 7$ ______
24. $8m + 6 = m - 8$ ______
25. $6 - 9y = -12y - 6$ ______
26. $11x - 4 = 5 + 8x$ ______
27. $2x - 7 = 3 - 3x$ ______
28. $4m + 6 = 8m - 10$ ______
29. $5a + 9 = 2a - 3$ ______
30. $14x - 3 = 7 + 9x$ ______

Write and solve an equation for each situation.

31. The product of 8 and a number, then added to 5, is 3 less than the product of 4 and the same number. ______

32. Noelle needs to have her computer fixed. One company charges a $25 estimation fee plus $45 per hour. Another company charges a $10 estimation fee plus $60 per hour. Find the number of hours for which the two costs would be the same. ______

33. The difference of a number from 8 is 3 more than the product of the same number and 5. ______

NAME ______________________ CLASS ____________ DATE ____________

Practice Masters Level B

3.4 Solving Multistep Equations

Solve and check each equation.

1. $12x - 7 = 8 + 3x + 12$ ____________
2. $8 - 8h = 7 - 3h + 16$ ____________
3. $7y - 4 = -3y + 11$ ____________
4. $3 - 7a = -9 - 4a + 3$ ____________
5. $1.5x - 5.3 = 2.5x - 2.7$ ____________
6. $9z - 10 = -3 + 2z + 7$ ____________
7. $13 - 3y = -8 + 4y - 7$ ____________
8. $2 - 18y - 6 = -7y + 3 + 10y$ ____________
9. $6 + 7a = -3 + 4a - 9$ ____________
10. $4x - 3 = -11 - 2x$ ____________
11. $11m + 15 = 4 + 7m - 5$ ____________
12. $-8x - 2 = -20 + 6x + 6$ ____________
13. $3y - 7 = 6 - 6y + 14$ ____________
14. $-4 - 6h - 5 = -3h + 3$ ____________
15. $9m - 11 = 2 - 2m + 9$ ____________
16. $5 - 4a + 3 = -6a - 8$ ____________
17. $10x - 4 = 15 + 6x - 7$ ____________
18. $7a - 16 = 3 + 12a + 6$ ____________
19. $-20 + 4y - 6 = 5 - 3y + 4$ ____________
20. $1.8 + 2.2h + 3.7 = -2.5 - 1.8h + 0.5$ ____________
21. $3.3h - 7.2 = 3.5 + 0.8h + 3.8$ ____________
22. $-x + 5 = -2x + 8 - 4x$ ____________
23. $0.25z - 0.75 = -0.25x + 3.25$ ____________
24. $1 + 2n + 3 = 4n - 5$ ____________
25. $3f + 9f - 3 = 7 - f$ ____________
26. $1.25 - 0.3k + 0.5 = 1.3 + 0.5k$ ____________
27. $-4x + 1 = 1 - x$ ____________
28. $5p + 8 = 3p + 10 - 2$ ____________
29. $3.2b - 6 = 2.1b + 27$ ____________
30. $-4x - 2 = -3 - 3x$ ____________

Write and solve an equation for each situation.

31. Sherry spent the same amount of money at the grocery store as she did at the drug store. At the grocery store, she spent \$3.40 on steak, \$1.19 on bread, and bought apples that cost \$0.15 each. At the drug store, she spent \$4.99 on shampoo and bought pencils that cost \$0.10 each. How many pencils and how many apples did Sherry buy? ____________

32. One less than the product of a number and negative 3 is the same as the sum of the same number and 11. What is the number? ____________

33. The area of a rectangle decreased by 4 is the same as the sum of -5 and twice the area of the rectangle. Find the area of the rectangle. ____________

NAME ____________________ CLASS __________ DATE __________

Practice Masters Level C

3.4 Solving Multistep Equations

Solve and check each equation.

1. $a - 16 = \frac{a}{4} + 2$ ________

2. $\frac{b}{5} + 8 = \frac{b}{2} + 10$ ________

3. $5 - \frac{c}{6} = 7 + \frac{c}{3}$ ________

4. $4d + \frac{3}{4} + \frac{5d}{4} = \frac{1}{2} + 2d$ ________

5. $12 + 4f = -2 + \frac{5f}{3}$ ________

6. $3 - \frac{g}{3} = \frac{g}{6} - 8$ ________

7. $\frac{h}{4} + 3 = -5 - \frac{h}{4}$ ________

8. $f - \frac{2}{5} + \frac{3f}{10} = \frac{1}{5} - 2f$ ________

9. $3k - 4 = \frac{k}{2} + 3.5$ ________

10. $12 + \frac{m}{4} = \frac{m}{8} - 4$ ________

11. $2 - \frac{n}{10} = 12 - \frac{n}{5}$ ________

12. $p + \frac{3}{5} + \frac{9d}{10} = 3d - \frac{1}{2}$ ________

13. $-q + 7 = \frac{2q}{3} - 3$ ________

14. $\frac{r}{10} + 2 = \frac{r}{5} - 3$ ________

15. $7 + \frac{s}{14} = -3 + \frac{s}{7}$ ________

16. $t + \frac{2}{3} + \frac{5t}{6} = 4t - \frac{5}{12}$ ________

17. $-3 + 3w = 2 - \frac{3w}{4}$ ________

18. $\frac{x}{3} - 6 = \frac{x}{2} + 4$ ________

19. $4 - \frac{y}{2} = 5 + \frac{y}{6}$ ________

20. $z - \frac{1}{4} - \frac{3z}{8} = z + \frac{1}{2}$ ________

Write and solve an equation for each situation.

21. In his first 4 algebra tests, Sam earned an 85, 92, 75, and 62. Determine how many points he must earn on the fifth test in order to have an average of 80. ________

22. The difference between 6 and the quotient of a number and 2 is equal to the sum of 8 and the quotient of the same number and 4. Find the number. ________

NAME ______________________ CLASS ____________ DATE ____________

Practice Masters Level A

3.5 Using the Distributive Property

Solve each equation.

1. $5s + 4 + s = -20$ ________
2. $9(c-2) = 18$ ________
3. $12a - 7 - 5a = 14$ ________
4. $6y - 4 = 2y + 12$ ________
5. $(3x - 2)7 = 5x + 2$ ________
6. $4m + 12 = m - 3$ ________
7. $-2y(3 + 4) = 8y + 12$ ________
8. $4(2 + 3r) = 4$ ________
9. $7c + 5 = 2(3c - 4)$ ________
10. $10d = (8d - 6)2$ ________
11. $3a - 9 = 3(9 - 3a)$ ________
12. $18x = 2(3x + 6)$ ________
13. $6(h - 2) = -6$ ________
14. $11(h - 2) = 6h + 3$ ________
15. $12b - 6 = 6(2 - b)$ ________
16. $7(2d - 3) = 6d + 3$ ________
17. $23 - 2a = 3(a - 4)$ ________
18. $-16 = 8(x - 3)$ ________
19. $-8g + 6 = (6g + 8)2$ ________
20. $34 = -2(4 - 3x)$ ________
21. $10a - 4 = 3(6a - 4)$ ________
22. $7(4 - 2g) = -7g$ ________
23. $2y - 12 = -4(2y - 2)$ ________
24. $6x = -3(2x - 4)$ ________
25. $2(x + 4) = 4(5 - 3x)$ ________
26. $-5h = 2(-2h + 6)$ ________
27. $-4y = 4(5y + 6)$ ________
28. $2 = -(4l - 2)$ ________
29. $-2(f + 1) = -2$ ________
30. $6(d - 1) = 3(d + 5)$ ________
31. $-2g - 10 + 3g = -4$ ________
32. $x = 3(5 + 2x)$ ________

Write an equation for each situation. Do not solve.

33. Mrs. Tyler has two rectangular flower gardens. If the perimeters of the two gardens are equal, find the value of x.

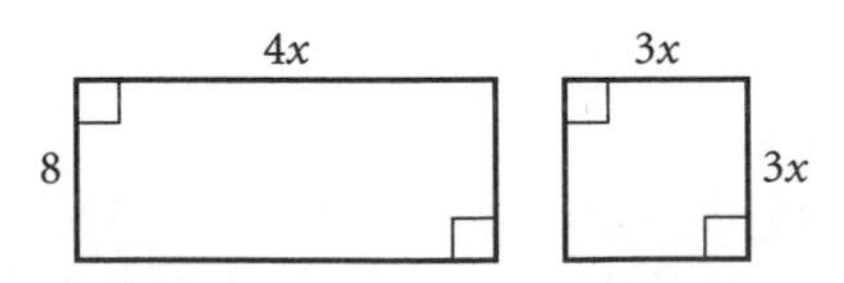

____

34. The product of 3 and the sum of a number and 5 is the same as the product of -2 and the difference of the same number from 3. Find the number. ________

Practice Masters Level B

3.5 Using the Distributive Property

Solve each equation.

1. $3(2d - 7) = (2 + 4d)2$ ______________
2. $10 + 3(4 - 6a) = -(6a + 12)$ ______________
3. $-3(2x + 4) = 3(x - 6)$ ______________
4. $6 + 2(5 - 3a) = 2a(2 + 8)$ ______________
5. $4 - (2x - 3) = 3(x + 1)$ ______________
6. $2(4a + 3) = 8 + 6(2a - 3)$ ______________
7. $5(x - 7) = 2(2 - 2x) - 3$ ______________
8. $-2 - 4(2y - 1) = (6 - 2y)3$ ______________
9. $22 + 3(a + 6) = -4(3a + 5)$ ______________
10. $-4(7y + 9) = (3 - 5y)3 - 4y$ ______________
11. $-8(2y - 4) = 9(4 - 2y)$ ______________
12. $3(x + 5) + 2 = -11 - 4x$ ______________
13. $8z + 2(3 + 2z) = 3(6 + 2z)$ ______________
14. $5d - 8 = 3(d + 2)$ ______________
15. $4a + 6 - 2a = 3(a - 3)$ ______________
16. $9y - 3(y + 4) = (4 + y)3$ ______________
17. $1.5(-2x - 4) = (3.5 + 4x)2 - 2$ ______________
18. $6.5(z - 2) = -2.5(3z + 3) - 6.5$ ______________
19. $9(y + 1) + 6 = 2(y + 3)$ ______________
20. $11(2m + 2) - 6 = 4m + 4(3m - 2)$ ______________
21. $-2g + 4(4 + g) = 2g - 4(6 - 3g)$ ______________
22. $3(1.5 + 2.5x) = -6.5 + 5.5x - 2.5(4 + 2x)$ ______________
23. $4z + 3(z + 1) = -22 + 4(3z - 5)$ ______________

Write and solve an equation for each situation.

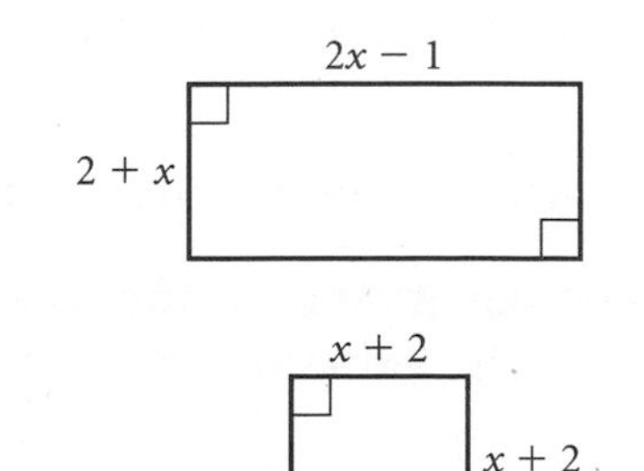

24. The perimeters of the two rectangular pools shown are equal. Find the value of x and the perimeter of the pools.

____

25. The product of 3 and the sum of a number and 5, then added to 6, is the same as the product of -2 and the difference of the same number from 3, then minus 5. Find the number. ______________

NAME ______________________ CLASS ______________ DATE ______________

Practice Masters Level C

3.5 Using the Distributive Property

Solve each equation.

1 $a + 4(2a + 1) = 3(3a - 2) + 10a$ ______________

2. $3(4h - 2) - 4h = 11(3 - 4h) + 39h$ ______________

3 $3x - 5 + x = 6 + 4x - 2$ ______________

4. $4(2g + 3) - 2g + 3 = 2g + 2 + 3(2g + 2)$ ______________

5 $4(k - 4) - 8(k + 4) = 5(3k - 4) - 9(3k + 4)$ ______________

6. $8(1 - m) + 2(4 + m) = 3(6 - 3m) - 5(m - 2)$ ______________

7. $4.8x + 0.2(x - 2) = 3.3 - 0.5(2 - 4x)$ ______________

8. $2(5 + z) + 1 = 11 - 4z + 6z$ ______________

9. $0.2(3n + 4) - 5n = 2.2(4 - 2n) - 3.4$ ______________

10. $2(b - 4) - 6(b + 8) = 12(b + 4) - (8b + 6)$ ______________

11. $\frac{1}{2}(4 - 10f) + 5 = \frac{2}{3}(6f - 3)$ ______________

12. $2(a + 3) - a = 10 + a - 4$ ______________

13. $2c - 3(c - 5) = 4(2c + 1) - 3c$ ______________

14. $2(3d + 1) - 9d = 8(5 + 2d)$ ______________

15. $\frac{3}{2}(10x - 4) - 5x = 7 - \frac{1}{5}(5 - 10x)$ ______________

16. $2(3j + 2) - 4(2j + 3) = 3(2j - 1) - 6(j + 5)$ ______________

17. $-3d + 5 + d = 3 - 2d + 5$ ______________

Write and solve an equation for each situation.

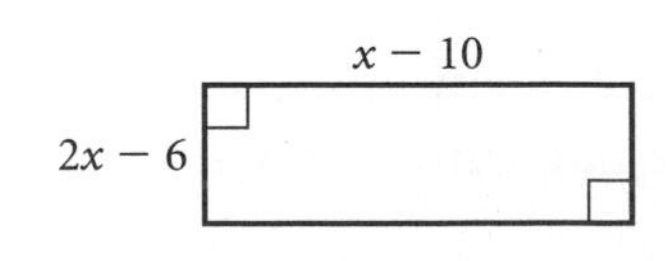

18. The perimeters of the two rectangles shown are equal. Find the value of x and the perimeter.

____

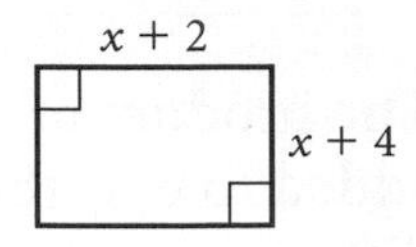

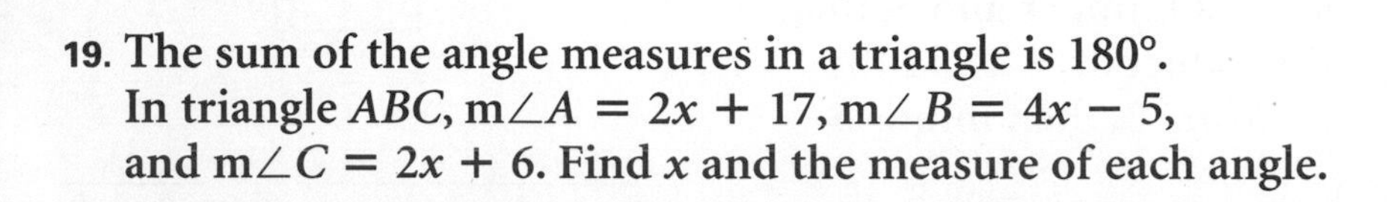

19. The sum of the angle measures in a triangle is 180°. In triangle ABC, $m\angle A = 2x + 17$, $m\angle B = 4x - 5$, and $m\angle C = 2x + 6$. Find x and the measure of each angle.

____

NAME ______________________ CLASS ______________ DATE ______________

Practice Masters Level A

3.6 Using Formulas and Literal Equations

Use the formula $\frac{9}{5}C + 32 = F$ to convert each temperature from degrees Celsius to degrees Fahrenheit.

1. 122°C ______________
2. 64.4°C ______________
3. 5°C ______________
4. 0°C ______________
5. 10°C ______________
6. 25°C ______________

Use the formula $A = bh$ to find the length of the missing base, b, for parallelograms with the given dimensions.

7. $A = 24, h = 8$ ______________
8. $A = 105, h = 15$ ______________
9. $A = 56, h = 4$ ______________
10. $A = 48, h = 12$ ______________
11. $A = 10, h = 1$ ______________
12. $A = 18, h = 2$ ______________

Solve each equation for the indicated variable.

13. $S + L = Q$, for L ______________
14. $S = L - Q$, for L ______________
15. $Q + S = L$, for Q ______________
16. $d + f = g$, for f ______________
17. $d = g - f$, for f ______________
18. $d + f = -g$, for d ______________
19. $d - f = g$, for d ______________
20. $m = nr$, for n ______________
21. $5z = y$, for z ______________
22. $hj = k$, for j ______________
23. $\frac{m}{r} = s$, for m ______________
24. $\frac{d}{k} = -5$, for d ______________
25. $R_1 + R_2 = R_T$ for R_1 ______________
26. $-J + K = L$, for K ______________
27. $ab = c$, for b ______________
28. $-3g = h$, for g ______________
29. $m - n = p$, for n ______________
30. $\frac{-b}{c} = d$, for b ______________
31. $t - s = -w$, for s ______________
32. $g - h = -k$, for g ______________
33. $ts = -w$, for t ______________
34. $-cd = -f$, for d ______________
35. $-b - c = -d$, for c ______________
36. $-2a = -b$, for a ______________

NAME ______________________ CLASS __________ DATE __________

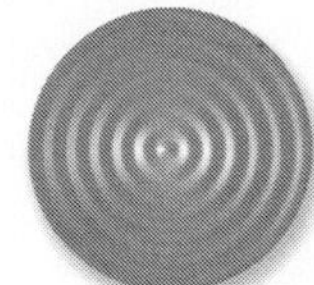

Practice Masters Level B

3.6 Using Formulas and Literal Equations

Use the formula $A = \frac{bh}{2}$ to find the length of the missing base, b, for triangles with the given dimensions.

1. $A = 3, h = 2$ ______
2. $A = 85, h = 10$ ______
3. $A = 20, h = 5$ ______
4. $A = 300, h = 25$ ______
5. $A = 42, h = 7$ ______
6. $A = 81, h = 9$ ______
7. $A = 56, h = 4$ ______
8. $A = 30, h = 12$ ______
9. $A = 102, h = 17$ ______
10. $A = 12, h = 6$ ______

Solve each equation for the indicated variable.

11. $y = mx - 5$, for m ______
12. $y = 2x + 2b$, for b ______
13. $2q + 3r = m$, for r ______
14. $3a - 5b = 4$, for a ______
15. $7m + 7n = p$, for m ______
16. $qx - 3 = d$, for x ______
17. $bx + c = 6$, for x ______
18. $q + p = 5m$, for q ______
19. $3y + 2x = 1$, for x ______
20. $gh + 5 = m$, for h ______
21. $5x + 2y = 3$, for y ______
22. $-2x + 3y = -5$, for x ______
23. $\frac{b}{j} + h = t$, for b ______
24. $\frac{n}{m} + 3 = 5$, for n ______
25. $\frac{b + k}{j} = t$, for k ______
26. $bk - p = q$, for k ______
27. $3m - n = k$, for n ______
28. $\frac{-b}{c} + 5 = d$, for b ______
29. $t - 4s = -w$, for s ______
30. $-4x - 6y = -3z$, for x ______

31. The perimeter of a rectangular frame is 24 inches and the length is twice the width. Use the formula $P = 2l + 2w$ to find the length and the width. ______

NAME ______________________ CLASS __________ DATE __________

Practice Masters Level C

3.6 Using Formulas and Literal Equations

Use the formula $A = \frac{1}{2}h(b_1 + b_2)$ to find the length of the missing base for trapezoids with the given dimensions.

1. $A = 100, h = 5, b_1 = 12$ ______
2. $A = 75, h = 4.5, b_1 = 10\frac{1}{3}$ ______
3. $A = 56, h = 8, b_1 = 12$ ______
4. $A = 25, h = 3, b_1 = 5\frac{2}{3}$ ______
5. $A = 240, h = 20, b_1 = 14$ ______
6. $A = 400, h = 10, b_1 = 60$ ______
7. $A = 400, h = 20, b_1 = 28$ ______
8. $A = 14, h = 2, b_1 = 4$ ______
9. $A = 100, h = 2.5, b_1 = 25$ ______
10. $A = 60, h = 5, b_1 = 10$ ______

Solve each equation for the indicated variable.

11. $\frac{1}{2}g + h = \frac{-2}{3}$, for g ______
12. $\frac{h}{2} + m = \frac{-2k}{3}$, for h ______
13. $2.5b + 3.1r = p$, for r ______
14. $\frac{1}{2}g + \frac{3}{4}h = m$, for h ______
15. $7.5m - 7.2 = p$, for m ______
16. $0.3x - 3.2 = 1.1d$, for x ______
17. $\frac{g + h}{4} + 5 = l$, for g ______
18. $1.5m + 3.2 = 5q$, for q ______
19. $0.3y + 0.2x = 0.1$, for x ______
20. $\frac{k - m}{2} - 2 = b$, for k ______

21. The circumference of a circle is 60 centimeters. Use the formula $C = 2\pi r$ to find the radius r of the circle. Use 3.14 as an approximation for π, and round your answer to the nearest hundredth.

____

22. A trapezoid has an area of 45 square inches. The shorter base measures 7 inches and the height is 6 inches. Use the formula $A = \frac{1}{2}h(b_1 + b_2)$ to find the length of the longer base.

____

NAME ______________________ CLASS __________ DATE __________

Practice Masters Level A

4.1 Using Proportional Reasoning

List the means and extremes of each given proportion.

1. $\frac{12}{7} = \frac{48}{28}$ ______________

2. $\frac{9}{32} = \frac{54}{192}$ ______________

3. $\frac{14}{5} = \frac{98}{35}$ ______________

4. $\frac{18}{45} = \frac{126}{315}$ ______________

5. $\frac{10}{50} = \frac{12}{60}$ ______________

6. $\frac{36}{73} = \frac{144}{292}$ ______________

Determine whether each proportion is true.

7. $\frac{10}{6} = \frac{20}{12}$ ______________

8. $\frac{15}{50} = \frac{180}{500}$ ______________

9. $\frac{4}{14} = \frac{32}{112}$ ______________

10. $\frac{23}{7} = \frac{115}{35}$ ______________

11. $\frac{18}{14} = \frac{36}{42}$ ______________

12. $\frac{54}{126} = \frac{324}{630}$ ______________

13. $\frac{1}{4} = \frac{10}{40}$ ______________

14. $\frac{17}{22} = \frac{51}{60}$ ______________

15. $\frac{27}{63} = \frac{3}{7}$ ______________

Solve each proportion. Round to the nearest hundredth if necessary.

16. $\frac{12}{n} = \frac{60}{50}$ ______________

17. $\frac{32}{13} = \frac{224}{x}$ ______________

18. $\frac{33}{5} = \frac{m}{80}$ ______________

19. $\frac{y}{16} = \frac{27}{72}$ ______________

20. $\frac{5}{70} = \frac{p}{28}$ ______________

21. $\frac{68}{z} = \frac{272}{35}$ ______________

22. $\frac{360}{24} = \frac{150}{x}$ ______________

23. $\frac{140}{y} = \frac{105}{60}$ ______________

24. $\frac{120}{4} = \frac{x}{9}$ ______________

25. $\frac{520}{x} = \frac{1200}{20}$ ______________

26. $\frac{75}{x} = \frac{5}{150}$ ______________

27. $\frac{y}{22} = \frac{4}{64}$ ______________

Practice Masters Level B

4.1 Using Proportional Reasoning

Determine whether each proportion is true.

1. $\frac{124}{81} = \frac{372}{324}$ ______

2. $\frac{13.5}{29} = \frac{81}{174}$ ______

3. $\frac{42.9}{31} = \frac{128.7}{96.1}$ ______

4. $\frac{22.3}{16.4} = \frac{178.4}{131.2}$ ______

5. $\frac{463}{561} = \frac{1852}{2244}$ ______

6. $\frac{612}{879} = \frac{2080}{2197}$ ______

Solve each proportion. Round to the nearest hundredth if necessary.

7. $\frac{x}{14} = \frac{15}{20}$ ______

8. $\frac{18}{y} = \frac{90}{12.5}$ ______

9. $\frac{162}{94} = \frac{z}{658}$ ______

10. $\frac{458}{817} = \frac{916}{n}$ ______

11. $\frac{42.8}{21} = \frac{m}{189}$ ______

12. $\frac{58.6}{p} = \frac{410.2}{122.5}$ ______

13. $\frac{q}{28.8} = \frac{4.3}{72}$ ______

14. $\frac{201.9}{x} = \frac{1615.2}{547.2}$ ______

15. $\frac{r}{3204} = \frac{612}{801}$ ______

16. $\frac{1400}{w} = \frac{1050}{600}$ ______

Solve.

17. If Alexa buys 3 shirts for \$42.00, how much would 8 shirts cost at the same price? ______

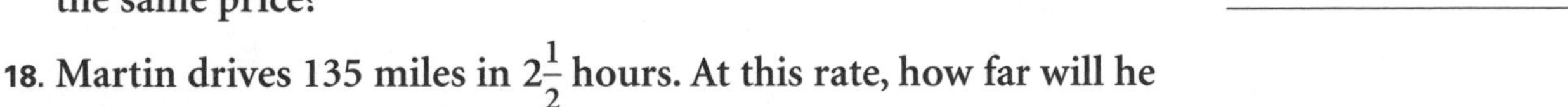

18. Martin drives 135 miles in $2\frac{1}{2}$ hours. At this rate, how far will he drive in 4 hours? ______

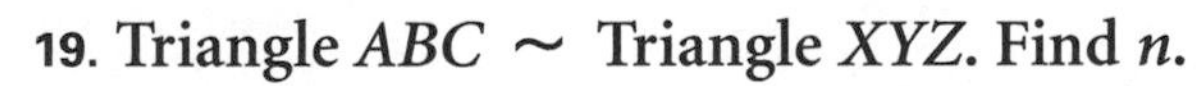

19. Triangle $ABC \sim$ Triangle XYZ. Find n.

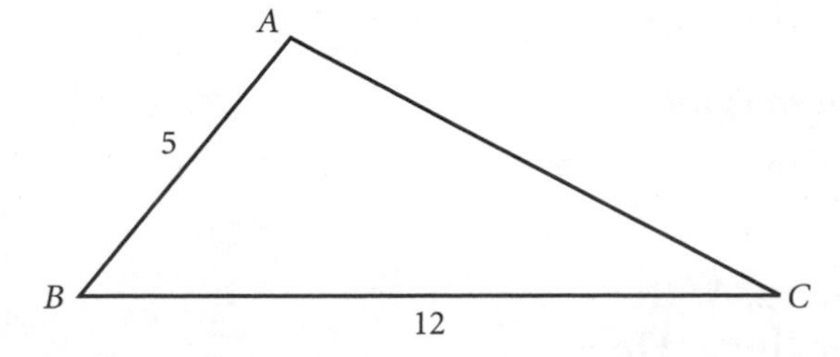

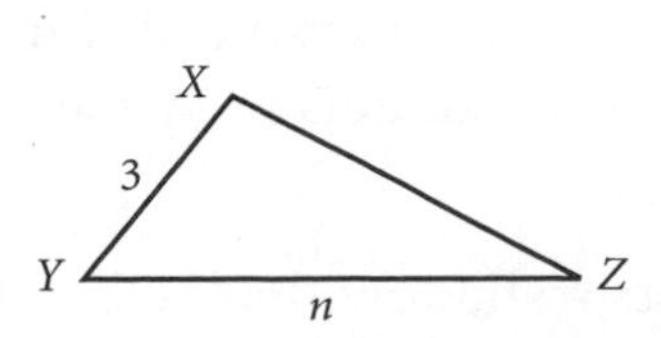

______

NAME ______________________ CLASS ____________ DATE ____________

Practice Masters Level C

4.1 Using Proportional Reasoning

Determine whether each proportion is true.

1. $\frac{68}{142} = \frac{272}{568}$ ____________
2. $\frac{184}{89} = \frac{1288}{623}$ ____________
3. $\frac{56}{163} = \frac{336}{815}$ ____________
4. $\frac{83.5}{12.6} = \frac{250.5}{378}$ ____________
5. $\frac{45.7}{28.7} = \frac{60.8}{36.2}$ ____________
6. $\frac{563}{312.1} = \frac{1126.5}{936.3}$ ____________

Solve each proportion. Round to the nearest hundredth if necessary.

7. $\frac{63.7}{x} = \frac{7}{9.2}$ ____________
8. $\frac{y}{243} = \frac{51}{729}$ ____________
9. $\frac{85.6}{407} = \frac{n}{610.5}$ ____________
10. $\frac{631}{260} = \frac{126.2}{m}$ ____________
11. $\frac{98.32}{p} = \frac{393.28}{102}$ ____________
12. $\frac{q}{1400} = \frac{600}{1050}$ ____________
13. $\frac{r}{115.8} = \frac{287.35}{405.3}$ ____________
14. $\frac{6201}{477} = \frac{1378}{w}$ ____________

Solve.

15. Michael runs 4 miles in 40 minutes. At this rate, how long will it take him to run 7 miles? ____________
16. If Celeste buys 5 cans of soda for \$3.00, how much will 12 cans cost at the same price? ____________
17. Pentagon *ABCDE* ~ pentagon *JKLMN*. Find *y*. ____________

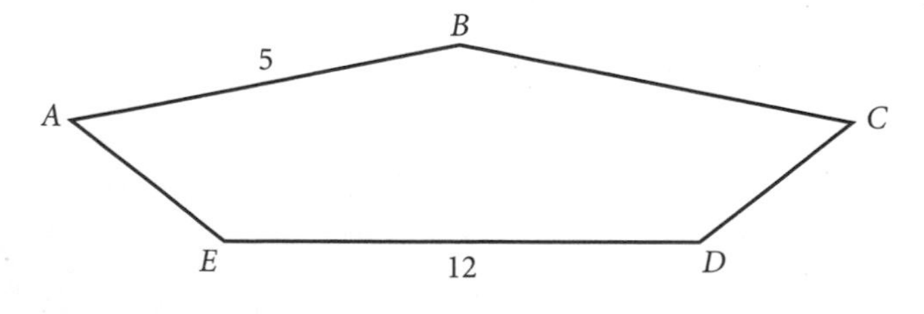

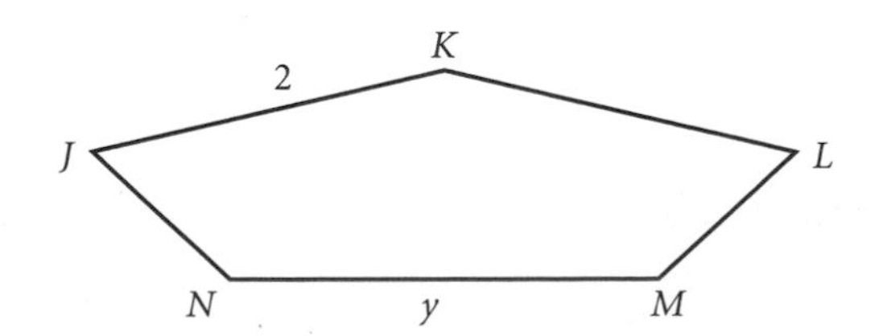

18. A recipe for 2 loaves of banana bread calls for 3 bananas and 2 cups of flour. How many bananas are needed to make 10 loaves of banana bread? ____________
19. A 90-foot building casts a shadow that is 58.5 feet long. At the same time, a fountain casts a shadow that is 9.75 feet long. How tall is the fountain? ____________

NAME ______________________ CLASS ____________ DATE ____________

Practice Masters Level A

4.2 Percent Problems

Write each percent as a decimal.

1. 4% ______ 2. 28% ______ 3. 67% ______

4. 150% ______ 5. 72% ______ 6. 230% ______

7. 55% ______ 8. 9% ______ 9. 32% ______

10. 1% ______ 11. 85% ______ 12. 100% ______

Write each decimal as a percent.

13. 0.12 ______ 14. 0.08 ______ 15. 1.85 ______

16. 2 ______ 17. 0.57 ______ 18. 3.6 ______

19. 0.7 ______ 20. 0.43 ______ 21. 0.03 ______

22. 0.26 ______ 23. 1.2 ______ 24. 0.98 ______

Write each percent as a fraction or mixed number in lowest terms.

25. 23% ______ 26. 50% ______ 27. 80% ______

28. 45% ______ 29. 64% ______ 30. 18% ______

31. 73% ______ 32. 36% ______ 33. 14% ______

34. 35% ______ 35. 81% ______ 36. 99% ______

Find each answer.

37. What is 20% of 80? ______ 38. What is 45% of 160? ______

39. Find 32% of 50. ______ 40. Find 60% of 185. ______

41. 35 is 10% of what number? ______ 42. 42 is 50% of what number? ______

43. 88 is 25% of what number? ______ 44. 7 is 35% of what number? ______

45. What percent of 100 is 43? ______ 46. What percent of 20 is 12? ______

47. What percent of 95 is 19? ______ 48. What percent of 250 is 200? ______

NAME ______________________ CLASS ____________ DATE ____________

Practice Masters Level B

4.2 Percent Problems

Write each percent as a decimal.

1. 14% ____________
2. 60% ____________
3. 150% ____________
4. 16.5% ____________
5. 245% ____________
6. 0.25% ____________
7. 781% ____________
8. 36% ____________
9. 89% ____________

Write each decimal as a percent.

10. 0.50 ____________
11. 1.2 ____________
12. 0.01 ____________
13. 1 ____________
14. 4.58 ____________
15. 0.68 ____________
16. 0.005 ____________
17. 0.89 ____________
18. 0.012 ____________

Write each percent as a fraction or mixed number in lowest terms.

19. 85% ____________
20. 43% ____________
21. 124% ____________
22. 8.5% ____________
23. 24% ____________
24. 260% ____________

Find each answer.

25. What is 55% of 600? ____________
26. What percent of 120 is 18? ____________
27. What percent of 60 is 96? ____________
28. Find 42% of 80. ____________
29. What is 260% of 15? ____________
30. What percent of 250 is 60? ____________
31. Find 3.5% of 88. ____________
32. 50 is 40% of what number? ____________
33. Georgia's bill for dinner was $25.00. She left the waitress a tip of $5.60. What percent of the bill did the waitress receive in tips? ____________
34. A suit is marked down to $67.40 from the original price of $84.00. By what percent has the original price been reduced? ____________
35. A magazine reports that 36% of its 4500 readers own a DVD player. How many of its readers own a DVD player? ____________
36. If there is a 6% sales tax on a sweatshirt that costs $22.00, what will be the total cost of the sweatshirt? ____________

Practice Masters Level C

4.2 Percent Problems

Write each percent as a decimal.

1. $\frac{1}{2}\%$ ______________ 2. 2500% ______________ 3. 8.756% ______________

Write each decimal as a percent.

4. 2.687 ______________ 5. 0.015 ______________ 6. 0.64 ______________

Write each percent as a fraction or mixed number in lowest terms.

7. 350% ______________ 8. 84% ______________ 9. 10.5% ______________

10. 44% ______________ 11. 24.4% ______________ 12. 167% ______________

Find each answer.

13. What is 125% of 650? ______________ 14. What is 33.4% of 45? ______________

15. What percent of 96 is 24? ______________ 16. 88 is 44% of what number? ______________

17. What percent of 65 is 260? ______________ 18. 173.6 is 140% of what number? ______________

19. A blouse originally priced at $36.00 is on sale for 25% off. What is the sale price of the blouse? ______________

20. If there is a 4.5% sales tax on the blouse in Exercise 19, find the total sale price of the blouse. ______________

21. This week the grocery store has ground beef marked down by 30%. Joseph has a coupon that entitles him to an additional 10% off his purchase. If the original price of the ground beef is $2.80 per pound, what will Joseph pay per pound of ground beef? ______________

22. A car dealership is advertising cars for 20% off the list price. Donna has chosen a car that is on sale for $14,016.00. What was the original price of the car? ______________

23. A survey was conducted of 2500 high-school students to find out their favorite subject. The results are as shown in the table.

History	Algebra	English	Spanish	Art
422	643	395	458	582

a. What percent of the students surveyed chose algebra? ______________

b. What percent of the students surveyed chose history or art? ______________

NAME ______________________ CLASS __________ DATE __________

Practice Masters Level A

4.3 Introduction to Probability

One number cube was rolled 10 times with the outcomes shown in the table below.

Trial	1	2	3	4	5	6	7	8	9	10
Outcome	4	1	1	5	2	3	6	3	6	3

Use the data above to find each experimental probability.

1. rolling a 4 ______________
2. rolling an even number ______________
3. rolling a 3 ______________
4. rolling a 1 ______________

Two number cubes were rolled 100 times. Based on the results below find the experimental probability of each outcome.

5. An odd number appeared on at least one cube 73 times. ______________
6. A sum of 8 appeared 12 times. ______________
7. A sum greater than 5 appeared 79 times. ______________

Two coins were tossed 15 times with the outcomes shown in the table below.

Trial	1	2	3	4	5	6	7	8	9	10	11	12	13	14	15
Coin 1	H	H	T	H	T	T	T	H	H	T	T	H	T	H	H
Coin 2	T	H	H	T	T	T	H	T	H	T	H	H	H	T	T

Use the data above to find each experimental probability.

8. At least one coin shows heads. ______________
9. Both coins show the same side of the coin. ______________
10. Neither coin shows heads. ______________

A survey of 200 students was conducted to find out the most popular sport. The results of the survey are shown in the table below.

Sport	Soccer	Basketball	Bowling	Track
Number of students	73	54	37	36

Use the data above to find each experimental probability.

11. A student chose basketball. ______________
12. A student chose bowling or track. ______________

NAME ______________________ CLASS __________ DATE __________

Practice Masters Level B

4.3 *Introduction to Probability*

Two number cubes were rolled 250 times. Based on the results below, find the experimental probability of each outcome.

1. A sum less than 10 appeared 203 times. ______________
2. A 4 appeared on at least one cube 67 times. ______________
3. An even sum appeared 130 times. ______________
4. An odd number appeared on at least one cube 198 times. ______________

Three coins were tossed 10 times with the outcomes shown in the table.

Trial	1	2	3	4	5	6	7	8	9	10
Coin 1	H	H	T	H	T	H	T	T	T	H
Coin 2	H	T	T	H	T	H	H	T	H	H
Coin 3	T	H	T	T	H	H	T	H	H	T

Use the results above to find each experimental probability.

5. All three coins show the same side of the coin. ______________
6. At least one coin shows tails. ______________
7. Two coins show heads. ______________
8. All three coins show tails. ______________
9. Two coins show tails. ______________

One red number cube and one blue number cube were rolled 5 times with the outcomes shown in the table.

Trial	1	2	3	4	5
Red cube	6	3	1	4	3
Blue cube	2	4	6	3	5

Use the results above to find each experimental probability.

10. A 3 appears on the red cube. __________
11. A 3 appears on the blue cube. __________
12. A sum of 8 or less is rolled. __________
13. A 2 appears on the red cube. __________
14. A sum of 5 is rolled. __________
15. A sum of 7 is rolled. __________
16. An even number appears on the blue cube. ______________
17. An even number appears on at least one cube. ______________

NAME ______________________ CLASS ____________ DATE ____________

Practice Masters Level C

4.3 Introduction to Probability

Two 10-sided die were rolled 40 times. Based on the results below, find each experimental probability.

1. A sum of 10 appeared 11 times. ______________________

2. A pair of threes was rolled 2 times. ______________________

3. A sum of at least 14 appeared 19 times. ______________________

The spinner shown at the right was spun 50 times. A summary of the outcomes is shown in the table below.

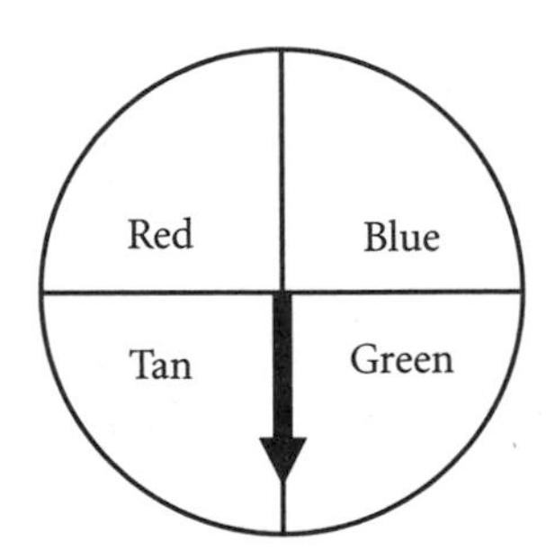

Color	Number of times landed on color
Red	10
Blue	12
Tan	16
Green	12

Use the results above to find the experimental probability that the spinner lands on:

4. tan ____________ 5. yellow ____________ 6. blue or green ____________

7. not red ____________ 8. red or tan ____________ 9. blue or orange ____________

Three coins were tossed 5 times with the outcomes shown in the table below.

Trial	1	2	3	4	5
Coin 1	H	T	H	H	T
Coin 2	T	T	H	T	H
Coin 3	T	T	T	H	H

Use the results above to find each experimental probability.

10. All three coins show heads. ____________ 11. At least one coin shows heads. ____________

12. All three coins show the same side. ______ 13. Exactly one coin shows heads. ____________

14. The result of a math class experiment to find the probability of rolling a sum of 6 on two number cubes was 16%. If the class performed 50 trials, how many favorable outcomes occurred? ____________

NAME ______________________ CLASS __________ DATE __________

Practice Masters Level A

4.4 Measures of Central Tendency

Find the mean, median, mode(s), and range for each set of data.

1. 11, 5, 8, 10, 11, 15, 7, 21

   mean ________ median ________ mode(s) ________ range ________

2. 27, 20, 33, 22, 26, 31, 30

   mean ________ median ________ mode(s) ________ range ________

3. 19, 14, 18, 19, 10, 7, 3, 10

   mean ________ median ________ mode(s) ________ range ________

4. 105, 107, 121, 110, 132, 127

   mean ________ median ________ mode(s) ________ range ________

5. 60, 71, 65, 72, 60, 58, 55

   mean ________ median ________ mode(s) ________ range ________

6. 1, 17, 15, 20, 10

   mean ________ median ________ mode(s) ________ range ________

The Southtown Library conducted a survey to determine how many books teenagers read in a month. The results are shown below. Use this data for Exercises 7 and 8.

Number of books read in a month				
4	1	1	5	6
3	2	3	2	3

7. Make a frequency table for the data.

Number	
Frequency	

8. Find the indicated measures of central tendency.

   mean ________ median ________ mode(s) ________ range ________

NAME ______________________ CLASS ____________ DATE ____________

Practice Masters Level B

4.4 Measures of Central Tendency

Find the mean, median, mode(s), and range for each set of data.

1. 12, 36, 24, 18, 33, 16, 35, 22

   mean ________ median ________ mode(s) ________ range ________

2. 58.5, 62, 41, 48.6, 61, 51.4

   mean ________ median ________ mode(s) ________ range ________

3. 107, 124.4, 117.3, 108, 117.3, 126, 105

   mean ________ median ________ mode(s) ________ range ________

4. 225.8, 200, 221.6, 231.8, 201.1, 260, 210.3, 244.5, 207

   mean ________ median ________ mode(s) ________ range ________

5. 0.25, 1.3, 0.88, 0.7, 1.3, 0.85, 0.25, 1.1

   mean ________ median ________ mode(s) ________ range ________

Kevin kept track of the minutes he ran each day for a month. The results are shown in the table. Use this information for Exercises 6 and 7.

Number of Minutes Run Each Day														
22	25	22	24	21	20	21	25	22	21	23	23	20	26	25
20	26	23	22	22	20	24	22	21	25	26	20	21	22	24

6. Make a frequency table for the data.

Minutes	
Frequency	

7. Find the indicated measures of central tendency.

   mean ________ median ________ mode(s) ________ range ________

8. Julia has test scores of 90, 96, 89, 94, and 90 so far this quarter. She would like to have a test average of 93. If there is one test remaining, what must Julia score to earn a 93 average? ________

NAME ______________________ CLASS ____________ DATE ____________

Practice Masters Level C

4.4 Measures of Central Tendency

1. Catherine says that $\frac{1}{2}$ of her employees earn more than $45,000 per year and $\frac{1}{2}$ earn less. Which measure of central tendency is she using? ____________

2. There are 4 breeds of dogs living in the neighborhood: Cocker Spaniel, Poodle, Dalmatian, and Collie. Which measure of central tendency is appropriate for these data? ____________

3. The monthly car payments for the last 6 customers at a bank were $325, $350, $404, $350, $350, and $412. Which measure best summarizes these data? Explain. ____________

4. The profits of the 6 schools that sold calendars were $220, $375, $550, $657, $1020, and $5300. Decide whether the mean or median is the better measure of central tendency for these data. Explain. ____________

Preston and Haley keep track of their golf scores for a month. The results are shown in the tables below. Use this information for Exercises 5–9.

Preston's Golf Scores				
78	80	80	82	80
79	79	80	81	83
78	79	80	80	82

Haley's Golf Scores				
80	82	81	80	79
79	79	85	83	80
83	79	81	79	83

5. Make a frequency table for Preston's scores.

Score	
Frequency	

6. Make a frequency table for Haley's scores.

Score	
Frequency	

7. Find the indicated measures of central tendency for Preston's scores.

mean ________ median ________ mode(s) ________ range ________

8. Find the indicated measures of central tendency for Haley's scores.

mean ________ median ________ mode(s) ________ range ________

9. In golf, the goal is to earn a low score.

a. Which measure of central tendency depicts Preston as the better golfer? ____________

b. Which measure of central tendency depicts Haley as the better golfer? ____________

c. Explain your choices. ____________

NAME ______________________ CLASS ____________ DATE ____________

Practice Masters Level A

4.5 Graphing Data

The bar graph below shows the weekly rental distribution at a video rental store.

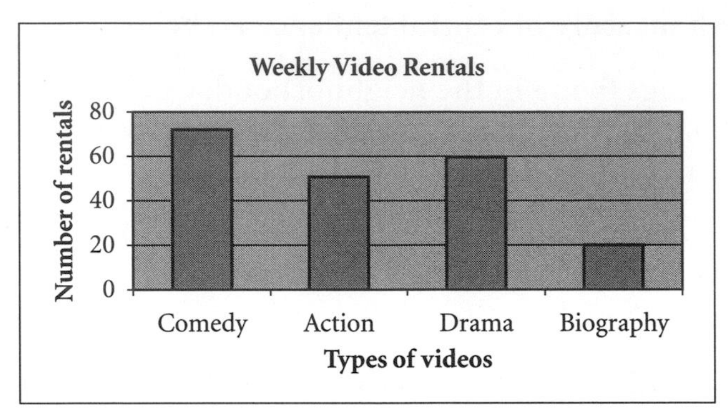

1. a. What type of movie is rented most often? ____________

   b. Approximately how many are rented per week? ____________

2. a. What type of movie is rented least often? ____________

   b. Approximately how many are rented per week? ____________

3. What are the approximate total weekly rentals? ____________

A survey was conducted of 150 people to determine their favorite ice cream flavor. The results are shown in the circle graph at the right.

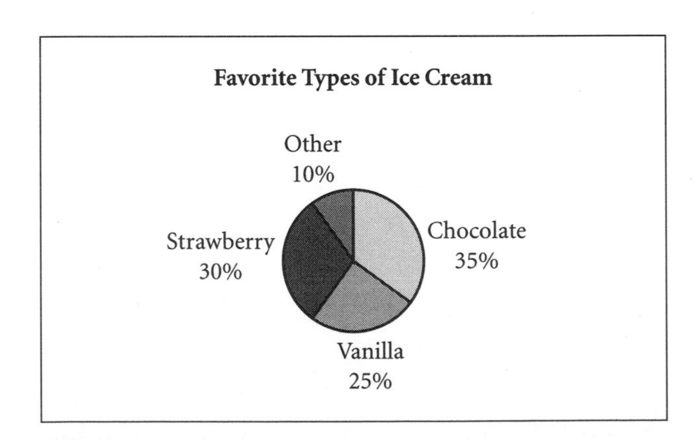

4. Of the 150 people surveyed, how many chose strawberry? ____________

5. Of the 150 people surveyed, how many chose vanilla? ____________

6. What is the most popular flavor among the people surveyed? ____________

7. Suppose you asked 80 people whether they are oldest, youngest, or in the middle of their family. Thirty-five people told you that they are the oldest, 25 said they are the youngest, and 20 people are in the middle of their families. Make a circle graph that displays the data.

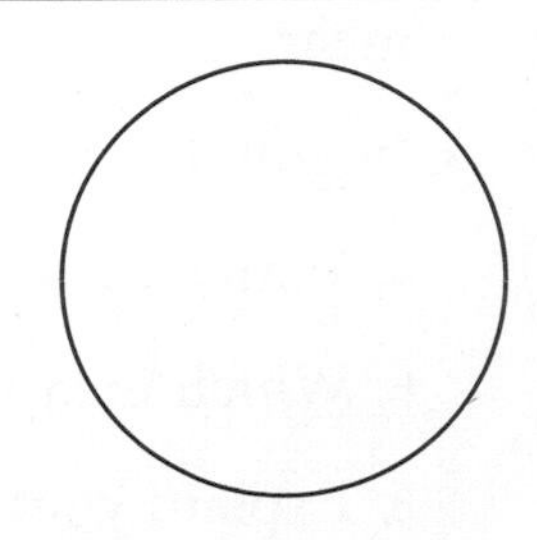

NAME ______________________ CLASS __________ DATE __________

Practice Masters Level B

4.5 Graphing Data

The line graphs below show the monthly rainfall for the cities of Springtown and Tropicville.

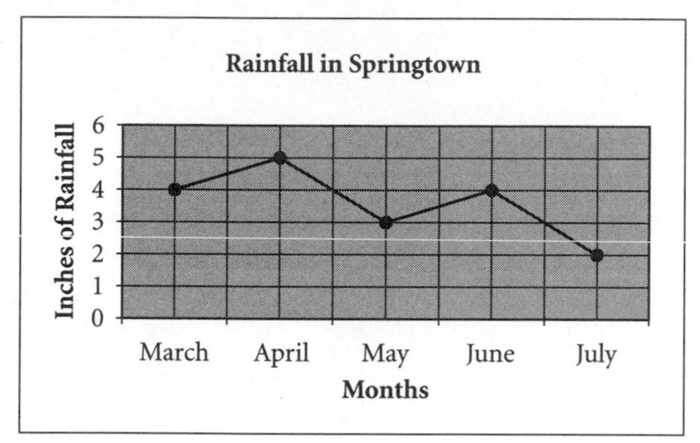

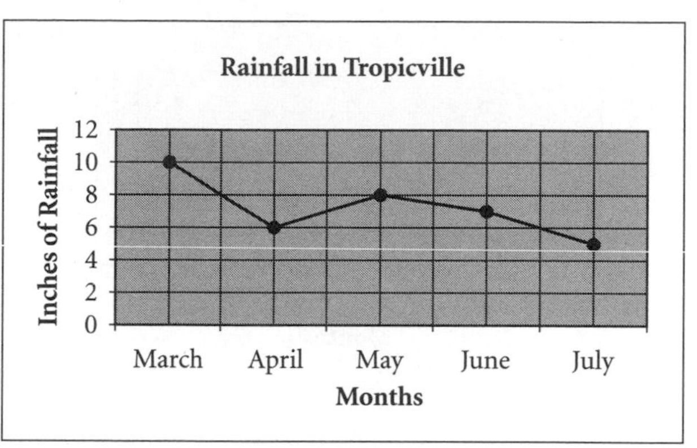

1. What was Springtown's rainfall in May? ______________

2. Which city had the highest rainfall? ______________ In what month? ______________

3. Which city had the lowest rainfall? ______________ In what month? ______________

4. Why might it be misleading to display these two graphs together? ______________

The bar graph shows the number of hours per week that different age groups watched TV in 1990 and 2000.

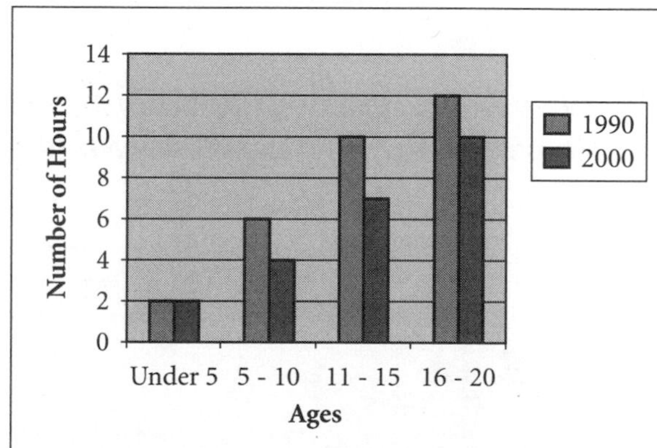

5. How many hours per week did the 5–10 age group watch TV in 1990?

____

6. What information about TV viewing habits can be learned from this graph?

____

____

7. A survey was conducted to determine how people planned to go on vacation. Make a circle graph that shows the percent of people in each category.

Type of travel	Car	Bus	Plane	Boat	Camper
Number	140	55	135	75	45

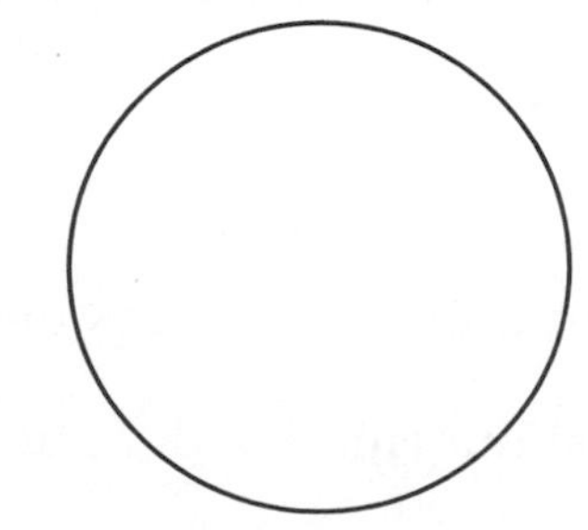

NAME ______________________ CLASS ______________ DATE ______________

Practice Masters Level C

4.5 Graphing Data

Kate has graphed her electric bill costs for the last 6 months. The results are shown in the line graph below.

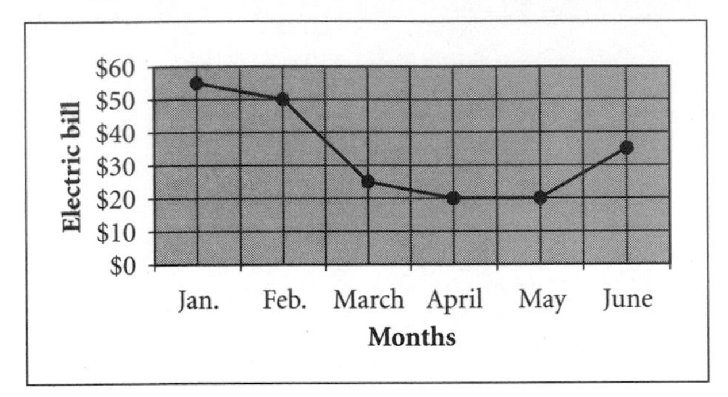

1. a. During what month was Kate's electric bill the highest? ______________

   b. Approximately how much was it? ______________

2. Based on the information in the graph, can you predict Kate's electric bill for July? If so, what will it be? ______________

3. Can you explain a possible reason why Kate's bill was high in January, then dropped and began to rise again in June? ______________

4. Use the information given in the line graph to make a bar graph about Kate's electric bill costs. Use the space provided beside the line graph.

A survey was conducted in order to learn the type of pets people own. The results are shown in the bar graph below.

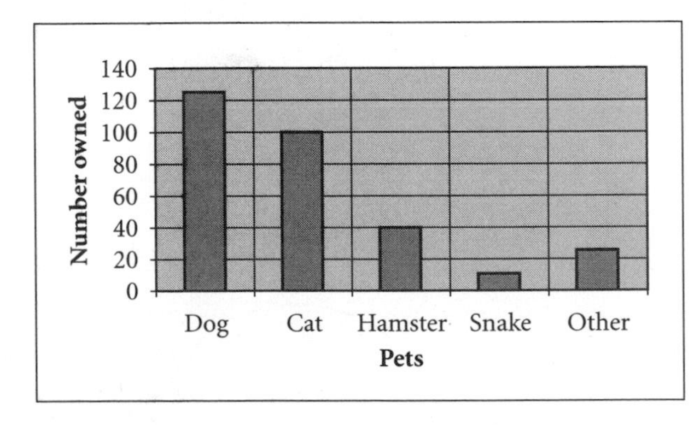

5. Make a circle graph of the data shown in the bar graph.

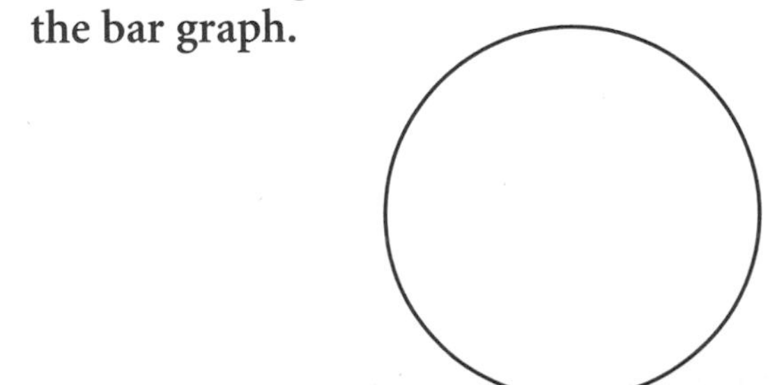

6. What information does the bar graph better display? ______________

7. What information does the circle graph better display? ______________

NAME ______________________ CLASS ______________ DATE ______________

Practice Masters Level A

4.6 Other Data Displays

Construct a stem-and-leaf plot for each data set.

1. 20, 12, 11, 12, 15, 18, 31, 24, 27, 10, 26, 18, 25

Stems	Leaves

2. 45, 55, 57, 61, 83, 48, 45, 66, 56, 63, 42, 42, 80, 60, 55

Stems	Leaves

3. 23, 35, 33, 21, 33, 19, 38, 20, 26, 15, 16, 35, 37, 11

Stems	Leaves

4. 89, 87, 77, 89, 101, 103, 76, 77, 81, 80, 82, 83, 75, 69

Stems	Leaves

Construct a box-and-whisker plot for each data set.

5. 35, 30, 42, 31, 45, 50, 48, 52, 63, 48 45, 55, 36 ______________

6. 81, 87, 70, 95, 88, 87, 110, 93, 83, 88, 100, 86, 100 ______________

7. 115, 118, 125, 122, 111, 141, 131, 135, 112, 124, 101 ______________

8. 1, 5, 12, 7, 2, 2, 10, 24, 15, 18, 7, 9, 11, 5, 15, 6, 10 ______________

Construct a histogram for each data set.

9. 12, 14, 25, 18, 27, 35, 35, 15, 36

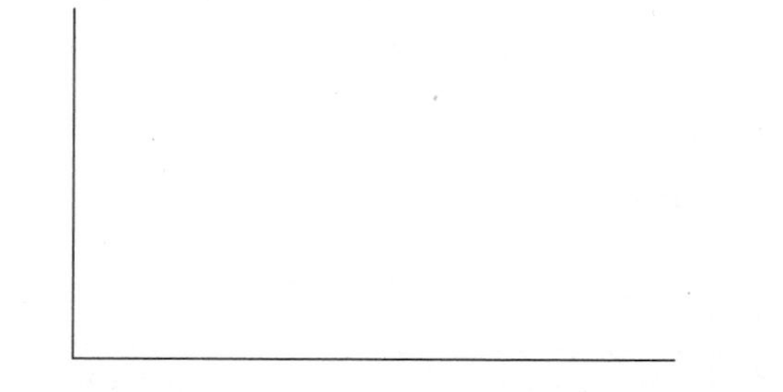

10. 88, 75, 78, 62, 79, 81, 82, 87, 72, 89, 80, 84

For Exercises 11–14, use the stem-and-leaf plot at the right.

Stems	Leaves
2	2, 2, 5, 7, 8
3	0, 5, 5, 9
4	1, 2, 7, 8, 9, 9

11. What is the minimum number in the data? ______

12. What is the range? ______________

13. What is the median? ______________

14. What numbers are repeated? ______________

NAME ______________________ CLASS ____________ DATE ____________

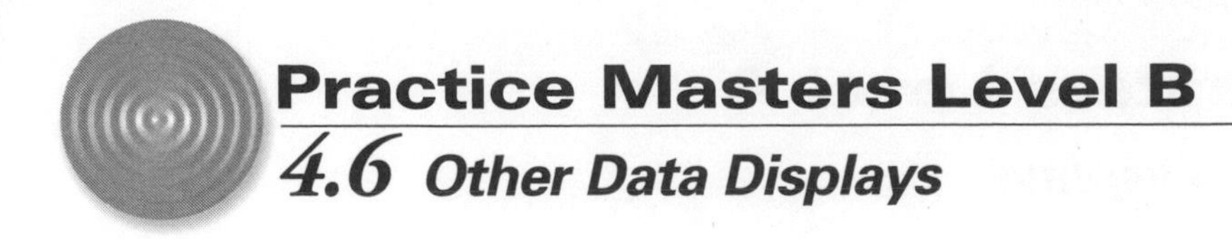

Practice Masters Level B

4.6 Other Data Displays

Use the box-and-whisker plot below for Exercises 1–4.

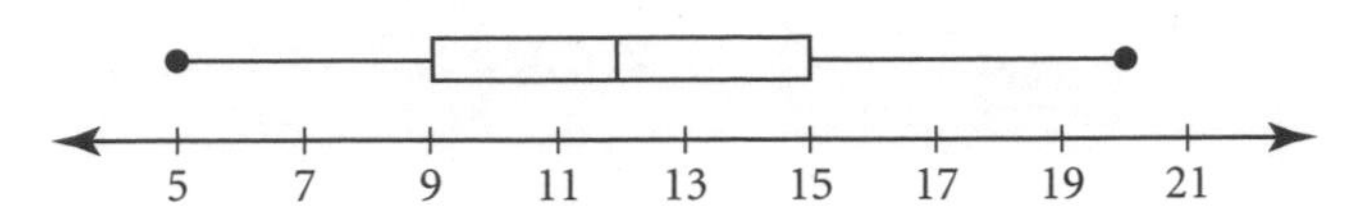

1. What is the lower quartile? ____________

2. The greatest number of the data: ____________

3. What is the median? ____________

4. What is the range? ____________

5. Can you determine if there is a mode for this data? ____________

Coach Barnett kept the following table of his basketball team's scores. Use this information for Exercises 5–9.

45	64	67	48	54	72	45	56	54
50	47	60	62	75	50	35	42	58

6. Construct a stem-and-leaf plot for the data.

Stems	Leaves

7. Construct a histogram of the data.

8. What is the median of the data? ____________

9. What is the mode of the data? ____________

The data below are the scores from the most recent test in Mr. Russell's class. Use this information for Exercises 10–15.

88	80	85	75	74	93	95	71
96	85	88	91	88	99	76	

10. Construct a box-and-whisker plot of the data. ____________

11. What is the lowest score? ____________

12. The upper quartile of the scores: ____________

13. What is the range of scores? ____________

14. What is the median of the scores? ____________

15. Does the box-and-whisker plot show the mean of the data? ____________

NAME ______________________ CLASS ____________ DATE ____________

Practice Masters Level C

4.6 Other Data Displays

The following tables show test scores from two of Mrs. Turner's classes. Use this data for Exercises 1–10.

Morning-Class Scores				
88	87	70	85	81
86	92	91	85	78
95	91	87	86	73
82	78	82	92	93

Afternoon-Class Scores				
76	78	89	75	86
90	85	73	79	82
88	85	75	78	79
90	89	79	80	84

1. Construct a box-and-whisker plot for the scores from the morning class.

2. Construct a box-and-whisker plot for the scores from the afternoon class.

3. What is the lowest score for the morning class? ____________

4. What is the upper quartile for the afternoon class? ____________

5. Which is larger, the highest score for the afternoon class or the upper quartile of the morning class? ____________

6. Which is smaller the lowest score for the morning class or the lowest score for the afternoon class? ____________

7. Which class did better on the test? Explain. ____________

8. Construct a stem-and-leaf plot for the scores from the morning class.

Stems	Leaves

9. Construct a histogram for the scores from the afternoon class.

10. What information is displayed in a stem-and-leaf plot that is not shown in a histogram? ____________

NAME ______________________ CLASS __________ DATE __________

Practice Masters Level A

5.1 Linear Functions and Graphs

For each relation, (a) describe the domain, (b) describe the range, and (c) determine whether the relation is a function.

1. {(5, 10), (3, 8), (13, 18), (9, 14)}

   a. ______________________

   b. ______________________

   c. ______________________

2. {(0, 4), (1,4), (2,4), (3, 4), (4, 4)}

   a. ______________________

   b. ______________________

   c. ______________________

3. {(2, 6), (10, 4), (2, 13), (16, 0)}

   a. ______________________

   b. ______________________

   c. ______________________

4. {(22, 8), (15, 20), (8, 3), (10, 20), (31, 6)}

   a. ______________________

   b. ______________________

   c. ______________________

5. {(5, 0), (14, 1), (5, 2), (14, 3)}

   a. ______________________

   b. ______________________

   c. ______________________

6. {(10, 3), (8, 3), (30, 7), (34, 9)}

   a. ______________________

   b. ______________________

   c. ______________________

7. {(16, 2), (16, 10), (16, 16), (16, 25)}

   a. ______________________

   b. ______________________

   c. ______________________

8. {(45, 45), (20, 20), (12, 12)}

   a. ______________________

   b. ______________________

   c. ______________________

Complete each ordered pair so that it is a solution to $3x + y = 20$.

9. (2, ________) 10. (________, 8) 11. (5, ________)

Complete each ordered pair so that it is a solution to $2x + y = 32$.

12. (12, ________) 13. (________, 30) 14. (________, 2)

Complete each ordered pair so that it is a solution to $4x + y = 64$.

15. (2, ________) 16. (________, 20) 17. (15, ________)

Complete each ordered pair so that it is a solution to $5x - y = 12$.

18. (4, ________) 19. (________, 38) 20. (20, ________)

NAME ______________________ CLASS ____________ DATE ____________

Practice Masters Level B

5.1 Linear Functions and Graphs

For each relation, (a) describe the domain, (b) describe the range, and (c) determine whether the relation is a function.

1. $\{(8, 12.5), (10, 12.3), (15, 12.1)\}$
   a. ______________________
   b. ______________________
   c. ______________________

2. $\{(22, 18.3), (22, 25.5), (22, 29), (22, 31.4)\}$
   a. ______________________
   b. ______________________
   c. ______________________

Complete each ordered pair so that it is a solution to $2x - y = 10$.

3. (2, ________) 4. (________, 8) 5. (−1, ________)

Complete each ordered pair so that it is a solution to $-3x + y = 14$.

6. (−5, ________) 7. (________, 20) 8. (________, 35)

The table below shows the cost for film developing. A processing fee of $2.00 is charged per roll, and there is an additional fee per picture. Use the table for Exercises 9–12.

Number of pictures, x	12	15	24	36
Cost, in dollars, y	$3.20	$3.50	$4.40	$5.60

9. Write an equation for the function. ________
10. Write a set of ordered pairs for the function. ________
11. Identify the domain shown in the table. ________
12. Identify the range shown in the table. ________
13. The graph below shows a linear function. Complete the table of values, and determine an equation of the line. ________

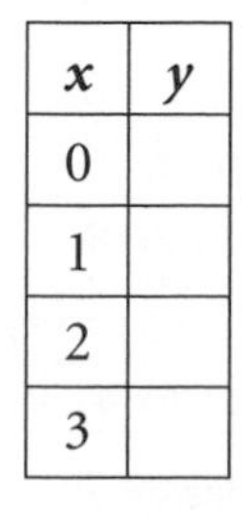

x	y
0	
1	
2	
3	

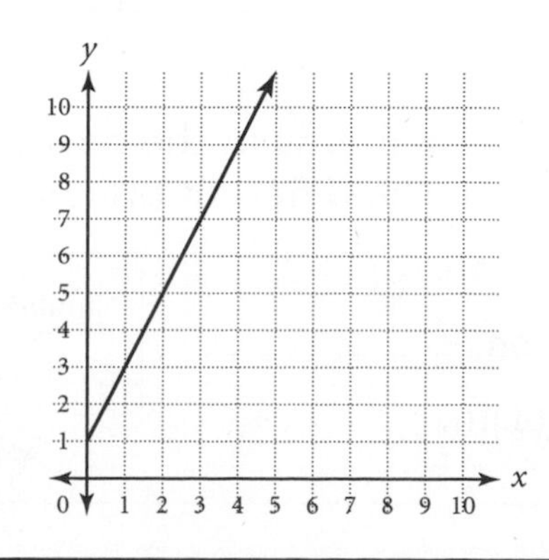

NAME ____________________ CLASS ____________ DATE ____________

Practice Masters Level C

5.1 Linear Functions and Graphs

The graph below shows the distance that George has traveled from his home on his way to college. Use this information for Exercises 1–6.

1. Identify the independent variable. ____________
2. Identify the dependent variable. ____________
3. Write an equation for this linear function. ____________
4. If George's college is 400 miles from his home, what is a reasonable domain for this function? ____________
5. Explain why this example is a function. ____________
6. The next time George drives to college, he drives faster. How would this change the equation you wrote for Exercise 3? ____________

y 300, 240, 180, 120, 60
Distance (in miles)
0 1 2 3 4 5
Time (in hours) x

7. Complete the table so the relation is a function.

Independent variable, x	5		12		21
Dependent variable, y		10		25	

8. Complete the table so the relation is *not* a function.

Independent variable, x	5		12		21
Dependent variable, y		10		25	

The cost, y, of renting a car for x days can be determined using the equation $y - 30x = 100$.

9. What is the cost of renting the car for 5 days? ____________
10. If the cost is $400, how many days was the car rented? ____________

Maryanne read the clothing advertisements in the newspaper. She then created a table displaying the different prices for jeans.

11. Identify the domain of the relation. ____________
12. Identify the range of the relation. ____________
13. Is this relation a function? Explain.

Number of jeans, x	2	3	2	4
Cost, in dollars, y	28	35	21	48

____

NAME ______________________ CLASS __________ DATE __________

Practice Masters Level A

5.2 Defining Slope

Find the slope of the lines graphed below.

1. ______________________

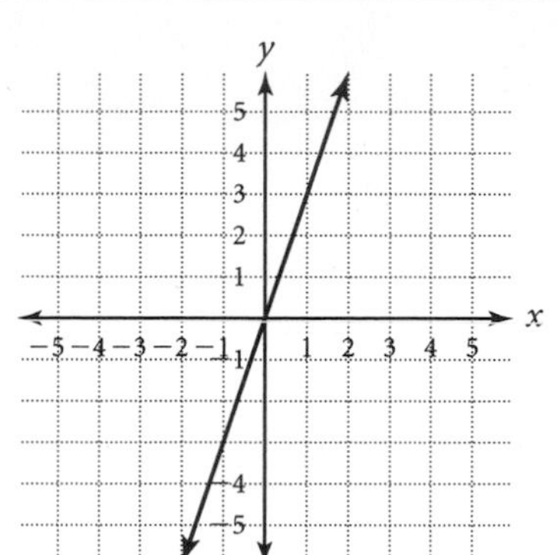

2. ______________________

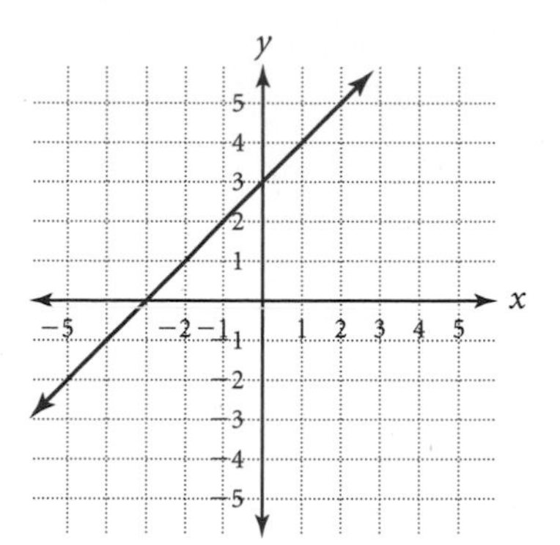

3. ______________________

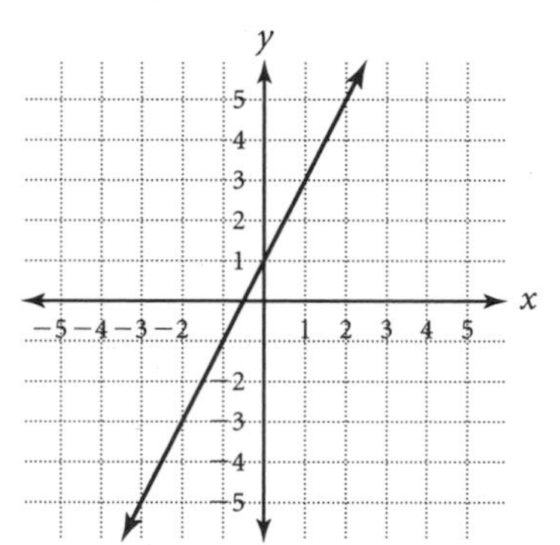

4. ______________________

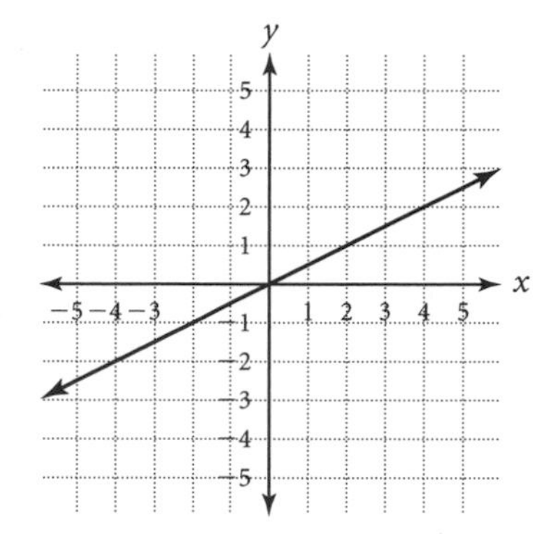

Find the slope for each rise and run.

5. rise: 6, run: 2 ______________________
6. rise: 4, run: 3 ______________________
7. rise: 4, run: 12 ______________________
8. rise: 16, run: 8 ______________________
9. rise: 8, run: 16 ______________________
10. rise: 20, run: 5 ______________________
11. rise: 1, run: 7 ______________________
12. rise: 5, run: 3 ______________________

Find the slope of the line that contains each pair of points.

13. $M(0, 0), N(1, 2)$ ______________________
14. $M(2, 4), N(5, 6)$ ______________________
15. $M(0, 8), N(5, 8)$ ______________________
16. $M(6, 4), N(7, 10)$ ______________________
17. $M(0, 0), N(10, 10)$ ______________________
18. $M(3, 5), N(3, 7)$ ______________________
19. $M(2, 5), N(7, 5)$ ______________________
20. $M(10, 2), N(10, 7)$ ______________________

NAME ______________________ CLASS ______________ DATE ______________

Practice Masters Level B

5.2 *Defining Slope*

Find the slope of the lines graphed below.

1. 

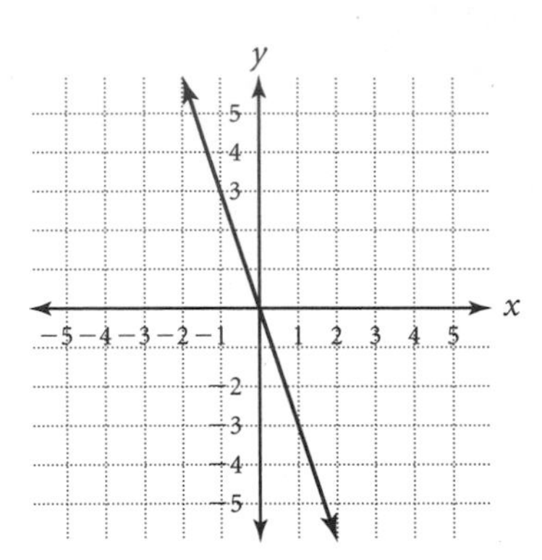

**2.** ______________________

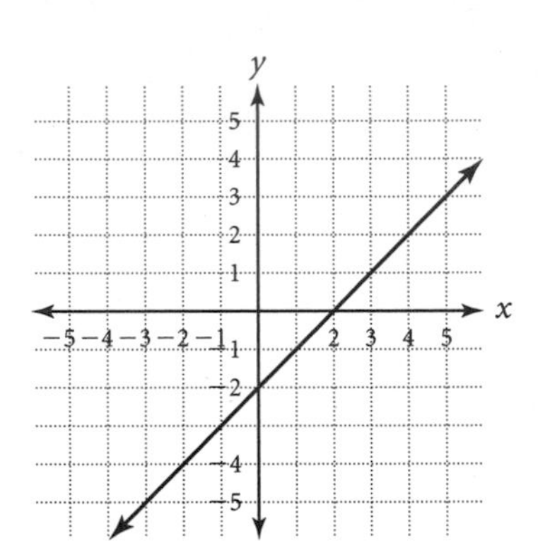

3. ______________________

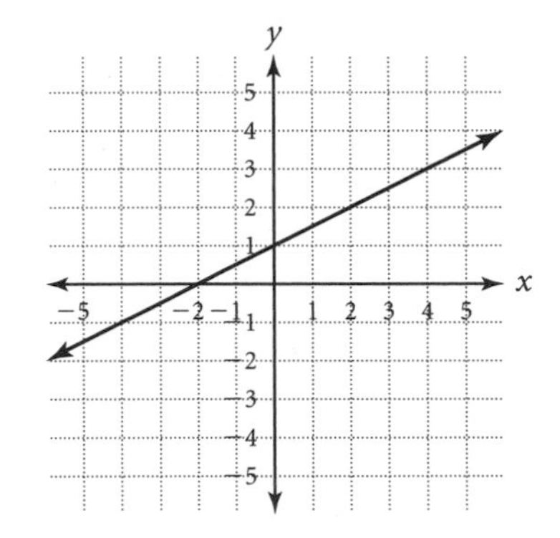

4. ______________________

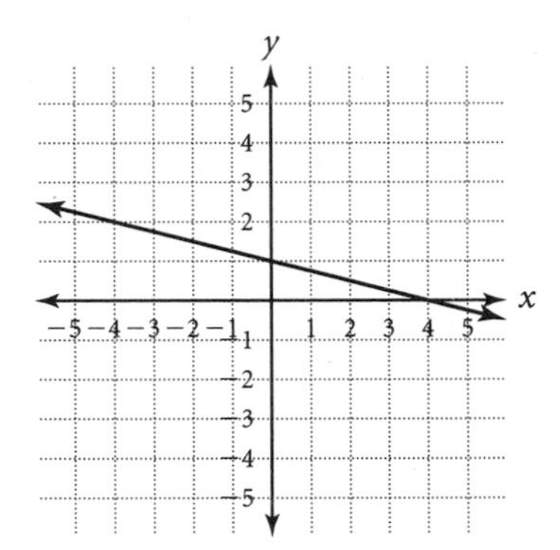

Find the slope for each rise and run.

5. rise: 10, run: 5 ________ **6.** rise: −2, run: 3 ________ **7.** rise: 3, run: −9 ________

8. rise: 20, run: 15 ________ **9.** rise: 12, run: −8 ________ **10.** rise: 0, run: 8 ________

Find the slope of the line that contains each pair of points.

11. $M(0, 0), N(3, 4)$ ______________ **12.** $M(5, 3), N(5, 8)$ ______________

13. $M(-4, 2), N(-4, -6)$ ______________ **14.** $M(-2, 2), N(1, -4)$ ______________

15. Wendy built a ramp that covers a distance of 30 centimeters across the bottom of the tank and rises 12 centimeters. What is the slope of the ramp? ______________

16. Josh is painting houses on the weekends. He is working on an area 14 feet high on the house, and the base of his ladder is 5 feet from the house. What is the slope of the ladder? ______________

NAME ______________________ CLASS __________ DATE __________

Practice Masters Level C

5.2 Defining Slope

Find the slope of the lines graphed below.

1. ______________________

2. ______________________

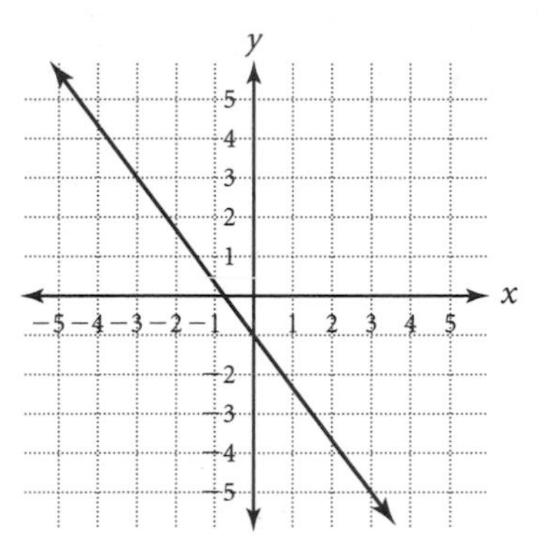

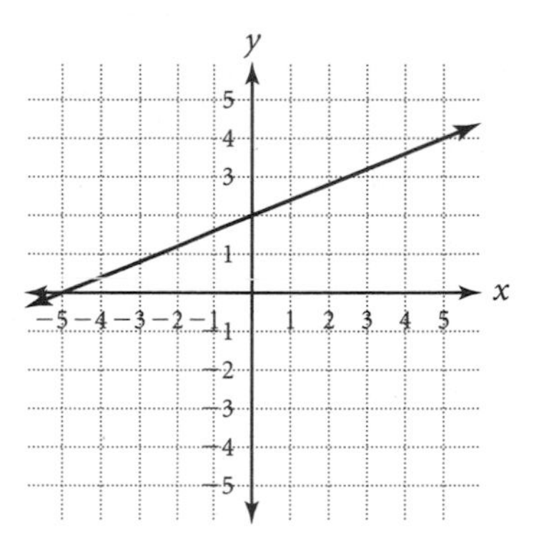

3. If one point on a horizontal line is $A(-9, 5)$, complete the coordinates of a second point on the line, $B(2,$ ________ $)$. ______________

4. If one point on a horizontal line is $J(a, b)$, use the appropriate variable to complete the coordinates of a second point on the line, $K(c,$ ________ $)$. ______________

5. If one point on a vertical line is $C(3, 4)$, complete the coordinates of a second point on the line, $D($ ________ $, -5)$. ______________

6. If one point on a vertical line is $P(a, b)$, use the appropriate variable to complete the coordinates of a second point on the line, $Q($ ________ $, d)$. ______________

Find the slope of the line that contains each pair of points.

7. $M(1, -4), N(-2, 2)$ ______________

8. $M\left(-3\frac{3}{10}, 2\frac{1}{2}\right), N\left(4\frac{2}{5}, -4\frac{3}{10}\right)$ ______________

9. $M\left(-1\frac{1}{5}, 6\frac{4}{5}\right), N\left(1\frac{1}{5}, 5\right)$ ______________

10. $M(-5, -2), N(-1, -6)$ ______________

11. A wheelchair ramp covers a flat distance of 12 feet and rises 1.5 feet. What is the slope of the ramp? ______________

12. Ken is doing an experiment using a ramp and a ball. He sets the ramp up so it covers a distance of 50 centimeters across the table, and it rises 20 centimeters. What is the slope of the ramp? ______________

13. Refer to Exercise 12. If the distance across the table is increased to 75 centimeters and the height remains 20 centimeters, will the new slope be less than or greater than the slope in Exercise 12? ______________

Practice Masters Level A

5.3 Rate of Change and Direct Variation

For the following values of x and y, y varies directly as x. Find the constant of variation, and write an equation of direct variation.

1. y is 12 when x is 3. ______________
2. y is 25 when x is 5. ______________
3. y is 18 when x is 2. ______________
4. y is 54 when x is 9. ______________
5. y is 64 when x is 4. ______________
6. y is 48 when x is 16. ______________
7. y is 32 when x is 4. ______________
8. y is 4 when x is 8. ______________

Make a table of values and graph each equation in the space provided.

9. $y = 3x$

x	y
0	
1	
2	
3	

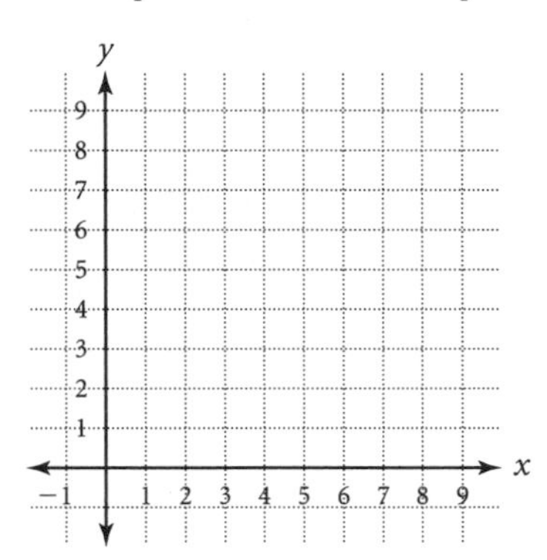

10. $y = 2x$

x	y
0	
1	
2	
3	

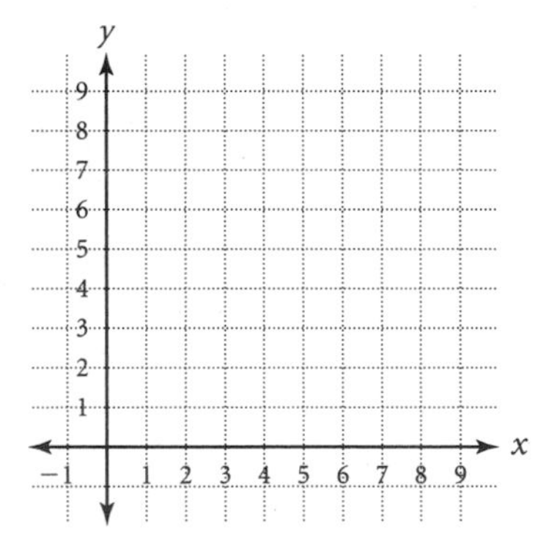

Danielle works at a shipping company. The number of boxes she sorts over time is shown in the graph at the right.

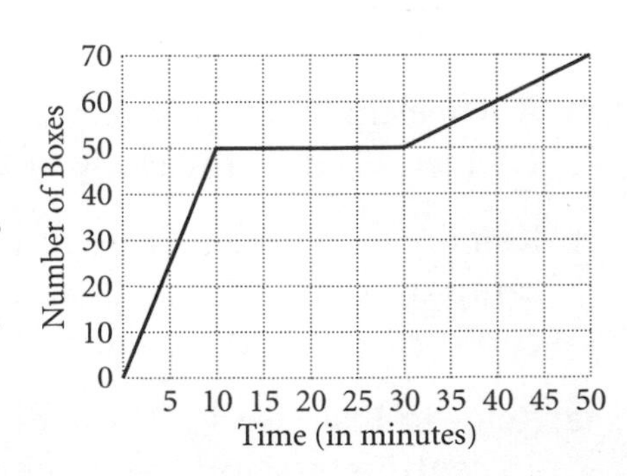

11. What does the horizontal segment of the graph represent?

____

12. What is her rate of box sorting during the first 10 minutes? During the last 20 minutes?

____

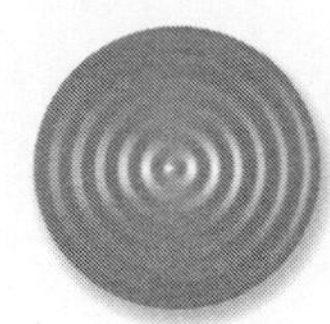

Practice Masters Level B

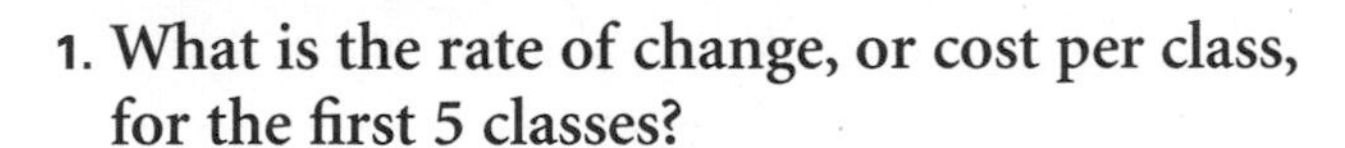

5.3 Rate of Change and Direct Variation

The graph at the right shows the cost of aerobics classes including an initial fee. The gym gives a discount on the cost per class after a member has attended 5 classes.

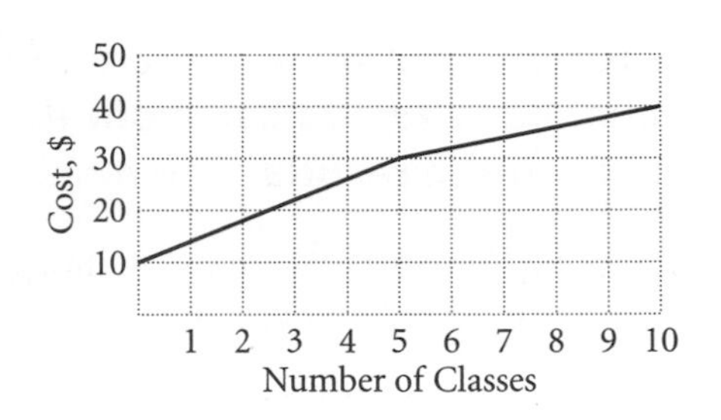

1. What is the rate of change, or cost per class, for the first 5 classes? ______________________

2. What is the rate of change, or cost per class for the last 5 classes? ______________________

For the following values of x and y, y varies directly as x. Find the constant of variation, and write an equation of direct variation.

3. y is 21 when x is 3. ______________
4. y is 13.5 when x is 4.5. ______________
5. y is 16 when x is 3.2. ______________
6. y is 13.2 when x is 4. ______________

For the following values of x and y, y varies directly as x. Find the missing value.

7. y is 50 when x is 12.5. Find x when y is 80. ______________
8. y is 63 when x is 9. Find y when x is 11. ______________
9. y is 41 when x is 5. Find y when x is 12. ______________
10. y is 85.4 when x is 14. Find x when y is 73.2. ______________
11. Suppose you are jogging at 6 miles an hour. Let x represent the time you spend jogging, and let y represent the distance that you jog. Write an equation of direct variation for this situation, and graph the equation. ______________

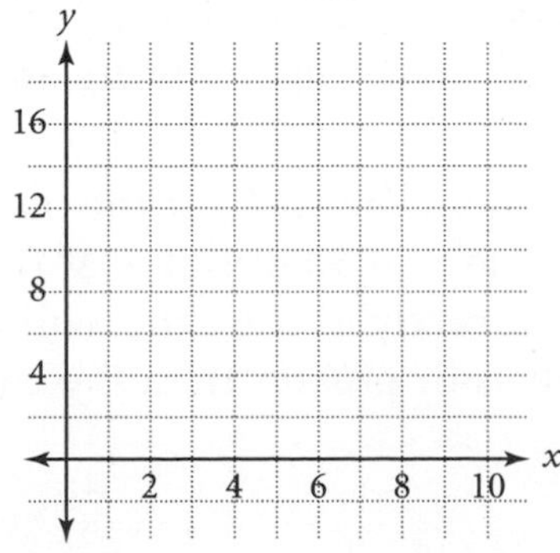

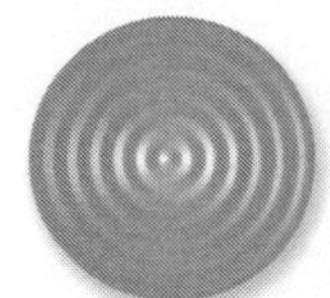

Practice Masters Level C

5.3 Rate of Change and Direct Variation

Aaron performed an experiment in class. He kept track of his distance as he walked from the marked starting place, and graphed the results. The results are shown at the right.

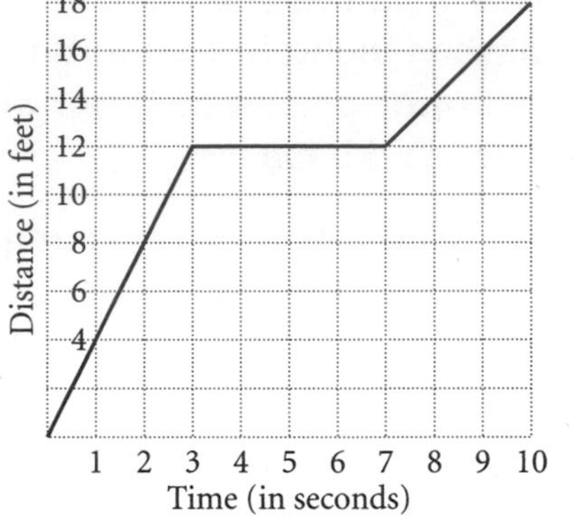

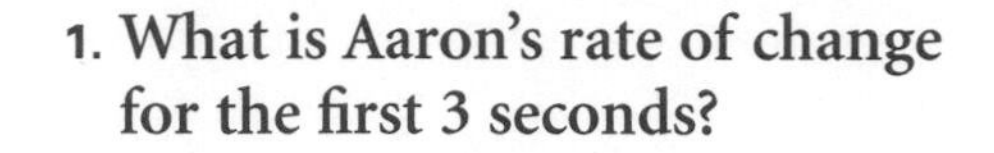

1. What is Aaron's rate of change for the first 3 seconds? ____________

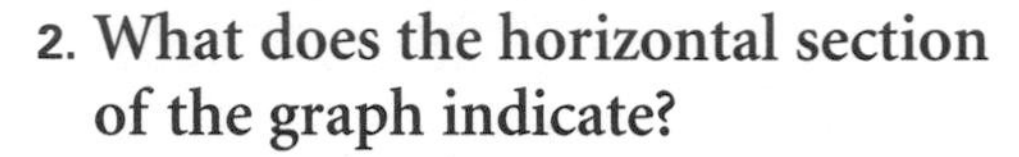

2. What does the horizontal section of the graph indicate? ____________

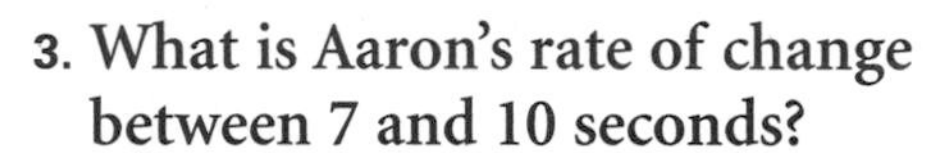

3. What is Aaron's rate of change between 7 and 10 seconds? ____________

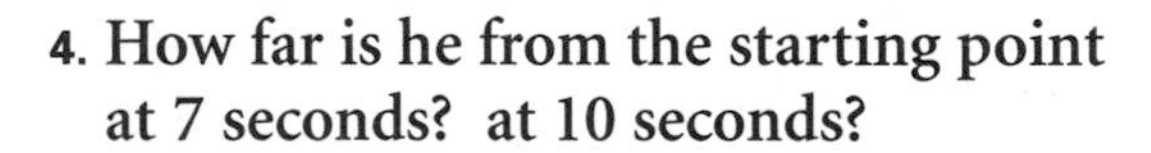

4. How far is he from the starting point at 7 seconds? at 10 seconds? ____________

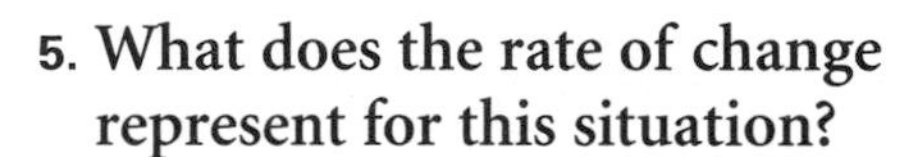

5. What does the rate of change represent for this situation? ____________

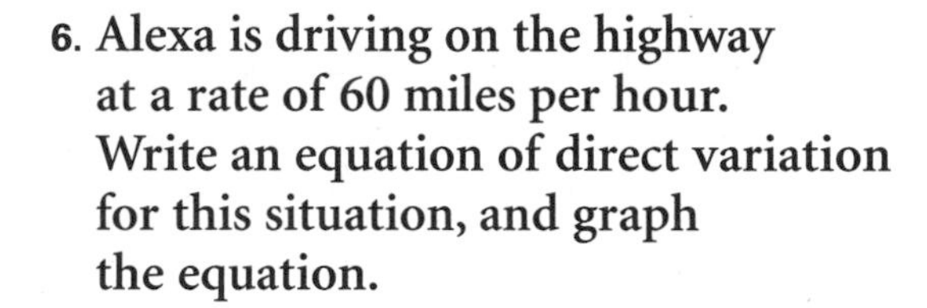

6. Alexa is driving on the highway at a rate of 60 miles per hour. Write an equation of direct variation for this situation, and graph the equation.

   ____

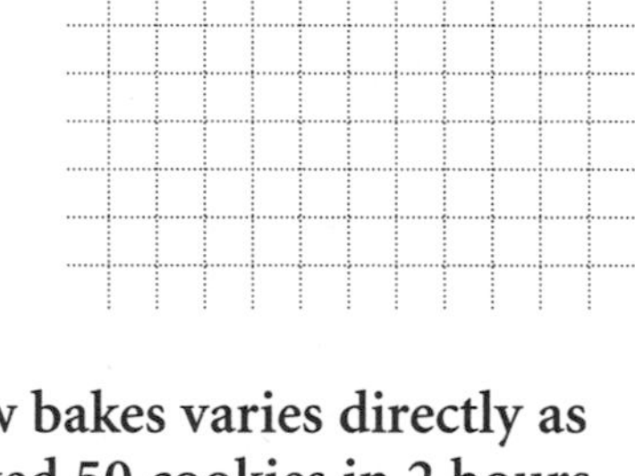

7. The number of cookies, y, that Matthew bakes varies directly as the time, x, that he bakes. Today he baked 50 cookies in 2 hours. Find the constant of variation for this situation, and write an equation of direct variation. ____________
8. Refer to Exercise 7. At his current rate, how long will Matthew need to bake to make 225 cookies? ____________
9. Bailey has a newspaper route and can deliver 40 papers in 50 minutes. The number of papers she delivers, y, varies directly as the time, x. Find the constant of variation and write an equation of direct variation. ____________
10. Suppose Bailey's friend, Sarah, helps her deliver the papers. Together they can deliver 60 papers in 50 minutes. If the number of papers delivered still varies directly as the time, will the new constant of variation be less than or greater than the constant of variation in Exercise 9? ____________

NAME ______________________ CLASS ______________ DATE ______________

Practice Masters Level A

5.4 The Slope-Intercept Form

Identify the *x*- and *y*-intercepts of each line.

1. $y = 6x + 12$ __________ **2.** $y = 5x + 10$ __________ **3.** $y = x - 7$ __________

4. $y = -4x + 8$ __________ **5.** $y = 10x + 50$ __________ **6.** $y = -8x + 4$ __________

Graph each equation by using the slope and *y*-intercept.

7. $y = \frac{2}{3}x + 4$

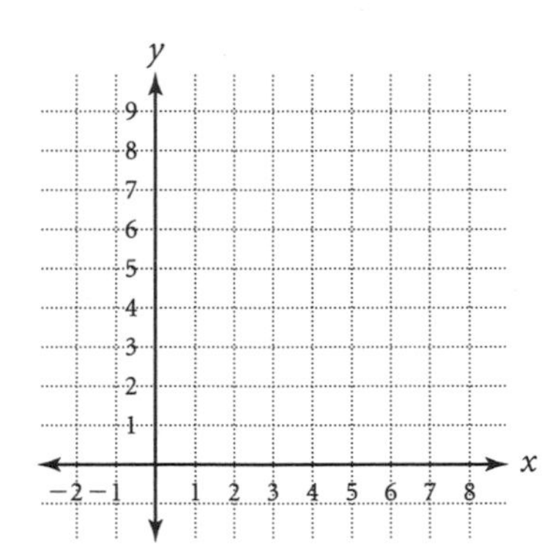

8. $y = -2x + 1$

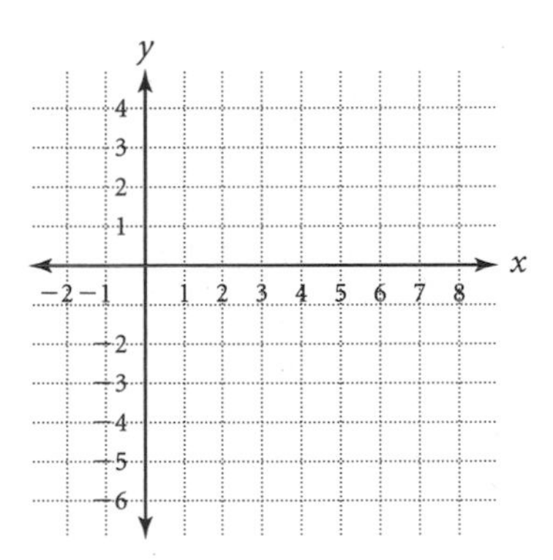

Write an equation in slope-intercept form for the line that contains each pair of points.

9. (2, 7), (5, 22) __________

10. (4, 11), (8, 14) __________

11. (6, 22), (2, 14) __________

Write an equation in slope-intercept form for each line graphed below.

12. ______________________

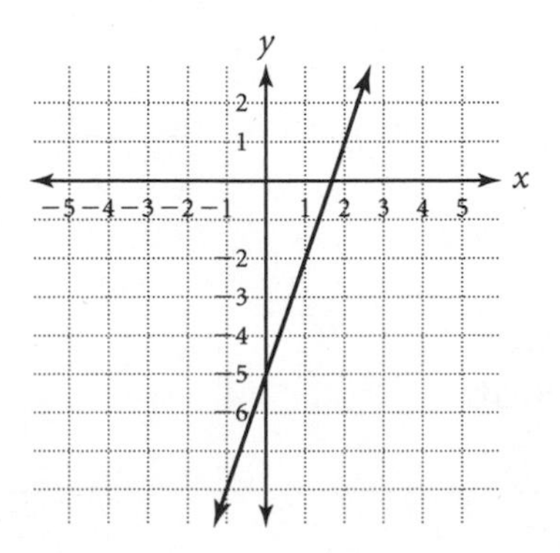

13. ______________________

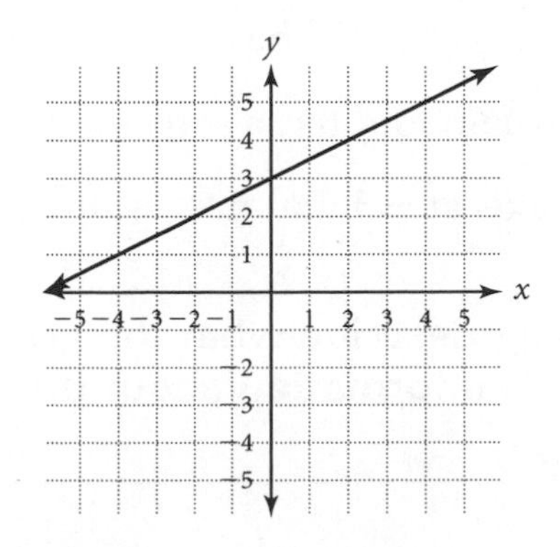

NAME ______________________ CLASS ____________ DATE ____________

Practice Masters Level B

5.4 The Slope-Intercept Form

Identify the *x*- and *y*-intercepts of each line.

1. $y = 3x + 1$ ____________
2. $y = \frac{2}{3}x + 6$ ____________
3. $y = \frac{3}{4}x - 3$ ____________

Graph each equation by using the slope and *y*-intercept.

4. $y = -\frac{3}{2}x + 4$ ____________
5. $y = \frac{1}{3}x - 2$ ____________

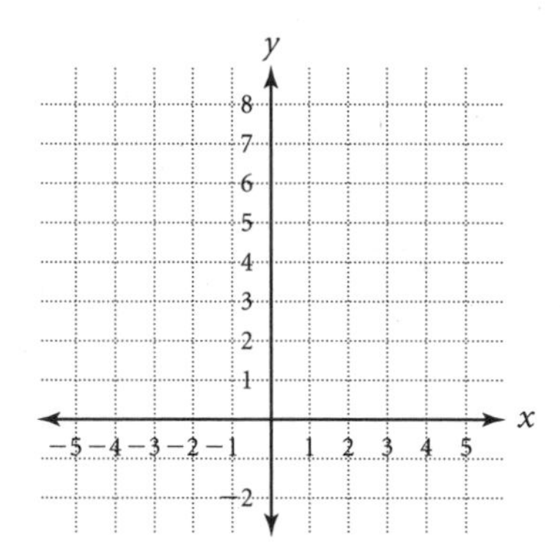

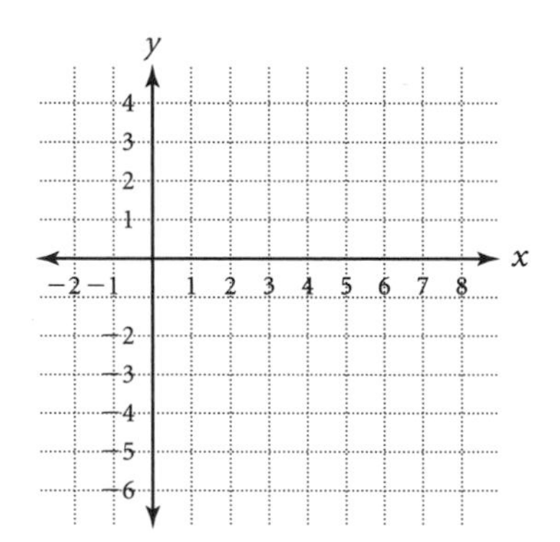

Write an equation in slope-intercept form for the line that contains each pair of points.

6. $(4, 2), (-6, 7)$ ____________
7. $(6, -9), (3, -5)$ ____________
8. $(-9, -1), (3, 7)$ ____________

Write an equation in slope-intercept form for the line that fits each description below.

9. crosses the *y*-axis at 3 and has a slope of $-\frac{1}{3}$ ____________
10. contains the origin and has a slope of 7 ____________
11. crosses the *y*-axis at -3 and has a slope of 6 ____________

Without graphing, describe what the graph of each equation looks like. Include information about the slope and *y*-intercept.

12. $y = -4x + 7$ ______________________
13. $y = \frac{4}{5}x - 3$ ______________________

NAME ______________________ CLASS ____________ DATE ____________

Practice Masters Level C

5.4 The Slope-Intercept Form

Match each equation with the appropriate description.

1. ________ $y = 6$
2. ________ $xy = 6$
3. ________ $x + y = 6$
4. ________ $y = 6x$
5. ________ $x = 6$

a. a line with a slope of -1 and a y-intercept of 6

b. a vertical line 6 units to the right of the origin

c. a horizontal line 6 units above the origin

d. something other than a straight line

e. a line through the origin with a slope of 6

6. Graph $y = \frac{2}{3}x - 4$ by using the slope and y-intercept.

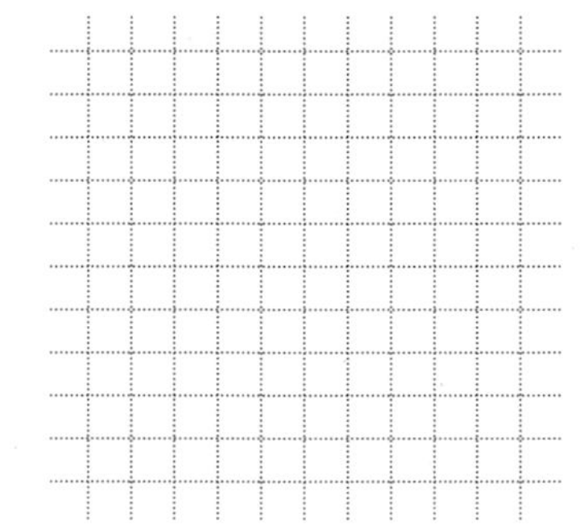

7. Write an equation in slope-intercept form for the line graphed below. ____________

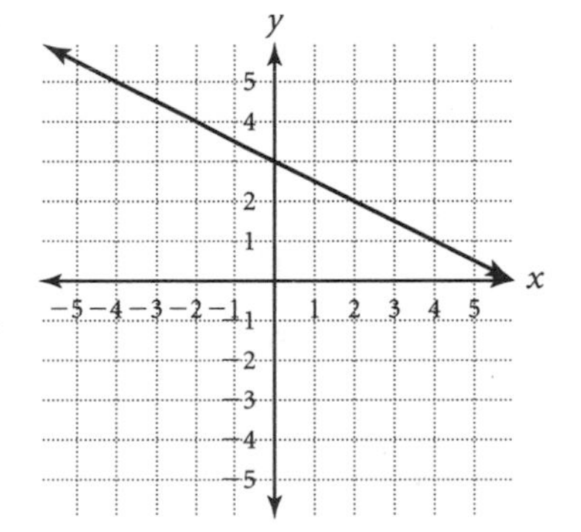

8. Nathaniel is setting up a jogging schedule for himself. He plans to jog 2 miles each day for the first week, and add 1.5 miles each week.

   a. Write an equation in slope-intercept form that relates the number of miles jogged each day, y, to the week number, x, of his schedule. ____________

   b. During what week of his jogging schedule will he jog 8 miles each day? ____________

9. Mohammed has \$80. He pays his brother \$3 a week to do his chores.

   a. Write an equation in slope-intercept form that relates the amount of money Mohammed has remaining, y, to the number of weeks, x. ____________

   b. How much money will Mohammed have left after 15 weeks? ____________

10. The sports boosters charge \$1.75 for each plate of cookies sold at the bake sale. Write an equation relating the total amount of money made, y, to the number of plates of cookies sold, x. ____________

NAME ______________________ CLASS ____________ DATE ____________

Practice Masters Level A

5.5 The Standard and Point-Slope Forms

Write each equation in standard form.

1. $6x = 4y + 20$ ______________
2. $-2x = 5y + 16$ ______________
3. $5x = -10y + 3$ ______________
4. $8x - 21y + 13 = 0$ ______________
5. $19x + y - 2 = 0$ ______________
6. $14y = 2x + 18$ ______________
7. $9y = 12x - 35$ ______________
8. $12x - 34y - 25 = 0$ ______________
9. $3x = -5y + 1$ ______________
10. $10x + 2y - 8 = 0$ ______________

Write an equation in point-slope form for the line that has the given slope and that contains the given point.

11. slope 7, $(1, 8)$ ______________
12. slope 2, $(4, 0)$ ______________
13. slope 4, $(7, 2)$ ______________
14. slope 5, $(6, 3)$ ______________
15. slope 3, $(8, 4)$ ______________
16. slope 10, $(5, 1)$ ______________

Find the *x*- and *y*-intercepts for the graph of each equation.

17. $5x + 2y = 10$ ______________
18. $3x + 2y = 12$ ______________
19. $6x + 10y = 30$ ______________
20. $2x + y = 14$ ______________
21. $x + 5y = 15$ ______________
22. $7x + 3y = 21$ ______________
23. $10x - 3y = 30$ ______________
24. $9x + 2y = 36$ ______________

Write an equation in point-slope form for the line that contains each pair of points.

25. $(5, 4), (7, 12)$ ______________
26. $(1, 3), (2, 8)$ ______________
27. $(7, 14), (4, 2)$ ______________
28. $(6, 4), (10, 20)$ ______________

Write an equation in standard form for the line that contains each pair of points.

29. $(2, 9), (1, 3)$ ______________
30. $(5, 22), (3, 12)$ ______________
31. $(9, 8), (4, 7)$ ______________
32. $(5, 8), (2, 2)$ ______________

NAME ______________________ CLASS ____________ DATE ____________

Practice Masters Level B

5.5 The Standard and Point-Slope Forms

Write each equation in standard form.

1. $8x - 3y + 12 = 0$ ____________
2. $3x = 9y + 15$ ____________
3. $-14x = -22y + 17$ ____________
4. $5y = 2x - 15$ ____________

Write each equation in point-slope form for the line that has the given slope and contains the given point.

5. slope $\frac{2}{3}$, (5, 6) ____________
6. slope $\frac{-1}{4}$, (7, −5) ____________
7. slope 5, (−2, 8) ____________
8. slope −2, (3, 1) ____________

Find the *x*- and *y*-intercepts for the graph of each equation.

9. $-6x + 2y = -24$ ____________
10. $\frac{1}{2}x - 4y = 16$ ____________
11. $15x - y = -45$ ____________
12. $-5x + \frac{2}{3}y = -8$ ____________

Use intercepts to graph each equation.

13. $-2x + 3y = 12$ ____________
14. $5x - 10y = -30$ ____________

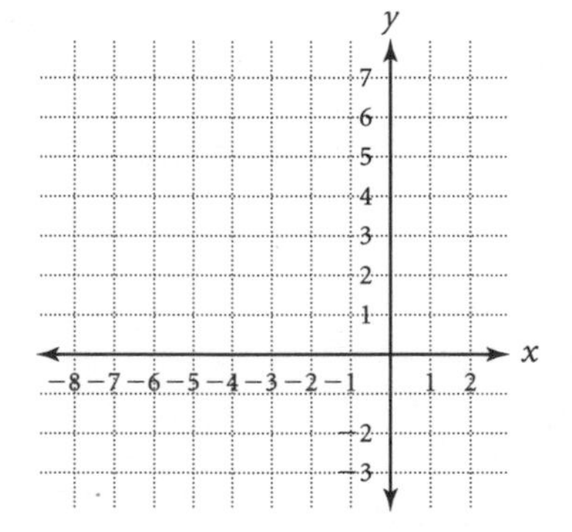

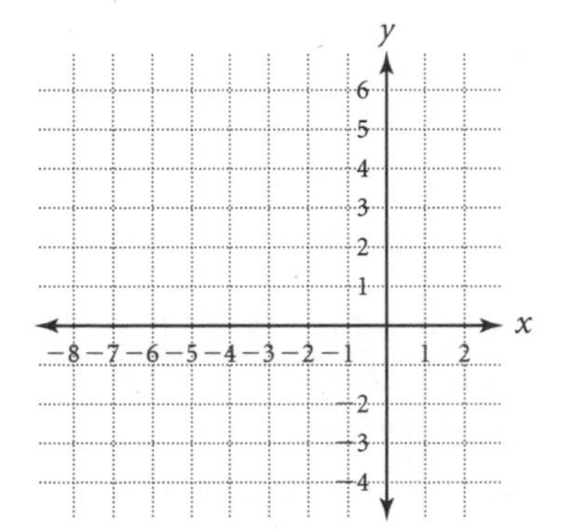

Write an equation in standard form for the line that contains each pair of points.

15. (3, 8), (6, 4) ____________
16. (−2, −1), (2, 2) ____________

17. Eric earns \$5 each week, and he adds this to money he has already saved. He has \$60 after 4 weeks of getting his allowance. Write an equation in point-slope form that relates the number of weeks, *x*, to his total amount of money, *y*. ____________

Practice Masters Level C

5.5 The Standard and Point-Slope Forms

Match each equation with the appropriate description.

1. ______ $-3x + 4y = 24$
2. ______ $y - 5 = 2(x - 8)$
3. ______ $y - 1 = 8(x - 2)$
4. ______ $4x + 0y = 8$
5. ______ $y - 3 = 5(x - 1)$

a. a vertical line 2 units right of the origin

b. can also be written as $y = 2x - 11$

c. has a y-intercept of 6

d. has a y-intercept of 8

e. has a slope of 8

Graph each equation by using the *x*- and *y*-intercepts.

6. $2x - 8y = 4$

7. $-x + 2y = 6$

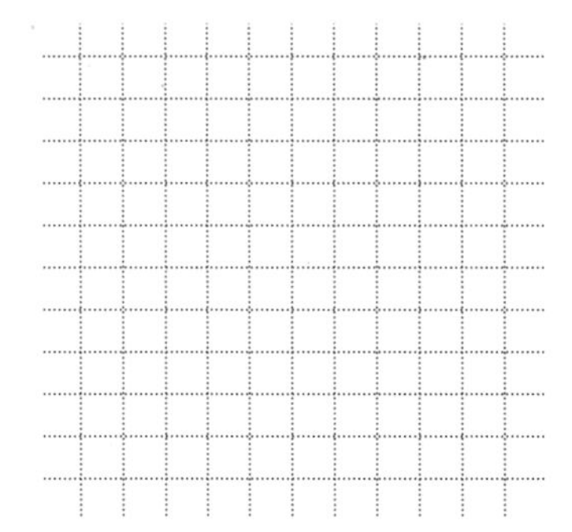

8. Jamal plans to lose weight for the state wrestling championships. He will lose 2 pounds each week. Suppose that after the fourth week of training he weighs 190 pounds. Write an equation in point-slope form that relates the number of weeks, x, to his weight, y. ________

9. Katie has a lemonade stand where she sells each glass of lemonade for 25 cents and each cookie for 50 cents. The number of glasses of lemonade sold is represented by x, the number of cookies sold is represented by y, and her total sales are \$15.

   a. Write an equation in standard form to model this situation. ________

   b. If Katie sells 20 glasses of lemonade, how many cookies must she sell to reach her total of \$15? ________

10. A certain recreation area has miniature golf and a driving range. A round of golf, x, costs \$5 and a bucket of golf balls, y, is \$7. On an average day the total money brought in is \$700.

    a. Write an equation in standard form to model this situation. ________

    b. If 50 buckets are purchased, how many games of golf are played? ________

NAME ______________________ CLASS __________ DATE __________

Practice Masters Level A

5.6 Parallel and Perpendicular Lines

Find the slope of each line.

1. $y = 2x + 13$ ______
2. $y = -3x + 8$ ______
3. $y = \frac{1}{2}x - 3$ ______
4. $4x + 2y = 16$ ______
5. $-12x + 4y = 20$ ______
6. $18 = 9x + 3y$ ______

Find the slope of a line that is perpendicular to the given line.

7. $y = 5x - 4$ ______
8. $y = -\frac{2}{3}x + 8$ ______
9. $y = \frac{4}{3}x + 21$ ______
10. $y = -2x - 10$ ______
11. $y = -4x + 17$ ______
12. $y = \frac{1}{2}x - 14$ ______

Determine whether each statement is true or false.

13. The line $y = 4x - 8$ is parallel to the line $y = 4x + 2$. ______
14. The line $y = 2x - 12$ is parallel to the line $y = 2x + 12$. ______
15. The line $y = 3x + 23$ is perpendicular to the line $y = \frac{1}{3}x - 12$. ______
16. The line $y = \frac{2}{3}x - 6$ is perpendicular to the line $y = -\frac{3}{2}x - 5$. ______
17. The line $y = 12$ is parallel to the line $y = 1$. ______
18. The line $x = 3$ is perpendicular to the line $x = 5$. ______

Write an equation in point-slope form that fits the description given.

19. contains the point (5, 8) and is parallel to $y = 5x + 12$

______

20. contains the point (6, 1) and is perpendicular to $y = 3x - 22$

______

21. contains the point (2, 5) and is parallel to $y = 8x + 10$

______

NAME ______________________ CLASS ____________ DATE ____________

Practice Masters Level B

5.6 Parallel and Perpendicular Lines

Find the slope of each line.

1. $y = 2x - 12$ ______________
2. $y = -\frac{2}{3}x + 1$ ______________
3. $10x + 5y = -15$ ______________
4. $\frac{4}{5}x - 16y = 30$ ______________

Find the slope of a line that is perpendicular to the given line.

5. $2x + 12 = y$ ______________
6. $y = \frac{2}{5}x - 8$ ______________
7. $6x - 2y = 8$ ______________
8. $-\frac{3}{4}x - 12y = 9$ ______________

Match each equation with the appropriate description.

9. ________ $y = 4x + 2$
10. ________ $y = -6x + 7$
11. ________ $y = -\frac{1}{2}x - 9$
12. ________ $y = \frac{1}{4}x - 3$
13. ________ $6x + 2y = 10$

a. parallel to $6x + y = 2$
b. parallel to $y = \frac{1}{4}x + 21$
c. parallel to $y = -3x + 8$
d. perpendicular to $y = 2x + 3$
e. perpendicular to $y = -\frac{1}{4}x + 8$

Write an equation in point-slope form, if possible, for the line that fits the description given.

14. contains the point $(2, 5)$ and is parallel to $y = 3x + 8$

____

15. contains the point $(4, 10)$ and is perpendicular to $y = \frac{1}{2}x - 5$

____

16. contains the point $(1, 9)$ and is parallel to $x = 8$

____

17. contains the point $(-3, -8)$ and is perpendicular to $y = -\frac{3}{4}x + 6$

____

Practice Masters Level C
5.6 Parallel and Perpendicular Lines

Find the slope of the line that is perpendicular to each given line.

1. $y = -4x + 6$ ______________
2. $y = -\frac{2}{3}x - 15$ ______________
3. $2x + 5y = 25$ ______________
4. $-8x + 2y = 12$ ______________

Match each equation with the appropriate description.

5. ________ $y = -4x + 10$
6. ________ $-x + 3y = -15$
7. ________ $y = 3x + 6$
8. ________ $x + 4y = 4$
9. ________ $y = 4$

a. perpendicular to $y = -3x + 4$
b. perpendicular to $y = \frac{1}{4}x - 8$
c. parallel to $y = \frac{1}{4}$
d. parallel to $y = 3x + 12$
e. parallel to $y = -\frac{1}{4}x - 5$

Write an equation in slope-intercept form for the line that fits each description.

10. contains the point $(-2, 5)$ and is parallel to $y = -3x - 7$ ______________
11. contains the point $(3, -4)$ and is perpendicular to $y = \frac{3}{4}x - 8$ ______________

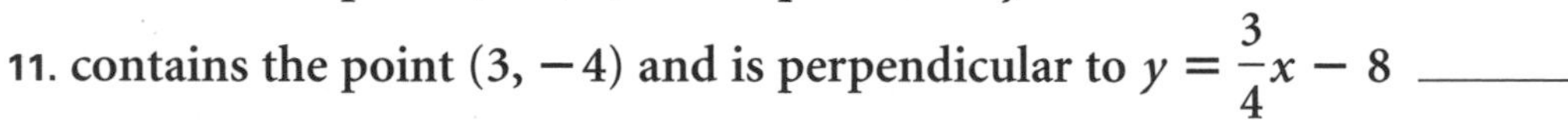

A city planner needs to map out two new roads. He has made a grid of the city that shows the present location of Road A to be $y = x - 1$.

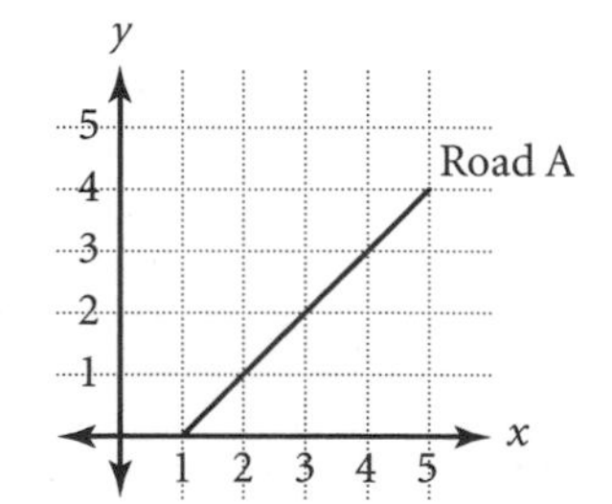

12. Road B must go through the coordinates (2, 4) and be constructed parallel to Road A. What is the equation of Road B?

____

13. Road C must intersect Road A at (2, 1) and be constructed perpendicular to Road A. What is the equation of Road C?

____

14. Will Road C also be perpendicular to Road B? ______________

15. Based on your answer to Exercise 14, can you make a true statement about a line that is perpendicular to one of two parallel lines?

____

NAME ______________________ CLASS ____________ DATE ____________

Practice Masters Level A

6.1 Solving Inequalities

State whether each inequality is true or false.

1. $8 > 4 + 2$ ________
2. $16 - 3 < 12$ ________
3. $-8 + 3 > -7$ ________
4. $-11 - 10 \geq 1$ ________
5. $-4 + 9 > 6$ ________
6. $13 < 5 - 19$ ________
7. $0 \leq -3 + 2$ ________
8. $6 - 8 \leq 4$ ________
9. $16 \geq 25 + 7$ ________
10. $-5 - 5 \geq 0$ ________
11. $-8 + 13 < 4$ ________
12. $19 \leq 14 + 4$ ________
13. $43 + 11 < 60$ ________
14. $-2 + 4 > -11$ ________
15. $30 - 7 > 23$ ________

Graph each inequality on a number line.

16. $x > 5$
17. $c < -2$
18. $d \leq 6$
19. $3 > k$
20. $n \geq 12$
21. $-3 < p$
22. $-2 > w$
23. $6 \geq t$
24. $h \geq -12$
25. $-5 < v$
26. $x < 7$
27. $0 > y$

Solve each inequality.

28. $x + 3 > 2$ ________
29. $y - 5 < 8$ ________
30. $b + 6 \leq 7$ ________
31. $y + (-12) \geq 9$ ________
32. $12 \geq g - 1$ ________
33. $f + 38 < 17$ ________
34. $h - 3 < -3$ ________
35. $t - 10 \leq 19$ ________
36. $y + 8 < -12$ ________
37. $4 + q > 13$ ________
38. $-7 + t > -8$ ________
39. $20 \leq x + 21$ ________
40. $h + 5 < 3$ ________
41. $v - 12 \geq 4$ ________
42. $m - 3 \leq -16$ ________
43. $-15 + c > -7$ ________
44. $63 + a > -1$ ________
45. $n - 15 < -7$ ________

NAME ______________________ CLASS ____________ DATE ____________

Practice Masters Level B

6.1 Solving Inequalities

Solve each inequality.

1. $x - 7 > -9$ ______________

2. $y + 0.3 < 1.6$ ______________

3. $-2.5 + v \geq -9.1$ ______________

4. $t - 4 < 2$ ______________

5. $\frac{1}{4} + r \leq \frac{5}{8}$ ______________

6. $w + 0.04 > 1.2$ ______________

7. $-\frac{2}{3} + p \leq \frac{8}{27}$ ______________

8. $12.5 > -9 + g$ ______________

9. $6.8 < b + 7.9$ ______________

10. $\frac{5}{4} + d > \frac{11}{12}$ ______________

11. $2.45 > -3.12 + q$ ______________

12. $t - \frac{32}{7} \geq \frac{5}{14}$ ______________

Match each expression with the correct inequality.

13. ______ $x < y$

14. ______ $x \leq y$

15. ______ $x \neq y$

16. ______ $y < x < z$

17. ______ $x \geq y$

18. ______ $y \leq x \leq z$

19. ______ $y > x > z$

a. x is less than z and greater than y.

b. x is less than y and greater than z.

c. x is less than y.

d. x is less than or equal to y.

e. x is greater than or equal to y and less than or equal to z.

f. x is greater than or equal to y.

g. x is not equal to y.

Write an inequality that describes the set of points on each number line.

20. 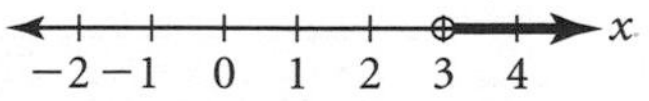

____

21. 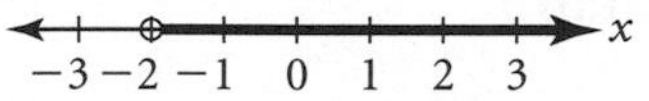

____

22. 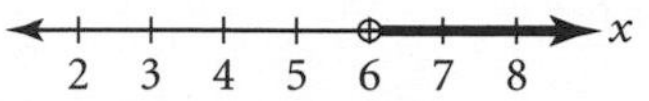

____

23. 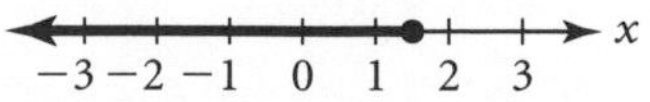

____

24. 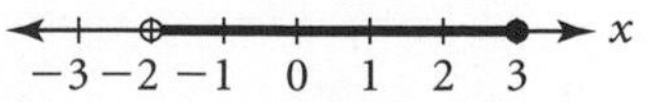

____

NAME ______________________ CLASS __________ DATE __________

Practice Masters Level C

6.1 *Solving Inequalities*

Write an inequality in terms of *x* to represent each situation.

1. a golf score that is at least 75 but lower than 110 ______________

2. a number that is no more than -14 ______________

3. a temperature that is above 32°F ______________

4. a car speed that is slower than 55 kilometer per hour ______________

5. a height taller than 20 centimeters but no more than 41 centimeters ______________

6. an age that is younger than 25 years ______________

Solve each problem.

7. A school bus has 42 seats available, 29 of which are occupied by students and teachers. Write and solve an inequality to describe the number of additional people who can be seated on the bus. ______________

8. An airplane has a total of 70 gallons of fuel. After 12 minutes the plane has used 11 gallons of fuel. Write and solve an inequality to describe the number of gallons of fuel remaining in the tanks. ______________

9. Tiffany has $149.50 in her savings account. If she spends $81.29 on groceries, $17.45 on gasoline, $14.95 on a haircut, and her bank requires a minimum balance of $20, write an inequality to describe how much money Tiffany has available to spend. ______________

10. If 3 less than 4 times a number is less than twice the number, what can the number be? ______________

11. If 5 more than 8 times a number is less than or equal to 6 more than 3 times the number, what is the largest possible number? ______________

12. An aquarium is to be maintained at a temperature of no more than 82°F and no less than 74°F. If the current temperature of the aquarium water is 79°F, what is the maximum temperature change the aquarium can undergo? ______________

13. A number divided by 7 is less than or equal to -4. What is the largest possible number? ______________

NAME ____________ CLASS ____________ DATE ____________

Practice Masters Level A

6.2 Multistep Inequalities

Write an inequality that corresponds to each statement.

1. y is greater than x. ____________
2. c is less than or equal to -12. ____________
3. Negative h is less than 1.9. ____________
4. z is less than or equal to 5 and greater than 0. ____________
5. n is between -14.3 and -18.9 inclusive. ____________

State whether each inequality is true or false.

6. $\frac{2}{9} > \frac{3}{7}$ ____________
7. $-3.35 \geq -3.53$ ____________
8. $\frac{23}{21} < \frac{21}{20}$ ____________
9. $1.14 < 1.41$ ____________
10. $\frac{11}{3} \leq \frac{14}{5}$ ____________
11. $3.77 > 3.707$ ____________
12. $0.1 \geq \frac{1}{5}$ ____________
13. $-\frac{2}{3} > -\frac{3}{4}$ ____________
14. $-8.1 > -\frac{21}{2}$ ____________
15. $0 > 0.001$ ____________
16. $\frac{1}{6} > \frac{1}{7}$ ____________
17. $-2.7 \geq -3.1$ ____________

Describe the steps needed to solve each inequality.

18. $3 + q < 10$ ____________
19. $v - 2 > 14$ ____________
20. $-8 + j \geq 1.9$ ____________
21. $\frac{k}{-3} \leq -2$ ____________
22. $7.8 < 3.9a$ ____________
23. $\frac{s}{3} > -\frac{8}{3}$ ____________

Solve each inequality.

24. $7y \leq 35$ ____________
25. $5x > 21$ ____________
26. $\frac{p}{9} > -\frac{2}{9}$ ____________
27. $\frac{r}{11} \geq 11$ ____________
28. $-6k < 36$ ____________
29. $-9t \leq -18$ ____________

NAME ______________________ CLASS __________ DATE __________

Practice Masters Level B

6.2 Multistep Inequalities

Describe the steps needed to solve each inequality.

1. $2y - 7 > 10$ ______________________

2. $\frac{x}{8} + 1 \leq 12$ ______________________

3. $-3p + 4 < p - 5$ ______________________

4. $-\frac{w}{6} + 9 \geq 3$ ______________________

5. $\frac{2}{3}m - 5 < 1$ ______________________

6. $4 > -(x + 1) - 7$ ______________________

Write an inequality that describes each situation.

7. 5 less than 2 times a number is smaller than 4. ______________________

8. 8 more than a number that is divided by 3 is at least 10. ______________________

9. A number divided by -7 is no more than the 3 times the number. ______________________

10. Twice the quantity of a number added to 6 is larger than the number divided by 10. ______________________

Solve each inequality.

11. $\frac{2}{5}y + \frac{3}{2} > \frac{1}{2}$ ______________________

12. $7.9z + 1.3 < 9.2$ ______________________

13. $-4 \geq -9h + 5$ ______________________

14. $2 + 3n > n - 5$ ______________________

15. $\frac{g}{3} + 10 < 12$ ______________________

16. $-\frac{p}{2.1} - 4 \geq 1.8$ ______________________

17. $u - \frac{1}{2} \leq \frac{1}{2} + 3u$ ______________________

18. $2(x - 1) > 10$ ______________________

19. $-3(y + 4) < -9$ ______________________

20. $h + 3 < 5 - h$ ______________________

21. $-u - 10 \geq u$ ______________________

22. $\frac{6}{5}x - \frac{1}{5} > 3$ ______________________

NAME ______________________ CLASS ____________ DATE ____________

Practice Masters Level C

6.2 *Multistep Inequalities*

Solve each inequality.

1. $-\frac{4}{5}t + 3 > t - 2$ ____________
2. $\frac{1}{5}(k - 1) \geq 2 + k$ ____________
3. $1.6m + 5 < m - 8.4$ ____________
4. $3(5 - h) \leq 7h$ ____________
5. $5u + 3(u + 1) > 4$ ____________
6. $3.1q - 2.7 > 1 - 2q$ ____________
7. $\frac{n}{6} - \frac{5}{12} \geq \frac{2}{3}n$ ____________
8. $-\frac{w}{4} + 1 \geq \frac{w}{5} - 3$ ____________
9. $5.3(2.1 - x) < 2 - x$ ____________
10. $5(c + 2) \leq 2(1 - c)$ ____________
11. $3.1b + 2.1 < 5.4 - b$ ____________
12. $\frac{1}{5}x + \frac{2}{7} \geq 7x$ ____________

Solve.

13. A banquet hall charges a flat rate of $250 plus $9 per person in attendance. If the Wilsons wish to spend no more than $450 for their retirement party, how many guests can they invite? ____________
14. Five times the quantity of a number increased by 3 is no more than 7 less than twice the number. What is the largest possible number? ____________
15. One-seventh the quantity of $\frac{2}{5}$ minus a number is at least $\frac{1}{6}$ the quantity of the number decreased by 3. What is the smallest possible number? ____________
16. A train travels at an average speed of 85 miles per hour. If a trip from Sue's house to Kentucky will take at least 4 hours, what is the minimum distance between Sue's house and Kentucky? ____________
17. A certain stock is trading at $53 per share. If an investor wishes to spend no more than $425 on the stock, what is the maximum number of shares she can buy? ____________
18. Tom earns $3.25 per hour at the car wash plus $1.50 for each car he washes. If he wishes to earn at least $100 on a busy weekend during which he works 12 hours, what is the minimum number of cars that he must wash? Which property of equality did you use to solve this problem? ____________
19. The length of rectangle is at least 6 centimeters longer than the width. If the rectangle is 12 centimeters wide, what is the minimum area of the rectangle? ____________

NAME ______________________ CLASS __________ DATE __________

Practice Masters Level A
6.3 Compound Inequalities

Write a compound inequality for each of the following.

1. h is greater than 4 AND h is less than 10. ______________

2. c is no more than 2 AND c is more than -7. ______________

3. d is at most 0 OR d is more than $\frac{2}{3}$. ______________

4. k is at least 5 OR k is at most 4. ______________

5. y is less than 9 AND y is greater than -12. ______________

6. x is no less than 0.25 AND x is no greater than 1. ______________

Graph each compound inequality.

7. $-7 < x < 2$

8. $m \leq -7$ OR $m > -4$

9. $w > 13$ AND $w \leq 15$

10. $x < -3$ OR $x \geq 1$

11. $17 \geq u > 14$

Solve each compound inequality.

12. $6 < x - 7 \leq 7$ ______________

13. $3 \leq t + 2 \leq 5$ ______________

14. $3g \leq 9$ OR $g + 2 > 40$ ______________

15. $12 > 4m > 8$ ______________

16. $3 < \frac{w}{2} \leq 4$ ______________

17. $0 > \frac{2}{5}h \geq -\frac{5}{2}$ ______________

18. $u - 1 > 3$ OR $\frac{u}{2} \leq 1$ ______________

19. $9 > \frac{k}{3} \geq \frac{1}{3}$ ______________

20. $-1.8 < y + 1.9 < 2.1$ ______________

21. $7r < 14$ OR $r - 3 > 0$ ______________

22. $\frac{1}{6} \leq \frac{p}{12} < \frac{5}{24}$ ______________

23. $2p > 4$ OR $\frac{p}{5} \leq \frac{3}{10}$ ______________

24. $0 \geq -n > -3.2$ ______________

25. $-j < -4$ OR $9j \leq 27$ ______________

NAME ______________________ CLASS __________ DATE __________

Practice Masters Level B

6.3 Compound Inequalities

Describe the steps needed to solve each compound inequality.

1. $5 < r - 4 \leq 7$ ______________________
2. $-2 < 3c - 1 < 0$ ______________________
3. $\frac{x}{4} \geq 7$ OR $x - 2 < 9$ ______________________
4. $5 > 6 - y > -2$ ______________________
5. $3n + 2 < 0$ OR $\frac{n}{5} > 3$ ______________________
6. $9 \leq 2v + 1 \leq 13$ ______________________

Graph each compound inequality.

7. $h > -\frac{3}{5}$ OR $h \leq -1$
8. $1.7 \leq b \leq 3.2$
9. $7.5 > c \geq 0$ 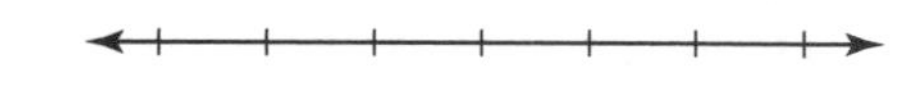
10. $f > -1.7$ AND $f \geq -4.1$
11. $m < 9$ OR $m \geq 10$

Solve each compound inequality.

12. $-4 < 3d - 2 \leq 7$ ______________
13. $1 \geq m - 2 \geq -4$ ______________
14. $2v - 1 > 1$ OR $\frac{v}{5} \leq -2$ ______________
15. $\frac{w}{2} > \frac{1}{2}$ OR $-w > 5$ ______________
16. $-1.5 < -y - 3 < 0$ ______________
17. $6a \leq 3$ OR $a - 4 > 2$ ______________
18. $14 \leq 3q + 5 < 17$ ______________
19. $4 \geq 3 - x > -12$ ______________
20. $1 - 2t > 0$ OR $\frac{t}{7} \geq 5$ ______________
21. $-17 < \frac{x}{4} + 5 < 2$ ______________

NAME ______________________ CLASS ______________ DATE ______________

Practice Masters Level C

6.3 Compound Inequalities

Solve and graph each compound inequality.

1. $3(v - \frac{2}{3}) > \frac{4}{5}$ OR $\frac{v}{8} < -\frac{7}{16}$ ______________

2. $14 - c \leq 3c + 6 \leq 28 - c$ ______________

Solve each compound inequality.

3. $-\frac{2}{7} < -\frac{x}{4} + 2 \leq \frac{4}{7}$ ______________

4. $3(2p + 6) < 18$ OR $-\frac{p}{9} < -2$ ______________

5. $5u - 4 > 2u + 6$ OR $22 \geq 11u$ ______________

6. $2g - 4 < 6g + 7 < 5 + 2g$ ______________

Solve.

7. Competitors in a target shooting contest earn 15 points for hitting each target. If Manuel needs at least 195 and at most 255 points to take over second place, what are the minimum and maximum targets he needs to hit to ensure second place? ______________

8. Water freezes at 32°F and boils at 212°F. Write a compound inequality to express the range of temperatures for which water is neither frozen nor boiling. ______________

9. Negative 7 is no more than twice the quantity of a number decreased by 9, which is no more than negative 4. What is the minimum value the number can be? What is the maximum value the number can be? ______________

10. If the value of a traded commodity falls below \$43 Mary will buy shares, and if it is at least \$71 she will sell her shares. The price of the commodity over the next week will be \$55 plus \$2 per point rise in the market. How many points must the market rise for Mary to sell her shares? ______________

11. In Exercise 10, how many points must the market drop over the week for Mary to purchase more shares? ______________

12. Either -7 times a number added to 3 is no more than 4, or twice the number added to $\frac{1}{3}$ is smaller than $\frac{2}{3}$. What is the minimum value the number can be? ______________

NAME ______________________ CLASS ____________ DATE ____________

Practice Masters Level A

6.4 Absolute-Value Functions

Find the absolute value (ABS) of each number.

1. 32 ________
2. -5.1 ________
3. $-\frac{4}{3}$ ________
4. 0.35 ________
5. -14 ________
6. $\frac{11}{23}$ ________
7. 19.21 ________
8. -0.1 ________
9. -5.4 ________
10. 0 ________
11. $-\frac{1}{21}$ ________
12. -100 ________
13. 0.7 ________
14. -7.03 ________
15. $-\frac{14}{3}$ ________
16. $\frac{1}{10{,}000}$ ________

Evaluate.

17. $|19 - 14|$ ________
18. $|5 - 8|$ ________
19. $|0 - 8|$ ________
20. $|4 + (-15)|$ ________
21. $|1.8 - 3.7|$ ________
22. $-|10 - 3|$ ________
23. $|-5 - 8|$ ________
24. $|7 - (-4)|$ ________
25. $|4 - 4|$ ________
26. $|2 - 19|$ ________
27. $-|23 - 7|$ ________
28. $|-1 - (-5)|$ ________
29. $|17 - 56|$ ________
30. $|60 - 1|$ ________
31. $|4.5 - 8.5|$ ________
32. $|7.1 - 2.8|$ ________
33. $-|0 - 5|$ ________
34. $|2 - (-8)|$ ________
35. $-|1.4 - 1.5|$ ________
36. $|69 - 70.5|$ ________
37. $|-8 - (-8)|$ ________
38. $|5 - 23|$ ________
39. $|23 - 5|$ ________
40. $\left|\frac{2}{4} - \frac{4}{5}\right|$ ________

Find the domain and range of each function.

41. $y = |x|$ ________________________
42. $y = |x| + 2$ ________________________
43. $y = |x - 1|$ ________________________
44. $y = |x + 5|$ ________________________
45. $y = 3 + |x|$ ________________________

NAME ______________________ CLASS ______________ DATE ______________

Practice Masters Level B

6.4 *Absolute-Value Functions*

Evaluate.

1. $\left|\frac{2}{9} - \frac{5}{18}\right|$ ________
2. $-|4.3 - 5.7|$ ________
3. $|1.7 - (-7.2)|$ ________
4. $|3.01 - 9.12|$ ________
5. $\left|-\frac{4}{7} - \frac{5}{8}\right|$ ________
6. $-\left|-\frac{9}{4} + \frac{7}{16}\right|$ ________
7. $-|6.7 - 48.1|$ ________
8. $|-4(2 - 9)|$ ________
9. $|1.4 - 3.7 + 8.1|$ ________
10. $|3.2(1.8 - 9.4)|$ ________
11. $\left|\frac{11}{9} - \frac{33}{4}\right|$ ________
12. $-|0 - 10(0.5)|$ ________
13. $-|5.01 - 5.1|$ ________
14. $|100.52 - 37.9|$ ________
15. $\left|-\frac{3}{8} - \frac{5}{7}\right|$ ________

Let $a = 3$, $b = -8$, and $c = 5$. Evaluate each of the following.

16. $|-a - b + c|$ ________
17. $-b|2c - a|$ ________
18. $|4a + b - c|$ ________
19. $|(a)(b)(c)|$ ________
20. $|-3ab - c|$ ________
21. $3a|a + b + c|$ ________
22. $b|3(b - c) - a|$ ________
23. $|2c - a + 3b|$ ________
24. $|a - c - (-a)|$ ________

Find the domain and range of each function.

25. $y = 7 - |x|$ ________
26. $y = 3|x - 4|$ ________
27. $y = (0.25)|x|$ ________
28. $y = -|x| - 9$ ________
29. $y = -|x - 5|$ ________
30. $y = \frac{1}{3}|x| + 2$ ________

Identify the transformation of $y = |x|$ for each of the following functions.

31. $y = |x - 1|$ ________
32. $y = |x + 13|$ ________
33. $y = |x - 1| - 9$ ________
34. $y = -|x| - 6$ ________

NAME ______________________ CLASS ____________ DATE ____________

Practice Masters Level C

6.4 Absolute-Value Functions

Identify the transformations of $y = |x|$ for each of the following.

1. $y = |x + 4| - 1$ ______________________

2. $y = -|x - 17|$ ______________________

3. $y = |x - 10| + 2$ ______________________

4. $y = -|x| + \frac{1}{3}$ ______________________

5. $y = -2|x - 9| - 4$ ______________________

6. $y = \frac{1}{3}|x + 3|$ ______________________

Write a function for each transformation of $y = |x|$ described.

7. The graph of $y = |x|$ is translated 11 units right and then 4 units up. ____________

8. The graph of $y = |x|$ is reflected across the x-axis and then translated 7 units left. ____________

9. The graph of $y = |x|$ is translated 2 units left and then reflected across the line $x = -4$. ____________

10. The graph of $y = |x|$ is translated 8 units up and then translated 6 units right and then reflected across the line $x = 3$. ____________

Solve.

11. Ms. Thompson's fifth grade class is having a contest to see who can come closest to guessing the number of pennies in a jar. Tom, Kim, Mike and Sue each guess 595, 1012, 402 and 774, respectively. If the jar holds 750 pennies, find the error and the absolute error for each student's guess.

____

12. In Exercise 11, who should be declared the winner? Justify your answer.

____

13. A river dam is opened for a short period of time to lower the water level. The water level, in feet, is modeled by the function $y = 22 + 3|t - 2|$, where t is in hours. How many hours will it take for the level to reach its lowest point? ____________

NAME ______________________ CLASS __________ DATE __________

Practice Masters Level A

6.5 Absolute-Value Equations and Inequalities

Solve each absolute-value equation. Check your answers.

1. $|x - 4| = 9$ ______________
2. $|x + 3| = 1$ ______________
3. $|x + 13| = 15$ ______________
4. $|x - 5| = 0$ ______________
5. $|x - 8| = 4$ ______________
6. $|x + 1.5| = 5.5$ ______________
7. $\left|x - \frac{1}{4}\right| = \frac{1}{2}$ ______________
8. $|x - 7.1| = 7$ ______________
9. $|x + 0.7| = 1.5$ ______________
10. $|x + 65| = 100$ ______________
11. $|x - 8| = 8$ ______________
12. $|x - 21.6| = 4.9$ ______________

For each guess and true value, find the error and absolute error.

13. value = 5, guess = 9 ______________
14. value = 1.8, guess = 1.3 ______________
15. value = $\frac{1}{3}$, guess = $\frac{1}{5}$ ______________
16. value = 100, guess = 95 ______________
17. value = 22, guess = 22.9 ______________
18. value = −32, guess = −29 ______________
19. value = $\frac{7}{9}$, guess = $\frac{7}{8}$ ______________
20. value = 1.75, guess = 1.25 ______________
21. value = −4, guess = −12 ______________
22. value = 0, guess = 0.02 ______________

Solve each absolute-value inequality. Graph each answer on a number line.

23. $|x - 2| < 5$ ______________

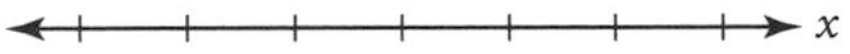

24. $|x + 3| > 7$ ______________

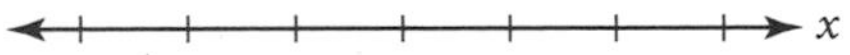

25. $|x - 9| < 3$ ______________

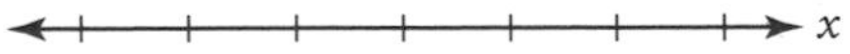

26. $|x - 1.5| \leq 3$ ______________

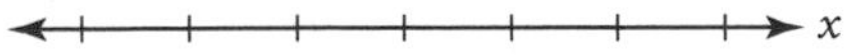

Solve each inequality. Check your answer.

27. $|x + 1| \leq 4$ ______________
28. $|x + 14| > 10$ ______________
29. $|x - 8| > 12$ ______________
30. $|x - 5.4| < 7.8$ ______________
31. $|x + 12| \leq 12$ ______________
32. $|x + 3| \geq 0$ ______________
33. $|x - 9| > 9$ ______________
34. $|x + 14| < 21$ ______________

NAME ______________________ CLASS __________ DATE __________

Practice Masters Level B

6.5 Absolute-Value Equations and Inequalities

Write an absolute-value inequality to represent each of the following.

1. all of the values on a number line that are within 8 units of 4 ________

2. all of the values on a number line that are closer than 2 units to 15 ________

3. all of the values on a number line that are within 5 units of 3 ________

4. all of the values on a number line that are at least 14 units more than 17 ________

5. all of the values on a number line that are closer than 22 units to 28 ________

Solve each equation if possible. Check your answers.

6. $|8x + 5| = -14$ ________

7. $\left|\frac{1}{3}x - \frac{4}{5}\right| = -6$ ________

8. $|-4(7 - x)| = 3$ ________

9. $|11x + 5| = 17$ ________

10. $|3 - 5x| = 0$ ________

11. $|0.25(x + 1)| = 19$ ________

12. $|-2(8 - x)| = 4$ ________

13. $|0.3x + 0.8| = 0.02$ ________

Solve by graphing on a coordinate plane.

14. $|2x + 7| > 9$ ________

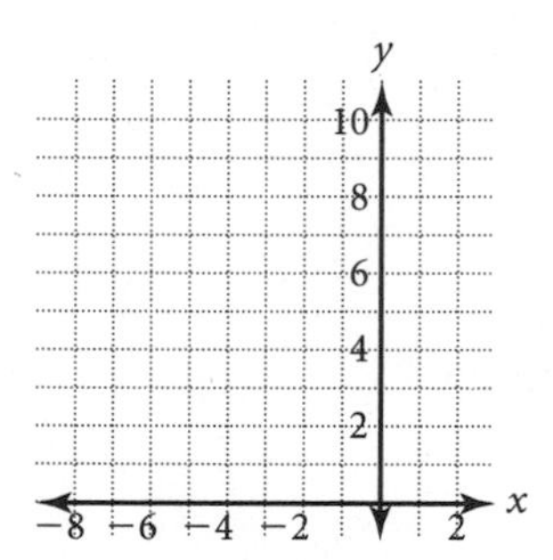

15. $|x - 5| \leq 3$ ________

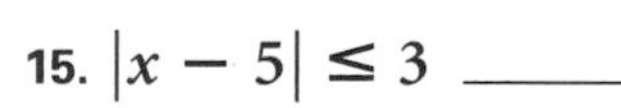

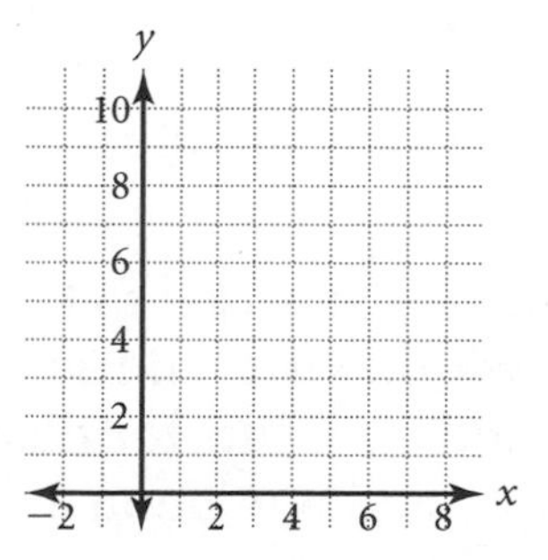

Solve each inequality. Check your answers.

16. $|4 - 11x| < 6$ ________

17. $|2x + 15| > 17$ ________

18. $|36(x - 1)| \geq 4$ ________

19. $\left|\frac{1}{3}x - \frac{4}{9}\right| < \frac{1}{9}$ ________

20. $\left|\frac{5}{11} - 2x\right| \leq 3$ ________

21. $|5 - 13x| > 25$ ________

NAME ____________________ CLASS __________ DATE __________

Practice Masters Level C

6.5 Absolute-Value Equations and Inequalities

For each absolute-value inequality, write a sentence that begins with "The distance between".

1. $|x + 4| < 8$ ____________________

2. $|x - 5| \geq 3$ ____________________

3. $|x - 3.1| \leq 9.2$ ____________________

4. $\left|x - \frac{1}{4}\right| < \frac{2}{7}$ ____________________

5. $|-(5 - x)| \leq 10$ ____________________

Solve each inequality. Check your answers.

6. $\left|-\frac{1}{3}\left(\frac{3}{2} - \frac{3x}{12}\right)\right| \leq \frac{4}{3}$ ____________________

7. $|13(4 - 7x) + 1| > 5$ ____________________

8. $|5x + 7(4 - x)| < 21$ ____________________

9. $|2x - 0.2(x - 5)| \leq 0.8$ ____________________

10. $|-13x - 5(10 - x)| \geq 10$ ____________________

11. $|0.01 - 3.6(x + 0.001)| < 0.1$ ____________________

12. $|0.7 + 0.23(1 - x)| < 0.9$ ____________________

13. $\left|\frac{2}{13}x + \frac{5}{26}\right| \leq \frac{9}{52}$ ____________________

Solve each problem.

14. A soda fountain is expected to dispense 21 ounces with an absolute error of no more than 0.75 ounces. Write and solve an absolute-value inequality describing the situation. What are the acceptable maximum and minimum values dispensed? ____________________

15. A jet-engine fan blade is expected to be within 0.001 centimeters of 1.1 centimeters in width. Write and solve an absolute-value inequality to describe the boundary values for the situation. Graph the acceptable maximum and minimum acceptable values for the width of the fan blade.

____

16. During a chemistry experiment the ambient temperature must be maintained within 2.5 degrees of 120 degrees Celsius. Write and solve an absolute-value inequality to describe the situation. Graph the maximum and minimum acceptable temperatures for the experiment.

____

NAME ______________________ CLASS ____________ DATE ____________

Practice Masters Level A

7.1 Graphing Systems of Equations

Determine whether the given point is a solution of the system of equations.

1. $(1, 3)$ $\begin{cases} y = 5x - 2 \\ y = -3x + 6 \end{cases}$ ____________

2. $(7, 5)$ $\begin{cases} y = 2x + 1 \\ y = x - 2 \end{cases}$ ____________

3. $(-1, 9)$ $\begin{cases} y = 14x + 5 \\ y = -7x + 2 \end{cases}$ ____________

4. $(0, 4)$ $\begin{cases} y = 4x + 4 \\ y = 15x + 4 \end{cases}$ ____________

5. $(1, 1)$ $\begin{cases} y = 9x - 8 \\ y = -14x + 15 \end{cases}$ ____________

6. $(2, 13)$ $\begin{cases} y = 8x - 4 \\ y = x + 10 \end{cases}$ ____________

7. $(5, -3)$ $\begin{cases} y = -x + 3 \\ y = 7x - 27 \end{cases}$ ____________

8. $(16, 3)$ $\begin{cases} y = 4x - 61 \\ y = -2x + 35 \end{cases}$ ____________

Write each equation in slope-intercept form.

9. $2x + 6y = -5$ ____________

10. $7y - 14x - 21 = 0$ ____________

11. $3 + 4y = 16x$ ____________

12. $-y - 5x = 25$ ____________

13. $8 = 12y + 6x$ ____________

14. $13 - 4y = 2x$ ____________

15. $6(x + 3y) = 36$ ____________

16. $18x + 18y = 36$ ____________

Solve by graphing.

17. $\begin{cases} y = x - 1 \\ y = 3x + 1 \end{cases}$ ____________

18. $\begin{cases} y = -x - 3 \\ y = 2x - 6 \end{cases}$ ____________

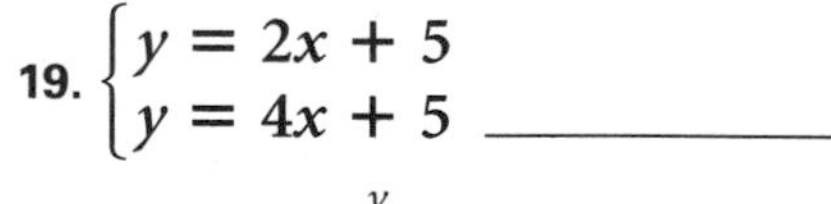

19. $\begin{cases} y = 2x + 5 \\ y = 4x + 5 \end{cases}$ ____________

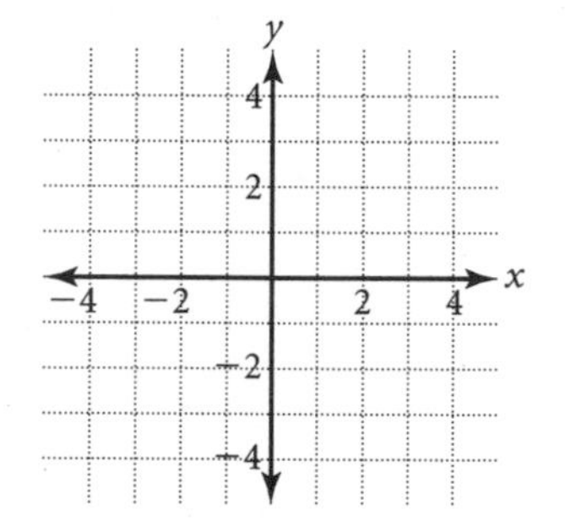

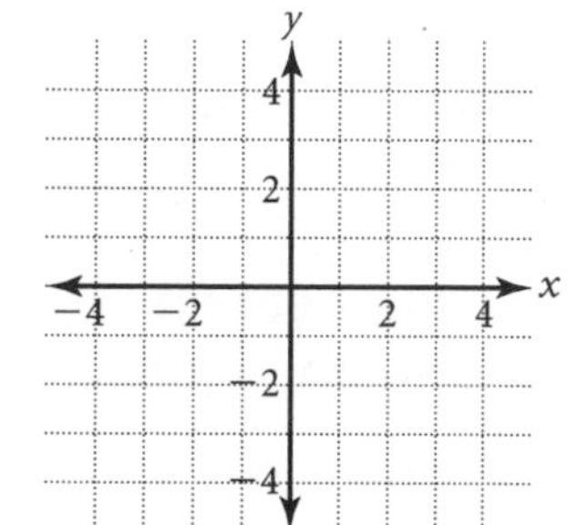

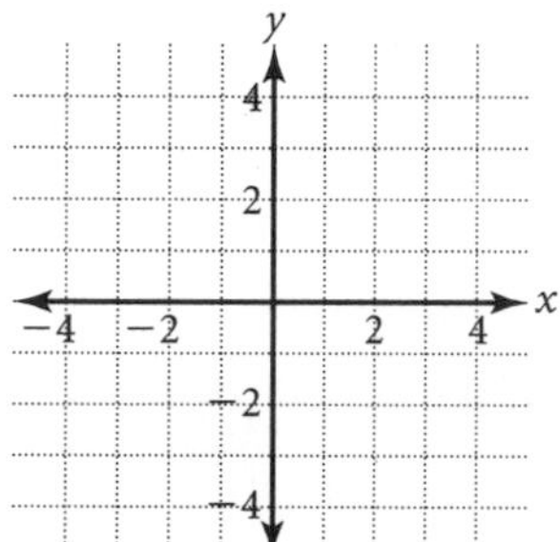

20. $\begin{cases} y = -x + 3 \\ y = 2x + 6 \end{cases}$ ____________

21. $\begin{cases} y = x - 2 \\ y = 3x - 4 \end{cases}$ ____________

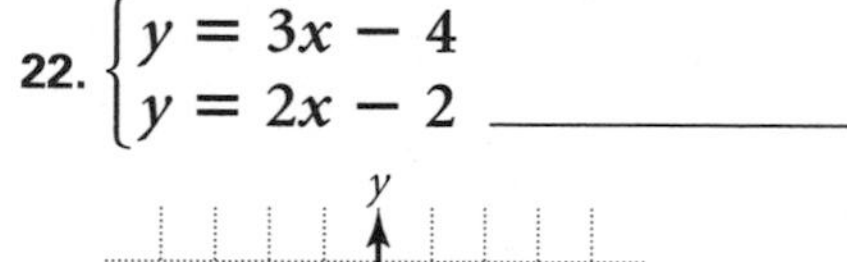

22. $\begin{cases} y = 3x - 4 \\ y = 2x - 2 \end{cases}$ ____________

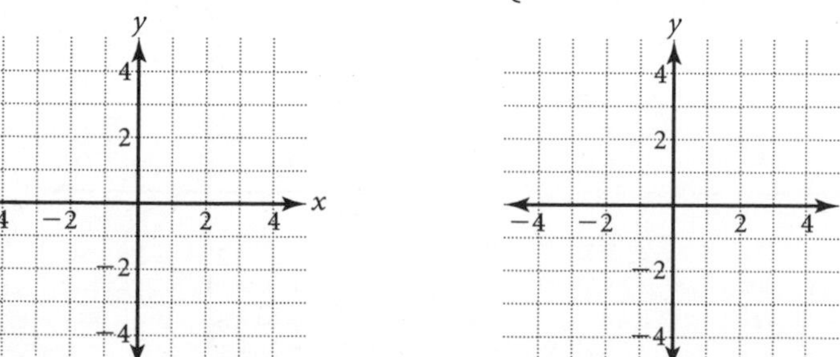

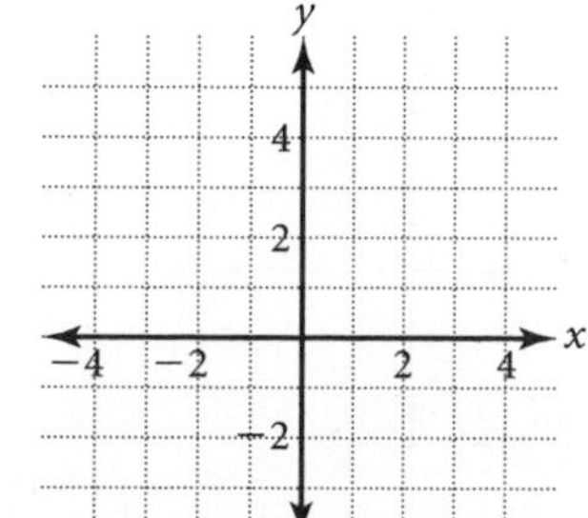

NAME ______________________ CLASS __________ DATE __________

Practice Masters Level B

7.1 Graphing Systems of Equations

Determine whether the given point is a solution of the system of equations.

1. $(-1, 9)$ $\begin{cases} 5y - 2x = -12 \\ -x - 3 = -12 - y \end{cases}$ ______

2. $(-3, -5)$ $\begin{cases} 10x + 1 = 5y - 4 \\ 0.5(x + 5) = 6 + y \end{cases}$ ______

3. $(9, -8)$ $\begin{cases} 3y - 5x = 7x + 6 \\ 2(x + y) = -\frac{y}{4}y \end{cases}$ ______

4. $(1, 0.25)$ $\begin{cases} 8x - 1 = 12y + 4 \\ 10y - x = x + 2y \end{cases}$ ______

5. $(0.1, 0.01)$ $\begin{cases} 10y - 4x = -3x \\ 8x - 0.5 = 3y \end{cases}$ ______

6. $(-4, 7)$ $\begin{cases} 3(x - 10) = 6y \\ 15 - 2y = 4x \end{cases}$ ______

7. $(0, 0)$ $\begin{cases} 6(y - x) = 10y \\ 3x + 2 = 2 - 9y \end{cases}$ ______

8. $(3, 12)$ $\begin{cases} y - 14x = -9x \\ -2(2x + y) = -3y \end{cases}$ ______

Write and graph a system of equations, to solve each problem.

9. Mandy scored 22 points in a basketball game. If she made 9 field goals, worth either 2 or 3 points, and no free throws, how many three-point field goals did she make? ______

10. Juan has \$67 in \$5 and \$1 bills. If he has 19 bills in all, how many \$1 bills does he have? ______

11. In a target shooting contest, contestants receive 5 points for hitting a red target and 10 points for hitting a blue target. If Tom scores a total of 135 points with 20 shots, how many red targets did he hit? ______

Solve by graphing. Round solutions to the nearest tenth. Check by substituting the solutions into the original equation.

12. $\begin{cases} 2x + 4y = 16 \\ 5y - 8x = -1 \end{cases}$ ______

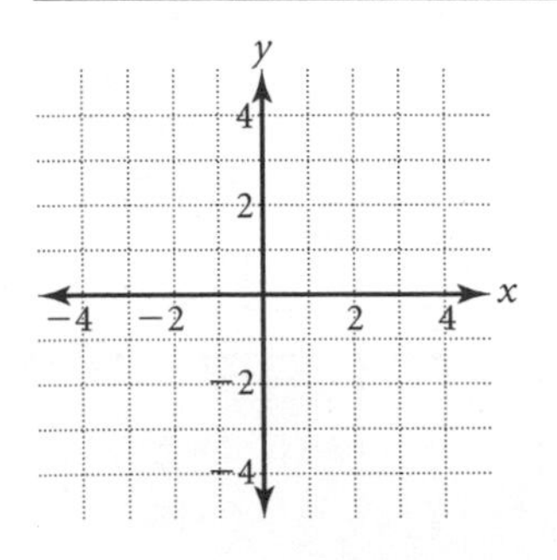

13. $\begin{cases} 4x + 9y = 2x - 7 \\ x - 10 = y - 12 \end{cases}$ ______

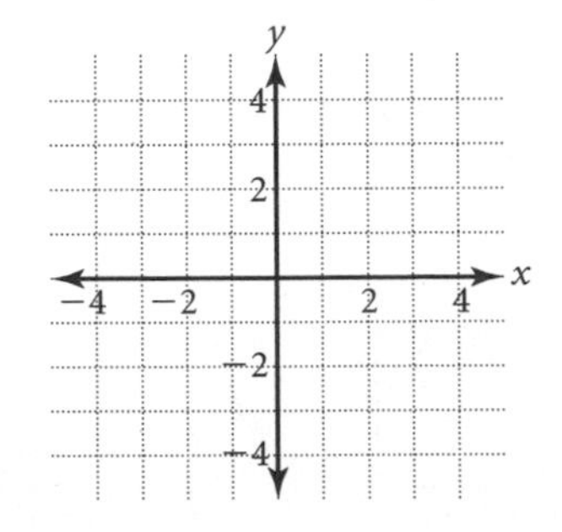

14. $\begin{cases} x = 5 \\ 7x + 5y = 11 \end{cases}$ ______

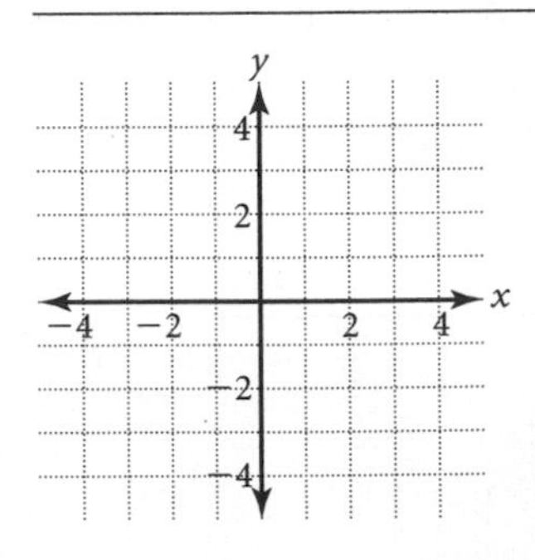

NAME ______________________ CLASS ____________ DATE ____________

Practice Masters Level C

7.1 Graphing Systems of Equations

Refer to points $D(-3, 6)$, $E(-1, 0)$ and $F(5, 6)$ for Exercises 1–4.

1. Write an equation for the line containing points D and E. ____________
2. Write an equation for the line containing points D and F. ____________
3. Write an equation for the line containing points E and F. ____________
4. Use graphing to solve the system of equations for lines DE and EF and then for lines DE and DF. ____________

For Exercises 5–9, write and solve a system of equations to solve the problem.

A retailer offers two options for satellite TV service. A customer may buy the dish for $150 and then pay $25 per month for service, or simply pay $35 per month for service and dish rental. Use this information for Exercises 5 and 6.

5. How many months will it take for the two plans to cost the same amount? ____________
6. If Tom chooses the dish-rental option and keeps the service for 24 months, how much money will he save? ____________

A deep-sea diver at the surface of the ocean begins descending at a rate of 120 feet per minute. Another diver who is 726 feet beneath the surface is rising at a rate of 60 feet per minute. Use this information for Exercises 7 and 8.

7. About how many minutes will it take for the divers to be at the same depth? ____________
8. About how many feet below the surface of the ocean will the two divers be when they meet? ____________
9. In a bike race, Pam has a 375-meter lead over Tonya. If Pam rides at an average rate of 7.9 meters per second and Tonya rides at an average rate of 8.3 meters per second, about how many minutes will it take for Tonya to catch up with Pam? ____________

NAME ______________________ CLASS ____________ DATE ____________

Practice Masters Level A

7.2 The Substitution Method

Determine whether the given point is a solution of the system of equations.

1. (2, 8) $\begin{cases} y = 3x + 2 \\ y = 9x - 10 \end{cases}$ ____________

2. (−4, 6) $\begin{cases} y = x - 2 \\ y = -7x + 22 \end{cases}$ ____________

3. (0, 9) $\begin{cases} 2y = 18x \\ 8x + y = 9 \end{cases}$ ____________

4. (−1, −3) $\begin{cases} 7x + 2y = -13 \\ 8 - 4y = 21 + x \end{cases}$ ____________

5. (7, 12) $\begin{cases} 6x - 2y = 22 \\ x - y = 5 \end{cases}$ ____________

6. (6, 5) $\begin{cases} -x - 6y = -36 \\ 3x - 3y = 3 \end{cases}$ ____________

7. (10, −2) $\begin{cases} 7y + x = 14 \\ x = 10 \end{cases}$ ____________

8. (3, 3) $\begin{cases} 9y - 3x = 6x \\ 4(x + y) = 24 \end{cases}$ ____________

Solve by using substitution, and check your answers.

9. $\begin{cases} y = 3x - 2 \\ x = 4 \end{cases}$ ____________

10. $\begin{cases} y = -2x + 7 \\ x = -1 \end{cases}$ ____________

11. $\begin{cases} y = 7x - 1 \\ y = -x + 14 \end{cases}$ ____________

12. $\begin{cases} y = 14x \\ y = -5x \end{cases}$ ____________

13. $\begin{cases} y = 2x + 9 \\ y = x + 5 \end{cases}$ ____________

14. $\begin{cases} y = -3x - 2 \\ x = 17 \end{cases}$ ____________

15. $\begin{cases} y = 13x - 1 \\ y = 10x + 29 \end{cases}$ ____________

16. $\begin{cases} y = 8x + 3 \\ y = -x + 3 \end{cases}$ ____________

17. $\begin{cases} y = x + 2 \\ y = -x - 2 \end{cases}$ ____________

18. $\begin{cases} y = 9x - 4 \\ x = 0 \end{cases}$ ____________

19. $\begin{cases} y = 6x + 7 \\ y = 10x + 11 \end{cases}$ ____________

20. $\begin{cases} y = -3x + 6 \\ y = 8x - 16 \end{cases}$ ____________

21. $\begin{cases} y = 8x - 2 \\ y = 5x - 0.5 \end{cases}$ ____________

22. $\begin{cases} y = 14x - 20 \\ y = -4x + 16 \end{cases}$ ____________

23. $\begin{cases} y = 2x + 8 \\ y = x - 7 \end{cases}$ ____________

24. $\begin{cases} y = 11x - 7 \\ x = 1 \end{cases}$ ____________

25. $\begin{cases} y = 9x - 4 \\ y = 13x + 2 \end{cases}$ ____________

26. $\begin{cases} y = -4x + 2 \\ y = 100x + 2 \end{cases}$ ____________

27. $\begin{cases} y = 13x + 17 \\ x = -1 \end{cases}$ ____________

28. $\begin{cases} y = 5x - 2 \\ y = 4x + 8 \end{cases}$ ____________

NAME ______ CLASS ______ DATE ______

Practice Masters Level B

7.2 The Substitution Method

Graph each system and estimate the solution. Then use the substitution method to get an exact solution.

1. $\begin{cases} y = -3x \\ y - x = 5 \end{cases}$ ______

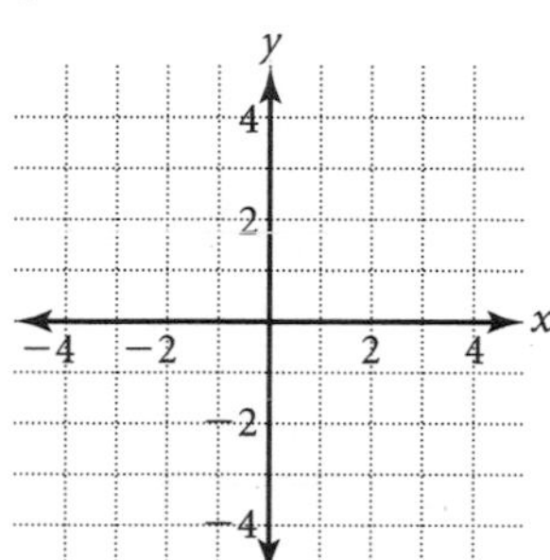

2. $\begin{cases} x - 2y = 4 \\ 8y + 24x = 5 \end{cases}$ ______

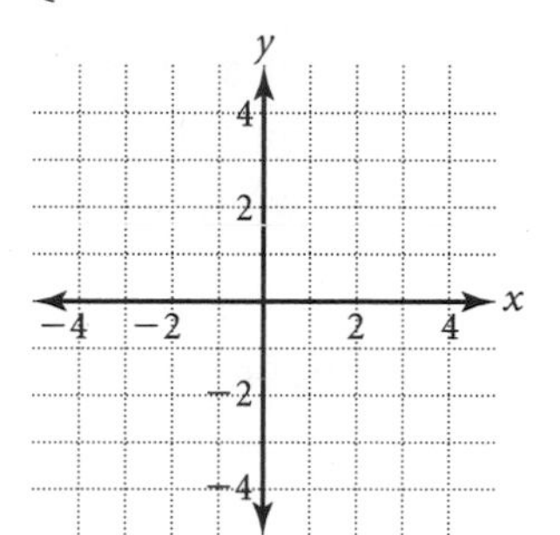

3. $\begin{cases} 2y - 6x = 10 \\ y = -x - 6 \end{cases}$ ______

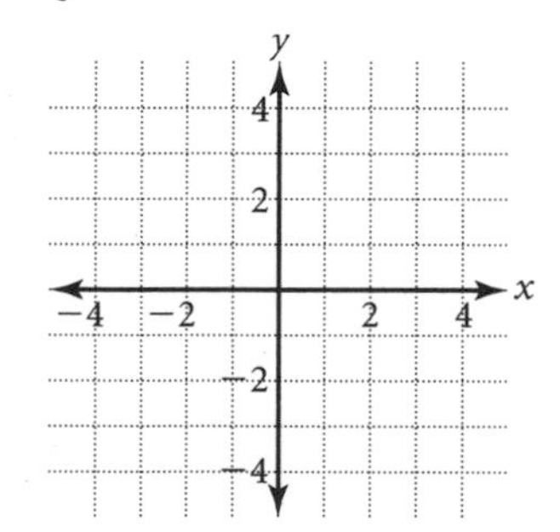

4. $\begin{cases} x + y = 1 \\ y + 4x = 2 \end{cases}$ ______

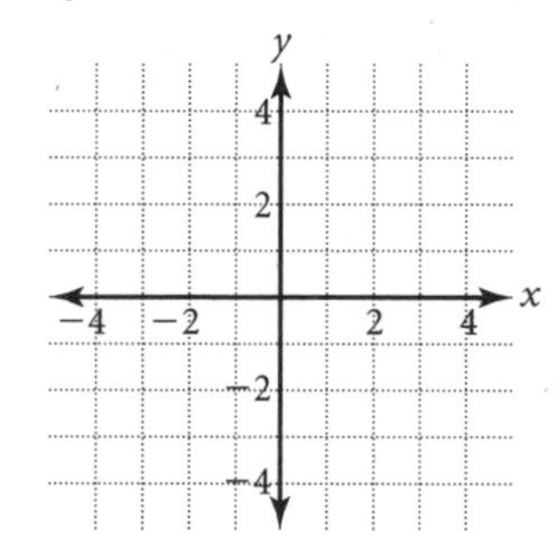

Solve by using substitution, and check your answers.

5. $\begin{cases} 4x + 16y = 2 \\ 2x = 4y \end{cases}$ ______

6. $\begin{cases} 2x + 3y = 24 \\ y + 7x = 46 \end{cases}$ ______

7. $\begin{cases} 8y - x = -4 \\ 5x - y = 59 \end{cases}$ ______

8. $\begin{cases} 3x - 4y = -1 \\ 11y + 2 = 13x \end{cases}$ ______

9. $\begin{cases} 6x + 3y = -2 \\ 9y + 10x = -2 \end{cases}$ ______

10. $\begin{cases} 7x - y = 10 \\ x = 2y - 6 \end{cases}$ ______

11. $\begin{cases} 10x - 14y = 62 \\ -7 - y = -2x \end{cases}$ ______

12. $\begin{cases} y - 6x = 22 \\ 4y + 17 = -11x \end{cases}$ ______

13. $\begin{cases} 0.5x - 2y = 1 \\ -3y + x = 4 \end{cases}$ ______

14. $\begin{cases} 9x + 9y = 9 \\ 2y = 14x + 20 \end{cases}$ ______

15. $\begin{cases} \frac{1}{3}y - x = \frac{1}{4} \\ 2y - 12x = -1 \end{cases}$ ______

16. $\begin{cases} 7y - 15x = 22 \\ 3x + 9y = 6 \end{cases}$ ______

NAME ______________________ CLASS ____________ DATE ____________

Practice Masters Level C
7.2 The Substitution Method

Write and solve a system of equations for each problem.

1. Twice a number is 4 more than 3 times another number. If 2 less than 7 times the first number is 1 more than 8 times the second number, what are the two numbers? ____________

2. Eight less than the opposite of a number is 5 more than 2 times another number. If 7 more than the first number is 1 less than 8 times the second number, what are the two numbers? ____________

3. The perimeter of Mrs. McCord's rectangular garden is 100 feet. If the length is 1.5 times longer than the width, what are the dimensions of the garden? ____________

A bungee jumper leaps from the top of a building that is 420 feet tall and falls at an average rate of 95 feet per second. At the time of the jump, the elevator is at a height of 130 feet and rising at a rate of 3.6 feet per second. Use this information for Exercises 4 and 5.

4. About how many seconds will pass until the jumper and the elevator are at the same height? ____________

5. At what height will the jumper and elevator be when they are at the same height? ____________

6. At a local baseball game, tickets cost \$4 for adults and \$2 for students. If there were a total of 94 people at the game and the ticket revenues were \$294, how many students and how many adults attended? ____________

7. A car lot is selling this year's model sports utility vehicle (SUV) for \$24,500 and last year's model for \$21,995. On a busy Saturday, the dealership sold 11 SUVs for a total of \$259,480. How many of each year's model SUVs were sold? ____________

A coal miner is 140 feet beneath the surface of the earth and ascending at an average rate of 2.1 feet per second. His replacement begins at the surface and descends at an average rate of 1.8 feet per second. Use this information for Exercises 8 and 9.

8. About how long will it be until the workers are at the same height? ____________

9. At what height, with respect to the surface, will the two workers be when they are at the same height? ____________

NAME ______________________ CLASS ____________ DATE ____________

Practice Masters Level A

7.3 The Elimination Method

In the following systems, which terms are opposites? Explain how you would solve each system.

1. $\begin{cases} 2x + 9y = 12 \\ -2x - 4y = 20 \end{cases}$ ______________________

2. $\begin{cases} 5c + 8d = -10 \\ 13c - 8d = 12 \end{cases}$ ______________________

3. $\begin{cases} 0.3v - 1.5b = 2.1 \\ 1.8v + 1.5b = 3.3 \end{cases}$ ______________________

4. $\begin{cases} 2m + n = 6 \\ 2m - n = 6 \end{cases}$ ______________________

5. $\begin{cases} 17x + 2y = 34 \\ -17x - 5y = 21 \end{cases}$ ______________________

6. $\begin{cases} \frac{1}{3}x - \frac{1}{5}y = 1 \\ 3x + \frac{1}{5}y = \frac{1}{9} \end{cases}$ ______________________

Solve each system of equations by elimination, and check your solution.

7. $\begin{cases} 3x + y = 5 \\ 2x - y = 0 \end{cases}$ ____________

8. $\begin{cases} 9y + 3x = 18 \\ 7y - 3x = -2 \end{cases}$ ____________

9. $\begin{cases} 7m + 3n = 10 \\ -7m - n = 2 \end{cases}$ ____________

10. $\begin{cases} 2p - 4t = -8 \\ 7t - 2p = 11 \end{cases}$ ____________

11. $\begin{cases} 6h - 5s = 13 \\ -5h + 5s = 2 \end{cases}$ ____________

12. $\begin{cases} 3k + j = 24 \\ 5j - 3k = -6 \end{cases}$ ____________

13. $\begin{cases} 0.2x - 1.3y = 1 \\ 2.3y - 0.2x = 3 \end{cases}$ ____________

14. $\begin{cases} m - 7n = 28 \\ 6m + 7n = 7 \end{cases}$ ____________

15. $\begin{cases} 8u + 6c = 20 \\ 7u - 6c = -5 \end{cases}$ ____________

16. $\begin{cases} 10h - 2g = 11 \\ 7h + 2g = 23 \end{cases}$ ____________

Solve each system of equations by using any method.

17. $\begin{cases} 7x - 0.5y = 4 \\ 3x + 0.5y = 16 \end{cases}$ ____________

18. $\begin{cases} 6m + 5n = -11 \\ 7m - 5n = -2 \end{cases}$ ____________

19. $\begin{cases} -5w + 8p = 22 \\ 9w - 8p = -6 \end{cases}$ ____________

20. $\begin{cases} 14g + 21b = 28 \\ -11g - 21b = 5 \end{cases}$ ____________

NAME ______________________ CLASS ______________ DATE ______________

Practice Masters Level B

7.3 The Elimination Method

Decide which is the best method for solving each system, and explain your decision.

1. $\begin{cases} 3x - 8y = 21 \\ 7x + 8y = -12 \end{cases}$ ______________

2. $\begin{cases} y = 6x - 1 \\ 3y + 5x = 12 \end{cases}$ ______________

3. $\begin{cases} 4y + 2x = 11 \\ 2y - 3x = -8 \end{cases}$ ______________

4. $\begin{cases} 3.21x - 9.7y = 1.01 \\ 5.53x + 2.09y = 0.7 \end{cases}$ ______________

Solve each system by elimination, and check your solution.

5. $\begin{cases} 5x - 3y = 12 \\ 2x + 1.5y = 3 \end{cases}$ ______________

6. $\begin{cases} 11m + 9n = -24 \\ 4m - 18n = -30 \end{cases}$ ______________

7. $\begin{cases} 3q + 4b = 17 \\ -2q + 6b = -20 \end{cases}$ ______________

8. $\begin{cases} 7u - 3v = 1 \\ 3u - 3v = -3 \end{cases}$ ______________

9. $\begin{cases} -x + 12y = 16 \\ 4x - 3y = 26 \end{cases}$ ______________

10. $\begin{cases} 2q + 5w = 18 \\ 3q - 4w = -19 \end{cases}$ ______________

11. $\begin{cases} 3h - 2g = 7 \\ 7g + 5h = 84 \end{cases}$ ______________

12. $\begin{cases} 7x - 4y = 23 \\ -3x + 5y = 0 \end{cases}$ ______________

Solve each system of equations by using any method.

13. $\begin{cases} y = 5x + 4 \\ 7x - y = 0 \end{cases}$ ______________

14. $\begin{cases} \frac{1}{3}p - \frac{4}{5}q = -1 \\ -3p - 0.6q = -30 \end{cases}$ ______________

15. $\begin{cases} -2m - 8 = -10n \\ 8n + 5 = 13m \end{cases}$ ______________

16. $\begin{cases} 7.1x + 3.5y = 1.9 \\ 1.5y - 3.21x = 8 \end{cases}$ ______________

17. $\begin{cases} 2c - 8d = -10 \\ 5d + 7c = 31 \end{cases}$ ______________

18. $\begin{cases} 7y - x = 12 \\ 10x = 8y + 4 \end{cases}$ ______________

19. $\begin{cases} k + p = 0 \\ 11k - 2p = -39 \end{cases}$ ______________

20. $\begin{cases} 13v - 2r = -10 \\ 6v + 5r = 25 \end{cases}$ ______________

21. $\begin{cases} 4x + 9y = 11 \\ 8y - 6x = 5 \end{cases}$ ______________

22. $\begin{cases} 17m - 3 = -41n \\ -n + 5m = 27 \end{cases}$ ______________

NAME ______________________ CLASS ____________ DATE ____________

Practice Masters Level C

7.3 The Elimination Method

Solve each system of equations by using any method.

1. $\begin{cases} \frac{2}{7}x - \frac{4}{5}y = -\frac{1}{5} \\ \frac{7}{27}y + \frac{2}{9}x = \frac{2}{9} \end{cases}$ ____________

2. $\begin{cases} \frac{5}{13}b - \frac{2}{11}c = 6 \\ \frac{9}{13}b + \frac{7}{11}c = 32 \end{cases}$ ____________

3. $\begin{cases} 0.001m - 0.006n = -0.59 \\ 0.03n + 0.07m = 3.7 \end{cases}$ ____________

4. $\begin{cases} 0.872w - 2.309q = 1.212 \\ 5.09q + 4.005 = 0.12w \end{cases}$ ____________

For Exercises 5–10, write and solve a system of equations by using any method. Check your answers.

5. Mary works at a factory assembling auto parts. She earns a fixed salary for the first 40 hours and then additional pay for any time over 40 hours. During one week, she works 47 hours and earns \$613.75. Another week she works 51 hours and earns \$678.75. What is Mary's weekly salary and overtime pay per hour? ____________

6. At a home and garden sale, Mark buys 3 rose bushes and 7 beds of petunias for a total of \$114.75. Cynthia buys 2 rose bushes and 6 beds of petunias for a total of \$89.50. How much does it cost for one rose bush? How much is a bed of petunias? ____________

Weather balloon A is at an elevation of 1400 feet and descending at a rate of 9.5 feet per second. Weather balloon B is released from the ground and rises at a rate of 12.8 feet per second. Use this information for Exercises 7 and 8.

7. About how many seconds will pass before the balloons are at the same height? ____________

8. At what elevation will the weather balloons be when they are at the same height? ____________

9. Marty's road-side grill sells hot dogs for \$1.25 and hamburgers for \$2.50. During a busy summer weekend, he sells a total of 177 hot dogs and hamburgers combined for \$363.75. How many hot dogs did he sell? How many hamburgers? ____________

10. Tina's parking garage charges a flat rate for the first hour, and then a per-minute-rate for every minute past 1 hour. Sam parks for 2 hours and 17 minutes and pays \$12.66. Tonya parks for 1 hour and 33 minutes and pays \$9.14. How much is the flat rate for the first hour? How much per minute after the first hour? ____________

NAME ______________________ CLASS ____________ DATE ____________

Practice Masters Level A

7.4 Consistent and Inconsistent Systems

Use a graphics calculator to determine whether each system is consistent or inconsistent.

1. $\begin{cases} y = 5x - 12 \\ 2y = 10x - 12 \end{cases}$ ____________

2. $\begin{cases} 3y = 14t - 9 \\ y = 9t + 10 \end{cases}$ ____________

3. $\begin{cases} m = 3n + 6 \\ m = 3n - 5 \end{cases}$ ____________

4. $\begin{cases} c = 14d - 3 \\ c = 15 - 14d \end{cases}$ ____________

5. $\begin{cases} 5g = 13h - 4 \\ 5g = 13h + 4 \end{cases}$ ____________

6. $\begin{cases} p = 8q - 3 \\ 2p = 16q + 4 \end{cases}$ ____________

7. $\begin{cases} y = x - 1 \\ y = 1 - x \end{cases}$ ____________

8. $\begin{cases} c + d = 4 \\ c - d = 4 \end{cases}$ ____________

9. $\begin{cases} 3x + y = 13 \\ y + 3x = 0 \end{cases}$ ____________

10. $\begin{cases} r = 10t - 32 \\ 2r = 5 - 10t \end{cases}$ ____________

Determine whether each consistent system is dependent or independent.

11. $\begin{cases} y = x + 2 \\ y = -x - 2 \end{cases}$ ____________

12. $\begin{cases} b = 3c - 4 \\ b = 4c - 3 \end{cases}$ ____________

13. $\begin{cases} m = 10(n - 2) \\ 2m = 20n - 40 \end{cases}$ ____________

14. $\begin{cases} q = 6w + 3 \\ 5q = 30w + 15 \end{cases}$ ____________

15. $\begin{cases} u = 4 - v \\ u = 4 + v \end{cases}$ ____________

16. $\begin{cases} p = 17t - 10 \\ 3p = 30 - 51t \end{cases}$ ____________

17. $\begin{cases} 2y = x + 10 \\ 3y = 1.5x + 15 \end{cases}$ ____________

18. $\begin{cases} 9r = 3d - 2 \\ 9r = -3d + 2 \end{cases}$ ____________

Solve each system algebraically. Identify each system as consistent and dependent, consistent and independent, or inconsistent.

19. $\begin{cases} y = x + 3 \\ y = x - 4 \end{cases}$ ____________

20. $\begin{cases} w = 3q - 18 \\ w = 7q + 4 \end{cases}$ ____________

21. $\begin{cases} m = 2n - 1 \\ 4m = 8n - 4 \end{cases}$ ____________

22. $\begin{cases} u = 13c + 4 \\ 2u = 26c - 8 \end{cases}$ ____________

23. $\begin{cases} g + h = 4 \\ g + h = -4 \end{cases}$ ____________

24. $\begin{cases} y = 3t + 12 \\ 1.5y = 4.5t + 18 \end{cases}$ ____________

25. $\begin{cases} k = 7h + 7 \\ -3k = 21h - 21 \end{cases}$ ____________

26. $\begin{cases} 10z = 8 + 6v \\ 5z = 4 + 3v \end{cases}$ ____________

NAME ____________________ CLASS ____________ DATE ____________

Practice Masters Level B

7.4 Consistent and Inconsistent Systems

For each equation, write another equation to create a system that is dependent.

1. $y = 3x - 1$ ____________
2. $m = 4n + 3$ ____________
3. $4y = 13t - 12$ ____________
4. $3c + 2d = 5$ ____________
5. $9p + 2k = 8$ ____________
6. $8q + 4r = 16$ ____________
7. $y + x = 3$ ____________
8. $3b - 8c = 2c$ ____________

For each equation, write another equation to create a system that is independent.

9. $y = -7x + 3$ ____________
10. $c + 2d = 6$ ____________
11. $7r - 14t = 28$ ____________
12. $m = 13n + 4$ ____________
13. $2y - 4t = -2$ ____________
14. $36(x + y) = 72$ ____________
15. $3r - 7t = 10$ ____________
16. $u = 12c + 3u$ ____________

Solve each system algebraically. Identify each system as consistent and dependent, consistent and independent, or inconsistent.

17. $\begin{cases} x + y = 12 \\ x + y = -12 \end{cases}$ ____________
18. $\begin{cases} 3c - d = 13 \\ 9c = 3d + 39 \end{cases}$ ____________
19. $\begin{cases} c + 10b = 12 \\ 2b = 10c - 4 \end{cases}$ ____________
20. $\begin{cases} 2 = 4m - 3n \\ m = 3n + 5 \end{cases}$ ____________
21. $\begin{cases} 11v + 5q = 12 \\ q = 5v - 12 \end{cases}$ ____________
22. $\begin{cases} 7y - 2x = 11 \\ 14y = 22 - 4x \end{cases}$ ____________
23. $\begin{cases} 7p - 20q = 9 \\ 7p - 9 = 20q \end{cases}$ ____________
24. $\begin{cases} -y + 21v = 2 \\ 3y - 63v = 6 \end{cases}$ ____________
25. $\begin{cases} 4m - 2n = -2 \\ n = 2m + 1 \end{cases}$ ____________
26. $\begin{cases} 8u + 2c = 20 \\ c - 5 = -4u \end{cases}$ ____________
27. $\begin{cases} 3g + 6h = 15 \\ 9h - 6 = 12g \end{cases}$ ____________
28. $\begin{cases} 19p - m = 1 \\ 3m - 1 = 5p \end{cases}$ ____________
29. $\begin{cases} 3x = t + 8 \\ 3t = x + 8 \end{cases}$ ____________
30. $\begin{cases} -h = 3g - 1 \\ 9g + 3h = 3 \end{cases}$ ____________

NAME ______________________ CLASS __________ DATE __________

Practice Masters Level C

7.4 Consistent and Inconsistent Systems

For Exercises 1–3, use the system $5x + 2y = 4$ and $10 - 3x = 2y + 2$.

1. Solve the system of equations. ____________

2. Classify the system of equations in terms of its consistency and dependency. ____________

3. What do you notice about the slopes of the two lines? ____________

For Exercises 4–6, use the system $2y - 3x = -5$ and $7.5x - 12.5 = 5y$.

4. Solve the system of equations. ____________

5. Classify the system of equations in terms of its consistency and dependency. ____________

6. What do you notice about the slopes of the two lines? ____________

For Exercises 7–9, use the system that consists of the line containing points $A(2, 3)$ and $B(-1, -6)$ and the line containing points $C(4, 14)$ and $D(0, 2)$.

7. Write and solve the system of equations. ____________

8. Classify the system of equations in terms of its consistency and dependency. ____________

9. Write an equation that will form a consistent system with line AB. ____________

10. One football player weighs 65 pounds more than another football player. Together, the two players weigh 505 pounds. How much does each football player weigh? ____________

11. One number is 39 more than another number. If the 3 times the larger number is 81 more than 6 times the smaller number, what are the two numbers? ____________

12. The equations for lines MN and NP form a consistent and dependent system. If the slope of line MN is 4 and the points $M(-7, p)$, and $P(2, 20 - p)$ are given, find the value of p. ____________

13. In a bike race, Juan has a 135 meter lead over Tom. If both racers average a speed of 2.1 meters per second over the remaining 900 meters of the race, prove by writing and solving an inconsistent system of equations that Tom will not be able to catch Juan. ____________

Practice Masters Level A

7.5 Systems of Inequalities

Graph each inequality. Determine if the given point is a solution.

1. $y > 3x + 2$; $(3, 16)$

____

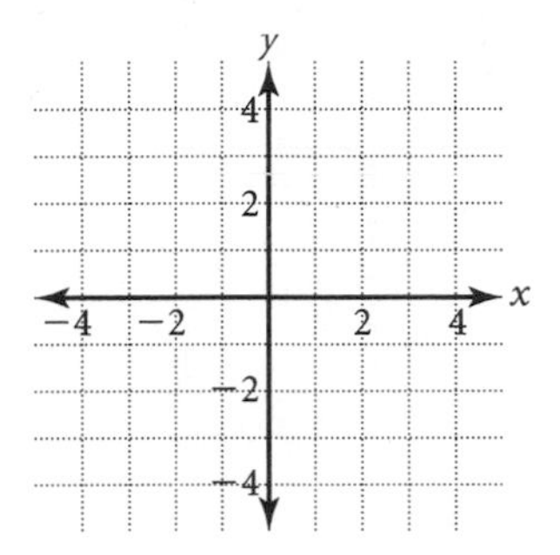

2. $y \leq -x + 4$; $(-1, 0)$

____

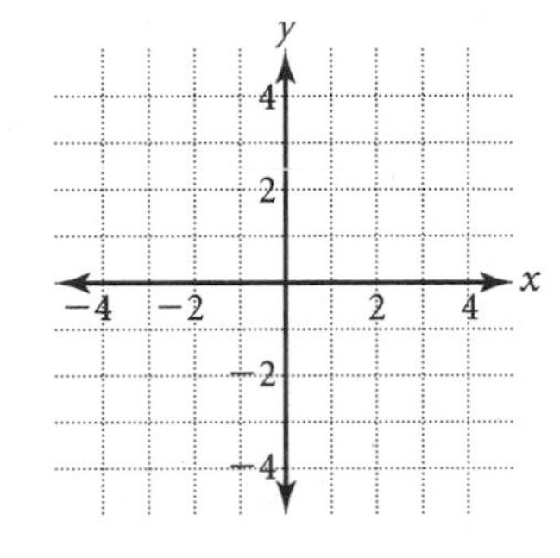

3. $y < 2x + 1$; $(4, -4)$

____

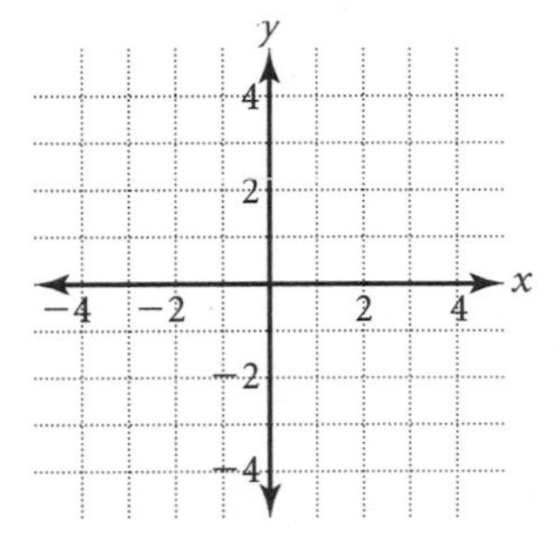

4. $y \geq -4x$; $(3, -4)$

____

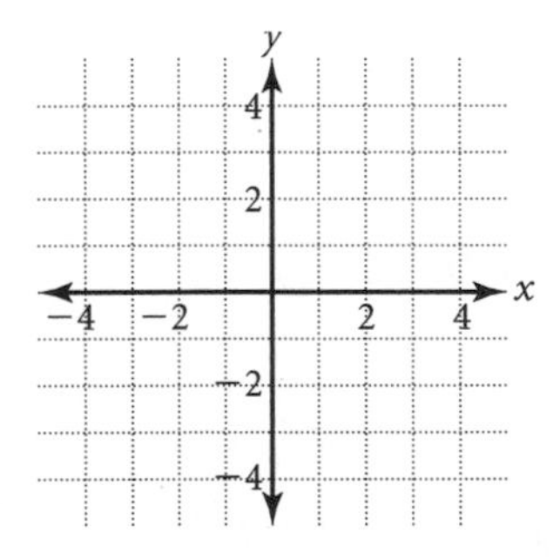

5. $y < -2x + 5$; $(0, 0)$

____

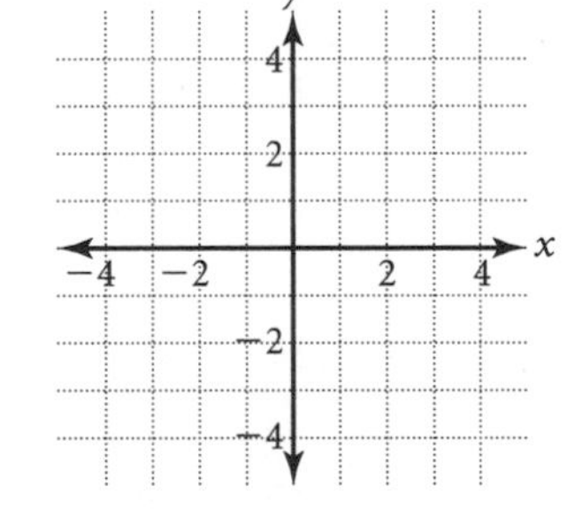

6. $y \geq 4x - 3$; $(-1, 1)$

____

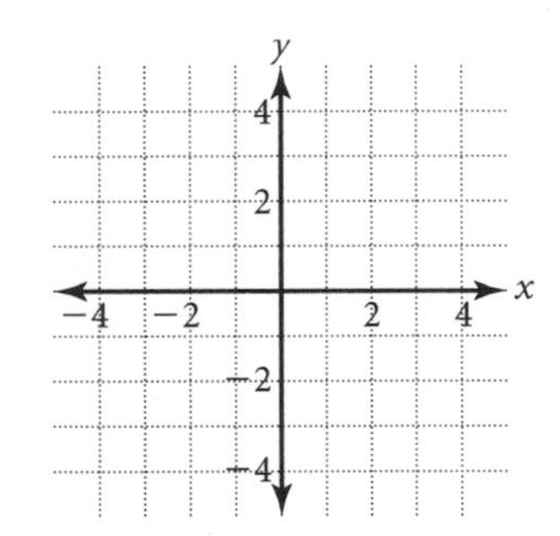

Solve by graphing.

7. $\begin{cases} y < 3x + 8 \\ y > -6x - 3 \end{cases}$

____

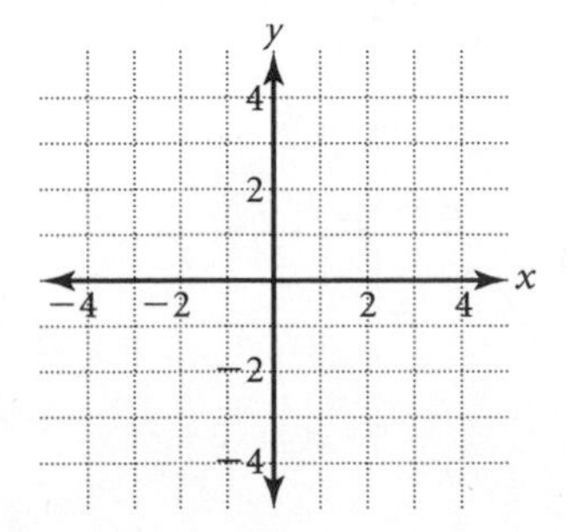

8. $\begin{cases} y > 4x - 7 \\ y < 12 \end{cases}$

____

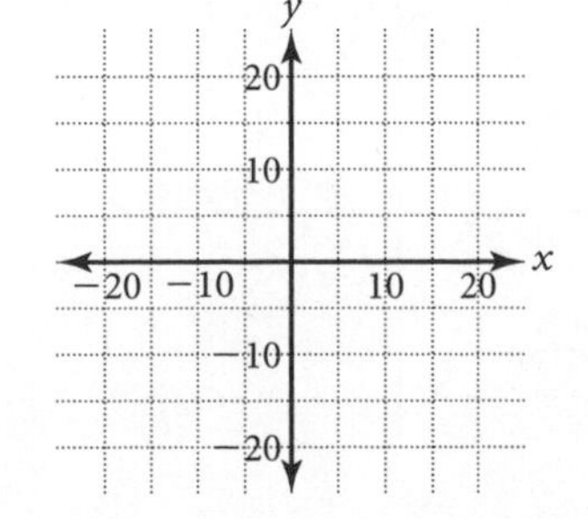

9. $\begin{cases} y \geq -1 \\ y < 2 - x \end{cases}$

____

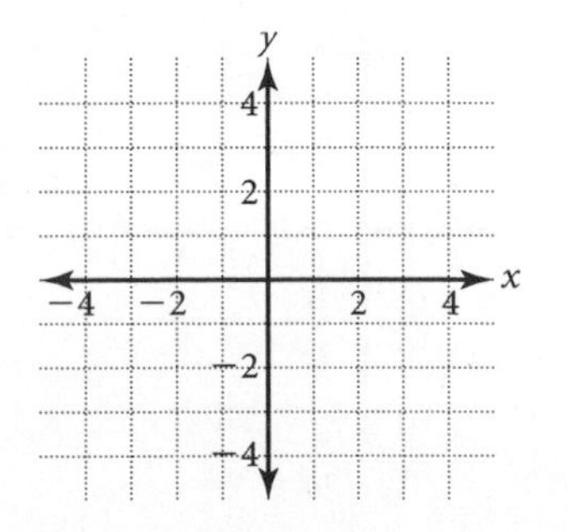

Practice Masters Level B

7.5 Systems of Inequalities

Graph each inequality.

1. $y > 2.5x + 4$

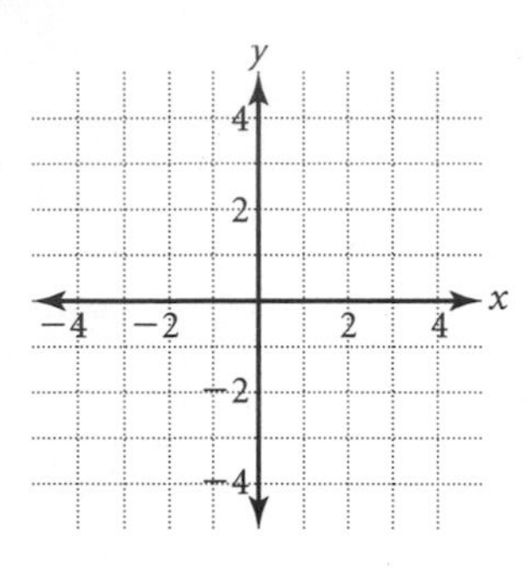

2. $y \leq -x + \frac{3}{4}$

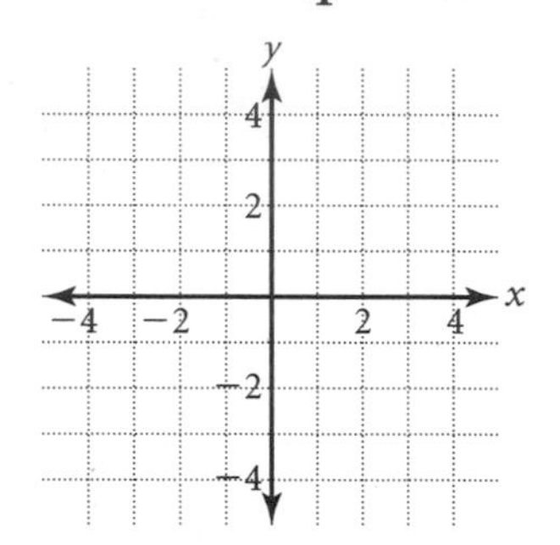

3. $y < \frac{2}{3}x + \frac{7}{12}$

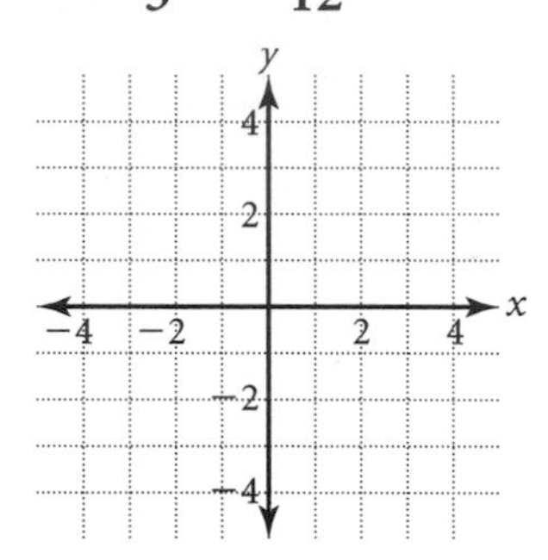

For each inequality, write another inequality to form a system that has the given point as a solution.

4. $(3, 4)$ $y < 2x + 1$ ______________

5. $(-8, 12)$ $y \geq 3x + 12$ ______________

6. $(0, 0)$ $3x - 2 < 2y$ ______________

7. $(9, -6)$ $4y + 2x \leq x - 4$ ______________

8. $(-5, -7)$ $8y - 8x \geq -16$ ______________

9. $(6, 13)$ $14x > 28$ ______________

10. $(19, -8)$ $9y + 3x < 4$ ______________

11. $(0, 53)$ $3x + 18y \geq 0$ ______________

12. $(10, 2)$ $13y - x > 9$ ______________

13. $(-5, 3)$ $7y \leq -3x + 6$ ______________

Solve by graphing. Choose a point from the solution and use it to check both inequalities.

14. $\begin{cases} 4x - 6y > 12 \\ 7y + 21 \leq 28x \end{cases}$

____

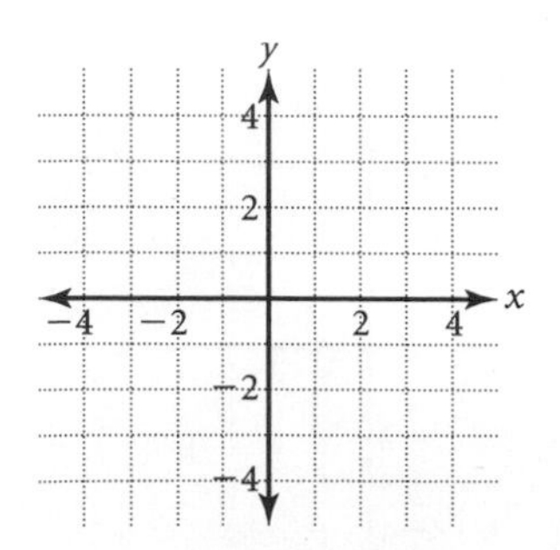

15. $\begin{cases} y \geq 7 \\ x \leq 10 \end{cases}$

____

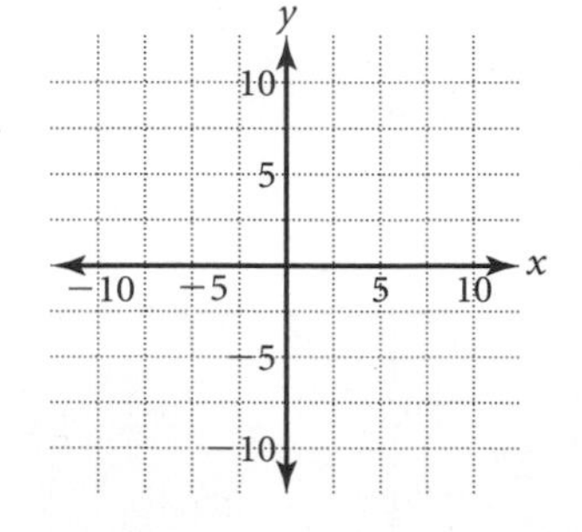

16. $\begin{cases} 4x + 7y < 0 \\ 3y - 9x > -36 \end{cases}$

____

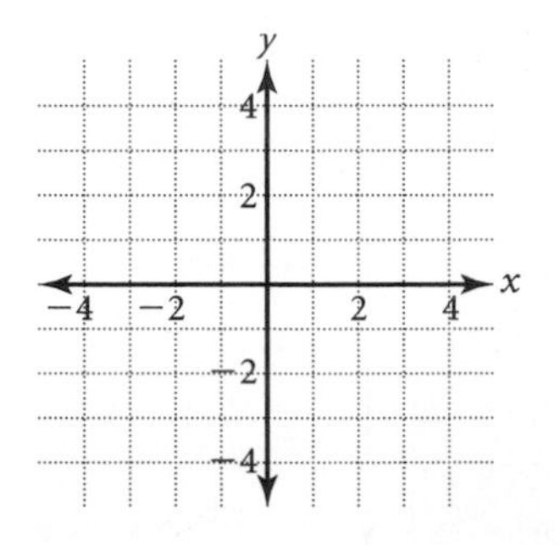

NAME ______________________ CLASS ____________ DATE ____________

Practice Masters Level C

7.5 Systems of Inequalities

1. Write a system of inequalities whose solution is the first quadrant of the coordinate plane, including all the points along the x-axis and the y-axis. ____________

2. How many inequalities would be needed to form a system whose solution area is the shape of a triangle? ____________

3. How many inequalities would be needed to form a system whose solution area is the shape of a rectangle? ____________

4. Write a system of inequalities whose solution is the triangle with vertices $A(-2, -5)$, $B(4, 7)$ and $C(4, -5)$. Include the sides of the triangle in the solution. ____________

5. Write a system of inequalities whose solution is the rectangle with corners $M(-8, 13)$, $N(5, 13)$, $O(-8, -19)$ and $P(5, -19)$. Include the sides of the rectangle in the solution if they are horizontal lines, but not if they are vertical lines. ____________

6. On an average day, an automobile factory produces no more than 50 sports utility vehicles (SUV). If the factory always makes about 3 times as many SUVs as trucks, what is the maximum number of trucks that the factory usually produces? ____________

James is planning on adding at least 15 new fish to his pond. He wants to add goldfish, which cost $12.50 each, and coy, which cost $17.75 each. He does not want to spend more than $380 total. Use this information for Exercises 7 and 8.

7. Write and solve a system of inequalities to represent the situation. ____________

8. Explain if it is feasible for James to purchase 10 goldfish and 10 coy.

____

Greg wants to earn at least $225 per week during the summer. He can earn $7.50 an hour working at a local restaurant and $6.25 an hour working at the park. Greg wishes to work no more than 40 hours per week. Use this information for Exercises 9 and 10.

9. Write and solve a system of inequalities to represent the situation. ____________

10. Does it make sense to consider any solutions that do not lie in the first quadrant? Explain.

____

NAME ______________________ CLASS ____________ DATE ____________

Practice Masters Level A

7.6 Classic Puzzles in Two Variables

Solve each problem. Check your answers.

1. A number is 2 more than twice another number. If the sum of the two numbers is 29, what are the two numbers? ____________

2. When a smaller number is subtracted from a larger number, the difference is 34. If the larger number is 30 less than 3 times the smaller number, what are the two numbers? ____________

3. One number divided by 6 is 37 less than another number. If the difference between the two numbers is 28, what are the two numbers? ____________

4. Mia has twice as many quarters as she has dimes. If she has $3.60 total, how many of each type of coin does she have? ____________

5. Hank has a total of $175 in $5 and $20 bills. The number of $5 bills he has is 3 more than 4 times the number of $20 bills. How many of each type of bill does Hank have? ____________

6. The tens digit of a 2-digit number is 1.5 times the units digit. If the sum of the digits is 5, what is the number? ____________

7. The hundreds digit of a 3-digit number is 1 more than twice the tens digit. The ones digit is the same as the tens digit, and the sum of all 3 digits is 13. What is the number? ____________

8. A mother is 23 years older than her son. In 7 years she will be twice as old as the daughter. How old is each now? ____________

9. Patrick is 5 years older than his nephew. In 2 years he will be 1.5 times older than his nephew. How old is each now? ____________

10. The difference between a grandmother's and a granddaughter's age is 44 years. If the grandmother's age is 13 less than 4 times the age of the granddaughter, how old is each now? ____________

11. The tens digit of a 2-digit number is 9 less than twice the units digit. If the sum of the 2 digits is 18, what is the number? ____________

12. Vicky is 5 years older than Pam. If the sum of their ages is 67 years, how old is each? ____________

13. Leo is 38 years older than Adam. Five times Adam's age, when decreased by 5 is 1 more than Leo's age. How old is each? ____________

Practice Masters Level B

7.6 Classic Puzzles in Two Variables

Solve each problem. Check your answers.

1. An airplane leaves North Dakota and heads toward Washington, which is 1200 miles away. Heading west, the plane, flying into the wind, takes 4.8 hours to get to Washington. After refueling, the plane returns to North Dakota, flying with the wind, in 4 hours. Find the speed of the wind and the speed of the plane with no wind. ____________

2. A swimmer begins a trip upstream. He swims a total of 3 miles in 2.5 hours. He then turns around and swims with the current to his original starting position. If the return trip takes him only 0.3 hours, how fast is the current and how fast does he swim when there is no current? ____________

3. A spelunker dives into an underwater canyon with a severe upward flowing current. It takes her 21.95 seconds to descend 90 feet. In her ascent, she covers the same distance in only 6.04 seconds. How fast is the upward current? How fast can the diver swim if there is no current? ____________

4. The soft drink fund of the mathematics club has a total of $21.25 in $1 bills and quarters. If 5 times the number of $1 bills decreased by 6, is 20 fewer than twice the number of quarters in the fund, how many of each type does the club have? ____________

5. Milo has a total of $4.40 in nickels and pennies. If 10 more than the number of his pennies divided by 5 is 24 fewer than the number of his nickels, how many of each type of coin does he have? ____________

6. A sprinter runs the 100-meter dash, with the wind behind him, in 9.96 seconds. He then runs the same distance in the other direction in 11.79 seconds. About how fast is the wind blowing? About how fast does the sprinter run when there is no wind? ____________

7. The difference between the digits in a 2-digit number is 1. If 7 times the smaller digit divided by 4 is 3 less than twice the larger digit, what is the number? ____________

8. Harrison is 9 more than 4 times the age of his twin sons. In 11 years he will be 10 years more than the sum of the twins' age. How old are the twins and Harrison now? ____________

9. The hundreds digit of a 3-digit number is 2 less than the tens digit, and the ones digit is 2 less than the hundreds digit. If twice the sum of the digits decreased by 10 is 20, what is the number? ____________

NAME ______________________ CLASS ____________ DATE ____________

Practice Masters Level C

7.6 Classic Puzzles in Two Variables

Solve each problem.

1. Acid solution A is composed of 12.5% hydrochloric acid, and acid solution B is composed of 34% hydrochloric acid. How much of each solution are needed to obtain 86 ounces of a solution that contains 25% hydrochloric acid? ____________
2. A soft drink company mixes a solution that contains 3% sugar with a solution that contains 5% sugar. If the company needs a total of 200 gallons of a solution that contains 4.2% sugar, how much of each type should be mixed? ____________
3. The thousands digit of a number is 5 less than twice the tens digit, and the units digit is 1 more than twice the tens digit. If the sum of all 4 digits is 20, what is the number? ____________
4. A coin bank contains pennies, nickels, quarters and dollars. There are 7 fewer quarters than twice the number of dollars, and there is 1 more nickel than 3 times the number of pennies. If the total amount of money in the bank is $15.76, how many of each denomination are there? ____________
5. The ten-thousands digit of a 5-digit number is 2 more than the thousands digit. The hundreds digit of the number is 4 more than twice the tens digit, and the units digit is 3 times the tens digit. If the difference between the sum of the ten-thousands and thousands digits and the sum of the hundreds, tens and units digits is 2, what is the number? ____________
6. A chemist has a solution that contains 50% sodium and a solution that contains 80% sodium. If he wants to obtain 2000 liters of a solution that contains 56% sodium, how much of each type of solution should be mixed? ____________
7. Nancy is 16 years older than Mark, and Mark is 8 years older than Claire. Ten years ago, Nancy was 21 years older than 4 times Claire's age and Mark was 7 years older than twice Claire's age. How old is each now? ____________
8. Write and solve your own classic puzzle in two variables.

____

____

____

____

NAME ______________________ CLASS __________ DATE __________

Practice Masters Level A

8.1 Laws of Exponents: Multiplying Monomials

For Exercises 1–8, identify the base and the exponent.

1. 6^3 ________
2. 4^2 ________
3. 8^5 ________
4. 10^5 ________
5. 5^5 ________
6. 2^7 ________
7. 12^2 ________
8. 20^1 ________

Evaluate.

9. 3^5 ________
10. 15^1 ________
11. 9^3 ________
12. 4^3 ________
13. 7^2 ________
14. 10^3 ________
15. 24^1 ________
16. 8^2 ________

Simplify each product. Leave the product in exponent form.

17. $5^2 \cdot 5^3$ ________
18. $8^4 \cdot 8^7$ ________
19. $9^3 \cdot 9^3$ ________
20. $10^5 \cdot 10^2$ ________
21. $7^4 \cdot 7^3$ ________
22. $10^3 \cdot 10^9$ ________
23. $2^5 \cdot 2^4$ ________
24. $4^2 \cdot 4^6$ ________
25. $6^3 \cdot 6^2$ ________
26. $10^2 \cdot 10^2$ ________
27. $3^3 \cdot 3^5$ ________
28. $5^4 \cdot 5^4$ ________
29. $10^5 \cdot 10^4$ ________
30. $6^2 \cdot 6^7$ ________
31. $7^2 \cdot 7^6$ ________
32. $4^5 \cdot 4^5$ ________

Simplify each product.

33. $(9x^2)(3x^3)$ ________
34. $(5x^4)(2x^2y^5)$ ________
35. $(3y^4)(4x^5y^5)$ ________
36. $(7a^3)(-6a^6b^3)$ ________
37. $(-5c^7)(2c^2d^3)$ ________
38. $(9t^4)(4t^3v^8)$ ________
39. $(-a^5)(4a^2)$ ________
40. $(8x^7)(-3y^5z^6)$ ________
41. $(7n^3)(-m^8n^6)$ ________
42. $(6f^2)(5f^2g^6)$ ________
43. $(-12r^3)(2r^4s^4)$ ________
44. $(3a^3)(5b^2c^8)$ ________
45. $(-x^4)(4y^5)$ ________
46. $(9h^3)(-4h^2i^6)$ ________
47. $(10w^2)(3u^6v^9)$ ________
48. $(6b^2)(-4a^5b^4)$ ________
49. $(-5s^2)(6s^5t^4)$ ________
50. $(8p^9)(3n^7q^5)$ ________
51. $(3d^4e^2)(5d^2e^3)$ ________
52. $(2x^4y^2)(-7x^3y^5)$ ________
53. $(7a^2c^5)(8b^6e^7)$ ________
54. $(4r^2s^6)(-5r^5s^3)$ ________

NAME ______________________ CLASS ______________ DATE ______________

Practice Masters Level B

8.1 Laws of Exponents: Multiplying Monomials

For Exercises 1–4, identify the base and the exponent.

1. 5^7 ______________ 2. 6^6 ______________ 3. 2^{12} ______________ 4. 7^4 ______________

Evaluate.

5. 10^9 ______________ 6. 4^7 ______________ 7. 12^4 ______________ 8. 9^5 ______________

Simplify each product. Leave the product in exponent form.

9. $9^5 \cdot 9^9$ __________ 10. $6^7 \cdot 6^7$ __________ 11. $10^6 \cdot 10^{12}$ __________ 12. $5^5 \cdot 5^{10}$ __________

13. $4^7 \cdot 4^6$ __________ 14. $8^5 \cdot 8^7$ __________ 15. $10^{10} \cdot 10^{10}$ __________ 16. $12^4 \cdot 12^5$ __________

17. $2^7 \cdot 2^4$ __________ 18. $3^8 \cdot 3^3$ __________ 19. $10^x \cdot 10^y$ __________ 20. $6^a \cdot 6^b$ __________

Simplify each product.

21. $(-x^6)(7x^3)$ ______________ 22. $(9x^4)(-5x^6y^7)$ ______________

23. $(-10a^3b^5)(6a^4b^7)$ ______________ 24. $(14x^3y^5)(-4x^7z^9)$ ______________

25. $(5r^4s^3)(-7r^5s^9)$ ______________ 26. $(8a^4c^5)(9b^3c^8)$ ______________

27. $(-4w^4)(7w^6)(2x^8)$ ______________ 28. $(3a^2)(4b^5)(-4a^2b^2)$ ______________

29. $(-9x^2)(3x^4)(-3x^2)$ ______________ 30. $(4f^5)(-2e^3)(5e^4f^6)$ ______________

31. $(-2a^2)(-2b^4)(2b^3)$ ______________ 32. $(-3x^5)(-5y^4)(-4x^4y^6)$ ______________

33. $(-7f^3g^6)(4h^2)(-5g^3)$ ______________ 34. $(8b^2)(3b^4)(4a^5c^6)$ ______________

35. $(-3r^2)(-6t^8)(-4s^8)$ ______________ 36. $(2x^6)(5y^7)(9x^3z^5)$ ______________

37. $(2a^7)(12a^3)(-c^9)$ ______________ 38. $(5x^3z^2)(-8y^7)(4x^2y^2)$ ______________

Use the formula $V = lwh$ to find the volume of each rectangular prism.

39. $l = 2x, w = 3y, h = 2z$ ______________ 40. $l = x, w = 5x, h = 4y$ ______________

41. $l = 4ab, w = 2ac, h = 2bc$ ______________ 42. $l = 6fgh, w = 2gh, h = 3fh$ ______________

43. $l = 3xyz, w = 3xy, h = 5xyz$ ______________ 44. $l = 7ab, w = 6b, h = 2bc$ ______________

NAME ______________________ CLASS ____________ DATE ____________

Practice Masters Level C

8.1 Laws of Exponents: Multiplying Monomials

Simplify each product.

1. $(-6a^3)(5a^2)(4a^4)$ ____________
2. $(2x^3)(5y^3)(-7x^4z^5)$ ____________
3. $(-4a^4b^3)(3a^5)(-2b^5)$ ____________
4. $(5c^4d^5)(-7c^3e^5)(-4d^5e^3)$ ____________
5. $(-a^5)(-b^7)(-2a^9b^5)$ ____________
6. $(-8r^4s^7)(-3r^2s^5)(-5r^6t^8)$ ____________
7. $(-2x^a)(2y^b)(-4x^c)$ ____________
8. $(4a^xb^x)(-3a^y)(7b^z)$ ____________
9. $(5r^a)(2r^a)(-10r^b)$ ____________
10. $(6c^fd^g)(8c^g)(-7d^{3g})$ ____________

For Exercises 11–14, find the value of *x*.

11. $5^x \cdot 5^4 = 5^8$ ____________
12. $7^x \cdot 7^3 = 7^{12}$ ____________
13. $9^6 \cdot 9^x = 9^7$ ____________
14. $6^9 \cdot 6^x = 6^{18}$ ____________

15. The width of a cube is represented by $(a + b)^x$. Find an expression for the volume of the cube. ____________
16. The length of a rectangular prism is represented by $3y$, the width is represented by $2x$ and the height is represented by $4x$. If the volume of the cube is 96 cubic units, find the possible values of x and y. ____________
17. The volume of a rectangular prism is represented by $70x^6y^3z^4$. The length is represented by $5xy$ and the width is represented by $2x^3y^2$. Find an expression for the height of the rectangular prism. ____________
18. Millie had a bacteria culture that doubled in size every hour. She started with 2 milligrams of the culture at 8:00 A.M. How many milligrams will she have at 7:00 P.M.? ____________
19. Find a such that $(2x^2y^3)(4x^2y^a)(ax^4y^5) = -24x^4y^5$. ____________
20. Find x such that $(5x^3)(4x^4) = 43{,}740$. ____________
21. Find a such that $(x^{4a})(x^{3a})(x^7) = x^{24}$. ____________
22. Find a and b such that $y^{a+b} = y^{10}$ and $y^b = y^3$. ____________
23. Find m and n such that $x^{m+n} = x^6$ and $x^{m-n} = x^4$. ____________
24. Find a positive number that gets smaller when raised to a power. ____________

NAME ____________ CLASS ________ DATE ________

Practice Masters Level A

8.2 Laws of Exponents: Powers and Products

Simplify and find the value of each expression when possible.

1. $(4^2)^3$ ______
2. $(5^3)^2$ ______
3. $(2^6)^3$ ______
4. $(1^8)^{12}$ ______
5. $(y^1)^3$ ______
6. $(10^4)^2$ ______
7. $(k^2)^2$ ______
8. $(0^6)^6$ ______
9. $(b^2)^4$ ______
10. $(12^1)^2$ ______
11. $(9^1)^3$ ______
12. $(7^2)^2$ ______

Simplify each expression.

13. $(z^4)^3$ ______
14. $(t^{12})^6$ ______
15. $(a^3)^2$ ______
16. $(b^7)^1$ ______
17. $(w^3)^5$ ______
18. $(k^{13})^2$ ______
19. $(xy)^4$ ______
20. $(j^4)^z$ ______
21. $(h^w)^5$ ______
22. $(ab)^9$ ______
23. $(m^2n)^2$ ______
24. $(kp^5)^7$ ______
25. $(c^2d^3)^4$ ______
26. $(s^{10}p)^2$ ______
27. $(h^2g^3)^1$ ______
28. $(xy^9)^3$ ______
29. $(z^2y)^m$ ______
30. $(u^5v^5)^5$ ______
31. $(m^pn^q)^2$ ______
32. $(g^ph^2)^7$ ______
33. $(x^2y^3)^4$ ______

Verify the Power-of-a-Product Property by finding the value of each expression in two different ways.

34. $(3 \cdot 5)^2$ ______
35. $(2 \cdot 1)^9$ ______
36. $(5 \cdot 3)^3$ ______
37. $(8 \cdot 1)^3$ ______
38. $(3 \cdot 4)^4$ ______
39. $(10 \cdot 1)^2$ ______
40. $(3 \cdot 2)^5$ ______
41. $(3 \cdot 3)^2$ ______
42. $(5 \cdot 2)^4$ ______
43. $(2 \cdot 0)^7$ ______
44. $(6 \cdot 3)^1$ ______
45. $(9 \cdot 7)^2$ ______

Evaluate each monomial for $a = 3$, $b = 5$, and $c = 2$.

46. a^2b ______
47. $4c^3$ ______
48. $-8ac^2$ ______
49. b^2c^3 ______
50. a^cb ______
51. $(ac)^2$ ______
52. $(a^2b)^a$ ______
53. $(abc)^2$ ______
54. a^2c^4 ______
55. $(a^3c)^4$ ______
56. $(3ab)^3$ ______
57. $9c^2$ ______

NAME ______________________ CLASS ______________ DATE ______________

Practice Masters Level B

8.2 Laws of Exponents: Powers and Products

Simplify and find the value of each expression when possible.

1. -3^2 ______
2. $(-r)^2$ ______
3. $(-1)^{501}$ ______
4. $(-g^2)^4$ ______
5. $-(3^2)$ ______
6. $(-7^1)^4$ ______
7. $-(w^2)^2$ ______
8. $(-8^2)^2$ ______
9. $-(-4^3)^2$ ______
10. -10^2 ______
11. $(-4^4)^3$ ______
12. $(-p^5)^3$ ______

Evaluate each monomial for $a = -2$, $b = 3$, and $c = -5$.

13. a^4 ______
14. $(a^2c)^3$ ______
15. a^2c^4 ______
16. $-a^b$ ______
17. $(ac)^4$ ______
18. a^2b^3 ______
19. $-5c^5$ ______
20. $(bc)^2$ ______
21. $(abc)^3$ ______
22. $(2ca^2)^2$ ______
23. $(-a^3)^3$ ______
24. $(a^b)^5$ ______

Simplify each expression.

25. $(-y^4x^3)^2$ ______
26. $(g^3)^6(g^2)^5$ ______
27. $(3c^2)^3(-4bc^2)^2$ ______
28. $(-6x^2y^5)^4(xy)^2$ ______
29. $(-13h^5p^8)^2$ ______
30. $(-5t^4v^9)^3$ ______
31. $(xy)(xy)^2(xy)^3$ ______
32. $(2m^5)^3(-3mn)^5$ ______
33. $(4cd^3)^3(-5c^2d)^2$ ______
34. $(12gh^2)(gh)^6$ ______
35. $(9a^2b)^2(-3ab)^5$ ______
36. $8(-5q^3)^3$ ______
37. $(-4a)^4(ab^2)^{11}$ ______
38. $(y^2)^3(x^6y^9)^4$ ______
39. $(3b^2)^2(-\mathrm{b})^5$ ______
40. $-15(ut^3)^8$ ______
41. $(4gh^2)^4(-hk^5)^7$ ______
42. $(-3n^2p^3)^5(np)^2$ ______
43. $(-1)^{794}a^8$ ______
44. $(-cd^3)^5(c^2d)^4$ ______
45. $(a^2bc^8)^2(-cd^3)^3$ ______
46. $(-3m^2n^6)^5(n^7p^3)^6$ ______
47. $(-a)^{200}(-b)^{201}$ ______
48. $(v^4w^8)^2(-w^3u^5)^2$ ______

Practice Masters Level C

8.2 Laws of Exponents: Powers and Products

Simplify each expression.

1. $\left(\dfrac{1}{2x^3y^4}\right)^5$ ______________

2. $\left(\dfrac{-3}{4a^3b^6c^2}\right)^3$ ______________

3. $\left(\dfrac{-1}{7m^5n^3p}\right)^4$ ______________

4. $\left(\dfrac{1}{4q^5w^{11}}\right)^5$ ______________

5. $\left(\dfrac{-1}{2u^2v^4}\right)^7$ ______________

6. $\left(\dfrac{-27a^7b^3d^{10}}{81}\right)^3$ ______________

7. $(-3u^3v^8w^2)^3\,(u^7v^2w^3)^4\,(u^3v^4w^5)^3$ ______________

8. $(-c)^{103}\,(b^{12}c^5d^{10})^4\,(2b^2c^9d^{15})^2$ ______________

9. $(10x^8y^3z^{12})^3\,(-x^{13}y^{22}z^4)^5$ ______________

10. $-(6m^2n^6p^8)^2\,(-m^3np^5)^3\,(-mn^2p^4)^5$ ______________

Solve.

11. The height of a box is $12c^2b^4$ and its width and length are given by $3c^3b^2$ and $2c^8b^4$, respectively. What is the volume of the box?

____

12. In Exercise 11, if $b = 1$ and $c = 1$, and the measurements of the box are in centimeters, what is the volume of the box?

____

13. The base of a right triangle is $\left(\dfrac{4m^2n^4}{2m^3n^2}\right)^3$ meters and the height of the triangle is $\left(\dfrac{-21m^5n^3}{7m^2n^8}\right)^4$. What is the area of the triangle?

____

14. Mrs. Stewart is planting a garden in her backyard. She wants the garden to have an area of $12x^2y^8z^6$ square meters. If the length of the garden is $4xy^5z^3$ meters, how wide is it?

____

15. In Exercise 14, if mulch costs \$13.75 per $6x^2y^8z^6$ square meters, how much will it cost for Mrs. Stewart to mulch her entire garden?

____

NAME ______________________ CLASS ____________ DATE ____________

Practice Masters Level A

8.3 Laws of Exponents: Dividing Monomials

Find the value of each expression.

1. $\frac{7^{11}}{7^9}$ ______
2. $\frac{12^4}{12^2}$ ______
3. $\frac{13^{14}}{13^{16}}$ ______
4. $\frac{25^3}{25}$ ______
5. $\frac{10^6}{10^3}$ ______
6. $\frac{14^7}{14^8}$ ______
7. $\frac{9^7}{9^3}$ ______
8. $\frac{2^{42}}{2^{48}}$ ______
9. $\frac{30^4}{30^6}$ ______
10. $\frac{15^8}{15^6}$ ______
11. $\frac{50^5}{50^2}$ ______
12. $\frac{8^7}{8^4}$ ______

Use the Quotient-of-Powers to simplify each quotient. Assume that the conditions of the property are satisfied.

13. $\frac{x^m}{x^n}$ ______
14. $\frac{c^8}{c^4}$ ______
15. $\frac{k^{p+1}}{k}$ ______
16. $\frac{h^{13}}{h^d}$ ______
17. $\frac{b^{8u}}{b^{6u}}$ ______
18. $\frac{n^4}{n^{k+2}}$ ______
19. $\frac{y^5}{y^q}$ ______
20. $\frac{(mn)^4}{(mn)^3}$ ______
21. $\frac{v^{u-4}}{v^{u-5}}$ ______
22. $\frac{b^c}{b^d}$ ______
23. $\frac{g^5}{g^{h-1}}$ ______
24. $\frac{y^x}{y^{-x}}$ ______
25. $\frac{f^4}{f^w}$ ______
26. $\frac{c^{3+d}}{c^d}$ ______
27. $\frac{n^{12}}{n^3}$ ______

Simplify each expression. Assume that the conditions of the Quotient-of-Powers Property are met.

28. $\frac{a^2b^4}{ab^2}$ ______
29. $\frac{6mn}{3m}$ ______
30. $\frac{h^5g^7}{2h^6g^9}$ ______
31. $\frac{(xy)^c}{(xy)^d}$ ______
32. $\frac{u^3v^7}{u^2v^9}$ ______
33. $\frac{k^4h}{k^3g^2}$ ______
34. $\frac{f^3p^7}{3fp^4}$ ______
35. $\frac{12m^8}{2m^4}$ ______
36. $\frac{169v}{13v}$ ______
37. $\frac{5xy}{25x^2y}$ ______
38. $\frac{-8gh^7}{4gh}$ ______
39. $\frac{14f^3}{7d^2}$ ______

NAME ______________________ CLASS __________ DATE __________

Practice Masters Level B

8.3 Laws of Exponents: Dividing Monomials

Simplify each expression. Assume that the conditions of the Quotient-of-Powers Property are met.

1. $\left(\frac{c}{d}\right)^5$ ______
2. $\left(\frac{g^2}{f^5}\right)^3$ ______
3. $\left(\frac{7y}{2x^3}\right)^2$ ______
4. $\left(\frac{2a^4}{3b^7}\right)^4$ ______
5. $\left(\frac{14uv^2}{7u^3v^2}\right)^6$ ______
6. $\left(\frac{-gh^4}{g^3h}\right)^7$ ______
7. $\left(\frac{9w^4}{q^6}\right)^2$ ______
8. $\left(\frac{5m^4}{n^7}\right)^{2p}$ ______
9. $\left(\frac{g^3f^6}{2f^3}\right)^4$ ______
10. $\left(\frac{2t^3}{5v^2}\right)^3$ ______
11. $\left(\frac{7x^2y}{28x^5y^3}\right)^2$ ______
12. $\left(\frac{3z^2y}{xyz}\right)^{3v}$ ______
13. $\left(\frac{-3uv^3}{9u^2v^5}\right)^3$ ______
14. $\left(\frac{(xy)^4}{x^3y^2}\right)^7$ ______
15. $\left(\frac{k^hp^5}{g^2}\right)^{2f}$ ______

Evaluate each quotient given $a = 3$, $b = -4$, and $c = 6$.

16. $\frac{b^5}{b^2}$ ______
17. $\frac{a^2bc^4}{ab^3c^5}$ ______
18. $\frac{b^4c}{b^5c^3}$ ______
19. $\frac{b^ac^2}{ab^4c}$ ______
20. $\frac{2abc}{4a^2b}$ ______
21. $\frac{(abc)^4}{a^3b^5c^2}$ ______
22. $\frac{c^{2a}b^2}{c^6b}$ ______
23. $\frac{c^2(2ab)^3}{c^3a^2}$ ______
24. $\left(\frac{ab}{c}\right)^3$ ______

Find each quotient. Assume that the conditions of the Quotient-of-Powers Property are met.

25. $\frac{65b^4}{13b^2}$ ______
26. $\frac{-13xy^3}{169x^2y^8}$ ______
27. $\frac{-m^4n^7}{-3m^7n^5}$ ______
28. $\frac{6a^5b^2c}{3a^4b^9c^3}$ ______
29. $\frac{-5z^{11}}{25yz^3}$ ______
30. $\frac{x^5yz^6}{x^2y^7z^2}$ ______
31. $\frac{-5.04w^{12}(u^3v)^4}{0.56(w^5u)^{11}v^7}$ ______
32. $\frac{144m^3(np)^5q}{-4m(n^3p^2)^7q^2}$ ______
33. $\frac{256(x^4y^3z^7)^2v^{11}}{-1024(xy^3z^2v^5)^2}$ ______
34. $\frac{535g^2h^7(f^7p)^4}{107(g^5hf^2)^5p^3}$ ______

NAME ______________________ CLASS __________ DATE __________

Practice Masters Level C

8.3 *Laws of Exponents: Dividing Monomials*

Find each quotient. Assume that the conditions of the Quotient-of-Powers Property are met.

1. $\left(\dfrac{13u^2v(u^4v^2w^7)^3uw^6}{-26(uv^4)^3v^3(uw^2)^2}\right)^5$ __________

2. $\left(\dfrac{-3x^7y^4(x^2yz^5)^2}{2y(x^7z)^3(xy^4)^3}\right)^4$ __________

3. $\left(\dfrac{-10mn(m^3n^5p)^2}{m^5n^3p(m^3p^7)^4}\right)^4$ __________

4. $\left(\dfrac{16q(q^4w^9v)^5}{4q^5w^{15}(qv^7)^2}\right)^5$ __________

5. $\left(\dfrac{-19uvw(uvw)^4u^2}{38u^2v^4(u^2v^4)^3w^7}\right)^3$ __________

6. $\left(\dfrac{-3a^7b^3d^{10}}{9a(b^7d^3)^8a^5(b^2d)^4}\right)^2$ __________

7. $\left(\dfrac{6a^3b(ab^3c^2)^4}{12a(a^5b^2c^7)^4}\right)^2 \cdot \left(\dfrac{28a^4(bc^5)^3ab^2c}{-56a(a^7bc^{12})^2b^7c^2}\right)^3$ __________

8. $\left(\dfrac{17u^3v^5(u^2w^7)^3uw^9}{34uw^7(v^3w)^7u^5v}\right)^3 \cdot \left(\dfrac{81(u^7v)^2w^{12}(uw^2)^2}{-27u^3(uvw^8)^2uv^{12}}\right)^3$ __________

9. In a city with a population of approximately 6.25×10^5 families, there are roughly 2×10^6 cars. Simplify the expression $\dfrac{2 \times 10^6}{6.25 \times 10^5}$ to find the average number of cars per family in the city. __________

10. The length of a side of a cube is $\dfrac{24a^3(bc^2)^4}{12ab(b^3c^3)^3}$ centimeters. What is the volume of the cube? __________

11. In Exercise 10, what is the surface area of the cube? Recall that the surface area of a cube is found by calculating the area of one of the sides and multiplying by 6. __________

12. The diameter of a circle is $\dfrac{2x^2(yz^5)^3}{x^7(y^2z^3)^2}$ meters. What is the area of the circle? Recall that the area of a circle is given by the formula $A = \pi r^2$, where r is the radius of the circle. __________

NAME ______________________ CLASS __________ DATE __________

Practice Masters Level A

8.4 Negative and Zero Exponents

Evaluate each expression.

1. 4^{-2} ______
2. $(-3)^{-4}$ ______
3. -2^3 ______
4. $\frac{7^{13}}{7^{15}}$ ______
5. 5^{-3} ______
6. $\frac{1^{15}}{1^{23}}$ ______
7. $-(-2)^3$ ______
8. $\frac{10^8}{10^{12}}$ ______
9. -8^2 ______
10. $\frac{15}{15^2}$ ______
11. 6^{-2} ______
12. $\frac{13^{-4}}{13^{-3}}$ ______
13. 12^{7-7} ______
14. $\left(\frac{34^8}{34^3}\right)^0$ ______
15. $\frac{12^3}{12^1}$ ______

Write each of the following without negative or zero exponents.

16. x^{-3} ______
17. p^{-5} ______
18. w^{t-t} ______
19. $(mn)^0$ ______
20. $5w^{-2}$ ______
21. $13f^0$ ______
22. $8b^{-12}$ ______
23. 12^0 ______
24. c^{-6} ______
25. $\frac{y^{-5}}{2}$ ______
26. $5k^{-7}$ ______
27. $(5k)^{-7}$ ______
28. $14cd^{-2}$ ______
29. $\frac{p^0}{q^0}$ ______
30. d^5w^{-6} ______
31. $a^{-4}b^0$ ______
32. $x^{-1}y^{-5}$ ______
33. $\frac{g^0}{h^{-9}}$ ______

Evaluate each expression for $a = 4$, $b = 2$, and $c = -3$.

34. a^c ______
35. $\frac{c^b}{c^a}$ ______
36. $a^{-b}b^c$ ______
37. $\frac{a^{-3}}{a^{-1}}$ ______
38. $b^{2c}b^4$ ______
39. $\frac{a^5}{a^{-c}}$ ______
40. $\frac{(ab)^c}{a^{-4}}$ ______
41. $\frac{a^{-2}c^2}{b}$ ______
42. $(abc)^{-2}$ ______
43. $ab^{-2}c^3$ ______
44. $b^{-3}a^2$ ______
45. $\frac{ab}{a^{-1}b^2c}$ ______
46. a^0b^c ______
47. $(a^cb^2)^0$ ______
48. $c^{-3}a^b$ ______

NAME ______________________ CLASS ____________ DATE ____________

Practice Masters Level B

8.4 Negative and Zero Exponents

Write each of the following without negative or zero exponents.

1. $(u^{-8}v^{2})^{0}$ ____________
2. $-m^{3}n^{-4}$ ____________
3. $(-3b^{2}c)^{-3}$ ____________
4. $\dfrac{-9q^{-8}}{3w^{-3}}$ ____________
5. $(2p^{2}q^{-7})^{-4}$ ____________
6. $\left(\dfrac{3t}{4s^{5}}\right)^{-1}$ ____________
7. $\dfrac{10h^{0}}{2h^{-7}}$ ____________
8. $(x^{-6}y^{-2})^{-2}$ ____________
9. $\dfrac{8^{0}}{(gh^{2})^{0}}$ ____________
10. $(-3v^{-5})^{-3}$ ____________
11. $\left(\dfrac{2a^{3}}{3b^{-2}}\right)^{-3}$ ____________
12. $w^{-4}w^{5}$ ____________
13. $\dfrac{3y^{-12}}{z^{-2}}$ ____________
14. $(-2m^{-2})^{-3}$ ____________
15. $\dfrac{12p^{-11}}{4p^{-11}}$ ____________

Evaluate each expression for $x = -2$, $y = 6$, and $z = -1$.

16. $\dfrac{x^{3}y^{-2}}{z}$ ____________
17. $y^{3}x^{-y}$ ____________
18. $\dfrac{z^{13}y^{x}}{y}$ ____________
19. $y^{-3}x^{4}z^{2}$ ____________
20. $\dfrac{xyz}{x^{-1}z^{-5}}$ ____________
21. $\left(\dfrac{y^{3}}{x^{4}z}\right)^{-3}$ ____________
22. $\dfrac{x^{3}z^{5}}{y^{-2z}}$ ____________
23. $(x^{-4}yz^{2})^{-2}$ ____________
24. $\dfrac{-3z^{-4}}{y^{x}}$ ____________

Write each of the following without negative or zero exponents.

25. $\dfrac{-3c^{-5}d^{4}f^{-1}}{-6d^{-5}d^{-4}f}$ ____________
26. $(-13q^{-7}w^{5})^{-2}$ ____________
27. $\dfrac{-x^{-9}x^{8}}{-25x^{4}x^{-4}}$ ____________
28. $\dfrac{21p^{0}b^{12}}{-42p^{-5}b^{12}}$ ____________
29. $\left(\dfrac{9m^{4}n^{7}}{81m^{-1}n^{3}}\right)^{0}$ ____________
30. $(8u^{6}v^{2}w^{-5})^{-1}$ ____________
31. $\dfrac{8a^{2}b^{4}c^{-3}}{a^{-3}b^{4}c^{6}}$ ____________
32. $\dfrac{12xy^{-5}z^{8}}{6x^{-1}y^{5}z^{3}}$ ____________
33. $(-g^{-4}f^{8}h^{12})^{-3}$ ____________
34. $\dfrac{99a^{-10}b^{2}cd^{-5}}{-11a^{-9}bc^{4}d^{-5}}$ ____________
35. $\left(\dfrac{r^{-4}s^{12}t^{0}}{4rs^{19}t^{-5}}\right)$ ____________
36. $(-4p^{5}qw^{-10})^{-2}$ ____________

Practice Masters Level C

8.4 Negative and Zero Exponents

Simplify and write each expression with positive exponents only.

1. $\left(\dfrac{10x^{-6}z^{2}}{2x^{3}y^{-3}}\right)^{3}$ ____________

2. $\left(\dfrac{-3a^{-3}b^{6}c}{6a^{5}b^{-2}c^{-2}}\right)^{2}$ ____________

3. $\left(\dfrac{7a^{-12}b^{2}c^{0}}{7a^{5}b^{3}c}\right)^{0}$ ____________

4. $\left(\dfrac{15x^{19}y^{-5}z^{-5}}{15(x^{-2}y)^{-8}yz}\right)^{6}$ ____________

5. $\left(\dfrac{-u^{-5}vw^{12}}{2u^{2}v^{-8}w^{13}}\right)^{-1}$ ____________

6. $\left(\dfrac{4m^{3}n^{-8}(np)^{-1}}{-8mn(m^{4}np)^{7}}\right)^{-4}$ ____________

7. $\left(\dfrac{13f^{-4}g^{9}h^{-1}}{-13(fg)^{-2}h^{5}}\right)^{3}\cdot\left(\dfrac{25f^{2}g^{-8}h^{4}}{5fg^{11}(fh^{2})^{-6}}\right)^{-2}$ ____________

8. $\left(\dfrac{-3u^{3}v^{-8}w^{2}(u^{-1}w^{3})^{-2}}{-9(uw^{5})^{4}v^{-8}w}\right)^{0}\cdot\left(\dfrac{9uvw^{6}(u^{-7})^{2}}{174u^{10}v^{12}w^{-4}}\right)^{0}$ ____________

9. $\left(\dfrac{12m^{-5}(n^{5}p)^{-4}q^{7}}{144(m^{-3}n^{2})^{-2}pq^{7}}\right)^{2}\cdot\left(\dfrac{5mn^{6}p^{-5}(n^{3}q^{10})^{-4}}{60m^{-3}n^{7}}\right)^{-2}$ ____________

10. $\left(\dfrac{-10xy^{13}(x^{-5}z^{4})^{2}}{10x^{7}(x^{13}y)^{5}z^{-8}}\right)^{-1}\cdot\left(\dfrac{9x^{-4}y(x^{6}z^{-7})^{-1}}{27y^{5}(x^{5}yz^{-2})^{-8}}\right)^{2}$ ____________

11. A circular swimming pool has a radius of $\dfrac{2m^{3}n^{-4}p}{8(mm)^{-2}p^{2}}$ meters and a height of $\dfrac{m^{-5}np^{2}}{27mn^{-2}p^{-1}}$ meters. What is the volume of the pool?
HINT: Find the area of the circle and multiply by the height. ____________

12. In Exercise 11, if $m = 7$, $n = 1$ and $p = 5$, what is the volume of the pool to the nearest tenth? ____________

13. Explain why it would not make sense to define $x^{0} = 0$, for any nonzero number x, and why it does make sense to define $x^{0} = 1$.

____

____

____

____

NAME ______________________ CLASS ______________ DATE ______________

Practice Masters Level A

8.5 Scientific Notation

Write each number in scientific notation.

1. 500,000 ______
2. 40,000,000 ______
3. 100,000,000,000,000 ______
4. 235,000,000 ______
5. 170,000,000,000 ______
6. 0.0000006 ______
7. 0.000077 ______
8. 0.00000000001 ______
9. 30,000,000 ______
10. 0.00000085 ______
11. 0.00978 ______
12. 0.0412 ______

Write each number in decimal notation.

13. 9×10^9 ______
14. 6×10^4 ______
15. 1.8×10^5 ______
16. 2.071×10^7 ______
17. 2×10^{-8} ______
18. 4.9×10^6 ______
19. 3.001×10^8 ______
20. 4×10^{-1} ______
21. 2.74×10^2 ______
22. 8.6×10^{-3} ______
23. 2.07×10^9 ______
24. 3×10^{-3} ______

Perform the following computations. Write your answers in scientific notation.

25. $(4 \times 10^8)(2 \times 10^3)$ ______
26. $(3 \times 10^2)(3 \times 10^5)$ ______
27. $(5 \times 10^3)(2 \times 10^6)$ ______
28. $(7 \times 10^5)(4 \times 10^9)$ ______
29. $(5.6 \times 10^4) - (2.1 \times 10^4)$ ______
30. $(7.8 \times 10^{14}) + (3.7 \times 10^{14})$ ______
31. Assume the answer to Exercise 30 represents the distance, in miles, between two stars. For your science class, you are to write a report about these two stars. Explain the benefits of using scientific notation in your report. ______

NAME ______________________ CLASS ______________ DATE ______________

Practice Masters Level B

8.5 Scientific Notation

Write each number in scientific notation.

1. 8,000,000,000,000 ______________
2. 22 ______________
3. 38,500,000,000 ______________
4. 40,080,000 ______________
5. 950 ______________
6. 2.7 ______________
7. 0.000304 ______________
8. 0.0000006819 ______________
9. 512,000,000,000,000,000,000 ______________
10. 70 ______________
11. 0.000010 ______________
12. 0.00000000000037 ______________

Write each number in decimal notation.

13. 2×10^{15} ______________
14. 8×10^{4} ______________
15. 7.6×10^{10} ______________
16. 8.075×10^{16} ______________
17. 3.2×10^{-4} ______________
18. 9.0×10^{-6} ______________
19. 3.001×10^{-8} ______________
20. 4.50×10^{-4} ______________
21. 7.19×10^{5} ______________
22. 1.06×10^{-8} ______________
23. 4.3×10^{9} ______________
24. 2×10^{-1} ______________

Solve for *x*. Write your answers in scientific notation.

25. $\dfrac{x}{2 \times 10^{3}} = (3 \times 10^{4})$ ______________
26. $\dfrac{x}{4 \times 10^{5}} = (6 \times 10^{8})$ ______________
27. $\dfrac{x}{5 \times 10^{9}} = (8 \times 10^{12})$ ______________
28. $(4 \times 10^{7}) + x = (7 \times 10^{7})$ ______________
29. $(4 \times 10^{5}) + x = (8 \times 10^{9})$ ______________
30. $(6 \times 10^{8}) - x = (8 \times 10^{8})$ ______________

NAME ______________________ CLASS ____________ DATE ____________

Practice Masters Level C

8.5 Scientific Notation

Solve for *x*. Write your answers in scientific notation.

1. $\dfrac{x}{1 \times 10^{15}} = (6 \times 10^{12})$ ____________

2. $\dfrac{x}{5 \times 10^{16}} = (8 \times 10^{20})$ ____________

3. $\dfrac{x}{2.4 \times 10^{10}} = (5 \times 10^{10})$ ____________

4. $\dfrac{x}{3.08 \times 10^{11}} = (2.1 \times 10^{30})$ ____________

5. $\dfrac{x}{7.5 \times 10^{33}} = (8 \times 10^{31})$ ____________

6. $\dfrac{x}{8.6 \times 10^{8}} = (4.5 \times 10^{15})$ ____________

7. $(2 \times 10^{8})x = (6 \times 10^{8})$ ____________

8. $(3.8 \times 10^{12})x = (1.9 \times 10^{21})$ ____________

9. $(7.5 \times 10^{10})x = (1.5 \times 10^{6})$ ____________

10. $(2.4 \times 10^{-8})x = (8.1 \times 10^{-6})$ ____________

11. $(5 \times 10^{-9})x = (1 \times 10^{4})$ ____________

12. $(1 \times 10^{14})x = (4 \times 10^{12})$ ____________

13. $(3 \times 10^{15}) + x = (8 \times 10^{15})$ ____________

14. $(8 \times 10^{9}) + x = (3 \times 10^{10})$ ____________

15. $(4.5 \times 10^{8}) + x = (1 \times 10^{11})$ ____________

16. $(1.2 \times 10^{6}) + x = (4 \times 10^{8})$ ____________

17. $(8 \times 10^{4}) + x = (3.7 \times 10^{7})$ ____________

18. $(2 \times 10^{-2}) + x = (7 \times 10^{3})$ ____________

19. An astrologer, specializing in black holes, concentrates his attention in one particular area of space for his research. To determine the distance, d, that a constellation is from Earth in this region, he uses the formula $d - (1 \times 10^{21}) = c$. If c equals 4.5×10^{19} miles, what is d? ____________

20. To determine the difference between an infrared wavelength and an ultraviolet wavelength, a physicist subtracted 5.2×10^{-8} meters from 1×10^{-6} meters. What was the result? ____________

NAME ________________ CLASS ________ DATE ________

Practice Masters Level A

8.6 Exponential Functions

Graph each of the following on the same coordinate axes.

1. $y = \left(\frac{1}{9}\right)^x$
2. $y = \left(\frac{1}{3}\right)^x$
3. $y = 3^x$
4. $y = 9^x$

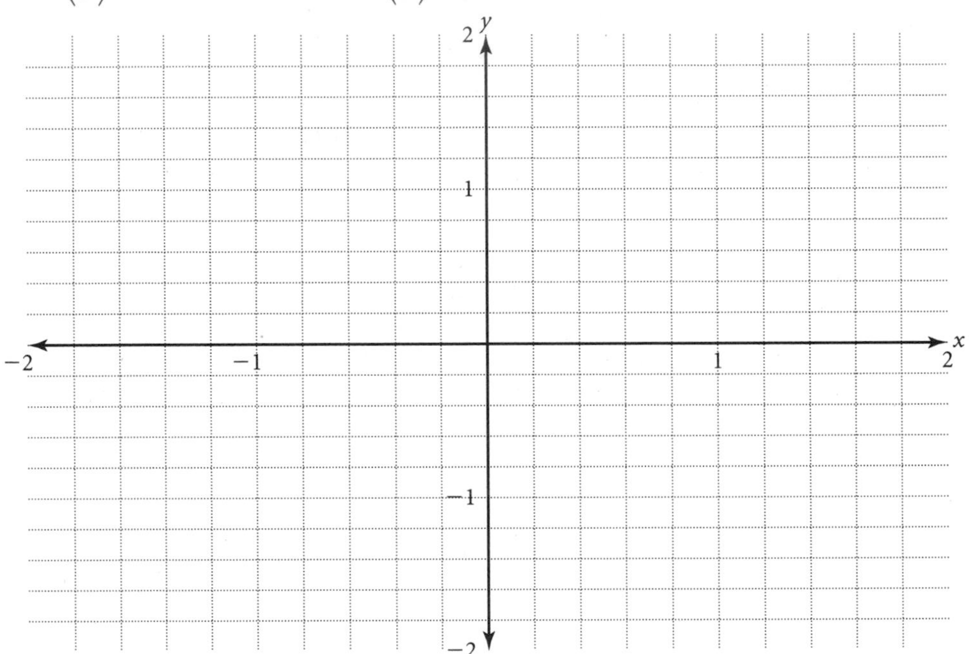

5. At what point do all the graphs from Exercise 1–4 intersect? ________

6. The graphs of all exponential functions having a base > 1 rises towards the __?__ and a base < 1 rises towards the __?__. ________

Graph each of the following.

7. $y = 5^x$

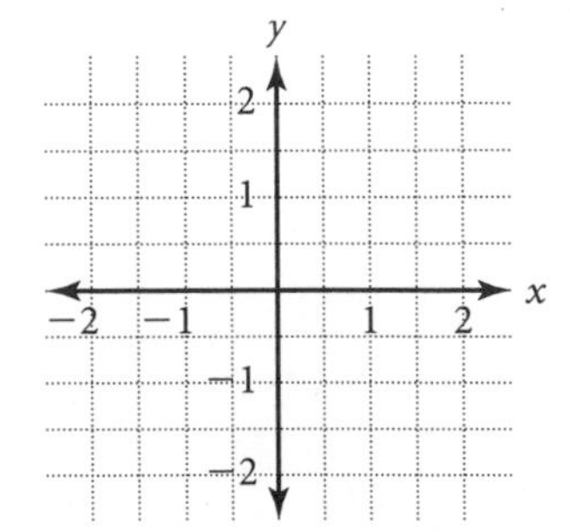

8. $y = 1.3^x$

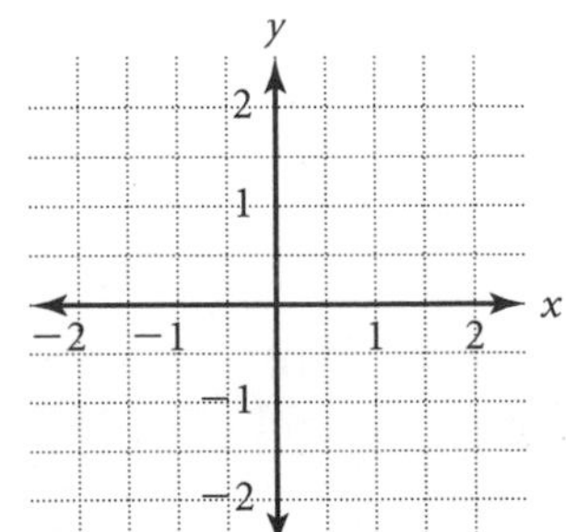

9. $y = 2.1^x$

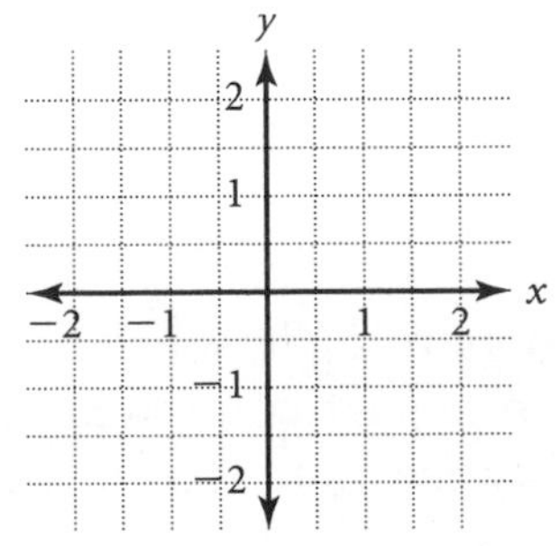

10. A telephone tree starts with the president of a club. The president phones two members. These members phone two other members. Each of those members phone two other members, and so on, until everyone is called.Write an equation, using x and y, for this situation. ________

NAME ______________________ CLASS __________ DATE __________

Practice Masters Level B

8.6 Exponential Functions

Graph each of the following on the same coordinate axes.

1. $y = -\left(\frac{1}{4}\right)^x$
2. $y = \left(\frac{1}{4}\right)^x$
3. $y = 4^x$
4. $y = -4^x$

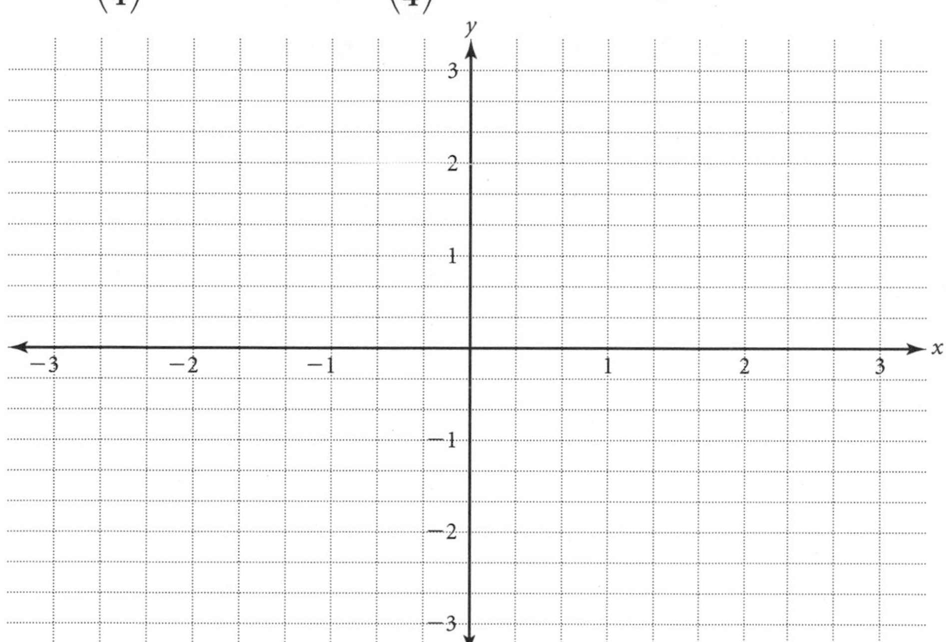

5. Each graph crosses the y-axis at $x =$ __?__. ________
6. Which quadrants do the graphs of all positive exponential functions lie in? ________
7. Which quadrants do the graphs of all negative exponential functions lie in? ________
8. The graph of $y = 4^x$ is the image of __?__ reflected over the x-axis. ________

Graph each of the following.

9. $y = 0.5^x$
10. $y = \left(\frac{1}{3}\right)^x$
11. $y = 1.75^x$

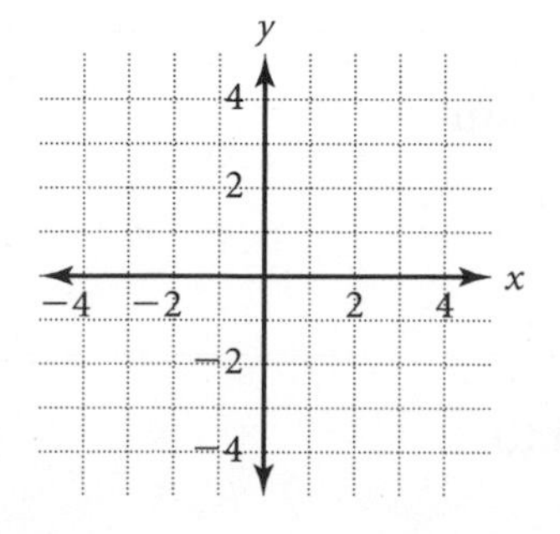

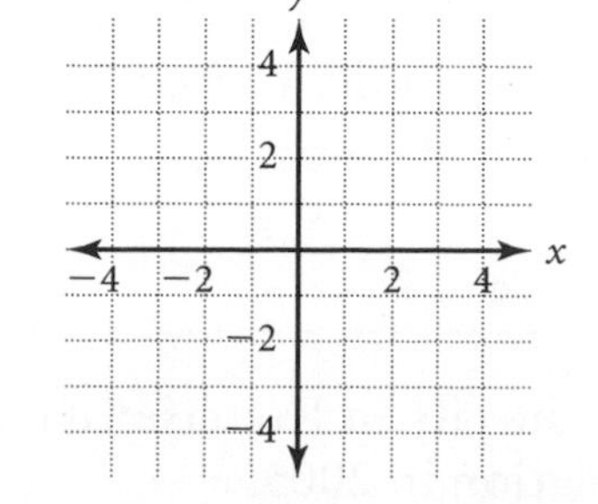

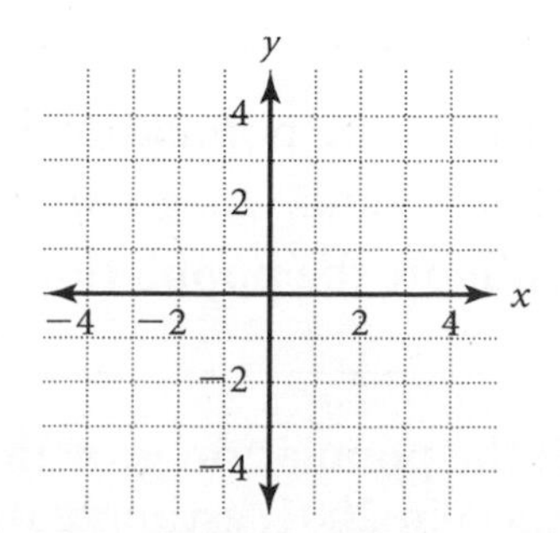

NAME ______________________ CLASS ____________ DATE ____________

Practice Masters Level C

8.6 Exponential Functions

Graph each of the following on the same coordinate axes.

1. $y = -\left(\frac{1}{2}\right)^x$
2. $y = \left(\frac{1}{2}\right)^x$
3. $y = 2^x$
4. $y = -2^x$
5. $y = 2^{x+4}$
6. 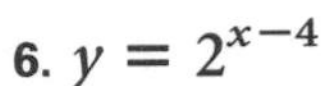 $y = 2^{x-4}$

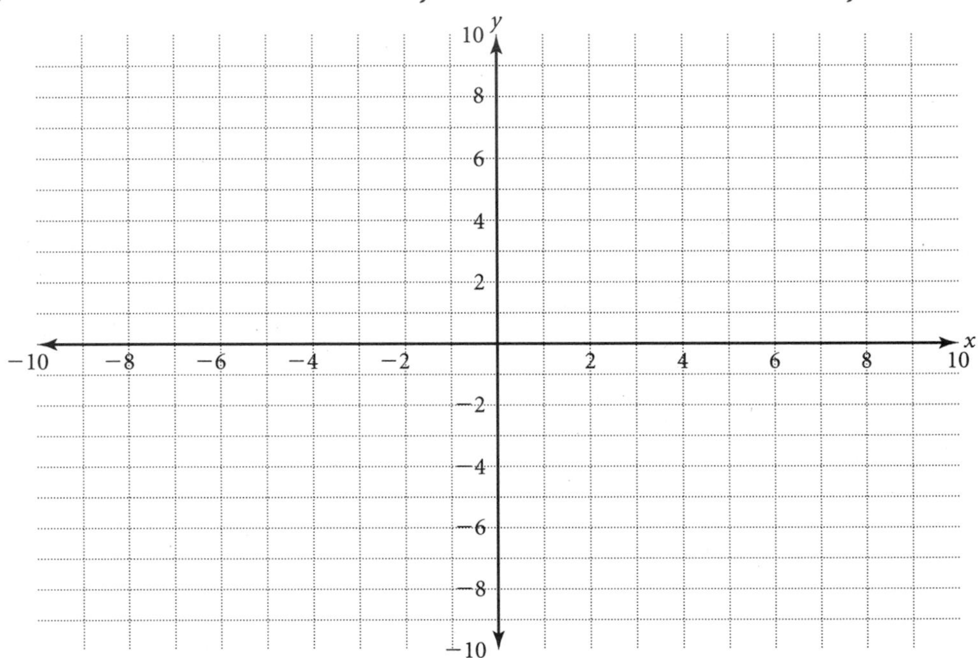

7. Multiplying the function rule of an exponential function by -1 reflects its graph across the __?__ -axis. ____________

8. Suppose a function is in the form of $y = b^x$. Its image is $y = b^{-x}$. Describe the transformation.

____

9. How does adding or subtracting a number to the exponent affect the graph?

____

10. In 1995 the population of a small city in the United States was about 123,000 and was growing at a rate of $33\frac{1}{3}\%$ per year. At this rate, estimate the population in 2000.

____

11. If the population growth rate of the city in Exercise 10 changed to 8.5% in 2001, estimate the population in 2005.

____

NAME ______________________ CLASS ______________ DATE ______________

Practice Masters Level A

8.7 Applications of Exponential Functions

Use the following information for Exercises 1–8. Carbon-14 is used to determine the age of artifacts. It has a half-life of 5700 years. This means that in 0 years, $\left(\frac{1}{2}\right)^0$, or 100%, of the original amount of carbon-14 remains in the artifact. In 5700 years, $\left(\frac{1}{2}\right)^1$, or 50%, of the original amount of carbon-14 remains.

1. Suppose that 40% of the original carbon remains in a piece of pottery. Between which two ages should you base an estimate of its age? ____________
2. Estimate the age of the pottery in Exercise 1. ____________
3. Suppose that 75% of the original carbon remains in a piece of pottery. Between which two ages should you base an estimate of its age? ____________
4. Estimate the age of the pottery in Exercise 3. ____________
5. Suppose that $\frac{1}{5}$ of the original carbon remains in a piece of pottery. Between which two ages should you base an estimate of its age? ____________
6. Estimate the age of the pottery in Exercise 5. ____________
7. Suppose that 2% of the original carbon remains in a piece of pottery. Between which two ages should you base an estimate of its age? ____________
8. Estimate the age of the pottery in Exercise 7. ____________

Use the general growth formula, $P = A(1 + r)^t$, for Exercises 9–11.

9. In 3 years you want to have $2000 in your savings account to buy a new computer. The current interest rate for a savings account is 4% compounded annually. How much do you need to deposit in a savings account now in order to have enough to buy the computer in 3 years? ____________
10. The value of a condominium has increased in value 3% per year for 5 years. If the condominium is worth $128,500 now, what was its value 5 years ago? Round your answer to the nearest hundred dollars. ____________
11. What will the condominium, in Exercise 10, be worth in 3 more years, if the annual percent increase remains the same? Round your answer to the nearest hundred dollars. ____________

NAME ________________ CLASS ________ DATE ________

Practice Masters Level B

8.7 Applications of Exponential Functions

Use the following information for Exercises 1–5.
A substance has a half-life of 72 hours.

1. Estimate the percent of the original substance remaining after one week? ________
2. Approximately what fraction of the original substance remains after 2 days? ________
3. To the nearest tenth of a percent, estimate how much of the original substance remains after 30 days? ________
4. In about how many days will the substance have about 10% of its original amount remaining? ________
5. In about how many days will the substance have about 70% of its original amount remaining? ________

Use the general growth formula $P = A(1 + r)^t$ for Exercises 6–9.

6. In 2 years you want to have $2500 in your savings account to buy a new computer. The current interest rate for a savings account is 3.5% compounded annually. How much do you need to deposit in a savings account now in order to have enough to buy the computer in 2 years? ________
7. Suppose in Exercise 6 the savings account interest rate is 4%, rather than 3.5%. Will the initial deposit amount be more than or less than the amount calculated in Exercise 6? How much more or less? ________
8. A house has increased in value 4.5% per year since it was purchased 4 years ago. The house is now worth $149,800. What was the original purchase price? Round your answer to the nearest hundred dollars. ________
9. If the house increases in value by the same percentage each year, what will be its value in 5 years? Round your answer to the nearest hundred dollars. ________

Use this information for Exercises 10–12.
The value of a car decreased from $38,500 to $28,875 in one year.

10. What is the percent decrease in value? ________
11. If the value of the car is decreasing exponentially, what will be its value in 4 years, to the nearest dollar? ________
12. What is the age of the car when its value is about 10% of its original value? ________

NAME ______________________ CLASS ____________ DATE ______________

Practice Masters Level C

8.7 Applications of Exponential Functions

Use the following information to answer Exercises 1–3.
When a substance was 40 years old, it had about 6.25% of its original amount remaining. When it was only 30 years old, it had about 12.5% of its original amount remaining.

1. What is the half-life of the substance? ____________
2. In approximately how many years will the substance contain about 20% of its original substance? ____________
3. To the nearest tenth of a percent, about how much of its original substance will remain after 100 years? ____________

Use this information for Exercises 4–8.
Suppose you deposited $10,000 into an account on January 1, 2000.

4. If you deposited all the money into an account earning 2.5% annual interest, how much money will be in the account after 5 years? ____________
5. If you deposited all the money into an account earning 5% annual interest, how much money will be in the account after 5 years? ____________
6. Suppose you deposited part of the money at 2.5% annual interest and the rest at 5% annual interest. After 5 years, you had a total of $12,400.63. How much did you deposit in the 5% account? ____________
7. In what year can you expect to have at least $20,000 if you deposit the entire amount in the 2.5% account? ____________
8. In what year can you expect to have at least $20,000 if you deposit the entire amount in the 5% account? ____________

Solve.

9. Tom is investing $3,500 in an account earning 4% annual interest. In another state, his cousin Marla will invest a certain amount earning 12% annual interest. How much should Marla invest if she wants to have the same amount in her account as Tom has in his, after 3 years? ____________
10. Hal invested some money in an account that earned 10.5% annual interest over 6 years. At the end of 6 years, he had $40,049.43 in his account. How much was his original investment? ____________

NAME ______________________ CLASS ____________ DATE ____________

Practice Masters Level A

9.1 Adding and Subtracting Polynomials

Write each polynomial in standard form.

1. $2a + 3 + a^2$ ______________
2. $n + 3 + 3n^2$ ______________
3. $-2x + x^2 + 3$ ______________
4. $-2 + y^2 + y$ ______________
5. $z^2 - 2z + 1$ ______________
6. $5 + z - 2z^2$ ______________

Find each sum.

7. $\begin{array}{r} 3x + 2 \\ +\ 2x + 1 \\ \hline \end{array}$ ______________

8. $\begin{array}{r} 10a + 3 \\ +\quad a - 3 \\ \hline \end{array}$ ______________

9. $\begin{array}{r} -2a + 1 \\ +\quad 5a + 1 \\ \hline \end{array}$ ______________

10. $\begin{array}{r} -7n - 2 \\ +\quad 7n + 2 \\ \hline \end{array}$ ______________

11. $\begin{array}{r} x^2 + 5x + 2 \\ +\qquad 2x - 1 \\ \hline \end{array}$ ______________

12. $\begin{array}{r} 3y^2 + 5y + 1 \\ +\qquad -6y - 4 \\ \hline \end{array}$ ______________

Find each difference.

13. $\begin{array}{r} 7b + 6 \\ -\ (2b + 1) \\ \hline \end{array}$ ______________

14. $\begin{array}{r} 10m + 6 \\ -\ (6m - 7) \\ \hline \end{array}$ ______________

15. $\begin{array}{r} -2t + 2 \\ -\ (5t + 7) \\ \hline \end{array}$ ______________

16. $\begin{array}{r} 7v - 1 \\ -\ (7v - 6) \\ \hline \end{array}$ ______________

17. $\begin{array}{r} 2r^2 + 5r + 3 \\ -\qquad (-2r - 1) \\ \hline \end{array}$ ______________

18. $\begin{array}{r} -2c^2 - 5c + 7 \\ -\qquad (-7c - 5) \\ \hline \end{array}$ ______________

Find each sum or difference.

19. $(3x - 2) + (2x + 5)$ ______________
20. $(-2t - 7) + (-t + 2)$ ______________
21. $(-z + 7) + (-z + 7)$ ______________
22. $(-3d - 7) + (-3d - 1)$ ______________
23. $(4n - 2) - 2n$ ______________
24. $(-v - 7) - (-2v)$ ______________
25. $(-2z + 7) - (3z + 11)$ ______________
26. $(-2t - 7) - (-7t - 2)$ ______________
27. $(4m^2 + 3) + 2m^2$ ______________
28. $(4m^2 + 3m) + (-2m^2)$ ______________

NAME ______________ CLASS ______________ DATE ______________

Practice Masters Level A

9.1 Adding and Subtracting Polynomials

Write each polynomial in standard form.

1. $-2b + 5 + b^2 - 3b^3$ ______________
2. $-5z^5 + 3z^3 - 3z^2 + 7$ ______________
3. $8 - 2r^3 + r^5 - 3r^2$ ______________
4. $-w^3 - 2w^6 + w^2 - w$ ______________
5. $5s^2 - 3s + 3 - s^7$ ______________
6. $-2x^3 - 5 + x - 2x^7$ ______________

Find each sum or difference.

7. $(4v^2 + 3v - 5) + (2v^2 - 4v + 5)$ ______________
8. $(-2z^2 - 3z - 3) + (2z^2 - 7z - 1)$ ______________
9. $(4b^5 - 3b^2 - 3b - 3) + (4b^4 - 6b^2 - \mathrm{b} + 5)$ ______________
10. $(-c^7 - c^5 - 3c - 3) + (c^7 - 6c^5 - 3c - 2)$ ______________
11. $(-5u^2 + 3u + 7) - (2u^2 - u + 1)$ ______________
12. $(-7h^2 - 4h + 7) - (7h^2 - 4h + 11)$ ______________
13. $(6k^5 - 4k^2 - 3k - 1) - (5k^4 - 2k^2 - k - 1)$ ______________
14. $(2y^7 - 7y^5 - 3y^2 - 3) - (2y^7 - 7y^5 - 3y - 2)$ ______________

Write a polynomial expression in standard form for each perimeter.

15.

2a + 3
2a
2a + 3

____

16.

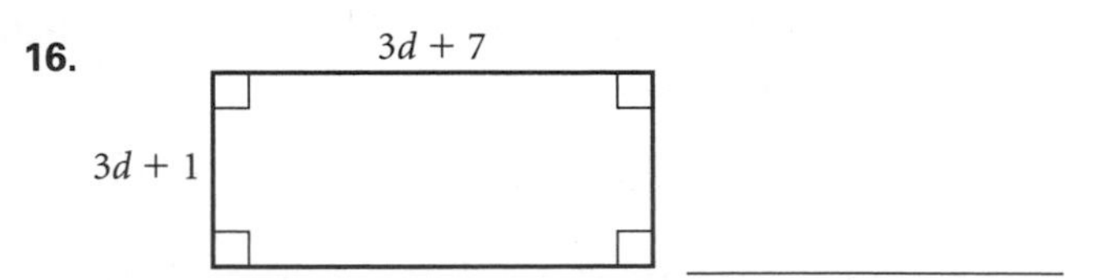

____

17.

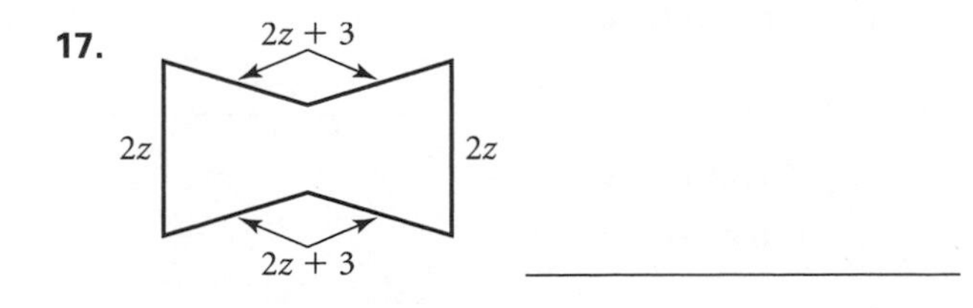

____

18.

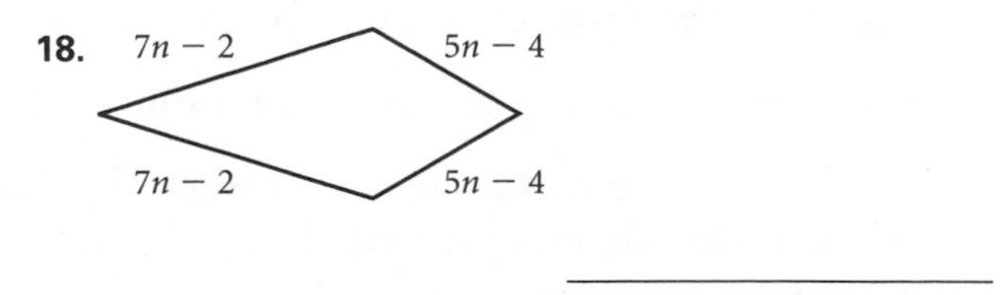

____

19.

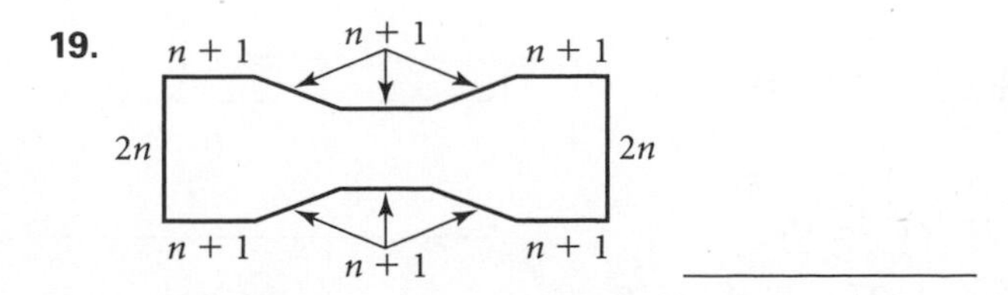

____

20.

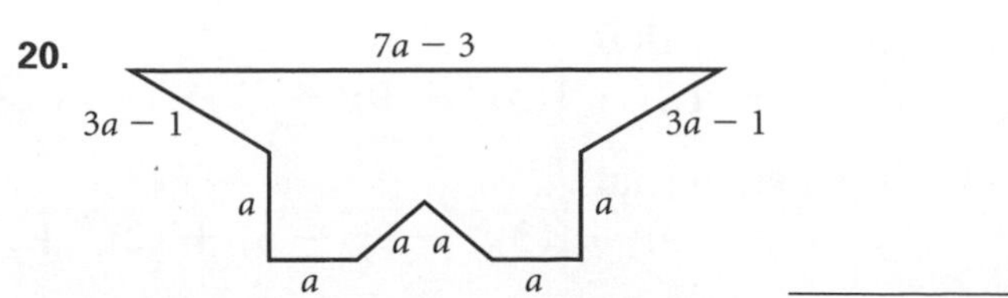

____

NAME ______________________ CLASS __________ DATE __________

Practice Masters Level C

9.1 Adding and Subtracting Polynomials

Write the result of each sum and difference in standard form.

1. $(-3a^2 + 4a - 5) - [(4a^2 - 7a - 5) + (2a^2 - 3a + 2)]$ ____________
2. $(7n^2 + 9n + 1) - [(-n^2 - n - 10) - (n^2 + 7n + 7)]$ ____________
3. $(6z^2 + z - 11) + [(9z^2 - 10z - 5) - (3z^2 + 9z + 1)]$ ____________
4. $(x^3 + x^2 - 11) - [(3x^3 - 9x^2 - 2) - (7x^2 + x + 10)]$ ____________
5. $-(2y^3 - 2y^2 - 1) + [(-y^3 + 2y^2 - 1) - (-3y^2 + 2y + 9)]$ ____________
6. $(m^3 + 2m^2 + 5) - [(2m^3 - 2m^2 - 3) +$
   $(-m^3 + 2m^2 - 1) - (m^2 + 2m^2)]$ ____________

Show that each quotient equals 1 or −1 for all allowable values of the variable.

7. $\dfrac{3m + 3 - (5m + 8)}{7m - 11 + (-9m + 6)}$ ____________

8. $\dfrac{(3n^2 + 3n) - (3n + 1)}{(3n^2 - 11n) + (11n - 1)}$ ____________

9. $\dfrac{(3a^3 + 3a^2) - (3a^3 + a)}{(-11a^3 - 3a^2) + (11a^3 + a)}$ ____________

10. The length of a rectangle is represented by $4a + 3b$ and its width is represented by $7a - 2b$. Find an expression for the perimeter of the rectangle. ____________
11. One carton of light bulbs contains n light bulbs. On one occasion, a customer bought 7 cartons and $\frac{1}{2}$ of an eighth carton. On another occasion, the same customer bought $\frac{1}{4}$ carton less than 6 cartons. Write an expression for the total number of bulbs purchased. ____________
12. The length of a rectangle is represented by $2a + 3$ and the width of the rectangle is represented by $3a + 7$. If the perimeter of the rectangle is 35 units, find a. ____________
13. Find x such that
    $(x^2 + x - 12) - [(3x^2 - 9x - 2) + (-2x^2 + x - 10)] = 0.$ ____________
14. Find y such that
    $(5y^2 + y + 9) - [(3y^2 - 9y - 6) + (2y^2 + y - 9)] = 0.$ ____________
15. Find a and b such that $(ay^2 + 2ay + 9) + (by^2 + 3by - 9) = 0.$ ____________

NAME ______________________ CLASS ____________ DATE ____________

Practice Masters Level A

9.2 Modeling Polynomial Multiplication

Write a product of binomial factors for each model. Use the model to find the product.

1.
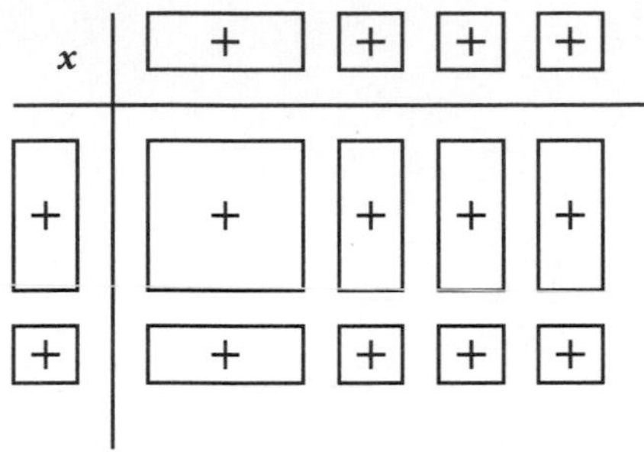

____

____

2.
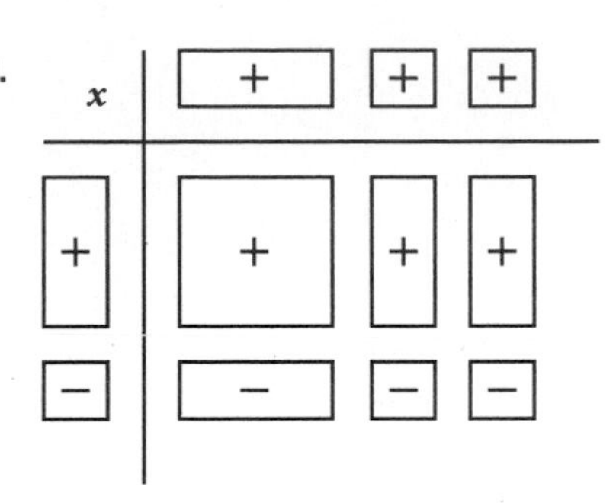

____

____

Model each product with tiles, and give the simplified product.

3. $(x + 2)(x + 3)$ ______________________

____

4. $(x + 3)(x - 1)$ ______________________

____

5. $(x + 1)(x - 4)$ ______________________

____

6. $(x - 2)(x - 2)$ ______________________

____

Find each product by using the rules for special products.

7. $(x + 3)(x - 3)$ ______________________

8. $(x + 1)^2$ ______________________

9. $(x + 5)(x - 5)$ ______________________

10. $(x - 3)^2$ ______________________

Use the Distributive Property to find each product.

11. $3(x + 5)$ ______________________

12. $7(x - 3)$ ______________________

13. $8(2x - 5)$ ______________________

14. $3(9x - 4)$ ______________________

15. $2(x^2 + 3)$ ______________________

16. $-5(7x^2 - 2x)$ ______________________

NAME ______________________ CLASS ____________ DATE ____________

Practice Masters Level B

9.2 Modeling Polynomial Multiplication

Model each product with tiles, and give the simplified product.

1. $(x + 3)(2x - 2)$ ______________
2. $(2x + 3)(2x + 1)$ ______________
3. $(2x - 3)(2x + 1)$ ______________
4. $(2x - 1)(x - 3)$ ______________

Find each product by using the rules for special products.

5. $(3x + 5)(3x - 5)$ ______________
6. $(x - 8)(x + 8)$ ______________
7. $(4x - 7)(4x - 7)$ ______________
8. $(9x + 2)(9x + 2)$ ______________
9. $(2x + 5)^2$ ______________
10. $(3x - 4)^2$ ______________
11. $(3x + 4)^2$ ______________
12. $\left(\frac{1}{2}x - 7\right)^2$ ______________

Use the Distributive Property to find each product.

13. $11(3x - 7)$ ______________
14. $-x(2x - 12)$ ______________
15. $9(2x^2 - 5x)$ ______________
16. $3x(9x^2 - 4x)$ ______________
17. $-2x(7 - x)$ ______________
18. $x(5 - x^2)$ ______________
19. The length of a rectangle is represented by $3a + 7b$ and its width is represented by $3a - 7b$. Find a simplified expression for the area of the rectangle. ______________
20. The side of a square is represented by $9x + 11y$. Find a simplified expression for the area of the rectangle. ______________

NAME ______________________ CLASS ______________ DATE ______________

Practice Masters Level C

9.2 Modeling Polynomial Multiplication

Find each product.

1. $(2x^2)(5x^3 - 7x + 5)$ ______________
2. $(x - 8x^3)(x + 8x^3)$ ______________
3. $(4x^2 - 5x)(4x^2 - 5x)$ ______________
4. $(11x - 2x^2)^2$ ______________
5. $(3x^3 + 4)^2$ ______________
6. $(-3x - 5x^2)(5x^2 - 3x)$ ______________

Construct a model with tiles that would give the following expression as its result.

7. $x^2 + 4x + 3$
8. $2x^2 + 3x - 2$

______________ ______________

9. The length of a rectangle is represented by $6a + 5b$ and its area is represented by $72a^2 - 50b^2$. Find an expression for the width. ______________

10. The side length of a square is represented by $8 + 2n$. If the area of this square is the same as the perimeter of another square with a side length of $n^2 + 48$, find the value of n. ______________

Samantha has volunteered to mow her grandmother's lawn. Grandma drew up some plans of her lawn so Samantha could see what area she was mowing. Use the diagram for Exercises 11–15. Simplify your expressions.

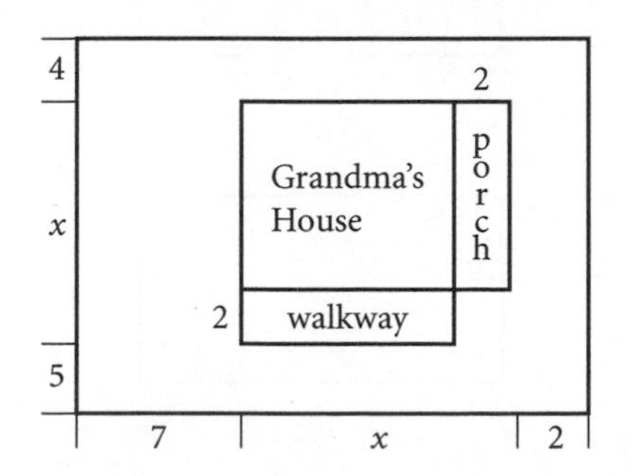

11. Write an expression for the dimensions of Grandma's house. ______________

12. Write an expression for the dimensions of the porch. ______________

13. Write an expression for the dimensions of the walkway. ______________

14. Write an expression for the area of the entire plot of land. ______________

15. Write an expression for the area of the lawn Samantha must mow. ______________

NAME ______________________ CLASS __________ DATE __________

Practice Masters Level A

9.3 *Multiplying Binomials*

Use the Distributive Property to find each product.

1. $9s(s + 6)$ ______________
2. $(x + 1)(x + 4)$ ______________
3. $(b - 2)(b - 5)$ ______________
4. $(c + 10)(c - 10)$ ______________
5. $(n + 4)(n - 4)$ ______________
6. $(y + 3)(y - 4)$ ______________
7. $(p + 5)(p - 9)$ ______________
8. $(w - 7)(w + 3)$ ______________
9. $(7f + 2)(7f + 2)$ ______________
10. $(3z + 8)(3z - 8)$ ______________

Use the FOIL method to find each product.

11. $(5z - 2)(2z + 3)$ ______________
12. $(w - 6)(w - 6)$ ______________
13. $(3x - 4)(3x + 2)$ ______________
14. $(7q + 2)(7q - 2)$ ______________
15. $(2c + 4)(c + 6)$ ______________
16. $(3w - 7)(2w - 1)$ ______________
17. $(5x + 2)(2x - 1)$ ______________
18. $(2y - 5)(y + 9)$ ______________
19. $(3x - 2)(2x + 5)$ ______________
20. $(x + 3)(5x - 7)$ ______________

21. Find the area of the rectangle in terms of b.

____

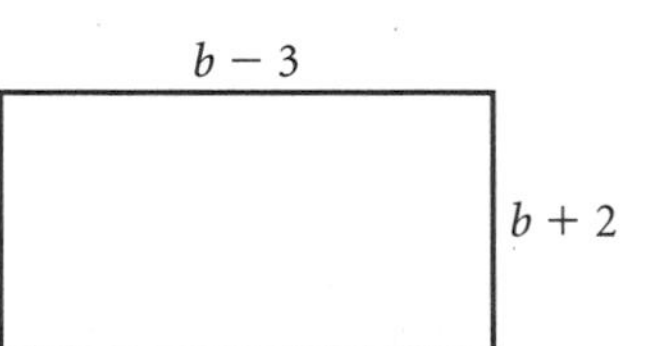

22. Find the area of the rectangle in terms of x.

____

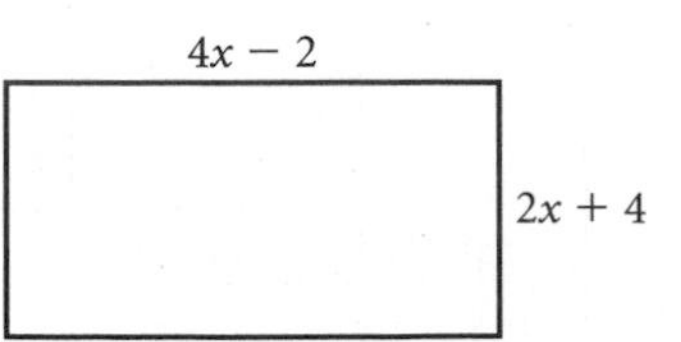

23. Find the area of a square with a side length of $5d - 2$.

____

NAME ______________________ CLASS ______________ DATE ______________

Practice Masters Level B

9.3 *Multiplying Binomials*

Use the Distributive Property to find each product.

1. $(x + y)(2x + y)$ ______________
2. $(a - 3b)(7a - 5b)$ ______________
3. $(-5d + c)(-6d - c)$ ______________
4. $(4n^2 - 3m)(4n^2 + 3m)$ ______________
5. $(s^3 - 4t)(s^3 + 4t)$ ______________
6. $(w^4 - w^3)(w^2 + w)$ ______________

Use the FOIL method to find each product.

7. $(6e - g)(8e + 3g)$ ______________
8. $(d^2 - 4d)(2d^2 - 5d)$ ______________
9. $\left(h + \frac{1}{3}\right)\left(h - \frac{1}{3}\right)$ ______________
10. $\left(k + \frac{2}{5}\right)\left(k + \frac{1}{10}\right)$ ______________
11. $(q^3 + 2q)(q^2 - 7q)$ ______________
12. $(8 - 5t^2)(r^2 - t)$ ______________
13. $(2.3k^2 + 10)(1.8k^2 - 4)$ ______________
14. Find the area, in terms of *d*, of a square tabletop whose sides are $4d - 6$ units long. ______________
15. Find the area, in terms of *d*, of a rectangular lawn whose width measures $2d + 5$ units long and whose length measures $8d - 3$ units long. ______________
16. Using the diagram of the rectangular frame, find an expression, in terms of *a*, for the area of the shaded region.

____

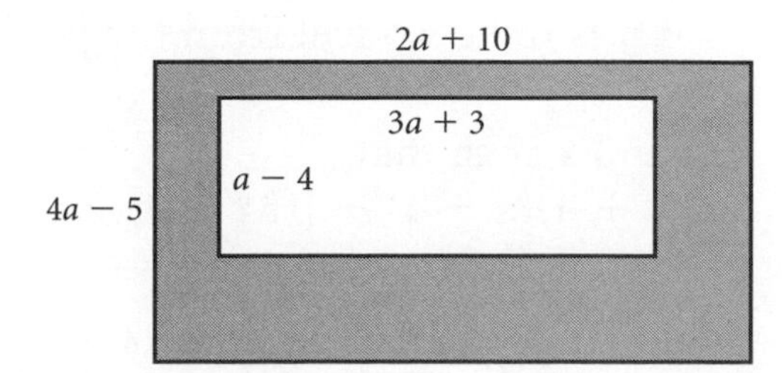

NAME ____________________ CLASS __________ DATE __________

Practice Masters Level C

9.3 Multiplying Binomials

Use the FOIL method or the Distributive Property to find each product.

1. $(-3a^2 + 4a)(7 - 2a)$ ____________________

2. $(1.3x^2 - 0.95y)(2.8x^2 + 1.2y)$ ____________________

3. $(11z^3 - 2z)(-6z + 5)$ ____________________

4. $\left(2x + \frac{1}{4}\right)\left(\frac{2}{3}x - \frac{5}{6}\right)$ ____________________

5. $(-5y^3 - 2x^2)^2$ ____________________

6. Find the value of a if the area of the unshaded region of the frame below is 156 square units.

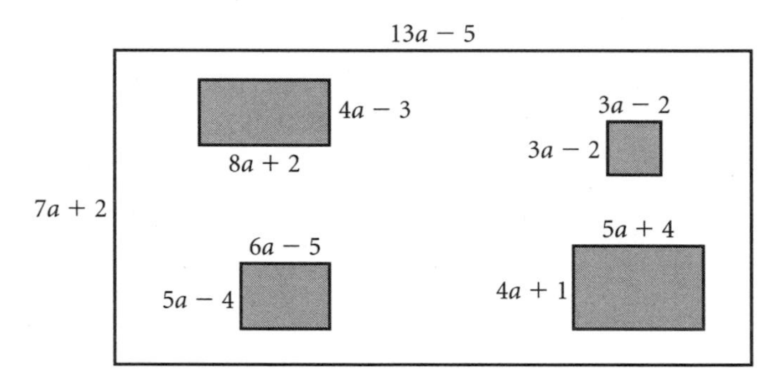

____

7. A rectangular pool is 25 feet long and 10 feet wide. It is surrounded by a fence that is w feet from all sides of the pool. Write and simplify an expression that represents the total fenced-in area. ____________________

8. Suppose the fence in Exercise 7 is 7 feet high and the cost of the fence is \$2.75 per linear foot. What is the total cost of the fence if it is to be 3.5 feet from the pool? ____________________

9. Find x such that
$(x + 4)(x - 4) - [(3x + 3)(2x - 7) - (5x - 3)(x - 2)] = 0.$ ____________________

10. Find y such that
$(8y - 5)(3y + 2) - [(5y - 2)(3y + 4) + (3y - 2)^2] = 0.$ ____________________

NAME ______________________ CLASS __________ DATE __________

Practice Masters Level A

9.4 Polynomial Functions

Determine whether each equation is true or false for *x*-values of −1, 0, and 1.

1. $(x + 3)(x + 4) = x^2 + 7x + 12$ ________
2. $(x - 4)(x + 5) = x^2 - x - 20$ ________
3. $x^2 - 25 = (x - 5)(x + 5)$ ________
4. $x^2 - 14x - 49 = (x - 7)^2$ ________

Complete the table to verify that each equation is an identity.

5. $(x - 8)(x + 4) = x^2 - 4x - 32$

x	$(x - 8)(x + 4)$	$x^2 - 4x - 32$
−3		
−2		
−1		
0		
1		
2		
3		

6. $(x - 3)^3 = x^3 - 9x^2 + 27x - 27$

x	$(x - 3)^3$	$x^3 - 9x^2 + 27x - 27$
−3		
−2		
−1		
0		
1		
2		
3		

7. $x^3 + 64 = (x + 4)(x^2 - 4x + 16)$

x	$x^3 + 64$	$(x + 4)(x^2 - 4x + 16)$
−3		
−2		
−1		
0		
1		
2		
3		

8. $x^2 + 12x + 36 = (x + 6)^2$

x	$x^2 + 12x + 36$	$(x + 6)^2$
−3		
−2		
−1		
0		
1		
2		
3		

Write the indicated equation for each geometric solid.

9. the volume of a cube with an edge of x inches ________
10. the volume of a cube with an edge of $3x$ inches long ________
11. the volume of a rectangular solid with a base length of 4 inches, a base width of 2 inches, and a height of x inches ________
12. the volume of a cylinder with a height of 6 meters and a radius of x meters ________

NAME ______________________ CLASS ______________ DATE ______________

Practice Masters Level B

9.4 Polynomial Functions

Create a table to verify that each equation is an identity by substituting integers from −2 to 2 for *x*.

1. $(2x + 1)^2 = 4x^2 + 4x + 1$

x	−2	−1	0	1	2
$(2x + 1)^2$					
$4x^2 + 4x + 1$					

2. $(3x - 2)(-2x + 5) = -6x^2 + 19x - 10$

x	−2	−1	0	1	2
$(3x - 2)(-2x + 5)$					
$-6x^2 + 19x - 10$					

3. $(x - 5)(x^2 + 5x + 25) = x^3 - 125$

x	−2	−1	0	1	2
$(x - 5)(x^2 + 5x + 25)$					
$x^3 - 125$					

4. $x^3 - 12x^2 + 48x - 64 = (x - 4)^3$

x	−2	−1	0	1	2
$(x - 4)^3$					
$x^3 - 12x^2 + 48x - 64$					

Write the indicated equation for each geometric solid.

5. the volume of a cube with an edge of $6x$ meters ______________
6. the surface area of a cube with an edge of x inches ______________
7. the surface area of a rectangular solid with base length of 9 inches, a base width of 5 inches, and a height of x inches. ______________
8. the volume of a cylinder with a height of 12 meters and a radius of $3x$ meters ______________
9. the volume of a cylinder with a height of 3.5 yards and a diameter of $4.6x$ yards ______________

Find the volume of the original container, the new volume, and the height of the new container.

10. The manufacturer of a frozen orange juice wants to increase the volume of its container by 25% by increasing the height. The current container is a cylinder with a radius of 3 centimeters and a height of 10 centimeters. ______________
11. The manufacturer of a canned juice wants to increase the volume of its can by 30% by increasing the height. The current container is a cylinder with a radius of 2.5 centimeters and a height of 7.5 centimeters. ______________

Practice Masters Level C

9.4 Polynomial Functions

Find the missing side of the equation, and create a table to verify that the expression makes the equation an identity.

1. $(2x + 3)(-4x - 5) =$ ______________

x	-2	-1	0	1	2

2. $(3x - 7)^2 =$ ______________

x	-2	-1	0	1	2

3. $(2x + 3)(x^2 - 2x + 1) =$ ______________

x	-2	-1	0	1	2

4. $(x - 6)(x^2 + 6x + 36) =$ ______________

x	-2	-1	0	1	2

For Exercises 5–9, write the indicated equation for each geometric solid.

5. the surface area of a cube with an edge of $4.7x$ meters ______________

6. the surface area of a rectangular solid with base length of $2x$ inches, a base width of 5 inches, and a height of 2 feet ______________

7. the volume of a cylinder with a height of 12 meters and a diameter of $3x$ meters ______________

8. the volume of a sphere with a diameter of $8.2x$ yards ______________

9. the volume of a box cut out from a 1-foot-by-2-foot piece of cardboard, where the corners cut on the cardboard to make the box are squares with x inches on each side ______________

10. The manufacturer of a frozen juice product wants to decrease the volume of a juice container by 15% by decreasing its height. The container is a cylinder with a radius of 3 centimeters and a height of 10 centimeters. Find the volume of the original container, and the height and the volume of the new container. ______________

11. The manufacturer of a drink container is comparing the sizes of two of its cylindrical containers. The first container has a radius of 4 centimeters and a height of x centimeters. The second container has a radius of 3 centimeters, but its height is twice as high as the first container. The second container is larger than the first container by what percentage? ______________

NAME ____________ CLASS ________ DATE ________

Practice Masters Level A

9.5 Common Factors

Factor each polynomial by finding the GCF.

1. $3x - 12$ ____________
2. $8z^2 - 4z$ ____________
3. $5x^2 - 5x - 20$ ____________
4. $q^6 - q^3$ ____________
5. $9x^2 + 36x + 15$ ____________
6. $12s^2 - 6s + 8$ ____________
7. $100 - 20d^3 + 10d$ ____________
8. $7b^4 + 7b^2$ ____________
9. $16t^2 + 32t$ ____________
10. $60c^3 - 45c^2 + 15c$ ____________
11. $2z^4 - z^3 + 5z^2$ ____________
12. $24s^4 - 15s^3 + 9s^2$ ____________

Write each polynomial as the product of two binomials.

13. $y(y + 2) + 3(y + 2)$ ____________
14. $7(w + 6) - x(w + 6)$ ____________
15. $(r + 15)t + (r + 15)3$ ____________
16. $k(k + 5) - 8(k + 5)$ ____________
17. $2x(x - 3) - 5(x - 3)$ ____________
18. $11(s - 9) - 3t(s - 9)$ ____________
19. $(6 + d)5 - e(6 + d)$ ____________
20. $4(p + 9) - t^2 (p + 9)$ ____________
21. $a(b + 2) - c(b + 2)$ ____________
22. $4(m + 3) - n(m + 3)$ ____________
23. $2f(3 - g) - 5(3 - g)$ ____________
24. $4x(x + 7) - 5(x + 7)$ ____________

Factor by grouping.

25. $3a + ax + 3b + bx$ ____________
26. $xy - y + 3x - 3$ ____________
27. $cd - 3c + 2d - 6$ ____________
28. $x^2 + 5x + 3x + 15$ ____________
29. $ax + 3x + ay + 3y$ ____________
30. $x^2 - 5x - 2x + 10$ ____________
31. $y^2 + 5y + 5y + 25$ ____________
32. $8x^2 - 6x - 12x + 9$ ____________
33. $7t^2 - 21t + 8rt - 24r$ ____________
34. $4m + 12 + 3m + 9$ ____________

NAME ______________________ CLASS ______________ DATE ______________

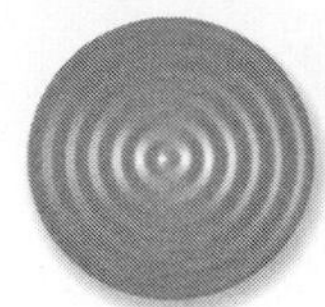

Practice Masters Level B

9.5 Common Factors

Factor each polynomial.

1. $3s^2 + 9s - 3$ ______________
2. $5v^4 - 15v^2 + 125v^3$ ______________
3. $12g^2h - 4gh^2 + 24g^2h^2$ ______________
4. $12x^3y^4 - 36x^2y^5 - 54x^4y^4$ ______________
5. $15d^5e^6 - 27d^4e^2 + 42d^3e^5$ ______________
6. $125f^6 + 25f^3 - 15$ ______________
7. $64s^8t^6 - 48s^5t^3 + 72s^6t^3$ ______________
8. $24q^5 - 36q^2$ ______________

Write each polynomial as the product of two binomials.

9. $9(w + 5) - r(w + 5)$ ______________
10. $r^2(r + 2w) - 5(2w + r)$ ______________
11. $(r + 8)t - (r + 8)3s$ ______________
12. $11(s - 5) - (s - 5)4a$ ______________
13. $x(3 - y) + (3 - y)$ ______________
14. $q(p + 9) - (p + 9)$ ______________
15. $3ab(6 - c) + 7(c - 6)$ ______________
16. $2k(k - 5) - 7(5 - k)$ ______________

Factor by grouping.

17. $8a^2 + 4ab + 2a + b$ ______________
18. $5x + 2zy + 10y + zx$ ______________
19. $5xa - 15xb - 3b + a$ ______________
20. $8xy + 9 - 12x - 6y$ ______________
21. $12 - 4a^2 - 15a + 5a^3$ ______________
22. $cd^2 + 12 + 3c + 4d^2$ ______________
23. $x^5 + y^5 + x^2y^3 + x^3y^2$ ______________
24. $6 + 15r + 2\pi + 5\pi r$ ______________
25. $6a^2y^2 - 20b^2x^2 + 10abxy - 12abxy$ ______________

For Exercises 26–28, solve by using factoring by grouping.

26. The area of a rectangular carpet is $a^2 + 8a + 3a + 24$ square units. What is the length and width of this carpet, in terms of a? ______________

27. The area of a rectangular field is $9r^2 + 15r + 15r + 25$ square yards. What is the length and width of this field, in terms of r? ______________

28. What is special about the shape of the field in Exercise 27? ______________

Practice Masters Level C

9.5 Common Factors

Find the missing factor.

1. $x^{a+5} - x^{a+1} = x^a(\_)$ ____________

2. $9x^{3n} + 24x^{5n+2} = 3x^{3n}(\_)$ ____________

3. $3e^{2n+3} - 12e^{5n-2} + 6e^n - 9e^{4n-3} = 3e^n(\_)$ ____________

Factor each polynomial.

4. $x^{n+1} - x^n$ ____________

5. $y^{n+5} + y^{n+2} - y^n$ ____________

6. $x^n(y - 1) - y^n(1 - y)$ ____________

7. $3a^x(x^a + 2) + 6(-2 - x^a)$ ____________

Factor by grouping.

8. $ax - 2a + bx - 2b - cx + 2c$ ____________

9. $2x^2 + 3z - 3x - 2xz - 2xy + 3y$ ____________

10. $xz - 15y - 5x + 3yz + 2z - 10$ ____________

11. $x^3 - y^3 + z^3 - x^2y - x^2z + xy^2 - xz^2 - y^2z + yz^2$ ____________

12. The area of a rectangle is $a^2 - 15 + 3a - 5a$ square units. Find its perimeter. ____________

13. The side of the square at the right is $3x + 7$ units long and the radius of the circular shaded region is $\frac{8x}{\sqrt{\pi}}$ units long. Find the area of the unshaded region in factored form. ____________

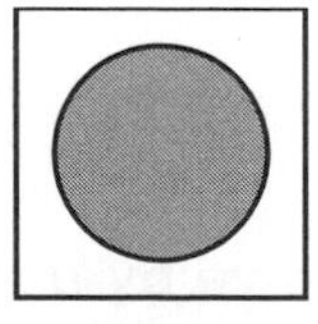

14. Find an expression for A that will make the expression $10x^3 + 4xy^3 + A - 6y^4$ factorable by grouping. Factor the resulting polynomial completely. ____________

NAME ______________________ CLASS ____________ DATE ____________

Practice Masters Level A

9.6 Factoring Special Polynomials

Factor each polynomial by using the rule for factoring a perfect-square trinomial.

1. $x^2 + 2xy + y^2$ ______________
2. $c^2 - 2cd + d^2$ ______________
3. $y^2 - 6y + 9$ ______________
4. $a^2 + 16a + 64$ ______________
5. $x^2 - 14x + 49$ ______________
6. $r^2 + 6r + 9$ ______________
7. $s^2 + 18s + 81$ ______________
8. $25 + 10t + t^2$ ______________

Factor by using the rule for factoring the difference of two squares.

9. $w^2 - x^2$ ______________
10. $9d^2 - c^2$ ______________
11. $25 - k^2$ ______________
12. $4f^2 - 49g^2$ ______________
13. $16y^2 - 81z^2$ ______________
14. $64s^2 - 25$ ______________
15. $25 - 36d^2$ ______________
16. $100a^2 - 9$ ______________

Factor each polynomial completely.

17. $x^2 - 8x + 16$ ______________
18. $x^2 - 4$ ______________
19. $36e^2 - 25d^2$ ______________
20. $25c^2 - 10c + 1$ ______________
21. $100s^2 - 60s + 9$ ______________
22. $c^2 - 24c^2 + 144$ ______________
23. $64q^2 - 49r^2$ ______________
24. $100 - 9d^2$ ______________
25. $4 - 12s + 9s^2$ ______________
26. $81q^2 - 144p^2$ ______________
27. $36q^2 - 12q + 1$ ______________
28. $9y^2 - 12y + 4$ ______________
29. $4q^2 + 20q + 25$ ______________
30. $121x^2 - 81y^2$ ______________
31. $64y^2 - 25x^2$ ______________
32. $49y^2 + 56y + 16$ ______________
33. $25s^2 - 30s + 9$ ______________
34. $x^2y^2 - z^2w^2$ ______________

NAME ______________________ CLASS __________ DATE __________

Practice Masters Level B

9.6 Factoring Special Polynomials

Find each product by using the difference of two squares.

1. $45 \cdot 55$ ______________
2. $63 \cdot 57$ ______________
3. $106 \cdot 94$ ______________
4. $28 \cdot 22$ ______________

Factor each polynomial completely.

5. $x^2 y^2 - 8xyz + 16z^2$ ______________
6. $x^4 - 9$ ______________
7. $16r^4 - 49s^4$ ______________
8. $x^4 + 6x^2 + 9$ ______________
9. $4x^6y^2 - 25z^2$ ______________
10. $49a^4 - 32a^2b + 81b^2$ ______________
11. $x^4 - 81$ ______________
12. $a^8 - b^8$ ______________

Find the missing term in each perfect-square trinomial.

13. $x^2 - 18x + ?$ ______________
14. $49y^2 + ? + 81$ ______________
15. $? - 90yz + 225z^2$ ______________
16. $49w^2 - 56w + ?$ ______________

Simplify the expression, and then factor completely.

17. $4x^2 - 8x + 75 - 2(-6x^2 + 16x + 25)$ ______________
18. $3x(18x - 4) + 25(4x^2 + x) - 10x(x + 1) - 3x - 121$ ______________
19. $5(x^4 - 3) - (x^4 + 1)$ ______________
20. The area of a rectangle is $16x^2 - 72x + 81$ square units. Find the length and width of the rectangle in terms of x. ______________
21. The area of a rectangle is $4x^2 - 68x + 289$ square units. Find the length and width of the rectangle in terms of x. ______________
22. The area of a circle is $(9x^2 + 24x + 16)\pi$ square units. Find the radius of the circle in terms of r. ______________

NAME ______________________ CLASS ____________ DATE ____________

Practice Masters Level C

9.6 *Factoring Special Polynomials*

Factor each polynomial completely.

1. $x^4 - \frac{1}{81}$ ______________________

2. $4x^5 - 36x$ ______________________

3. $x^2(y^2 - 9) + 4(9 - y^2)$ ______________________

4. $4(x - 2) - x^2(2 - x) + 4x(2 - x)$ ______________________

5. $(4x^2 - 25)^2 - 625$ ______________________

6. $(2y + 7)^2 - 81$ ______________________

7. $(x^4 - y^4)z^2 - 2w^2(y^4 - x^4) + w^2(x^4 - y^4)$ ______________________

8. $\frac{1}{36}p^2 - \frac{4}{25}q^2$ ______________________

9. $\frac{4}{81}s^2 - \frac{1}{3}s + \frac{9}{16}$ ______________________

10. $x^{2n}y^{4n} - 9$ ______________________

11. $a^{4x+6}b^{2x} - 16a^4b^4$ ______________________

Find the value of *A* that will make each equation true.

12. $25x^2 + 3Ay = (5x - 12y^2)(5x + 12y^2)$ ______________________

13. $Ax^{n-2}y^4 - 49x^{6n}y^{2n} = (16x^ny^2 - 7x^{3n}y^n)(16x^ny^2 + 7x^{3n}y^n)$ ______________________

Find the values of *A* that make the expression factorable. Factor the resulting polynomial completely.

14. $x^{2n} + A + 121y^6$ ______________________

15. $a^{2x}b^{4x} + A + 9a^4b^6$ ______________________

NAME ______________________ CLASS ____________ DATE ____________

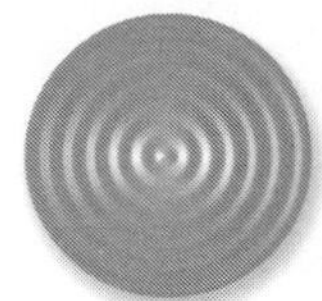

Practice Masters Level A

9.7 Factoring Quadratic Trinomials

Use algebra tiles to factor each trinomial. Make a sketch of your model.

1. $x^2 + 5x + 6$ ____________
2. $x^2 - 2x - 3$ ____________

Factor each trinomial. If a trinomial cannot be factored, write *prime.*

3. $x^2 + 5x + 4$ ____________
4. $x^2 + 6x + 5$ ____________
5. $c^2 - 9c + 18$ ____________
6. $s^2 - 12s + 3$ ____________
7. $y^2 - 2y - 3$ ____________
8. $h^2 + 2h - 3$ ____________
9. $r^2 - 5r - 6$ ____________
10. $a^2 + 3a - 28$ ____________
11. $w^2 - 19w - 36$ ____________
12. $h^2 - 10h + 24$ ____________
13. $z^2 - 9z + 20$ ____________
14. $x^2 + 9x - 21$ ____________
15. $q^2 - 8q + 15$ ____________
16. $e^2 + 14e - 32$ ____________
17. $t^2 - 13t - 48$ ____________
18. $e^2 + 12e + 32$ ____________
19. $s^2 - 21s - 100$ ____________
20. $y^2 + 10y - 75$ ____________

Factor each trinomial. In some cases you will need to factor out a common monomial term.

21. $2x^2 + 26x + 44$ ____________
22. $3y^2 + 21y + 36$ ____________
23. $2s^2 - 8s - 42$ ____________
24. $5y^2 + 45y - 50$ ____________
25. $3k^2 - 12k + 9$ ____________
26. $4y^2 + 12y - 160$ ____________
27. $6s^3 - 36s^2 - 96s$ ____________
28. $3y^2 + 48y - 108$ ____________
29. $x^3 - 6x^2 + 5x$ ____________
30. $x^3y - x^2y - 20xy$ ____________

NAME ____________ CLASS ____________ DATE ____________

Practice Masters Level B

9.7 Factoring Quadratic Trinomials

To factor quadratic trinomials of the form $ax^2 + bx + c$ where $a \neq 1$, list all of the factors of a and c. Then use the factors to write binomials and test and check each pair of binomials until the sum of the outer and inner products equals b.

Example: Factor $2x^2 + 3x - 5$.
The factors of 2: 2 and 1, −2 and −1
The factors of −5: −5 and 1, 5 and −1
Test the product of every combination of binomial factors until you find +3 as the b term.

Try: 2 and 1, −5 and 1
$(2x - 5)(x + 1) = 2x^2 - 3x - 5$
The b term is −3, not +3, so these are *not* the factors.

Try: 2 and 1, 5 and −1
$(2x + 5)(x - 1) = 2x^2 + 3x - 5$
The b term is +3, so stop testing. The factors result in the correct factorization of the trinomial.

Factor each trinomial. If a trinomial cannot be factored, write *prime*.

1. $r^2 + 16r - 36$ ____________
2. $6x^2 + 13x - 5$ ____________
3. $5c^2 + 12c + 7$ ____________
4. $2x^2 - x - 3$ ____________
5. $3h^2 + 19h + 20$ ____________
6. $2d^2 + 7d - 15$ ____________
7. $2g^2 - 17g + 36$ ____________
8. $4t^2 - t - 60$ ____________
9. $6w^2 - w - 35$ ____________
10. $6h^2 - 31h + 30$ ____________
11. $4z^2 + 19z - 12$ ____________
12. $4y^2 - 36y + 45$ ____________
13. $40 - x - 6x^2$ ____________
14. $6f^2 - \frac{3}{2}f + \frac{1}{12}$ ____________

Factor each polynomial completely. In some cases you will need to factor out a constant monomial term.

15. $45x^2 - 103x^2 + 12$ ____________
16. $6st^4 + 18st^2 - 168s$ ____________
17. $6x^2 + 34x + 40$ ____________
18. $64s^4 - 96s^2 + 36$ ____________
19. Find the dimensions of a box if its volume is $24x^3 + 4x^2 - 60x$. ____________

NAME ______________________ CLASS ____________ DATE ____________

Practice Masters Level C

9.7 Factoring Quadratic Trinomials

Use algebra tiles to factor each trinomial. Make a sketch of your model.

1. $2x^2 + x - 6$ ____________

2. $3x^2 - 8x + 4$ ____________

Factor each trinomial.

3. $32x^2 - 56x + 12$ ____________

4. $2x^2 + 5b - 3$ ____________

5. $2x^2 - 7x - 15$ ____________

6. $2x^6y^6 - 7x^3y^3 - 4$ ____________

Factor each polynomial completely.

7. $2xy^2 - 8xy - 90x + 3y^3 + 12y - 135$ ____________

8. $30a^2b + 25ab - 105b - 12a^2 - 10a + 42$ ____________

9. $3s^2t - 4s^2 - 3st + 4s - 60t + 80$ ____________

10. $36c^2d^2 + 216c - 81c^2 + 64d^2 - 96cd^2 - 144$ ____________

Find all the values of *A* that make the trinomial factorable. Write each set of binomials.

11. $2x^2 + Ax + 15$ ____________

12. $3x^2 + Ax - 20$ ____________

13. Determine the dimensions, in terms of x, of a rectangle if its area is $7x^2 + x - 8$. ____________

14. The volume of a rectangular prism is $30x^2y + 56xy - 64y$ cubic units.

    a. Find expressions for the dimensions of the prism. ____________

    b. Find expressions for the area of each unique face. ____________

15. Find the values of x and y that would classify a rectangular prism with a volume of $40x^2y + 90 + 130xy - 52x - 225y - 16x^2$ cubic units as a cube. ____________

NAME ______________________ CLASS ____________ DATE ____________

Practice Masters Level A

9.8 *Solving Equations by Factoring*

Identify the zeros of each function.

1. $y = (x - 3)(x - 4)$ ____________
2. $y = (x + 7)(x + 2)$ ____________
3. $y = (x + 4)(x - 6)$ ____________
4. $y = (x - 8)(x + 1)$ ____________
5. $y = (x - 10)(x - 7)$ ____________
6. $y = (x - 9)(x + 5)$ ____________
7. $y = x(x + 8)$ ____________
8. $y = (x + 11)(x - 12)$ ____________
9. $y = (x - 2)(x + 2)$ ____________
10. $y = (x - 15)(x - 15)$ ____________
11. $y = (2x - 1)(x + 3)$ ____________
12. $y = x(4x - 3)$ ____________

Solve by factoring.

13. $x^2 - 6x = 0$ ____________
14. $x^2 + 12x = 0$ ____________
15. $3x^2 - 12x = 0$ ____________
16. $-6x^2 - 36x = 0$ ____________
17. $x^2 + 9x + 20 = 0$ ____________
18. $x^2 - 7x + 12 = 0$ ____________
19. $x^2 + 8x + 16 = 0$ ____________
20. $x^2 - 4x + 4 = 0$ ____________
21. $x^2 + 3x + 2 = 0$ ____________
22. $x^2 - 6x + 5 = 0$ ____________
23. $x^2 - 11x + 28 = 0$ ____________
24. $x^2 - 3x - 10 = 0$ ____________
25. $x^2 - 5x - 36 = 0$ ____________
26. $x^2 - 2x - 15 = 0$ ____________
27. $x^2 - 9x + 18 = 0$ ____________
28. $x^2 - 3x - 40 = 0$ ____________
29. $6x^2 + x - 2 = 0$ ____________
30. $3x^2 - 2x - 1 = 0$ ____________
31. $2x^2 + 19x + 24 = 0$ ____________
32. $10e^2 - 26e + 12 = 0$ ____________
33. $x^2 - 10x - 75 = 0$ ____________
34. $4s^2 + 5s - 6 = 0$ ____________
35. $3x^2 - 4x + 1 = 0$ ____________
36. $2x^2 + 3x + 1 = 0$ ____________
37. $2x^2 - x - 1 = 0$ ____________
38. $3x^2 + 5x - 12 = 0$ ____________

Practice Masters Level B

9.8 Solving Equations by Factoring

Solve by factoring.

1. $x^2 - 6x = -9$ ______________
2. $3x^2 + 7x - 6 = 0$ ______________
3. $2x^2 + 7x - 15 = 0$ ______________
4. $4x^2 + x = 60$ ______________
5. $3x^2 + 14x = 15$ ______________
6. $x^2 - 16x = -64$ ______________
7. $x^2 - 25 = 0$ ______________
8. $10x^2 - 27x + 18$ ______________
9. $4x^2 + 19x + 12 = 0$ ______________
10. $x + 6x^2 = 40$ ______________
11. $6x^2 + 15x = 4x + 10$ ______________
12. $4x^2 - 6x = 6x - 9$ ______________
13. $6x^2 = -7x + 20$ ______________
14. $21 = -12x^2 + 37x$ ______________
15. $2x^2 = 5x + 88$ ______________
16. $25x^3 - 100x = 0$ ______________
17. $2x^2 - 50x = -x^2 + 2x + 36$ ______________
18. $2(4x^2 + 11) = 48x$ ______________
19. $6x(3x - 5) = 49 - (15 + 9x)2x$ ______________
20. $x^2 - \frac{1}{5}x = \frac{2}{25}$ ______________

21. The area of the rectangle at the right is 20 square inches. Find the value of x.

____

x

$x + 8$

22. The area of the square at the right is 49 square meters. Find the value of x.

____

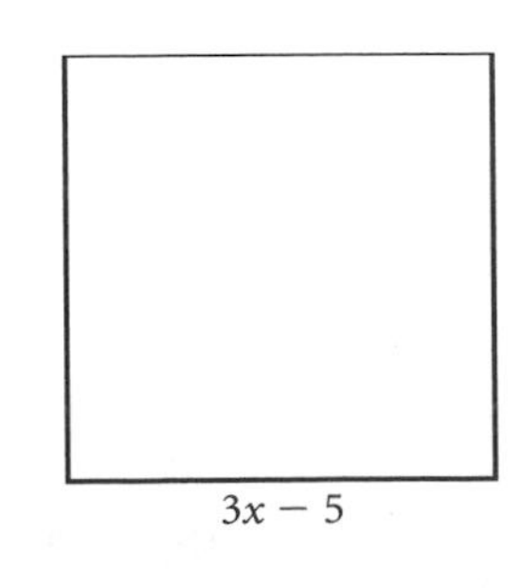

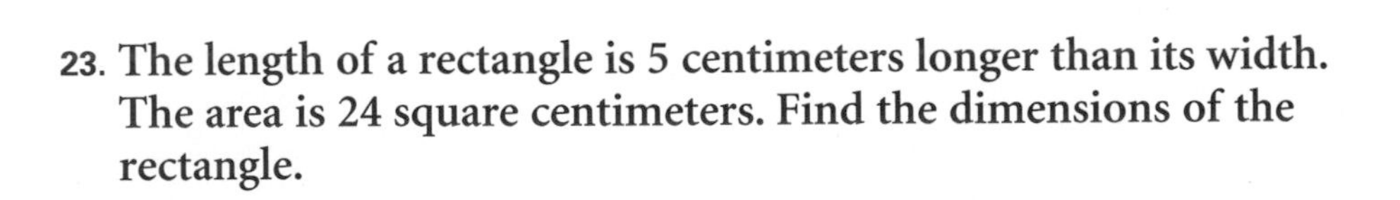

23. The length of a rectangle is 5 centimeters longer than its width. The area is 24 square centimeters. Find the dimensions of the rectangle.

____

24. The width of a rectangle is 7 inches shorter than its length. The area is 78 square inches. Find the dimensions of the rectangle.

____

NAME ______________________ CLASS ____________ DATE ____________

Practice Masters Level C

9.8 Solving Equations by Factoring

Solve by factoring.

1. $24x^3 - 8x^2 = -\frac{4}{6}x$ ______________________

2. $\frac{3}{16}x^2 + \frac{1}{24}x = \frac{5}{18}$ ______________________

3. $18x^3 - 32x + 45x^2 = 80$ ______________________

4. $(2x - 3)^2 - (x + 2)^2 = 17$ ______________________

Find an equation in the form $ax^2 + bx + c = 0$, where a, b and c are integers, that has the given pair of solutions.

5. $8, 3$ ______________________

6. $-2, -5$ ______________________

7. $-4, 9$ ______________________

8. $0, 7$ ______________________

9. $\frac{1}{3}, -10$ ______________________

10. $\frac{3}{5}, -\frac{2}{7}$ ______________________

11. The perimeter of a rectangular field is 64 yards and the area is 192 square yards. What are the dimensions of the field? ______________

12. Suppose that you want the length of your new garden to be 10 feet less than 3 times the width. The area of the garden must be 48 square feet. If fence material costs $2.50 per foot, how much money will it cost to enclose your garden? ______________

13. The base of a right triangle is 14 more than 8 times the height. If the area of the triangle is 15 square units, find its dimensions. ______________

14. The height of a ball after it has been thrown can be modeled by the function $h = -16t^2 + vt + s$, where h is measured in feet, t represents time in seconds, v represents initial velocity in feet per second, and s represents starting height in feet.
From a point 6 feet above the ground, a tennis ball is thrown up into the air, with an initial velocity of 10 feet per second. In how many seconds will the ball hit the ground? ______________

NAME ______________________ CLASS ____________ DATE ____________

Practice Masters Level A

10.1 Graphing Parabolas

Compare the graph of each function to the graph of $y = x^2$. Describe the vertical and horizontal translations of each vertex.

1. $y = x^2 + 5$ ______________________

2. $y = (x - 3)^2$ ______________________

3. $y = (x - 4)^2 + 3$ ______________________

4. $y = (x + 1)^2 - 10$ ______________________

5. $y = (x - 12)^2 - 7$ ______________________

6. $y = (x + 6)^2 + 8$ ______________________

Find the vertex and axis of symmetry for the graph of each function, and then sketch the graph.

7. $y = (x - 1)^2 + 2$

____

____

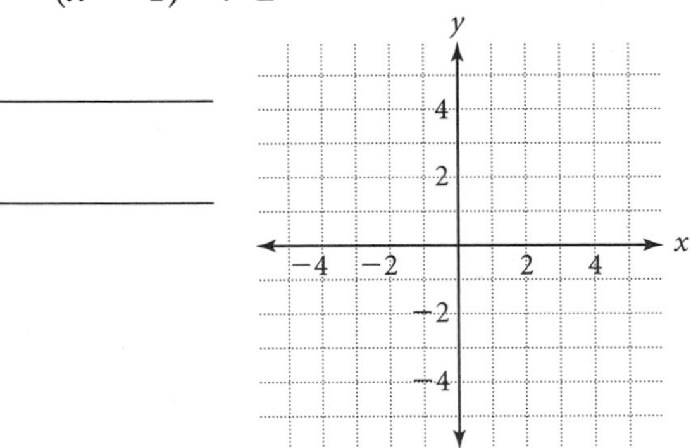

8. $y = (x + 2)^2 - 1$

____

____

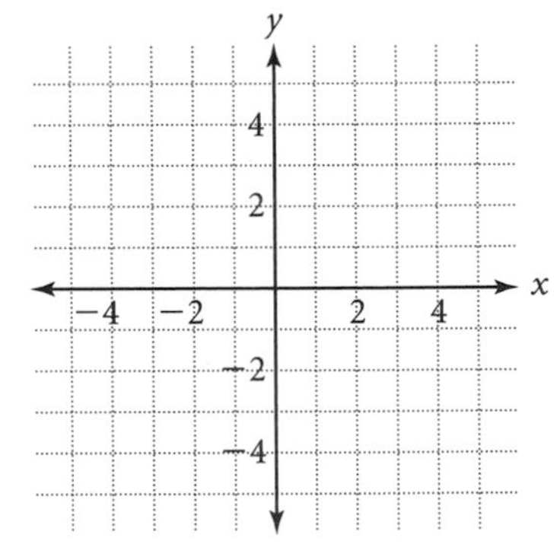

Use factoring to find the zeros of each function.

9. $y = x^2 - 5x - 24$ ______________

10. $y = x^2 - 13x + 36$ ______________

11. $y = x^2 + 8x + 15$ ______________

12. $y = x^2 + 3x - 18$ ______________

Use factoring to find the zeros of each function. Use the zeros to find the vertex of its graph.

13. $y = x^2 - 10x + 24$ ______________

14. $y = x^2 - 4x - 5$ ______________

NAME ______________________ CLASS ____________ DATE ____________

Practice Masters Level B

10.1 Graphing Parabolas

Compare the graph of each function to the graph of $y = x^2$. Describe the vertical and horizontal translations of each vertex.

1. $y = (x + 8)^2 + 2$ ______________________

2. $y = (x - 12)^2 - 9$ ______________________

3. $y = -x^2 - 3$ ______________________

4. $y = -(x + 5)^2 - 6$ ______________________

Find the vertex and axis of symmetry for the graph of each function, and then sketch the graph.

5. $y = -(x + 2)^2 + 3$

____

____

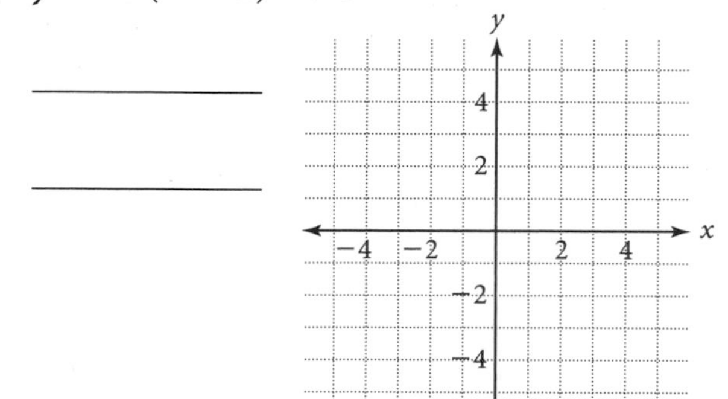

6. $y = (x + 1)^2 - 2$

____

____

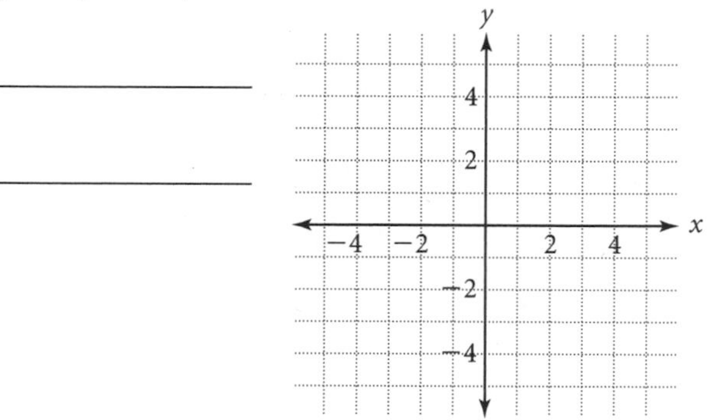

7. $y = 3(x - 1)^2 - 2$

____

____

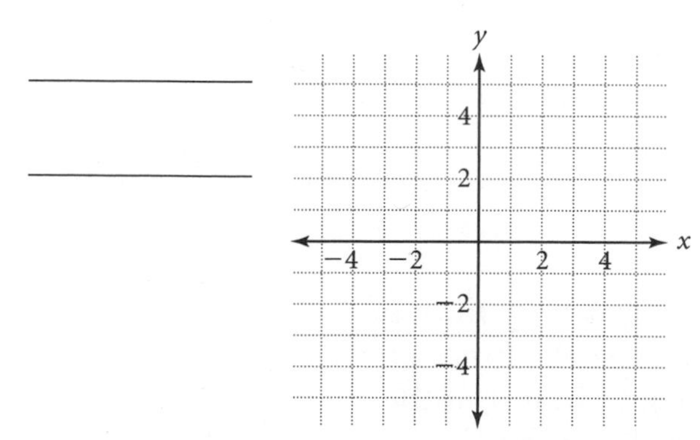

8. $y = \frac{1}{2}(x - 2)^2$

____

____

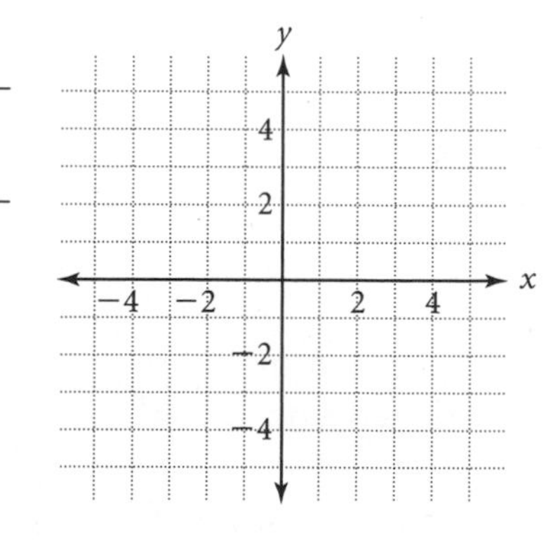

Use factoring to find the zeros of the function. Use the zeros to find the vertex of its graph.

9. $y = x^2 - 32x - 144$ ______________

____

10. $y = x^2 - x - 132$ ______________

____

NAME ______________________ CLASS __________ DATE __________

Practice Masters Level C

10.1 Graphing Parabolas

Compare the graph of each function to the graph of $y = x^2$. Describe the vertical and horizontal translations of each vertex.

1. $y = (3x - 5)^2 - 2$ ______________________

2. $y = -\frac{2}{3}(2x + 7)^2 + 1$ ______________________

Write an equation in vertex form of each graph described.

3. the graph of $y = x^2$ reflected vertically, shifted $\frac{5}{2}$ units to the right and $\frac{3}{2}$ units down ______________________

4. the graph of $y = x^2$ reflected vertically and shifted 5 units up ______________________

5. the graph of $y = x^2$ reflected vertically, shifted 4 units to the left, and shifted 2 units down ______________________

Use factoring to find the zeros of the function. Use the zeros to find the vertex of its graph. Graph the function.

6. $y = 4x^2 - 4x - 3$

____

____

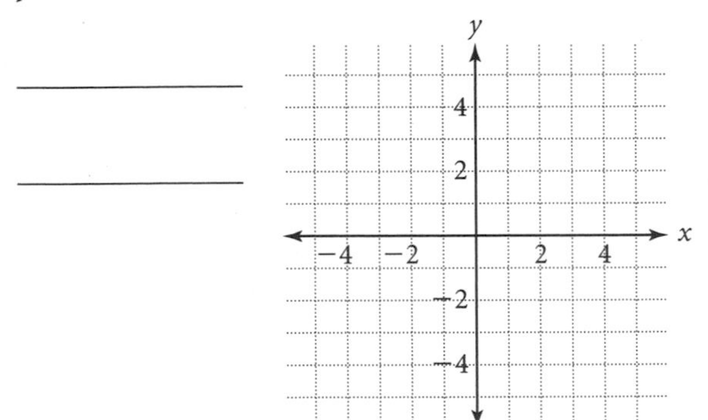

7. $y = 4x^2 - 24x + 35$

____

____

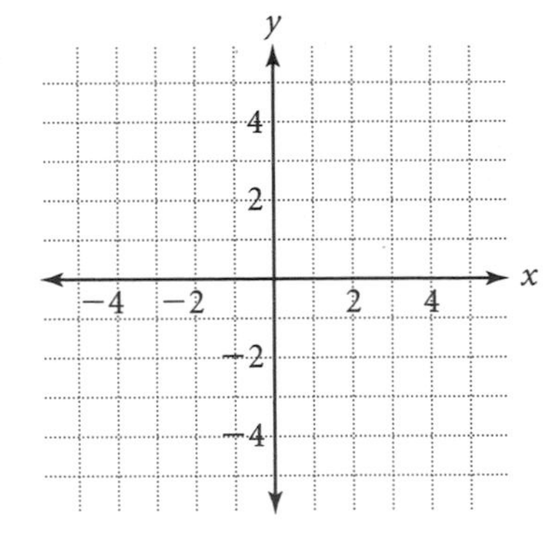

8. From an initial height of 5 feet, a tennis ball is thrown up into the air with an initial velocity of 16 feet per second. Use the function $h = -16t^2 + vt + s$, where h represents the height in feet of the ball t seconds after it is thrown, v is the initial velocity, and s is the starting height.

   a. What is the maximum height attained by the tennis ball? ______________________

   b. How long does it take for the tennis ball to reach its maximum height? ______________________

NAME ______________________ CLASS ____________ DATE ____________

Practice Masters Level A

10.2 Solving Equations by Using Square Roots

Find each positive square root. Round answers to the nearest hundredth when necessary.

1. $\sqrt{196}$ ________
2. $\sqrt{100}$ ________
3. $\sqrt{225}$ ________
4. $\sqrt{96}$ ________
5. $\sqrt{625}$ ________
6. $\sqrt{125}$ ________
7. $\sqrt{45}$ ________
8. $\sqrt{30}$ ________

Solve each equation. Round answers to the nearest hundredth when necessary.

9. $x^2 = 36$ ________
10. $x^2 = 49$ ________
11. $x^2 = 144$ ________
12. $x^2 = 289$ ________
13. $x^2 = 20$ ________
14. $x^2 = 400$ ________
15. $x^2 = 600$ ________
16. $x^2 = 32$ ________
17. $x^2 = \frac{9}{16}$ ________
18. $x^2 = \frac{1}{25}$ ________
19. $x^2 = \frac{64}{121}$ ________
20. $x^2 = \frac{81}{256}$ ________
21. $3x^2 = 48$ ________
22. $2x^2 = 1800$ ________
23. $4x^2 = 300$ ________
24. $9x^2 = 450$ ________
25. $(x - 3)^2 = 16$ ________
26. $(x + 2)^2 = 9$ ________

Find the vertex, axis of symmetry, and zeros of each function. Then sketch the graph of each.

27. $y = (x - 2)^2 - 1$

____

____

____

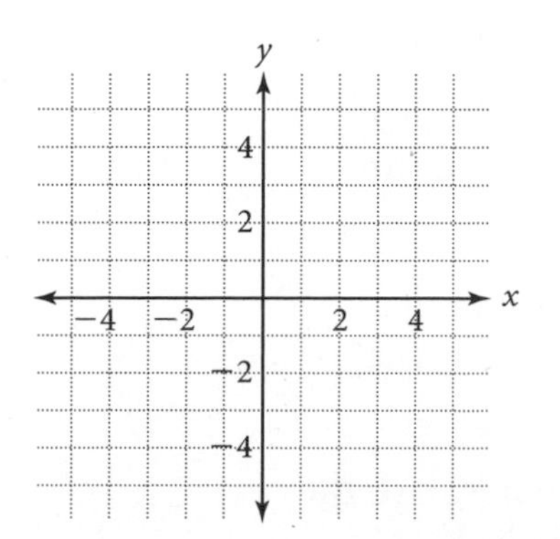

28. $y = (x + 1)^2 - 4$

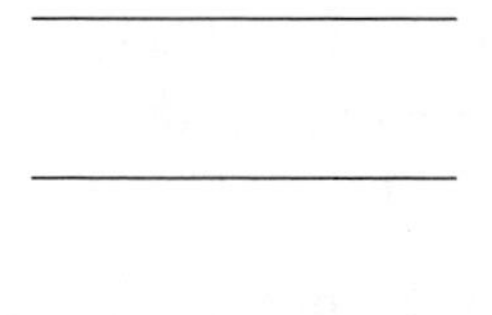

____

____

____

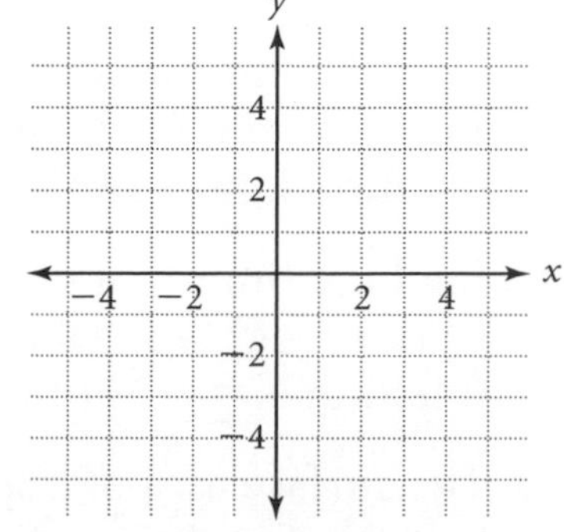

NAME ______________________ CLASS ______________ DATE ______________

Practice Masters Level B

10.2 Solving Equations by Using Square Roots

Solve each equation. Round answers to the nearest hundredth when necessary.

1. $x^2 - 361 = 0$ ______________
2. $x^2 - 300 = -75$ ______________
3. $3x^2 - 123 = 0$ ______________
4. $4x^2 + 15 = 24$ ______________
5. $36x^2 - 140 = 4$ ______________
6. $\frac{1}{2}x^2 + 5 = 12$ ______________
7. $3x^2 - 6 = 0$ ______________
8. $(x - 3)^2 - 16 = 0$ ______________
9. $(x + 10)^2 = 15$ ______________
10. $(x - 7)^2 - 28 = 0$ ______________
11. $2(x - 5)^2 - 27 = 23$ ______________
12. $\frac{3}{4}(x - 4)^2 - 12 = 0$ ______________
13. $3\left(x - \frac{2}{3}\right)^2 = \frac{1}{3}$ ______________
14. $4\left(x + \frac{1}{3}\right)^2 - \frac{4}{9} = 0$ ______________

Find the vertex, axis of symmetry, and zeros of each function. Then sketch the graph of each.

15. $y = 2(x - 2)^2 - 8$

____

____

____

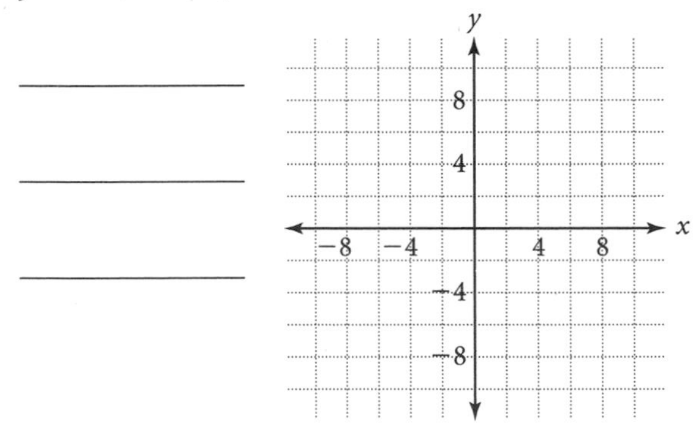

16. $y = -2(x + 2)^2 + 8$

____

____

____

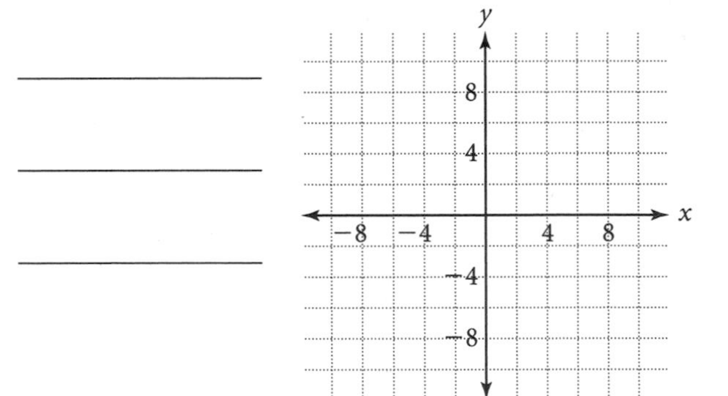

Solve. Round your answers to the nearest tenth.

17. Find the time, t, in seconds that it takes an object dropped from a height of 784 feet to reach the ground. Use the function $h = -16t^2 + 784$. ______________

18. Find the time, t, in seconds that it takes an object dropped from a height of 900 feet to reach the ground. Use the function $h = -16t^2 + 900$. ______________

19. The surface area of a sphere is 100 square inches. Use the formula $S = 4\pi r^2$ to find the length of the radius. ______________

Practice Masters Level C

10.2 Solving Equations by Using Square Roots

Find two different quadratic equations that have the solutions listed below. The equations should be in the form $ax^2 = k$ or $a(x - h)^2 = k$.

1. $-14, 14$ ______________
2. $2, -8$ ______________
3. $-\sqrt{21}, \sqrt{21}$ ______________
4. only 8 ______________

Solve each equation. Round answers to the nearest hundredth when necessary.

5. $0.6(x + 3)^2 - 100 = 0$ ______________
6. $\frac{3}{4}(x + 2)^2 - \frac{1}{2} = \frac{1}{2}$ ______________
7. $\frac{2}{3}(5x - 4)^2 - 12 = 3$ ______________
8. $4.5\left(\frac{1}{4}x - 25\right)^2 + 7 = 16$ ______________

Find the vertex, axis of symmetry, and zeros of each function. Then sketch the graph of each.

9. $y = -4(x + 2)^2 + 1$

____

____

____

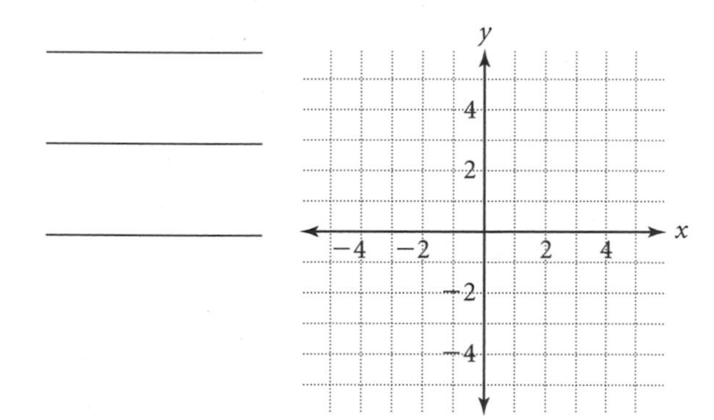

10. $y = -9(x - 2)^2 - 1$

____

____

____

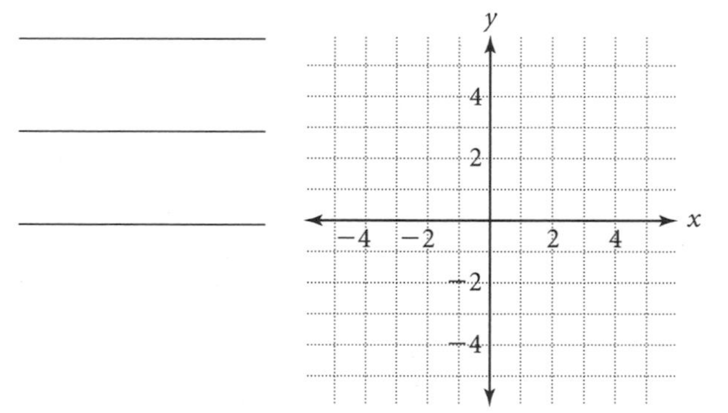

11. An object's height, h, in feet, t seconds after it is dropped is given by $h = -16t^2 + s$, where s represents the starting height in feet. Solve this equation for t. (Simplify the equation so there are no square roots in the denominator.) ______________

12. Find two different sets of values for x and y for which $4a(x - 2)^2 + 3xb - (16a + 2yb) = 0$ is true. ______________

13. Find two different sets of values for x and y for which $4a\left(3x - \frac{1}{2}\right)^2 + 4xb - \left(9a + \left(y + \frac{5}{8}\right)b\right) = 0$ is true. ______________

NAME ______________________ CLASS ____________ DATE ____________

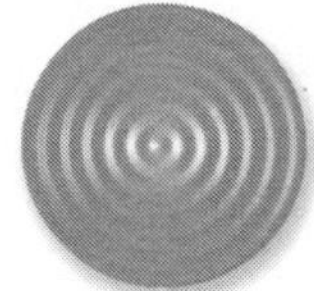

Practice Masters Level A

10.3 Completing the Square

Use algebra tiles to complete the square for each binomial. Draw a sketch of your algebra-tile model. Write each perfect-square trinomial in the form $x^2 + bx + c$.

1. $x^2 + 2x$ ____________

2. $x^2 - 4x$ ____________

Rewrite each function in the form $y = (x - h)^2 + k$.

3. $y = x^2 - 6x + 9 - 9$ ____________

4. $y = x^2 + 10x + 25 - 25$ ____________

5. $y = x^2 - 14x + 49 - 49$ ____________

6. $y = x^2 + 8x + 16 - 16$ ____________

Complete the square to create a perfect-square trinomial. Write the new expression in the form $(x - h)^2$.

7. $x^2 + 4x$ ____________

8. $x^2 - 12x$ ____________

9. $x^2 - 18x$ ____________

10. $x^2 + 22x$ ____________

Find the vertex of the parabola represented by each function.

11. $y = x^2$ ____________

12. $y = x^2 + 5$ ____________

13. $y = x^2 - 3$ ____________

14. $y = x^2 - 2$ ____________

Rewrite each function in the form $y = (x - h)^2 + k$ by completing the square. Find the vertex of its graph.

15. $y = x^2 + 7$ ____________

16. $y = x^2 - 2$ ____________

17. $y = x^2 - 4x$ ____________

18. $y = x^2 + 6x$ ____________

19. $y = x^2 + 2x$ ____________

20. $y = x^2 - 12x$ ____________

21. $y = x^2 - 20x$ ____________

22. $y = x^2 + 24x$ ____________

23. $y = x^2 - 10x + 1$ ____________

24. $y = x^2 + 4x + 2$ ____________

NAME ______________________ CLASS __________ DATE __________

Practice Masters Level B

10.3 Completing the Square

Complete the square for each binomial. Write the new perfect-square trinomial in the form $x^2 + bx + c$.

1. $x^2 + 28x$ ______________
2. $x^2 - 30x$ ______________
3. $x^2 - x$ ______________
4. $x^2 + 9x$ ______________
5. $x^2 - 50x$ ______________
6. $x^2 + 15x$ ______________
7. $x^2 - \frac{1}{2}x$ ______________
8. $x^2 + \frac{2}{5}x$ ______________

Rewrite each function in the form $y = (x - h)^2 + k$ by completing the square. Find the vertex of its graph.

9. $y = -8 + x^2$ ______________
10. $y = x^2 - 6x$ ______________
11. $y = x^2 - 3x$ ______________
12. $y = x^2 + 21x$ ______________
13. $y = x^2 + 6x + 4$ ______________
14. $y = x^2 - 10x - 5$ ______________
15. $y = x^2 + 18x + 50$ ______________
16. $y = x^2 + 3x + \frac{1}{4}$ ______________
17. $y = x^2 - 9x + \frac{13}{4}$ ______________
18. $y = x^2 - 11x + 5$ ______________
19. $y = 4 - 22x + x^2$ ______________
20. $y = \frac{1}{2} - 3x + x^2$ ______________

Rewrite each function in the form $y = (x - h)^2 + k$ by completing the square. Find the vertex of its graph, and graph the function.

21. $y = x^2 - 4x + 1$

____

____

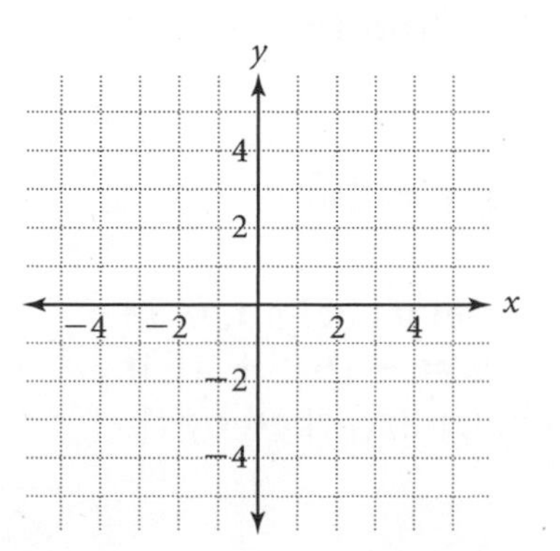

22. $y = x^2 + 6x + 7$

____

____

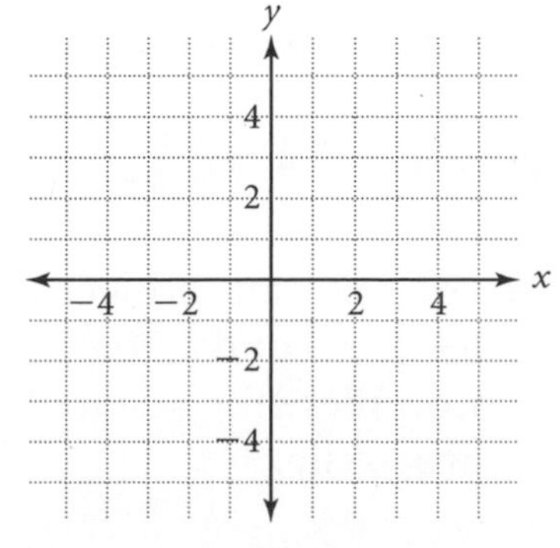

NAME ____________ CLASS ________ DATE ________

Practice Masters Level C

10.3 Completing the Square

Rewrite each function in the form of $y = (x - h)^2 + k$ by completing the square.

1. $y = x^2 - 19x$ ____________
2. $y = x^2 + 25x$ ____________
3. $y = x^2 - \frac{1}{2}x + 2$ ____________
4. $y = x^2 + \frac{1}{3}x + 2$ ____________
5. $y = -\frac{2}{3} - 3x + x^2$ ____________
6. $y = \frac{2}{9} + \frac{1}{3}x + x^2$ ____________
7. $4y = 4x^2 - 20x - 12$ ____________
8. $2y = 2x^2 + 6x + 11$ ____________
9. $3y = 3x^2 - 14x - 22$ ____________
10. $\frac{2}{3}y = \frac{2}{3}x^2 + 8x - 7$ ____________

Find the maximum (or minimum) value for the function by completing the square. Then graph the function.

11. $y = x^2 - 3x + 2$

____

____

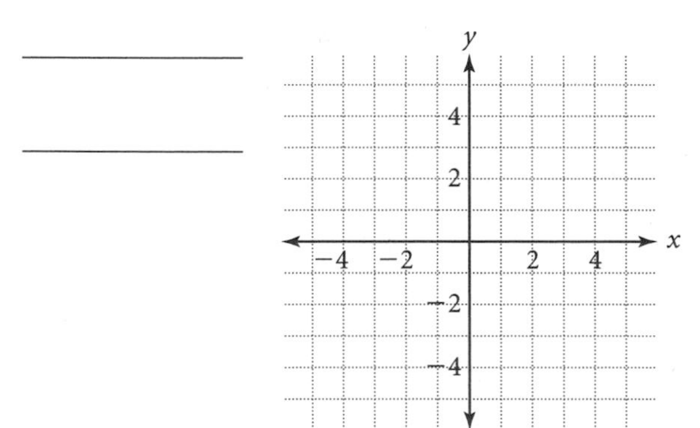

12. $y = x^2 + x + 2$

____

____

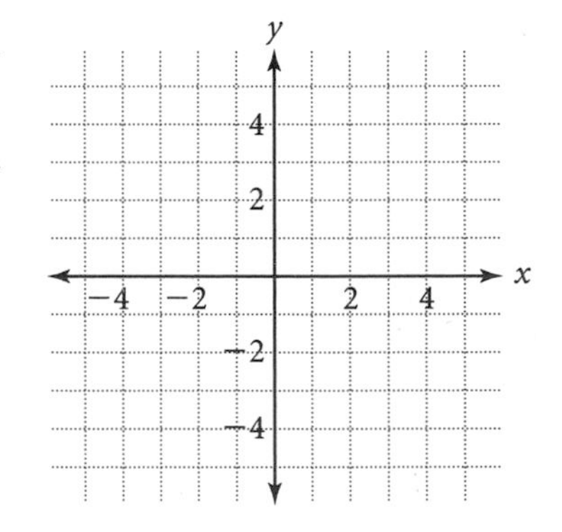

Rewrite each function in the form of $y = a(x - h)^2 + k$ by completing the square. Remember that completing the square requires the leading coefficient to be 1.

13. $y = 2x^2 + 8x$ ____________
14. $y = 3x^2 + 18x$ ____________
15. For a certain projectile, the relationship between its height, h, in feet, t seconds after it is projected is given by $h = -16t^2 + 144t$. Complete the square to find the maximum height reached by the projectile. ____________

NAME ______________________ CLASS ______________ DATE ______________

Practice Masters Level A

10.4 Solving Equations of the Form $x^2 + bx + c = 0$

Find the zeros of each function.

1. $y = x^2 - 4x - 5$ ______________
2. $y = x^2 + 4x - 12$ ______________

Solve each equation by factoring.

3. $x^2 - 18x + 45 = 0$ ______________
4. $x^2 + 6x - 27 = 0$ ______________
5. $x^2 - 22x - 75 = 0$ ______________
6. $x^2 + 5x - 24 = 0$ ______________
7. $0 = x^2 - 10x + 25$ ______________
8. $0 = x^2 + 9x - 36$ ______________
9. $x^2 - 49 = 0$ ______________
10. $x^2 - 13x - 48 = 0$ ______________

Solve each equation by completing the square. Round your answers to the nearest hundredth when necessary.

11. $x^2 + 2x = 3$ ______________
12. $x^2 + 8x = 9$ ______________
13. $x^2 - 12x = 64$ ______________
14. $x^2 + 20x = -99$ ______________
15. $x^2 + 4x = 8$ ______________
16. $x^2 - 4x = 1$ ______________

Solve each equation by factoring or by completing the square.

17. $d^2 - 6d = 0$ ______________
18. $r^2 - 6r = 2$ ______________
19. $s^2 - 4s + 1 = 0$ ______________
20. $q^2 - 7q - 12 = 0$ ______________
21. $s^2 - 7s = 60$ ______________
22. $s^2 - 3s = 40$ ______________

Find the points where the graphs of each system intersect.

23. $\begin{cases} y = 4 \\ y = x^2 \end{cases}$

____

____

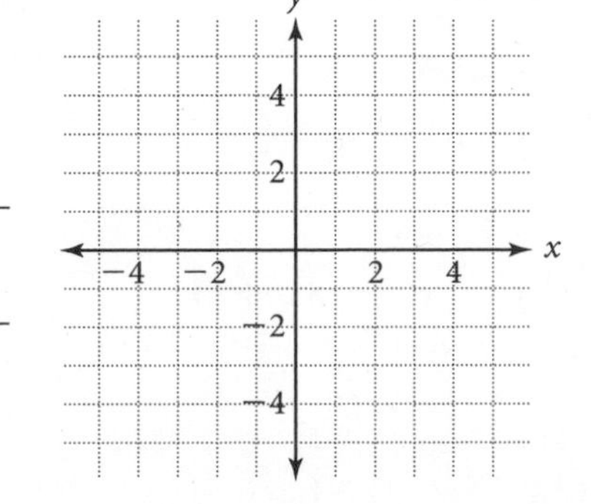

24. $\begin{cases} y = x^2 + 1 \\ y = 2 \end{cases}$

____

____

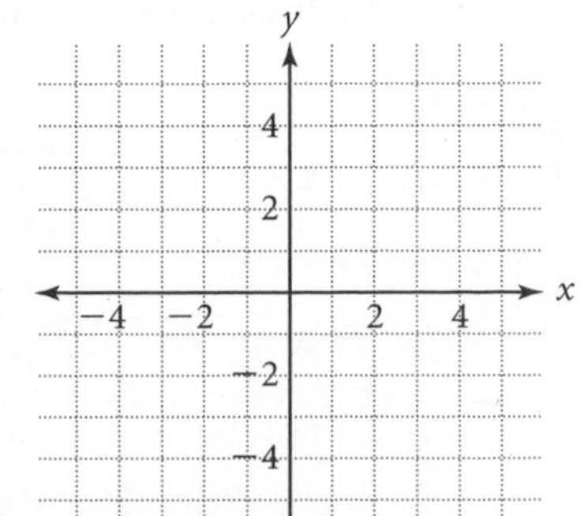

NAME ________________ CLASS ________ DATE ________

Practice Masters Level B

10.4 Solving Equations of the Form $x^2 + bx + c = 0$

Find the zeros of each function.

1. $y = 2x^2 - 5x - 12$ ________________
2. $y = 3x^2 - 22x + 24$ ________________

Solve each equation by factoring.

3. $6x^2 - 13x = 63$ ________________
4. $4x^2 - 121 = 0$ ________________
5. $7x^2 - 175 = 0$ ________________
6. $2x^2 + 23x = 39$ ________________
7. $3x^2 = 46x - 120$ ________________
8. $4x^2 - 66x + 90 = 0$ ________________

Solve each equation by completing the square. Round your answers to the nearest hundredth when necessary.

9. $x^2 + 2x = 8$ ________________
10. $x^2 + 6x - 5 = 0$ ________________
11. $x^2 - 10x = -12$ ________________
12. $x^2 - 14x - 3 = 0$ ________________
13. $x^2 - 3x = \frac{3}{4}$ ________________
14. $x^2 - 5x + 6 = 0$ ________________

Solve each equation by factoring or by completing the square.

15. $3d^2 - 9d = 0$ ________________
16. $r^2 - 8r = 2$ ________________
17. $2p^2 + 3p = 44$ ________________
18. $q^2 - 13q = 14$ ________________
19. $w^2 - w = \frac{3}{4}$ ________________
20. $4k^2 - 120k = -225$ ________________

Find the points where the graphs of each system intersect.

21. $\begin{cases} y = 6 \\ y = x^2 + x \end{cases}$ ________________
22. $\begin{cases} y = 5 \\ y = 2x^2 + 3x \end{cases}$ ________________
23. $\begin{cases} y = -2x - 3 \\ y = x^2 - x - 9 \end{cases}$ ________________
24. $\begin{cases} y = 2x - 4 \\ y = x^2 - 8x + 5 \end{cases}$ ________________

NAME ______________________ CLASS ____________ DATE ____________

Practice Masters Level C

10.4 Solving Equations of the Form $x^2 + bx + c = 0$

Find the zeros of each function.

1. $y = 36x^2 - 33x - 108$ ____________

2. $y = \frac{2}{9}x^2 - 2x - 20$ ____________

Solve each equation by factoring or completing the square. Round your answers to the nearest hundredth when necessary.

3. $\frac{1}{4}x^2 + \frac{3}{8}x = \frac{5}{4}$ ____________

4. $-16x^2 = -136x + 281$ ____________

5. $x^2 - 13x = -36$ ____________

6. $x^2 - 17x = 25$ ____________

Find the exact *x*-coordinates of the points where the graphs of each system intersect.

7. $\begin{cases} y = x^2 + 11 \\ y = 3x^2 - x - 4 \end{cases}$ ____________

8. $\begin{cases} y = 3x^2 - 5x - 2 \\ y = x^2 - 3x + 1 \end{cases}$ ____________

Solve by factoring or by completing the square. Round your answers to the nearest tenth when necessary.

9. Find 2 consecutive odd integers such that 3 times the square of the larger is 45 more than 10 times the smaller. ____________

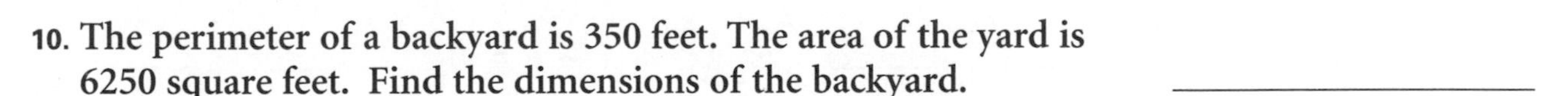

10. The perimeter of a backyard is 350 feet. The area of the yard is 6250 square feet. Find the dimensions of the backyard. ____________

11. The height of a rectangular prism is 2 feet. The base of this prism has a length that is 8 more than twice the width. The volume of the prism is $26\frac{1}{4}$ square feet. Find the length and the width of the base of the prism. ____________

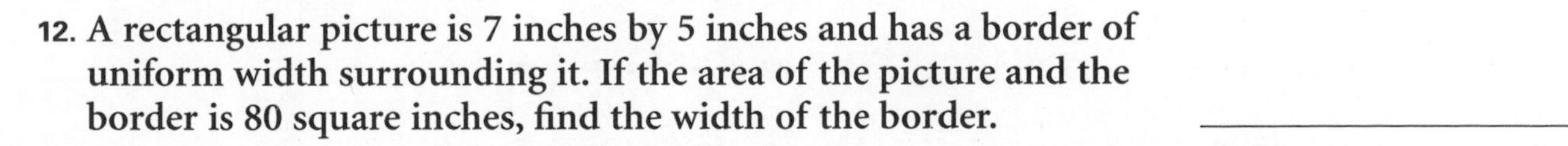

12. A rectangular picture is 7 inches by 5 inches and has a border of uniform width surrounding it. If the area of the picture and the border is 80 square inches, find the width of the border. ____________

NAME ______________________ CLASS ______________ DATE ______________

Practice Masters Level A

10.5 The Quadratic Formula

Identify *a*, *b*, and *c* in each quadratic equation.

1. $x^2 - 5x + 14 = 0$ ______________
2. $3x^2 - 4x + 1 = 0$ ______________
3. $2x^2 + 3x - 10 = 0$ ______________
4. $x^2 + 16 = 0$ ______________
5. $3x^2 - 15x = 0$ ______________
6. $7x^2 - x + 3 = 0$ ______________

Find the value of the discriminant, and determine the number of real solutions for each equation. Determine whether the equation can be factored.

7. $x^2 + 4x - 3 = 0$ ______________
8. $x^2 - 2x + 4 = 0$ ______________
9. $4x^2 + 4x + 1 = 0$ ______________
10. $3x^2 - 5x + 2 = 0$ ______________
11. $2x^2 - 9x + 11 = 0$ ______________
12. $x^2 - 36 = 0$ ______________
13. $5x^2 - 7x = 0$ ______________
14. $x^2 - 6x + 8 = 0$ ______________

Use the quadratic formula to solve each equation. Check by substitution.

15. $x^2 + x - 2 = 0$ ______________
16. $x^2 - 4x - 12 = 0$ ______________
17. $x^2 + 3x - 10 = 0$ ______________
18. $x^2 - 8x = 0$ ______________
19. $2x^2 - 7x + 6 = 0$ ______________
20. $3x^2 - 9x + 5 = 0$ ______________

Use the quadratic formula to solve each equation. Round your answers to the nearest hundredth when necessary.

21. $x^2 - 6x - 4 = 0$ ______________
22. $2x^2 + x - 6 = 0$ ______________
23. $x^2 + 5x - 3 = 0$ ______________
24. $x^2 - 4x + 2 = 0$ ______________

Choose any method to solve each quadratic equation. Round your answers to the nearest hundredth when necessary.

25. $x^2 - 21 = 0$ ______________
26. $x^2 + 3x - 28 = 0$ ______________
27. $d^2 - 3d - 1 = 0$ ______________
28. $s^2 - 3s + 1 = 0$ ______________
29. $w^2 - 18w + 45 = 0$ ______________
30. $3r^2 - 10r - 8 = 0$ ______________

NAME ______________________ CLASS ______________ DATE ______________

Practice Masters Level B

10.5 The Quadratic Formula

Find the value of the discriminant, and determine the number of real solutions for each equation. Determine whether the equation can be factored.

1. $3w^2 - 6w + 4 = 0$ ______________
2. $4r^2 + 10r - 7 = 0$ ______________
3. $8y^2 - 10y + 3 = 0$ ______________
4. $-9r^2 + 12r - 4 = 0$ ______________
5. $z^2 + 121 = 0$ ______________
6. $-7q^2 - 15q = 0$ ______________

Use the quadratic formula to solve each equation. Round your answers to the nearest hundredth when necessary.

7. $x^2 + 12x - 28 = 0$ ______________
8. $2x^2 + 4x = 7$ ______________
9. $3x^2 - 5x = 2$ ______________
10. $5x^2 = 21$ ______________
11. $3x^2 + 10x - 5 = 0$ ______________
12. $7x^2 - 14x = 2$ ______________
13. $6x^2 - x = 40$ ______________
14. $2x^2 = 5x + 9$ ______________
15. $x^2 - 3x = \frac{3}{4}$ ______________
16. $\frac{1}{2}x^2 - \frac{1}{4}x = 1$ ______________

Factor each expression by using the quadratic formula.

17. $2x^2 - 2x - 24$ ______________
18. $3x^2 + 9x - 30$ ______________
19. $2x^2 - 11x + 12$ ______________
20. $4x^2 + 24x + 35$ ______________

Choose any method to solve each quadratic equation. Round your answers to the nearest hundredth when necessary.

21. $3x^2 - 19x = -20$ ______________
22. $16s^2 - 225 = 0$ ______________
23. $y^2 - 7y = 10$ ______________
24. $2t^2 - 0.75t = 6$ ______________
25. $4p^2 = 25p + 21$ ______________
26. $45c^2 + 2 = 33c$ ______________
27. The length of a rectangle is 3 more than its width. If the area of this rectangle is 13.75 square feet, find its dimensions. ______________

NAME ______________________ CLASS ____________ DATE ____________

Practice Masters Level C

10.5 The Quadratic Formula

Choose any method to solve each quadratic equation. Round answers to the nearest hundredth when necessary.

1. $0.8x^2 - 3x = 10$ ____________

2. $\frac{2}{9}x^2 + 3\frac{1}{3}x = 12$ ____________

Find the values of *P* so that each quadratic equation has one unique solution.

3. $x^2 - 8x + P = 0$ ____________

4. $Px^2 + 10x + 5 = 0$ ____________

5. $3x^2 + Px + 3 = 0$ ____________

6. $Px^2 + Px - 1 = 0$ ____________

Determine an inequality for *P* that makes each quadratic equation have more than one solution.

7. $3x^2 - 4x + P = 0$ ____________

8. $Px^2 - \frac{3}{4}x + \frac{3}{16} = 0$ ____________

Solve by using the quadratic formula.

9. The length of a rectangle is 3 less than twice the width. If the area of the rectangle is $39\frac{3}{8}$ square units, find its dimensions. ____________

10. Can a rectangle that has a perimeter of 32 feet have an area of 124 square feet? Explain why or why not. ____________

    a. If it is possible, state the dimensions of the rectangle. If this is not possible, what must the area be in order to have only one solution? ____________

For Exercises 11–12, use the model $h = -16t^2 + vt + s$, where *h* is the height in feet, *v* is the initial velocity in feet per second, and *s* is the initial height in feet. Round your answers to the nearest hundredth.

11. An emergency flare is shot upward in the air from a tower that is 50 feet off the ground. The flare has an initial velocity of 75 feet per second. Find the time, in seconds, when the flare will be 10 feet from the ground. ____________

12. A basketball is thrown up in the air from an initial height of 5 feet. The ball has an initial velocity of 15 feet per second. Find the time, in seconds, when the ball will be 3 feet from the ground. ____________

NAME ______________________________ CLASS ______________ DATE ______________

Practice Masters Level A

10.6 Graphing Quadratic Inequalities

Solve each quadratic inequality by using the Zero Product Property. Graph each solution on a number line.

1. $x^2 - 9 < 0$ ______________

2. $x^2 - 16 \geq 0$ ______________

3. $x^2 - 4x + 3 \leq 0$ ______________

4. $x^2 + 3x - 10 > 0$ ______________

Graph each quadratic inequality.

5. $y \leq x^2$

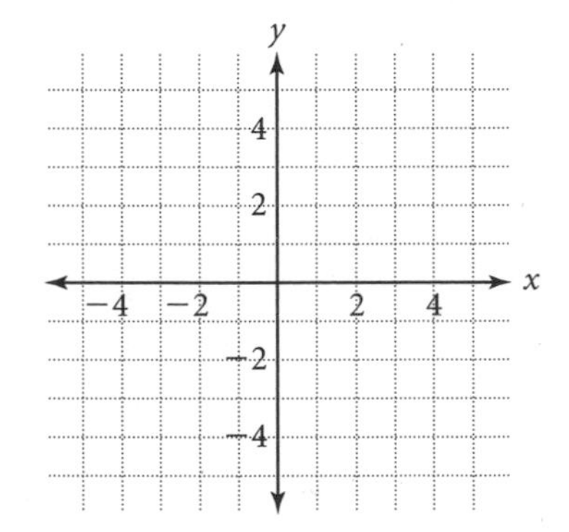

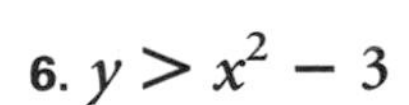

6. $y > x^2 - 3$

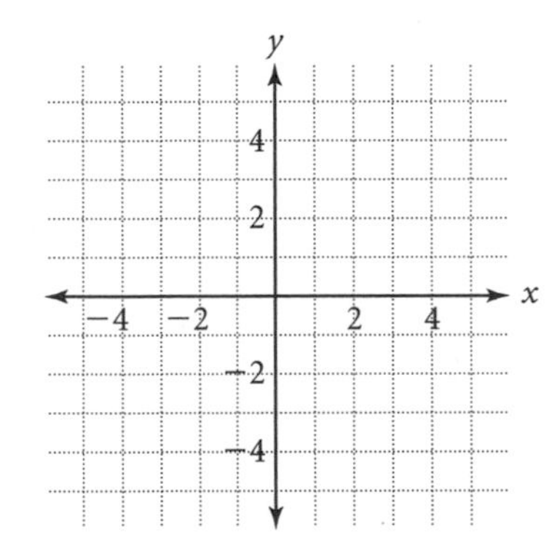

7. $y \geq -x^2$

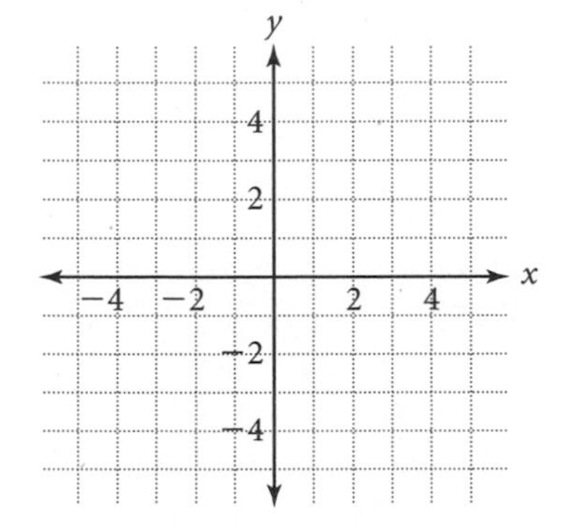

8. $y > x^2 + 2$

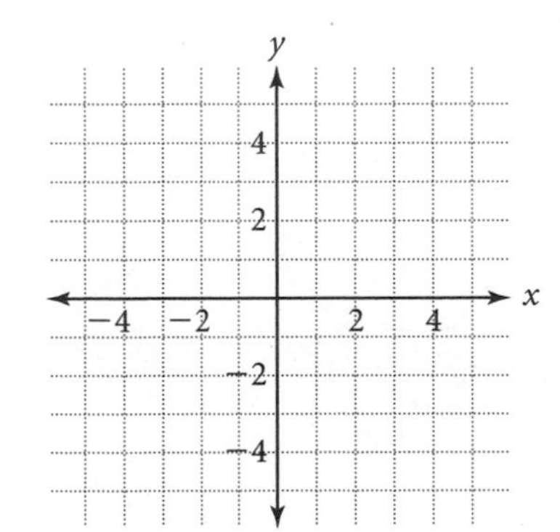

NAME ______________________ CLASS ______________ DATE ______________

Practice Masters Level B

10.6 Graphing Quadratic Inequalities

Solve each quadratic inequality by using the Zero Product Property. Graph each solution on a number line.

1. $x^2 - 9x < -18$ ______________________

2. $x^2 - 8x \geq -16$ ______________________

3. $x^2 \geq x + 20$ ______________________

4. $15 < 4x^2 + 4x$ ______________________

Graph each quadratic inequality.

5. $y \leq -2x^2 + 5$

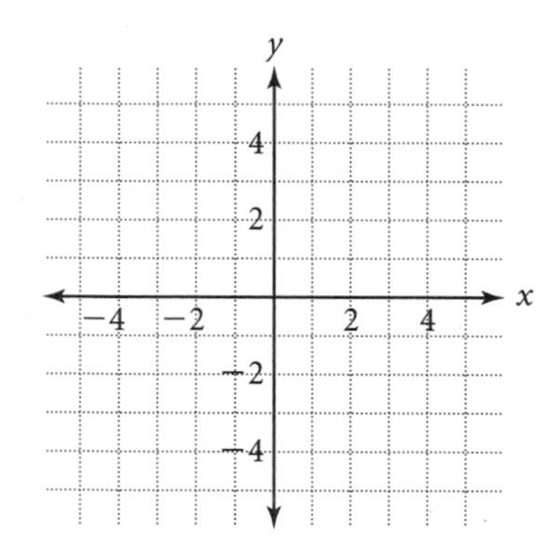

6. $y > 4 - x^2$

7. $y \geq x^2 + 2x + 1$

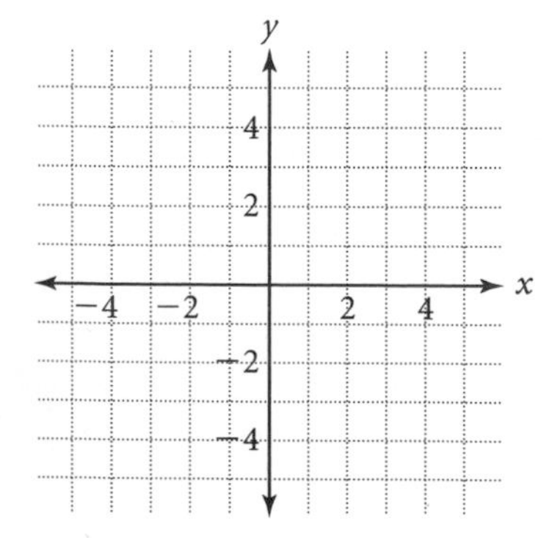

8. $y < x^2 + 4x - 1$

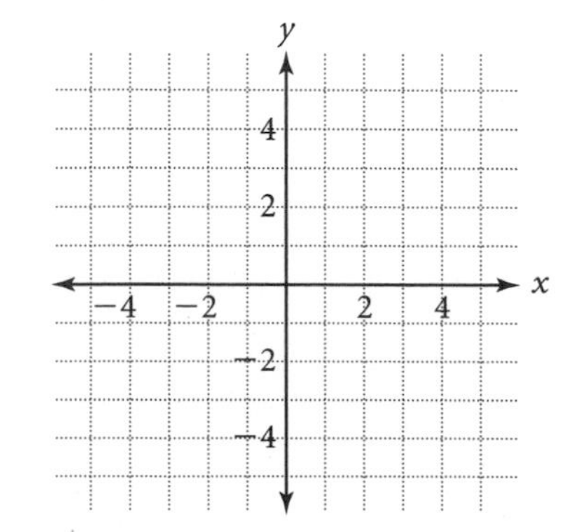

NAME ______________________ CLASS ____________ DATE ____________

Practice Masters Level C

10.6 Graphing Quadratic Inequalities

Solve each quadratic inequality by using the Zero Product Property. Graph each solution on a number line.

1. $4x^2 + 100 \geq 40x + 9$ ____________

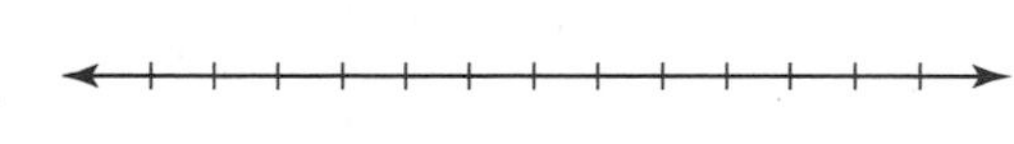

2. $-6x^2 + 6x \geq -2\frac{2}{3}$ ____________

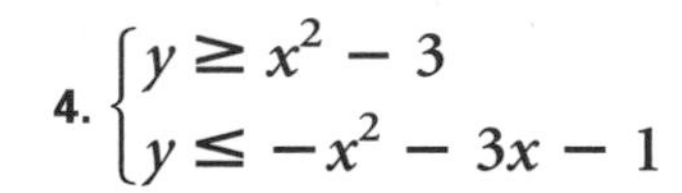

Graph each quadratic inequality. Shade the solution region. Label the intersection points of the two parabolas.

3. $\begin{cases} y \geq x^2 - 2 \\ y \leq -x^2 + 6 \end{cases}$

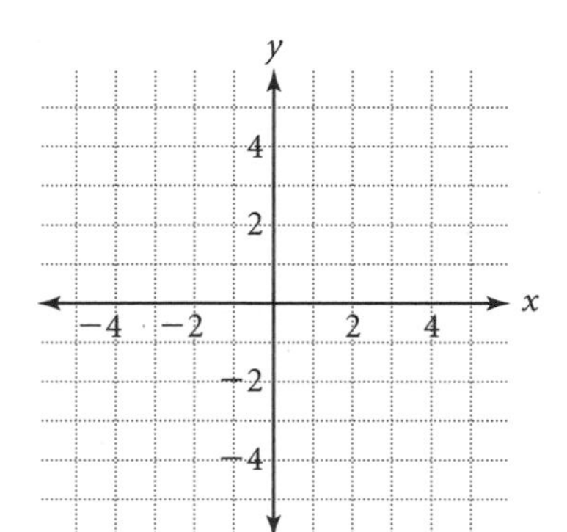

4. $\begin{cases} y \geq x^2 - 3 \\ y \leq -x^2 - 3x - 1 \end{cases}$

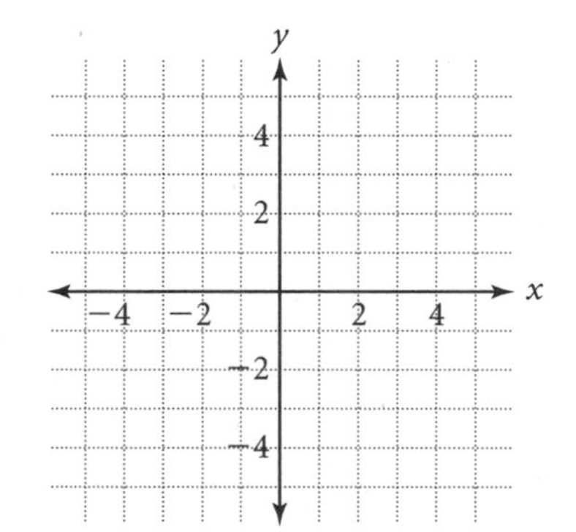

5. A rocket is shot upward into the air. Its height can be described by $h = -16t^2 + 96t$, where h is the rocket's height, in feet, after t seconds. Use a quadratic inequality to determine the following.

   a. During what time interval(s) will the height of the rocket be below 80 feet? ____________

   b. During what time interval(s) will the height of the rocket be above 80 feet? ____________

   c. After how many seconds will the rocket reach its maximum height? ____________

   d. What maximum height will the rocket reach? ____________

NAME ______________________ CLASS ______________ DATE ______________

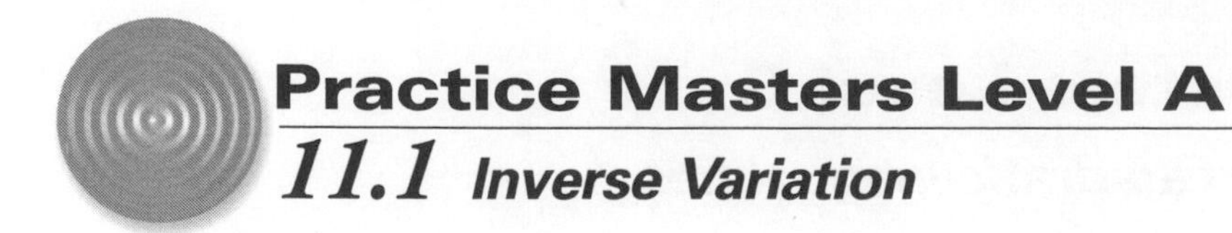

Practice Masters Level A

11.1 Inverse Variation

Determine whether each equation represents an inverse variation.

1. $3 + x = y$ ______
2. $3x = y$ ______
3. $\frac{3}{x} = y$ ______
4. $xy = 5$ ______
5. $a = \frac{1}{b}$ ______
6. $5 - y = x$ ______
7. $\frac{d}{2} = f$ ______
8. $\frac{m}{n} = \frac{3}{2}$ ______
9. $\frac{42}{z} = \frac{a}{8}$ ______
10. $\frac{2}{x} = \frac{y}{5}$ ______
11. $\frac{p}{5} = \frac{q}{3}$ ______
12. $3x = \frac{5}{y}$ ______

For Exercises 13–19, y varies inversely as x. Use $y = \frac{k}{x}$ to solve.

13. If $y = 4$, when x is 3, find x when $y = 12$. ______
14. If $y = 18$, when x is 24, find y when $x = 72$. ______
15. If $y = -6$, when x is -64, find x when $y = -8$. ______
16. If $y = 16$, when x is 12, find y when $x = 12$. ______
17. If $y = 12$, when x is -32, find x when $y = -16$. ______
18. If $y = 1$, when x is 2, find y when $x = 4$. ______
19. If $y = -3$, when x is 8, find x when $y = -2$. ______

For Exercises 20–26, y varies inversely as x. Use $x_1y_1 = x_2y_2$ to solve.

20. If $y = -15$, when x is 3, find x when $y = 125$. ______
21. If $y = 10$, when x is 21, find y when $x = 7$. ______
22. If $y = 4$, when x is -3, find x when $y = -6$. ______
23. If $y = -18$, when x is 5, find y when $x = 30$. ______
24. If $y = 13$, when x is 6, find x when $y = 26$. ______
25. If $y = 14$, when x is 5, find y when $x = 10$. ______
26. If $y = -16$, when x is -3, find x when $y = -12$. ______

NAME ______________________ CLASS ____________ DATE ____________

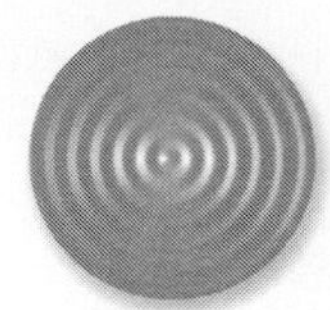

Practice Masters Level B

11.1 Inverse Variation

Determine which equation is *not* an inverse variation. There could be more than one answer.

1. a. $xy = 5$ b. $y = \frac{5}{x}$ c. $x = \frac{5}{y}$ d. $y = 5x$ ____________

2. a. $\frac{2}{x} = \frac{y}{5}$ b. $10 = xy$ c. $2y = 5x$ d. $10x = y$ ____________

Determine the value of the constant, *k*, in each inverse-variation equation.

3. $\frac{42}{z} = \frac{a}{8}$ ____________

4. $\frac{3}{x} = y$ ____________

5. $3x = \frac{5}{y}$ ____________

6. $\frac{1}{y} = x$ ____________

For Exercises 7–12, *y* varies inversely as *x*. Use $y = \frac{k}{x}$ to determine *k*.

7. $y = 2.5$, when x is 8. ____________

8. $y = -1\frac{1}{2}$, when x is $\frac{2}{3}$. ____________

9. $y = \frac{1}{2}$, when x is -10. ____________

10. $y = 3.71$, when x is 2.8. ____________

11. $y = 1.9$, when x is 7. ____________

12. $y = 3\frac{1}{3}$, when x is $\frac{2}{5}$. ____________

For Exercises 13–16, *y* varies inversely as *x*. Use $y = \frac{k}{x}$ or $x_1y_1 = x_2y_2$ to solve.

13. If $y = 8.8$, when x is 13.2, find x when $y = 6.4$. ____________

14. If $y = -3$, when x is -1, find y when $x = 2.5$. ____________

15. If $y = 15$, when x is 3, find x when $y = 13\frac{1}{3}$. ____________

16. If $y = 14$, when x is $\frac{4}{3}$, find y when $x = 14$. ____________

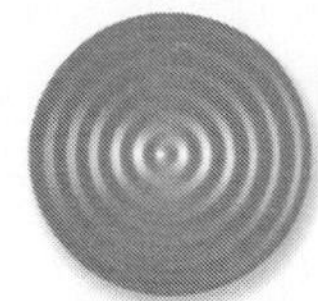

Practice Masters Level C

11.1 Inverse Variation

For Exercises 1–9, *y* varies inversely as *x*. Use $y = \frac{k}{x}$ or $x_1y_1 = x_2y_2$ to solve.

1. If $y = 45$, when x is 5, find x when $y = 9$. ____________
2. If $y = -2$, when x is -6, find y when $x = 1.5$. ____________
3. If $y = 12$, when x is -3, find x when $y = 4$. ____________
4. If $y = 10$, when x is 5, find y when $x = 5$. ____________
5. If $y = 10$, when x is 50, find y when $x = 20$. ____________
6. If $y = 8$, when x is 12, find x when $y = 6$. ____________
7. If $y = 12$, when x is -60, find y when $x = 15$. ____________
8. If $y = 15$, when x is 20, find y when $x = -30$. ____________
9. If $y = 5$, when x is 14, find y when $x = 10$. ____________

The graph of an inverse variation of the form $y = \frac{k}{x}$ or $xy = k$, where $x \neq 0$ and $k \neq 0$ is a hyperbola. The graphs of $xy = 1$, $xy = 2$, and $xy = 3$ are displayed at the right. Use the graph for Exercises 10–12.

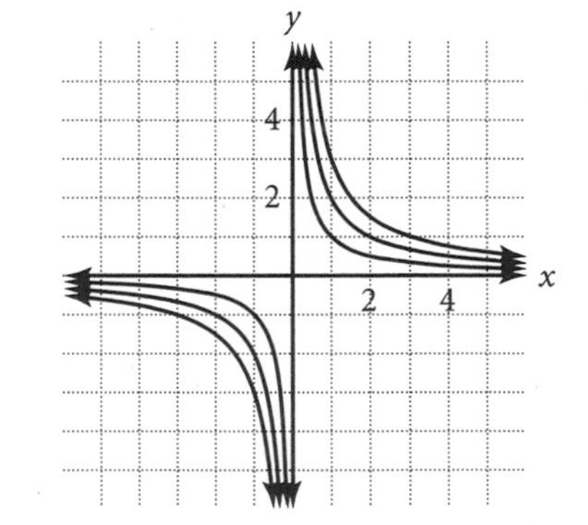

10. In which quadrants is the graph of $xy = k$ located if k is greater than zero? If k is less than zero?

____

11. What are the lines of symmetry of the graph of $xy = k$?

____

12. Determine the point in Quadrant I, on the graph of $xy = 1$, that is closest to the origin. ____________

13. The force, F, needed to pry open a crate, varies inversely as the length, l, of the crowbar used. When the length is 2 meters, the force needs to be 12 Newtons. What force would be needed if the crowbar is 1.6 meters long? ____________

NAME ______________________ CLASS ____________ DATE ____________

Practice Masters Level A

11.2 Rational Expressions and Functions

For what values of the variable is each rational expression undefined?

1. $\dfrac{2x + 8}{x}$ ____________

2. $\dfrac{3m - 4}{9m^2 - 12m}$ ____________

3. $\dfrac{5}{y - 4}$ ____________

4. $\dfrac{8z - 2}{3z + 1}$ ____________

5. $\dfrac{x^2 + 10x + 21}{x^2 - 2x - 15}$ ____________

6. $\dfrac{2m^2 + 4m}{3m + 5}$ ____________

7. $\dfrac{4x^2 + 3x}{3x^2}$ ____________

8. $\dfrac{x^3}{3x^2 + 2x}$ ____________

Evaluate each rational function or expression for $x = -1$ and $x = 2$. Write *undefined* if appropriate.

9. $y = \dfrac{1}{x - 2}$ ____________

10. $y = \dfrac{-2}{x - 2} - 4$ ____________

11. $y = -\dfrac{1}{x} + 2$ ____________

12. $y = \dfrac{-5}{x + 6} - 2$ ____________

13. $y = \dfrac{1}{3x + 2}$ ____________

14. $y = \dfrac{1}{x^2 - 1}$ ____________

15. $y = \dfrac{x^2 + 2}{-x^2 + 1}$ ____________

16. $y = \dfrac{x^2 - 5}{x^2 + 4}$ ____________

Graph each function. List any values of x for which the function is undefined.

17. $y = \dfrac{1}{3(x - 3)}$ ____________

18. $y = \dfrac{-4}{2(x + 2)}$ ____________

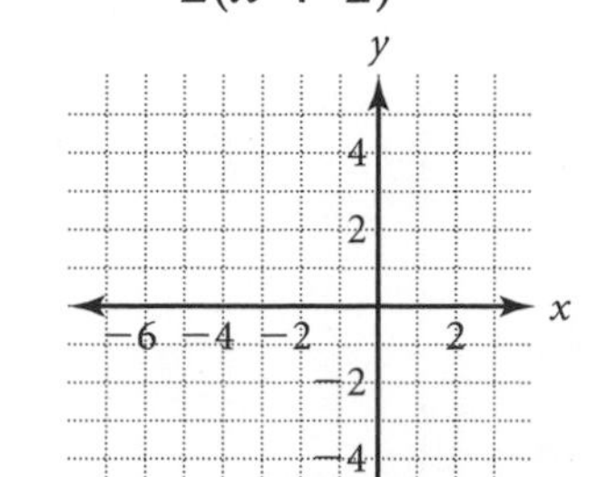

19. $y = \dfrac{3}{x - 2} + 4$ ____________

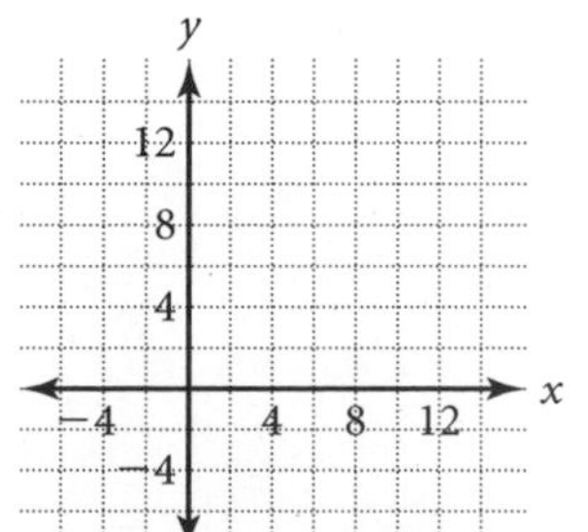

NAME ______________________ CLASS ____________ DATE ____________

Practice Masters Level B

11.2 Rational Expressions and Functions

1. Define in your own words a rational expression. ______________

For what values of the variables is each rational expression undefined?

2. $\dfrac{7}{9mn}$ ______________

3. $\dfrac{y}{5y + 15}$ ______________

4. $\dfrac{z - 5}{(z - 1)(z + 8)}$ ______________

5. $\dfrac{2x + 13}{x^2 + 2x - 48}$ ______________

6. $\dfrac{3}{pqr}$ ______________

7. $\dfrac{6x}{18x - 12}$ ______________

8. $\dfrac{3q + 8}{(q - 4)(q + 6)}$ ______________

9. $\dfrac{3y - 11}{y^2 - 10y - 56}$ ______________

Identify the domain of each function.

10. $y = \dfrac{2x}{x - 5}$ ______________

11. $y = \dfrac{x^2 - 25}{x - 5}$ ______________

12. $y = \dfrac{x}{2x^2 - 3x}$ ______________

13. $y = \dfrac{4}{2x - 1}$ ______________

14. $y = \dfrac{2x}{4x - 6}$ ______________

15. $y = \dfrac{3x^2}{3x^3 - 9x}$ ______________

Graph each function. List any values of *x* for which the function is undefined.

16. $y = \dfrac{1}{4x - 16}$ ______________

17. $y = \dfrac{2 - x}{5x}$ ______________

18. $y = \dfrac{6}{x} - 3$ ______________

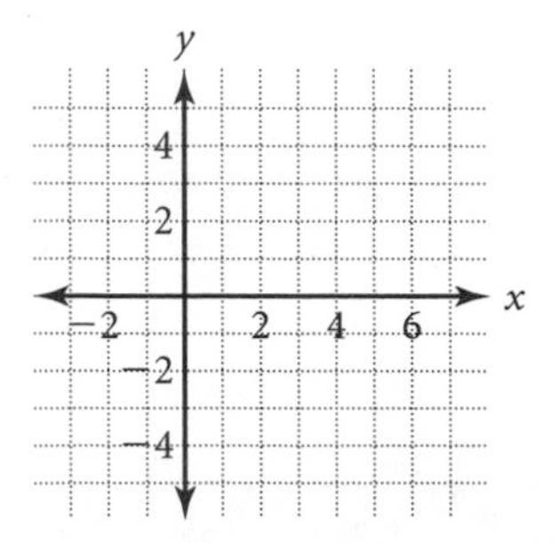

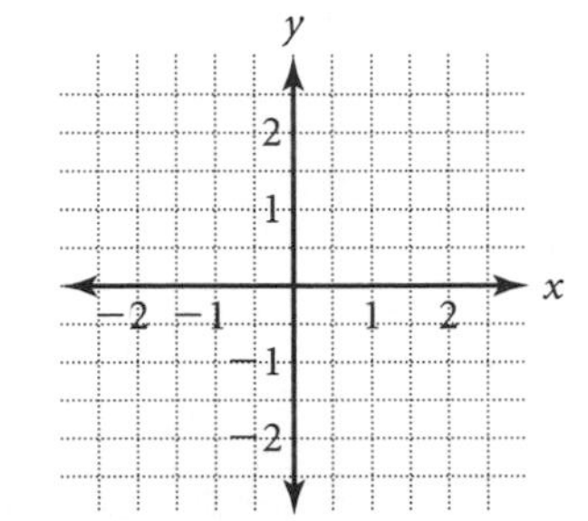

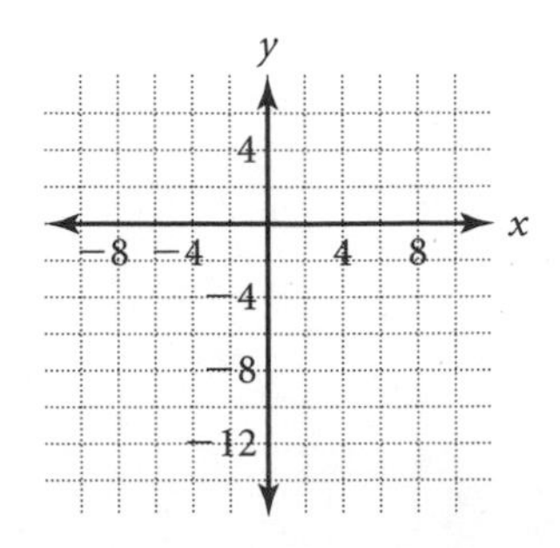

NAME ______________________ CLASS ____________ DATE ____________

Practice Masters Level C

11.2 Rational Expressions and Functions

1. Explain the difference between a rational expression and a rational function.

____

For what values of the variables is each rational expression undefined?

2. $\frac{3}{5xy}$ ____________

3. $\frac{z}{3z - 15}$ ____________

4. $\frac{z + 1}{(z - 2)(z + 6)}$ ____________

5. $\frac{3x - 18}{x^2 - 3x - 18}$ ____________

6. $\frac{-1}{mn}$ ____________

7. $\frac{2k}{4k - 6}$ ____________

Identify the domain of each function.

8. $y = \frac{3x}{x + 6}$ ____________

9. $y = \frac{x + 5}{x^2 - 25}$ ____________

10. $y = \frac{x}{x^2 - 2x}$ ____________

11. $y = \frac{-2x}{x - 1}$ ____________

Describe the transformations applied to the graph of $f(x) = \frac{1}{x}$ to create the graph of each rational function below. List any values for which the function is undefined.

12. $g(x) = \frac{3}{x} + 2$ ____________

13. $k(x) = \frac{1}{x - 3} - 1$ ____________

Graph each function. List any values of x for which the function is undefined.

14. $y = \frac{1}{-x + 3}$ ____________

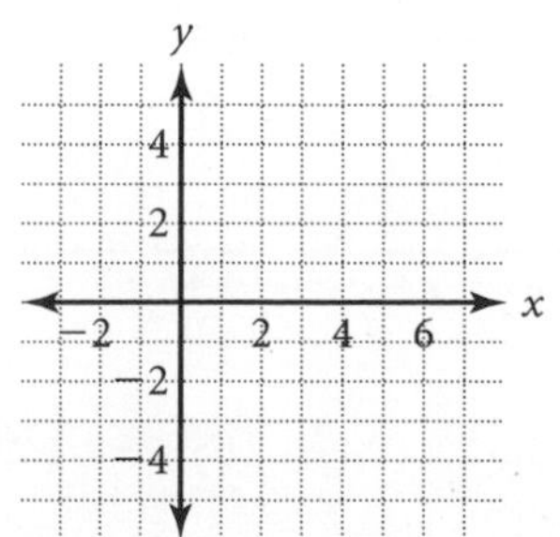

15. $y = \frac{x - 2}{x + 3}$ ____________

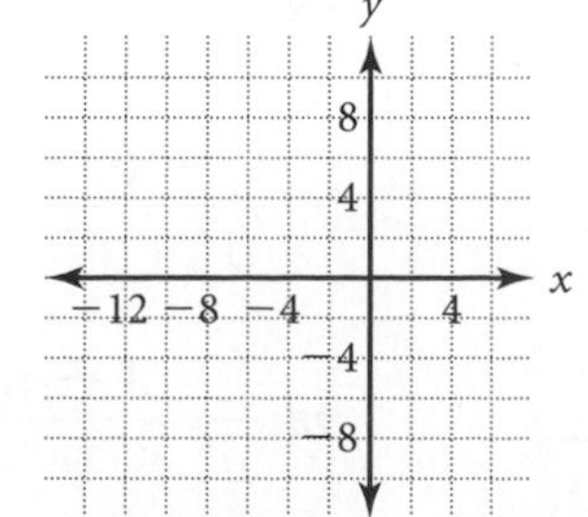

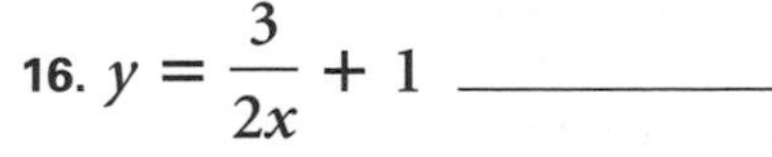

16. $y = \frac{3}{2x} + 1$ ____________

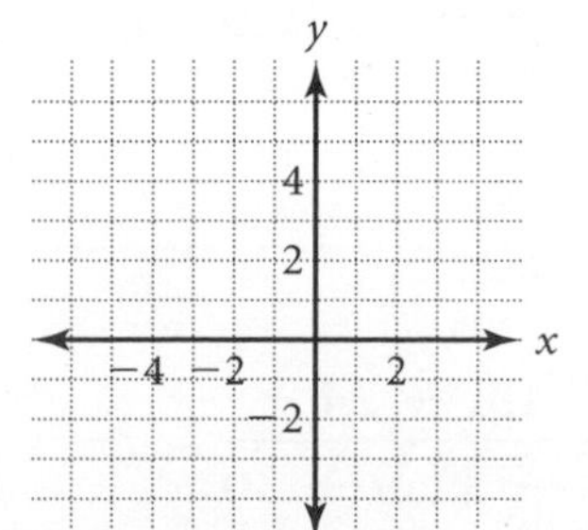

NAME ______________________ CLASS ____________ DATE ____________

Practice Masters Level A

11.3 Simplifying Rational Expressions

Name the common factors of the numerator and denominator.

1. $\dfrac{16}{20}$ ______________________

2. $\dfrac{x + 2}{(x + 3)(x + 2)}$ ______________________

3. $\dfrac{9}{15}$ ______________________

4. $\dfrac{3x + 9}{3x^2 - 27}$ ______________________

5. $\dfrac{4(x + 3)}{8x}$ ______________________

6. $\dfrac{x^2 - 4}{x^2 + 6x + 8}$ ______________________

7. $\dfrac{x + y}{(x + y)(x - y)}$ ______________________

8. $\dfrac{r + 2}{r^2 + 5r + 6}$ ______________________

For what values of the variable is each rational expression undefined?

9. $\dfrac{5x}{x - 2}$ ______________________

10. $\dfrac{r - 5}{r}$ ______________________

11. $\dfrac{6q}{q - 3}$ ______________________

12. $\dfrac{12}{3 - x}$ ______________________

13. $\dfrac{p - 4}{4 - p}$ ______________________

14. $\dfrac{2}{x(x^2 - 3x + 2)}$ ______________________

15. $\dfrac{(m - 1)(m + 1)}{(m + 1)(m + 1)}$ ______________________

16. $\dfrac{(b + 1)(b + 2)}{(b - 1)(b + 2)}$ ______________________

Simplify each expression and state any restrictions on the variables.

17. $\dfrac{15(x - 1)}{36(x - 2)}$ ______________________

18. $\dfrac{14t + 35}{2t + 5}$ ______________________

19. $\dfrac{5m + 15}{5}$ ______________________

20. $\dfrac{16 + 4x}{4x}$ ______________________

21. $\dfrac{c - 2}{c^2 - 4}$ ______________________

22. $\dfrac{-(b + 1)}{b^2 + 6b + 5}$ ______________________

23. $\dfrac{4m - 24}{m^2 - 11m + 30}$ ______________________

24. $\dfrac{3 - n}{n^2 + 2n - 15}$ ______________________

NAME ______________________ CLASS ____________ DATE ____________

Practice Masters Level B

11.3 Simplifying Rational Expressions

Name the common factors of the numerator and denominator.

1. $\frac{r^2 + 4r + 4}{r^2 + 7r + 10}$ ____________
2. $\frac{45x}{60x^2}$ ____________
3. $\frac{3z + 6}{z^2 - 3z - 10}$ ____________
4. $\frac{x^2 - 2x - 3}{x^2 - 5x + 6}$ ____________
5. $\frac{z^2 - 9}{5z - 15}$ ____________
6. $\frac{6x + 18}{3x}$ ____________

For what values of the variable is each rational expression undefined?

7. $\frac{m^2 + 3m + 2}{m^2 + m - 2}$ ____________
8. $\frac{l + 3}{l - 3}$ ____________
9. $\frac{12}{x^2 - 1}$ ____________
10. $\frac{2x}{10 - x}$ ____________
11. $\frac{g^2 - 1}{g^2 + 2g + 1}$ ____________
12. $\frac{x + 5}{5 - x}$ ____________

Simplify each expression and state any restrictions on the variables.

13. $\frac{8x}{4x^2}$ ____________
14. $\frac{10x}{5x^2}$ ____________
15. $\frac{6y + 5}{12}$ ____________
16. $\frac{2 + x}{x^2 - 5x - 14}$ ____________
17. $\frac{4 - 8b}{40b - 20}$ ____________
18. $\frac{a - 9}{9a - 81}$ ____________
19. $\frac{4a - 16}{a - 4}$ ____________
20. $\frac{4 - 3b}{9b - 12}$ ____________

21. The efficiency of a machine is the ratio of the work output to the work input. Both are measured in joules. A machine has a work output of $(4x^2 + 18x + 12)$ joules and a work input of $(2x + 1)$ joules. What is the efficiency of the machine? ____________

NAME ______________________ CLASS ____________ DATE ____________

Practice Masters Level C

11.3 Simplifying Rational Expressions

Name the common factors of the numerator and denominator.

1. $\frac{s - 5}{5 - s}$ ______________

2. $\frac{-15xy}{90x^2y^3}$ ______________

3. $\frac{16 - z^2}{z^2 - 8z + 16}$ ______________

4. $\frac{x^2 - 5x + 6}{x^2 - 6x + 9}$ ______________

For what values of the variable is each rational expression undefined?

5. $\frac{z^2 - 16}{z^2 + 2z - 8}$ ______________

6. $\frac{(k + 3)^2}{k - 3}$ ______________

7. $\frac{(j + 1)(j - 3)}{(2j + 3)(j + 4)}$ ______________

8. $\frac{4m}{2(m - 4)}$ ______________

9. $\frac{h^2 - 4}{h^2 + 2h + 1}$ ______________

10. $\frac{x + 7}{7 - x}$ ______________

Simplify each expression and state any restrictions on the variables.

11. $\frac{2y - 1}{4 - 8y}$ ______________

12. $\frac{5x + 10}{15}$ ______________

13. $\frac{x + 3}{x^2 - 10x - 39}$ ______________

14. $\frac{x^2 + 10x + 25}{x^2 + 11x + 30}$ ______________

15. $\frac{c^2 + c - 56}{c^2 + 16c + 64}$ ______________

16. $\frac{x^4 - 1}{x^2 - 2x + 1}$ ______________

17. $\frac{x^5 + 10x^4 + 25x^3}{x^3 + 4x^2 - 5x}$ ______________

18. $\frac{x^{2n} + 7x^n + 6}{x^{2n} + 9x^n + 8}$ ______________

19. Compare simplifying $\frac{30x^3}{2x}$ to simplifying $\frac{6x^2 - 24x - 30}{4x - 20}$. ______________

20. Write an algebraic expression for which the values $\frac{1}{2}$, $-\frac{1}{4}$, and 0 must be excluded. ______________

21. Given the equation $\frac{(x + y)^p}{(x + y)^q} = x^2 + 2xy + y^2$, write an equation that expresses the relationship between p and q. ______________

NAME ________________________ CLASS ____________ DATE ____________

Practice Masters Level A

11.4 Operations with Rational Expressions

Perform the indicated operations. Simplify, and state the restrictions on the variable.

1. $\frac{3}{x} + \frac{5}{x}$ ____________

2. $\frac{7}{z} - \frac{5}{z}$ ____________

3. $\frac{1}{x} \cdot \frac{5}{2}$ ____________

4. $\frac{4}{m} \div \frac{2}{m^2}$ ____________

5. $\frac{15}{2x} + \frac{3}{2x}$ ____________

6. $\frac{17}{3m} - \frac{5}{3m}$ ____________

7. $\frac{4}{2 + n} \cdot \frac{2 + n}{3 + n}$ ____________

8. $\frac{x - 2}{4} \div \frac{x^2 - 4}{3}$ ____________

9. $\frac{1}{y} + \frac{1}{2y}$ ____________

10. $\frac{3}{4n} - \frac{1}{n}$ ____________

11. $\frac{n + 4}{n + 3} \cdot \frac{n - 2}{n + 2}$ ____________

12. $\frac{x + 4}{2x - 8} \div \frac{x}{x - 4}$ ____________

13. $\frac{5}{x + 2} + \frac{10}{x + 2}$ ____________

14. $\frac{11}{p + 3} - \frac{6}{p + 3}$ ____________

15. $\frac{3p}{p^2 + 2p - 3} \cdot \frac{p + 3}{p - 1}$ ____________

16. $\frac{6}{g^2} \div \frac{5}{g}$ ____________

17. $\frac{10}{-2 + w} + \frac{12}{w - 2}$ ____________

18. $\frac{12}{q - 2} - \frac{5}{-2 + q}$ ____________

19. $\frac{y + 2}{y(y + 1)} \cdot \frac{y^3}{(y + 2)(y + 4)}$ ____________

In physics, then term work represents a force multiplied by the distance a body moves in the direction of the force applied. That is, *W* = *Fd*, where *W* represents work in joules, *F* represents force in Newtons, and *d* is the distance through which the force is applied. (1 joule = 1 Newton · 1 meter)

20. If a body moves 5 meters in the direction of a force of 12 Newtons, determine the work. ____________

21. If a body moves $\frac{x^2 + 13x + 42}{x + 6}$ meters in the direction of a force of $\frac{x + 4}{x^2 - 16}$ Newtons, determine the work. ____________

NAME ______________________ CLASS ____________ DATE ____________

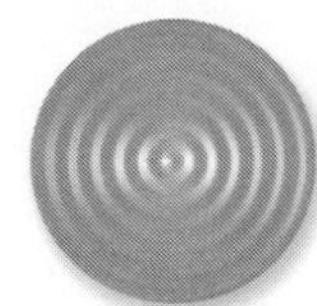

Practice Masters Level B

11.4 Operations with Rational Expressions

Perform the indicated operations. Simplify, and state the restrictions on the variables.

1. $\frac{3}{a+6} + \frac{2}{a-5}$ ______________

2. $\frac{t}{2(t-3)} - \frac{2}{3(t+3)}$ ______________

3. $\frac{m-5}{2(m+6)} \cdot \frac{4(m+6)}{8(m-5)}$ ______________

4. $\frac{5c^2d}{9ce^2} \div \frac{15c^2d^2}{18e}$ ______________

5. $\frac{5}{a+7} + \frac{4}{a-4}$ ______________

6. $\frac{r}{3(r+3)} - \frac{3}{4(r+3)}$ ______________

7. $\frac{7x+35}{3x^2-108} \cdot 6x+36$ ______________

8. $\frac{x^2-49}{x^2-x-42} \div 3x+21$ ______________

9. $\frac{4y+16}{y^2-16} + \frac{-4}{y+4}$ ______________

10. $\frac{11}{121-v^2} - \frac{v^2}{v+11}$ ______________

11. $\frac{a^2b}{c} \cdot \frac{2ac^3}{a}$ ______________

12. $\frac{8qr^2}{12qr} \div \frac{9qrs}{6s}$ ______________

13. $\frac{8x+32}{4x^2-100} \cdot 3x-15$ ______________

14. $\frac{10v-1}{100-v^2} - \frac{5v}{v+10}$ ______________

15. $\frac{wv^2}{v^2w} \cdot \frac{v^2w}{3v}$ ______________

16. $\frac{1}{5k} \div \frac{3}{20k^2}$ ______________

Before making a loan to a company, a bank conducts a ratio analysis to determine how the company is doing in the marketplace.

17. The current ratio, the ratio of current assets to current liabilities, is one measure of the liquidity of the company. If the current assets can be represented by the expression $\frac{2x^2+4x-30}{2x^2-18}$ and the liabilities by the expression $\frac{3x+15}{4x+12}$, determine the current ratio. ______________

18. Leverage is the firm's debt in relation to its equity. If the debt can be represented by the expression $\frac{2x^2-2x-4}{4x-8}$, and the equity by the expression $\frac{x^2-1}{4x-4}$, determine the debt to equity ratio. ______________

NAME ______________________ CLASS ____________ DATE ____________

Practice Masters Level C

11.4 Operations with Rational Expressions

Perform the indicated operations. Simplify, and state the restrictions on the variables.

1. $\frac{7}{6t^2} + \frac{12}{12t} - \frac{3}{t}$ ____________

2. $\frac{3}{8r^2} + \frac{8}{16r} - \frac{2}{r}$ ____________

3. $\frac{7}{g+2} - \frac{g+6}{g+2}$ ____________

4. $\frac{t^2 - t}{t^2 - 2t - 3} \cdot \frac{t^2 + 2t + 1}{t^2 + 4t}$ ____________

5. $\frac{pq^2}{r} \cdot \frac{3p^2r}{p}$ ____________

6. $\frac{4}{3y} \div \frac{2}{9y^2}$ ____________

7. $\frac{8}{3x} - \frac{2}{9x}$ ____________

8. $\frac{6x + 24}{2x^2 - 8} \cdot (3x + 6)$ ____________

9. $\frac{x^2 - 36}{x^2 + 4x - 12} \div (3x - 18)$ ____________

10. $\frac{5}{3z^2} - \frac{2}{z} + \frac{1}{6z}$ ____________

11. $\frac{t}{2t+8} - \frac{5}{3t+12}$ ____________

12. $\frac{n+7}{2n-6} \div \frac{10(n+7)}{5n-15}$ ____________

13. $\frac{5x^2 + 5x}{x^2 - 3x - 4} \cdot (3x - 12)$ ____________

14. $\frac{3}{x-5} \div \frac{1}{x^2 - 25}$ ____________

15. $\left(\frac{2}{x+2} - \frac{1}{x+5}\right)\left(\frac{2x+10}{x^2+8x}\right)$ ____________

16. $\frac{(2x+3)(x-5)}{4x} \div \frac{3(x^2 - 25)}{6x + 30}$ ____________

Write an expression for the perimeter of a triangle with sides of the given lengths.

17. $\frac{1}{x}, \frac{2}{x+1}, \frac{3}{x}$ ____________

18. $x, \frac{4x}{1-x}, \frac{x}{6x-1}$ ____________

19. $\frac{2}{x}, \frac{3}{x+1}, \frac{x}{3x-1}$ ____________

20. Give an example of the addition or subtraction of two rational expressions in which using the LCD produces an answer that is in simplest form, and an example that is not in simplest form.

____

____

NAME ______________________ CLASS __________ DATE __________

Practice Masters Level A

11.5 Solving Rational Equations

Solve each rational equation by using the lowest common denominator.

1. $\frac{16}{x} + 3 = 7$ ______________

2. $\frac{3x}{8} = \frac{2x - 3}{5}$ ______________

3. $\frac{4}{y} + 10 = 18$ ______________

4. $\frac{9}{32} = \frac{27}{x}$ ______________

5. $12 - \frac{48}{x} = 28$ ______________

6. $\frac{12}{x} - 3 = -7$ ______________

7. $\frac{5}{2a - 3} = 1$ ______________

8. $\frac{-5}{x} - 2 = -7$ ______________

9. $6 + \frac{16}{n} = 10$ ______________

10. $2 = \frac{9}{5 - 2x}$ ______________

11. $4 + \frac{10}{x} = 14$ ______________

12. $\frac{-6}{y} = \frac{2}{3}$ ______________

13. $1 - \frac{8}{x} = 3$ ______________

14. $6 + \frac{2}{x} = 5$ ______________

15. $3 - \frac{10}{x} = 8$ ______________

16. $\frac{1}{x} + \frac{3}{5} = 2$ ______________

17. $\frac{3}{2} = \frac{12}{x}$ ______________

18. $\frac{1}{2} + \frac{4}{x} = \frac{3}{8}$ ______________

19. $\frac{5}{x} + \frac{x - 1}{x} = 3$ ______________

20. $\frac{2}{3} = \frac{6}{x}$ ______________

Solve each rational equation by graphing.

21. $\frac{3}{x} + \frac{1}{2} = \frac{2}{x}$

____

22. $\frac{1}{3} + \frac{1}{x} = \frac{1}{3x}$

____

NAME ______________________ CLASS ____________ DATE ____________

Practice Masters Level B

11.5 Solving Rational Equations

Solve each rational equation by using the lowest common denominator.

1. $\frac{x}{x-2} = \frac{3}{x-2} + 5$ ____________

2. $\frac{d}{d+3} - \frac{4}{d-3} = \frac{d-27}{d^2-9}$ ____________

3. $\frac{2}{x-1} = 5 - \frac{1}{x-1}$ ____________

4. $\frac{k}{k-1} = 1 + \frac{k+2}{k+1}$ ____________

5. $\frac{j}{j+6} = 3 - \frac{6}{j+6}$ ____________

6. $\frac{x+5}{x^2+4x-5} = \frac{2x+8}{x^2+3x-4}$ ____________

7. $\frac{k}{k+2} - \frac{5}{k+2} = 8$ ____________

8. $\frac{4}{5q} - \frac{2}{7q} = \frac{1}{35}$ ____________

9. $\frac{m}{m-5} - \frac{13}{m-5} = 7$ ____________

10. $\frac{x-3}{x} + \frac{3}{4x} = 16$ ____________

11. $4 - \frac{6}{n} = 5$ ____________

12. $\frac{3d-8}{d-5} = \frac{d+6}{d-5} + 2$ ____________

Solve each rational equation by graphing.

13. $\frac{4}{x-3} - \frac{9}{x^2-3x} = \frac{1}{x}$

____

14. $\frac{x-4}{x-5} = \frac{x-6}{x-5}$

____

15. $\frac{4}{7x+x^2} = \frac{10}{x^2+8x}$

____

16. $\frac{10x-5}{x} = \frac{15}{x}$

____

17. $\frac{1}{x-1} + \frac{5}{(x+3)(x-1)} = \frac{2}{x+3}$

____

18. $\frac{1}{x+3} + 2 = \frac{-15}{x^2+x-6}$

____

NAME ______________________ CLASS ____________ DATE ____________

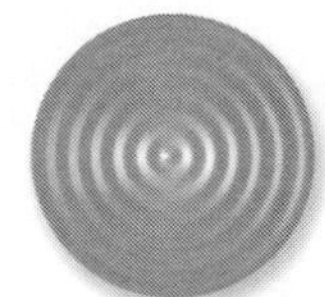

Practice Masters Level C

11.5 Solving Rational Equations

Solve each rational equation by using the lowest common denominator.

1. $\frac{4d - 6}{d - 11} = \frac{d + 8}{d - 11} + 2$ ____________

2. $\left(\frac{x - 3}{x + 6}\right)^2 \cdot \left(\frac{x + 6}{2x - 6}\right)^3 = 1$ ____________

3. $\left(\frac{x - 2}{x + 4}\right)^2 \div \left(\frac{3x - 6}{x + 4}\right)^3 = 1$ ____________

4. $2b + \frac{4}{5} = \frac{2}{3}$ ____________

5. $\frac{y}{y - 2} - \frac{14}{y - 2} = 5$ ____________

6. $\frac{k + 5}{k} - \frac{2}{5k} = 10$ ____________

Solve each rational equation by graphing.

7. $\frac{1}{x + 2} + \frac{2}{x^2 + 2x} = 1$ ____________

8. $\frac{x + 1}{x + 2} = \frac{x + 4}{x + 3}$ ____________

9. $\frac{(x - 1)}{(x - 1)^2} = \frac{3}{(x + 2)}$ ____________

Solve.

10. Luis' checking account contains $1500 more than Sergio's. If Sergio makes a deposit of $450, the ratio of the amount in Luis's account to the amount in Sergio's account will be 46:25. Determine the amount of money in each account. ____________

11. A jet can fly 550 miles per hour in calm air. With a tailwind, it can fly 2400 miles in the same time it can fly 2000 miles against the wind. How long does it take to travel 2400 miles with a tailwind? ____________

12. A motorboat can travel 12 miles downstream in $\frac{3}{4}$ of the time it takes to travel the same distance upstream. Determine the rate of the motorboat, in still water, if the rate of the current is 4 miles per hour. ____________

NAME ______________________ CLASS ______________ DATE ______________

Practice Masters Level A

11.6 Proof in Algebra

Complete each statement with the word or phrase that makes the sentence true.

1. In a 2-column proof, for each statement, there must be a ______________.

2. In the conditional statement, "*If* $a = b$, *then* $a + c = b + c$," the phrase, "*If* $a = b$" is called the ______________.

3. In the conditional statement, "*If* $a = b$, *then* $a + c = b + c$," the phrase, "*then* $a + c = b + c$" is called the ______________.

4. To prove a statement is false, you need to find a ______________.

5. The ______________ is the statement where the hypothesis and the conclusion of the original statement are interchanged.

6. The process of reasoning that a given principle is true because the special cases that you have seen are true is called ______________.

State a reason for each step in the proof of the following conjecture:

If $4m - 3 = 0$, then $m = \frac{3}{4}$.

Statement	Reason
$4m - 3 = 0$	Given or Hypothesis
$4m - 3 + 3 = 0 + 3$	7.
$4m + 0 = 0 + 3$	8.
$4m = 3$	9.
$\frac{4m}{4} = \frac{3}{4}$	10.
$1 \cdot m = \frac{3}{4}$	11.
$m = \frac{3}{4}$	12.

If $2n - 1 = 3$, then $n = 2$.

Statement	Reason
$2n - 1 = 3$	Given or Hypothesis
$2n - 1 + 1 = 3 + 1$	13.
$2n + 0 = 3 + 1$	14.
$2n = 4$	15.
$\frac{2n}{2} = \frac{4}{2}$	16.
$1 \cdot n = \frac{4}{2}$	17.
$n = 2$	18.

NAME ______________________ CLASS __________ DATE __________

Practice Masters Level B

11.6 Proof in Algebra

Match each definition with the word or phrase that makes the sentence true.

_____ 1. converse

_____ 2. hypothesis

_____ 3. conclusion

_____ 4. conditional statement

_____ 5. theorem

_____ 6. inductive reasoning

_____ 7. deductive reasoning

_____ 8. proof

a. the process of reasoning that a general principle is true because the special cases that you have seen are true

b. when you interchange the hypothesis and the conclusion of a conditional statement

c. a series of statements in a logical sequence which demonstrate that the conclusion is true whenever the hypothesis is true

d. the p portion of a conditional statement

e. the process of concluding that a special case of a proven general principle is true

f. "if p, then q" statements

g. the q portion of a conditional statement

h. the process of concluding that a special case of a proven general principle is true

Complete the 2-column proof for each conjecture. Let all variables represent real numbers.

9. If $-4x + 3 = 19$, then $x = -4$.

Statement	Reason

10. If $m = (cx - b) + cy$, then $m = c(x + y) - b$

Statement	Reason

NAME ______________________ CLASS ____________ DATE ____________

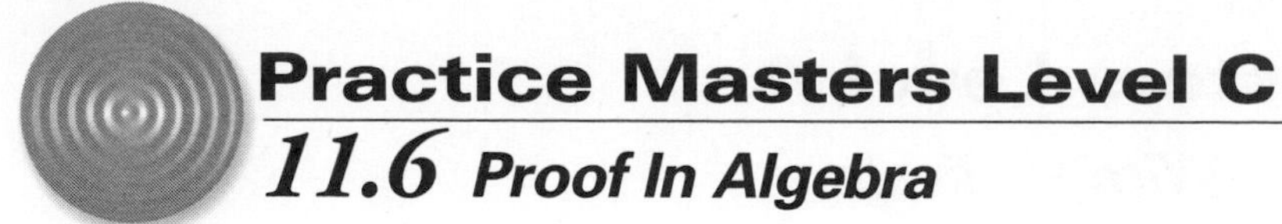

Practice Masters Level C

11.6 Proof In Algebra

Explain in your words the definition of each of the following:

1. converse ______________________

2. hypothesis ______________________

3. conclusion ______________________

4. conditional statement ______________________

5. theorem ______________________

6. inductive reasoning ______________________

7. deductive reasoning ______________________

Write a paragraph proof for each conjecture. Let all variables represent real numbers.

8. The set of integers is not closed under division.

9. The sum of two even integers is an even integer.

NAME ______________________ CLASS ____________ DATE ____________

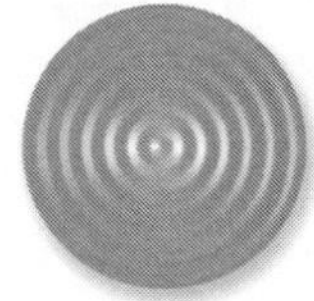

Practice Masters Level A

12.1 Operations With Radicals

Evaluate each square root.

1. $\sqrt{81}$ ______________
2. $-\sqrt{49}$ ______________
3. $\pm\sqrt{256}$ ______________
4. $\sqrt{900}$ ______________

Simplify.

5. $9\sqrt{5} + 2\sqrt{5}$ ______________
6. $3\sqrt{2} - 8\sqrt{2}$ ______________
7. $-4\sqrt{3} - 5\sqrt{3}$ ______________
8. $-5\sqrt{11} + 7\sqrt{11}$ ______________

Express in simplest radical form.

9. $\sqrt{8}$ ______________
10. $\sqrt{32}$ ______________
11. $\sqrt{75}$ ______________
12. $\sqrt{72}$ ______________

Simplify.

13. $(\sqrt{7})^2$ ______________
14. $(\sqrt{3})^2$ ______________
15. $(\sqrt{11})^2$ ______________
16. $(\sqrt{9})^2$ ______________

Simplify each radical by factoring.

17. $\sqrt{2}\sqrt{8}$ ______________
18. $\sqrt{3}\sqrt{27}$ ______________
19. $\sqrt{2}\sqrt{18}$ ______________
20. $\sqrt{6}\sqrt{24}$ ______________

Express in simplest radical form.

21. $\sqrt{2}\sqrt{6}$ ______________
22. $\sqrt{3}\sqrt{8}$ ______________
23. $\sqrt{5}\sqrt{15}$ ______________
24. $\sqrt{7}\sqrt{14}$ ______________
25. $\sqrt{\frac{72}{2}}$ ______________
26. $\frac{\sqrt{200}}{\sqrt{8}}$ ______________
27. $\frac{\sqrt{112}}{\sqrt{7}}$ ______________
28. $\sqrt{\frac{405}{5}}$ ______________

NAME ______________________ CLASS ______________ DATE ______________

Practice Masters Level B

12.1 Operations With Radicals

Find each square root. If the square root is irrational, approximate the value to the nearest hundredth.

1. $\sqrt{64}$ ______________
2. $-\sqrt{10}$ ______________
3. $\pm\sqrt{24}$ ______________
4. $\sqrt{225}$ ______________
5. $-\sqrt{0.123}$ ______________
6. $\sqrt{0.999}$ ______________

Simplify each of the following. Assume that all variables are nonnegative and that all denominators are nonzero.

7. $\sqrt{t^{12}v^8}$ ______________
8. $\sqrt{f^5g^7}$ ______________
9. $\sqrt{\dfrac{z^8}{a^6}}$ ______________
10. $\sqrt{\dfrac{x^5}{y^{16}}}$ ______________
11. $\sqrt{8a^5b^7}$ ______________
12. $\sqrt{16p^{12}f^{16}}$ ______________

Simplify.

13. $7\sqrt{7} + 8\sqrt{7} - 11\sqrt{7}$ ______________
14. $6\sqrt{4} - 3\sqrt{4} + 5$ ______________
15. $-3 - 5\sqrt{3} + 7$ ______________
16. $-19 - 6\sqrt{11} + 9 - \sqrt{11}$ ______________
17. $\sqrt{3} + \sqrt{2} - 9\sqrt{3}$ ______________
18. $\sqrt{3} - 2\sqrt{5} + \sqrt{7}$ ______________

If possible, perform the indicated operation, and simplify your answer.

19. $2(\sqrt{3} - 7)$ ______________
20. $-8(3 + \sqrt{5})$ ______________
21. $\sqrt{2}(\sqrt{5} - 14)$ ______________
22. $\sqrt{3}(\sqrt{3} - 5\sqrt{27})$ ______________
23. $4\sqrt{7}(2\sqrt{7} - 1)$ ______________
24. $-6\sqrt{2}(2\sqrt{2} - 3\sqrt{3})$ ______________
25. $(\sqrt{5} - 7)(\sqrt{5} + 3)$ ______________
26. $(4 + \sqrt{3})(2 - \sqrt{3})$ ______________
27. $(7 - \sqrt{2})(7 + \sqrt{2})$ ______________
28. $(\sqrt{5} - 4)^2$ ______________
29. $(-3 + \sqrt{7})^2$ ______________
30. $(\sqrt{2} + 3)(\sqrt{2} - 3)$ ______________

NAME ____________ CLASS ________ DATE ________

Practice Masters Level C

12.1 Operations With Radicals

Find each square root. If the square root is irrational, approximate the value to the nearest hundredth.

1. $\sqrt{200}$ ____________
2. $\pm\sqrt{99}$ ____________
3. $-\sqrt{0.169}$ ____________
4. $-\sqrt{\frac{25}{4}}$ ____________

If possible, perform the indicated operations, and simplify your answer.

5. $3\sqrt{28} + \sqrt{63}$ ____________
6. $3\sqrt{75} - 2\sqrt{18}$ ____________
7. $\dfrac{\sqrt{20} + \sqrt{30}}{\sqrt{10}}$ ____________
8. $\dfrac{\sqrt{12} - \sqrt{48}}{\sqrt{3}}$ ____________
9. $\dfrac{\sqrt{18} - \sqrt{72}}{3}$ ____________
10. $\dfrac{\sqrt{16} + \sqrt{25}}{\sqrt{9} - \sqrt{4}}$ ____________

Simplify.

11. $(4 + \sqrt{3})(2 + \sqrt{3})$ ____________
12. $(\sqrt{15} - 7)(2 - \sqrt{15})$ ____________
13. $(-3 + \sqrt{7})(-8 - \sqrt{2})$ ____________
14. $(6\sqrt{6} + \sqrt{10})(\sqrt{3} - 5\sqrt{2})$ ____________
15. $(2\sqrt{3} - 5)^2$ ____________
16. $-5(6 - \sqrt{7})^2$ ____________
17. $\sqrt{2}(\sqrt{2} - 4)^2$ ____________
18. $-7\sqrt{3}(-3 + \sqrt{5})^2$ ____________

Determine the side length of a square with the given area.

19. 225 square feet ____________
20. 125 square meters ____________

Decide whether each statement is true or false. Assume that $a > 0$, $b > 0$, and $c > 0$.

21. $\sqrt{a - b} = \sqrt{a} - \sqrt{b}$ ____________
22. $\sqrt{abc} = \sqrt{a}\sqrt{b}\sqrt{c}$ ____________
23. $\sqrt{\dfrac{ab}{c}} = \dfrac{\sqrt{a}\sqrt{b}}{\sqrt{c}}$ ____________
24. $\sqrt{ab + ac} = a\sqrt{b + c}$ ____________

NAME ______________________ CLASS ____________ DATE ____________

Practice Masters Level A

12.2 Square-Root Functions and Radical Equations

Solve each equation.

1. $\sqrt{x + 7} = 3$ ____________
2. $\sqrt{9 - x} = 8$ ____________
3. $\sqrt{x - 5} = 4$ ____________
4. $\sqrt{x + 15} = 5$ ____________
5. $\sqrt{24 + x} = 6$ ____________
6. $\sqrt{-x + 9} = 7$ ____________

Use graphing technology to solve each equation.

7. $\sqrt{4x} = 8$ ____________
8. $\sqrt{27x} = 9$ ____________
9. $\sqrt{2x} = 2$ ____________
10. $\sqrt{12x} = 6$ ____________

Solve each equation algebraically. Be sure to check your solution.

11. $\sqrt{2x + 7} = 5$ ____________
12. $\sqrt{7x - 4} = 1$ ____________
13. $\sqrt{11 - 9x} = 2$ ____________
14. $\sqrt{15 + 6x} = 3$ ____________
15. $\sqrt{7x + 4} = 6$ ____________
16. $\sqrt{9x - 1} = 7$ ____________

Solve each equation for *x*. Give answers in simplified form.

17. $x^2 = 40$ ____________
18. $x^2 = 72$ ____________
19. $x^2 = 75$ ____________
20. $x^2 = 63$ ____________
21. $x^2 = 500$ ____________
22. $x^2 = 176$ ____________
23. $x^2 = 1250$ ____________
24. $x^2 = 1053$ ____________
25. $9x^2 = 11$ ____________
26. $4x^2 = 7$ ____________
27. $2x^2 = 32$ ____________
28. $5x^2 = 45$ ____________
29. $4x^2 = 64$ ____________
30. $25x^2 = 21$ ____________
31. In which quadrant can you find the graph of the function $y = \sqrt{x}$? ____________

NAME ______________________ CLASS ____________ DATE ____________

Practice Masters Level B

12.2 Square-Root Functions and Radical Equations

Use graphing technology to solve each equation.

1. $\sqrt{x + 7} = x + 1$ ____________
2. $\sqrt{9 - x} = x - 3$ ____________
3. $\sqrt{2x + 1} = x - 1$ ____________
4. $\sqrt{12 + 3x} = x + 4$ ____________

Solve each equation, if possible. If there is no solution, write *no solution.*

5. $\sqrt{-1 + 2x} = 2 - x$ ____________
6. $\sqrt{3 - x} = 3 + x$ ____________
7. $\sqrt{12x + 37} = 4 + x$ ____________
8. $\sqrt{-6x - 1} = 3x - 2$ ____________
9. $\sqrt{x} = x - 2$ ____________
10. $\sqrt{3x} = 6 - x$ ____________
11. $\sqrt{4x} = x + 1$ ____________
12. $\sqrt{2x} = 4 - x$ ____________
13. $\sqrt{x^2 - 4x + 5} = x$ ____________
14. $\sqrt{x^2 + 3x + 1} = x$ ____________
15. $\sqrt{x^2 + 5x - 2} = x$ ____________
16. $\sqrt{x^2 - 3x - 3} = x$ ____________
17. $\sqrt{x^2 + 4x + 12} = x + 4$ ____________
18. $\sqrt{x^2 - 2x + 1} = x - 7$ ____________
19. $\sqrt{x^2 + 5x - 2} = 3 - x$ ____________
20. $\sqrt{x^2 + 6x + 7} = 2 + x$ ____________

Solve each equation for *x*. Give answers in simplified form.

21. $2x^2 - 128 = 0$ ____________
22. $7x^2 - 84 = 0$ ____________
23. $81x^2 - 18 = 0$ ____________
24. $16x^2 - 25 = 0$ ____________
25. $x^2 - 4x + 4 = 0$ ____________
26. $x^2 + 10x + 25 = 0$ ____________
27. $x^2 - 50x + 625 = 0$ ____________
28. $x^2 + 8x + 16 = 0$ ____________
29. $4x^2 - 12x + 9 = 0$ ____________
30. $9x^2 + 30x + 25 = 0$ ____________

31. Use graphing technology to graph the functions $y = \sqrt{7 - x}$ and $y = -\sqrt{7 - x}$ on the same screen. What is the domain and the range of the combined functions? ____________

NAME ______________________ CLASS __________ DATE __________

Practice Masters Level C

12.2 Square-Root Functions and Radical Equations

Solve each equation, if possible. If not possible, explain why.

1. $x^2 = 4$ ______________________
2. $|x| = -4$ ______________________
3. $\sqrt{x} = 4$ ______________________
4. $x^2 = -4$ ______________________
5. $|x| = 4$ ______________________
6. $\sqrt{x} = -4$ ______________________

Solve each equation algebraically. Be sure to check your solution(s).

7. $\sqrt{3x + 6} = x + 2$ ______________________
8. $\sqrt{5x - 6} = -x$ ______________________
9. $\sqrt{x^2 + x - 7} = x$ ______________________
10. $\sqrt{x^2 - 4x + 9} = x - 7$ ______________________

Use graphing technology to solve each equation.

11. $\sqrt{3x + 12} = x + 4$ ______________________
12. $\sqrt{6x + 7} = -x$ ______________________
13. $\sqrt{4x^2 - 8x - 6} = 2x - 5$ ______________________
14. $\sqrt{9x^2 + 6x + 3} = 2 + 3x$ ______________________

The motion of a pendulum can be modeled by $t = 2\pi\sqrt{\frac{\ell}{980}}$, where l is the length of the pendulum in centimeters, and t is the number of seconds required for one complete swing.

15. Determine the time in seconds that it takes a 10-centimeter pendulum to make one complete swing. ______________________

16. Determine the length in centimeters of a pendulum that takes 2 seconds to make one complete swing. ______________________

17. Determine the length in kilometers of a pendulum that takes 7 minutes to make one complete swing. ______________________

The height of a falling ball can be modeled by $v = -16t^2 + 48t + 65$, where v is the height above the ground in feet, and t is the time in seconds that the ball has traveled.

18. Determine the height in feet above the ground that this ball will be after traveling 3 seconds. ______________________

19. Determine the time in seconds that it will take this ball to get to a position that is 1 foot above the ground. ______________________

NAME ________________ CLASS ________ DATE ________

Practice Masters Level A

12.3 The Pythagorean Theorem

Decide whether each set of numbers can represent the side lengths of a right triangle.

1. 7, 12, 16 ________
2. 3, 4, 5 ________
3. 15, 36, 39 ________
4. 32, 60, 80 ________
5. 12, 15, 19 ________
6. 1, 1, 2 ________
7. 9, 12, 15 ________
8. 5, 6, 7 ________
9. 10, 24, 26 ________
10. 15, 20, 25 ________

For Exercises 11–16, the tables give the lengths of various sides of right triangles. Complete the tables. Use a calculator and round each answer to the nearest tenth.

	Leg	Leg	Hypotenuse
11.	9	12	____
13.	16	___	34
15.	10	50	____

	Leg	Leg	Hypotenuse
12.	6	8	____
14.	___	11	12
16.	10	___	26

Solve.

17. What is the length, in simplified form, of the hypotenuse of a right triangle whose legs are 8 and 12 units long? ________

18. What is the length, in simplified form, of the hypotenuse of a right triangle whose legs are 21 and 28 units long? ________

19. What is the length, in simplified form, of the hypotenuse of a right triangle whose legs are 14 and 21 units long? ________

20. A baseball diamond is a square with sides of 90 feet. To the nearest foot, how far does a catcher at home plate need to throw the ball to second base? ________

21. A football field is a rectangle that is 100 feet by 50 feet. To the nearest foot, what is the diagonal length of the field? ________

22. A local youth group is constructing a right triangular pennant. If the triangular shape of the pennant has legs of 12 inches and 18 inches, how long is the hypotenuse to the nearest inch? ________

NAME ______________________ CLASS __________ DATE __________

Practice Masters Level B

12.3 The Pythagorean Theorem

Decide whether each set of numbers can represent the side lengths of a right triangle.

1. 30, 40, 50 ______________
2. 30, 60, 90 ______________
3. 24, 45, 51 ______________
4. 27, 36, 45 ______________
5. 19, 150, 151 ______________
6. 26, 168, 170 ______________
7. 51, 1600, 1601 ______________
8. 41, 850, 851 ______________
9. 31, 480, 481 ______________
10. 41, 840, 841 ______________

For Exercises 11–16, the tables give the lengths of various sides of right triangles. Copy and complete the tables. Use a calculator and round each answer to the nearest tenth.

	Leg	Leg	Hypotenuse
11.	___	35	48
13.	27	___	50
15.	45	47	______

	Leg	Leg	Hypotenuse
12.	11	60	______
14.	25	___	100
16.	___	56	57

Solve.

17. What is the length, in simplified form, of the hypotenuse of a right triangle whose legs are 33 and 44 units long? ______________

18. What is the length, in simplified form, of the hypotenuse of a right triangle whose legs are 27 and 31 units long? ______________

19. What is the length, in simplified form, of the hypotenuse of a right triangle whose legs are 42 and 56 units long? ______________

20. Kelley is replacing the rectangular windows in her house. If a window measures 30 centimeters by 50 centimeters, what is its diagonal length to the nearest centimeter? ______________

21. The cover of a rectangular book has a diagonal length of 13.5 inches and a 9-inch side. What is the measure of the other side of the book to the nearest tenth of an inch? ______________

22. Tia wants to divide a 14-foot square room in half by placing a divider diagonally across the room. To the nearest hundredth of a foot, how long should the divider be? ______________

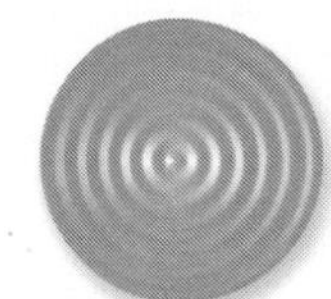

Practice Masters Level C

12.3 The Pythagorean Theorem

Decide whether each set of numbers can represent the side lengths of a right triangle.

1. 24, 301, 302 ______________
2. 25, 312, 313 ______________
3. 39, 760, 761 ______________
4. 49, 1200, 1201 ______________

For Exercises 5–10, the tables give the lengths of various sides of right triangles. Copy and complete the tables. Use a calculator and round each answer to the nearest tenth.

	Leg	Leg	Hypotenuse
5.	44	___	79
7.	___	48	60
9.	79	81	______

	Leg	Leg	Hypotenuse
6.	___	180	181
8.	33	___	67
10.	85	___	151

Solve.

11. A popular brand of laptop computer has a 16-inch screen when measured diagonally. If the screen is 13 inches wide, what is its length to the nearest inch? ______________

12. In order to commute from home to work, Rob drives 8 miles straight north before turning straight west. If direct distance is 15 miles, how far must he drive west, to the nearest tenth of a mile? ______________

13. If a manufacturer plans to make a 75-inch television, measured diagonally, that is 45 inches wide, what must be its height? ______________

14. Jenny is creating a 2-foot square kite. To the nearest hundredth of a foot, how long are the diagonal supports? ______________

Ed wants to create a right triangular garden in his yard. The garden is to have a 17-foot diagonal length and a 15-foot side length.

15. What must be the length of the other side? ______________

16. How much fencing is needed to enclose the garden? ______________

17. If fencing costs $3.79 per foot, how much will it cost to enclose the garden? ______________

NAME ______________________ CLASS ______________ DATE ______________

Practice Masters Level A

12.4 The Distance Formula

Find the distance between each pair of points. Round answers to the nearest hundredth.

1. $A(1, 2), B(3, 4)$ ______________
2. $C(-3, 4), D(2, -7)$ ______________
3. $E(-5, -9), F(-1, -13)$ ______________
4. $G(6, -12), H(-8, 3)$ ______________
5. $I(13, 10), J(11, 20)$ ______________
6. $K(7, 9), L(-5, -7)$ ______________
7. $M(2, 9), N(4, 10)$ ______________
8. $O(1, 5), P(10, -2)$ ______________
9. $Q(-3, 11), R(-5, 19)$ ______________
10. $S(12, 21), T(25, 52)$ ______________
11. $U(-1, 1), V(2, -8)$ ______________
12. $W(1, 1), X(2, 2)$ ______________
13. $Y(-3, -12), Z(2, 1)$ ______________
14. $A(-5, -1), Z(-9, -7)$ ______________
15. $B(8, 5), Y(12, 13)$ ______________
16. $C(10, -4), X(8, -7)$ ______________
17. $D(-8, -1), W(-1, 8)$ ______________
18. $E(15, 2), V(9, -6)$ ______________
19. $F(7, 10), T(9, 10)$ ______________
20. $G(12, -2), U(12, 7)$ ______________

What is the midpoint of a segment with the following endpoints?

21. $H(1, 2), S(1, 5)$ ______________
22. $I(7, 2)$ $R(9, 1)$ ______________
23. $J(0, 5), Q(3, 9)$ ______________
24. $K(-6, 12), P(2, 8)$ ______________
25. $L(5, -6), O(11, -4)$ ______________
26. $M(-4, -6), N(-2, -12)$ ______________

Given the vertices of $\triangle PQR$, decide which of the following terms apply to $\triangle PQR$: scalene (no sides equal), isosceles (two sides equal), equilateral (three sides equal), or right.

27. $P(0, 4), Q(5, 7), R(5, 1)$ ______________
28. $P(-4, 0), Q(0, 3), R(6, -5)$ ______________
29. $P(8, 5), Q(4, 4)$ $R(9, 1)$ ______________
30. $P(0, 0), Q(2, 5), R(7, 0)$ ______________
31. $P(-5, -2), Q(-1, -2), R(-3, 0)$ ______
32. $P(-3, -5), Q(-7, -2), R(-7, -8)$ ______
33. $P(-2, -1), Q(2, -1), R(2, 2)$ ______________
34. $P(0, -2), Q(0, 2), R(6, 0)$ ______________
35. $P(3, 1), Q(5, 1), R(4, 4)$ ______________
36. $P(1, -2), Q(4, -2), R(5, 3)$ ______________

NAME ______________________ CLASS ____________ DATE ____________

Practice Masters Level B

12.4 The Distance Formula

Find the distance between each pair of points. Round answers to the nearest hundredth.

1. $H(19, 21), S(17, 25)$ ____________
2. $I(76, 132), R(79, 128)$ ____________
3. $J(0, 5), Q(3, 9)$ ____________
4. $K(-6, 12), P(2, 7)$ ____________
5. $L(5, -6), O(11, -3)$ ____________
6. $M(-4, -6), N(-3, -9)$ ____________

Identify the coordinates of the midpoint of each segment.

7. $\overline{AB}$, where $A(3, 6)$ and $B(10, 6)$ ____________
8. $\overline{CD}$, where $C(1, 8)$ and $D(7, 2)$ ____________
9. $\overline{EF}$, where $E(-9, -6)$ and $F(-10, -3)$ ____________
10. $\overline{GH}$, where $G(11, 13)$ and $H(7, 18)$ ____________
11. $\overline{IJ}$, where $I(-4, 6)$ and $J(1, -4)$ ____________
12. $\overline{KL}$, where $K(-1, 1)$ and $L(1, -1)$ ____________
13. $\overline{MN}$, where $M(1, 2)$ and $N(3, 4)$ ____________
14. $\overline{OP}$, where $O(0, 0)$ and $P(-5, 6)$ ____________

In Exercises 15–19, the location of two points is given relative to the origin, (0, 0). Calculate the coordinates of the midpoint of the segment connecting each point and the origin.

15. Point A is 3 units west and 2 units south of the origin. ____________
16. Point B is 8 units east and 6 units south of the origin. ____________
17. Point C is 16 units east and 7 units north of the origin. ____________
18. Point D is 76 units west and 80 units south of the origin. ____________
19. Point E is 23 units east and 15 units south of the origin. ____________
20. Given the vertices of $\triangle ABC$, $A(3, 6)$, $B(9, -3)$, $R(-15, -6)$, determine if the triangle is a right triangle. Explain how you know. ____________

NAME ______________________ CLASS __________ DATE __________

Practice Masters Level C

12.4 The Distance Formula

Identify the coordinates of the midpoint of each segment.

1. $\overline{UV}$, where $U(12, 15)$ and $V(17, -19)$ ______________________
2. $\overline{WX}$, where $W(-2, 1)$ and $X(13, -5)$ ______________________
3. $\overline{YZ}$, where $Y(-9, -11)$ and $Z(2, 6)$ ______________________
4. $\overline{ST}$, where $S(-7, 6)$ and $T(2, -9)$ ______________________

The midpoint of $\overline{PQ}$ is M. Calculate the missing coordinates.

5. $P(-6, 8), Q(4, -2), M(?, ?)$ __________
6. $P(?, 5), Q(-3, ?), M(2, 7)$ __________
7. $P(2, ?), Q(?, 6), M(1, -1)$ __________
8. $P(9, -2), Q(?, ?), M(-3, 4)$ __________
9. $P(?, ?), Q(9, 4), M(3, 0)$ __________
10. $P(6, ?), Q(-3, 10), M(?, 1)$ __________
11. $P(-6, -1), Q(?, 4), M(2, ?)$ __________
12. $P(?, ?), Q(9, -2), M(0, 0)$ __________
13. $P(-9, 7), Q(?, -11), M(-3, ?)$ __________
14. $P(5, ?), Q(1, -7), M(?, -9)$ __________
15. $P(3, 0), Q(0, -6), M(?, ?)$ __________
16. $P(12, ?), Q(?, 15), M(-1, 3)$ __________
17. $P(?, ?), Q(8, -1), M(4, 1)$ __________
18. $P(?, -2), Q(-3, 0), M(1, ?)$ __________
19. A small plane sitting on a runway 5 miles west and 2 miles north of town is scheduled to fly to a second runway 17 miles east and 25 miles north of the same town. To the nearest mile, what is the distance between the two runways? ______________________
20. A police car that is 4 miles east and 1 mile south of the station must report to an accident located 5 miles west and 2 miles north of that station. To the nearest tenth of a mile, what is the direct distance between the police car and the accident? ______________________
21. Given the vertices of $\triangle ABC$, $A(0.5, 1)$, $B(2, 3)$, $R(3.5, 1)$, determine whether the triangle is a right triangle. Explain how you know.

    ____

22. True/False: A triangle with side lengths $\sqrt{a}$, $\sqrt{b}$, and $\sqrt{a + b}$ is always a right triangle. ______________________

NAME ______________________ CLASS ____________ DATE ____________

Practice Masters Level A

12.5 Geometric Properties

Write the equation of a circle with its center at the origin and the given radius.

1. radius $= 2$ ____________
2. radius $= 7$ ____________
3. radius $= 12$ ____________
4. radius $= 25$ ____________
5. radius $= 3.2$ ____________
6. radius $= 5.7$ ____________
7. radius $= 14.6$ ____________
8. radius $= 36.1$ ____________

Find the equation of a circle with the given radius, *r*, and the given center, *C*.

9. $r = 2, C(3, 5)$ ____________
10. $r = 5, C(7, 9)$ ____________
11. $r = 3, C(9, 1)$ ____________
12. $r = 4, C(10, 15)$ ____________
13. $r = 7, C(-1, 3)$ ____________
14. $r = 9, C(4, -6)$ ____________
15. $r = 1, C(-8, 9)$ ____________
16. $r = 16, C(7, -1)$ ____________
17. $r = 5, C(-2, -5)$ ____________
18. $r = 4, C(-7, -2)$ ____________

From each equation of a circle, give the center and the radius.

19. $(x - 2)^2 + (y - 3)^2 = 25$ ____________
20. $(x - 7)^2 + (y - 9)^2 = 9$ ____________
21. $(x - 1)^2 + (y - 8)^2 = 4$ ____________
22. $(x - 5)^2 + (y + 1)^2 = 121$ ____________
23. $(x + 6)^2 + (y - 2)^2 = 81$ ____________
24. $(x + 49)^2 + (y + 64)^2 = 1$ ____________

25. Test the Triangle Midsegment Theorem for the triangle shown.

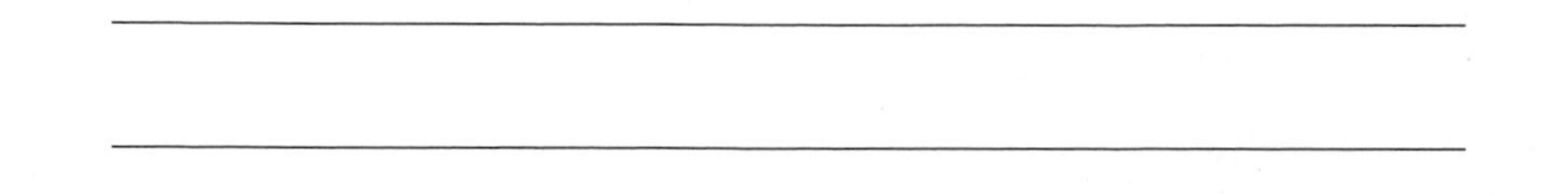

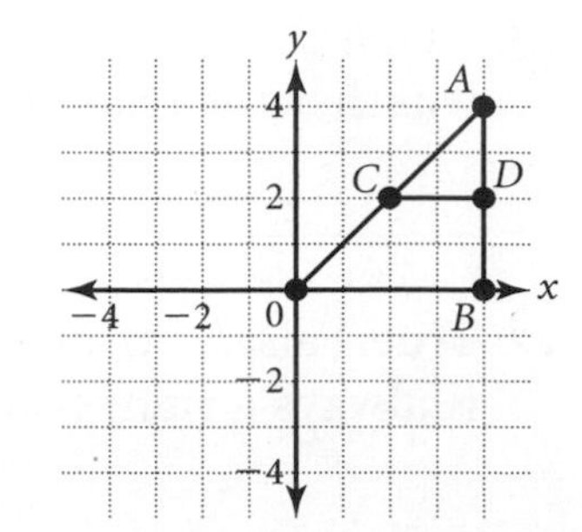

NAME ______________________ CLASS __________ DATE __________

Practice Masters Level B

12.5 Geometric Properties

Write the equation of a circle with its center at the origin and the given radius.

1. radius $= \sqrt{5}$ ______________
2. radius $= \sqrt{17}$ ______________
3. radius $= \sqrt{57}$ ______________
4. radius $= \sqrt{101}$ ______________
5. radius $= 2\sqrt{3}$ ______________
6. radius $= 5\sqrt{7}$ ______________
7. radius $= 4\sqrt{2}$ ______________
8. radius $= 11\sqrt{11}$ ______________

Find the equation of a circle with the given radius, *r*, and the given center, *C*.

9. $r = 11, C(-12, -13)$ ______________
10. $r = 2\sqrt{5}, C(6, -7)$ ______________
11. $r = 2, C(0, -5)$ ______________
12. $r = 5\sqrt{6}, C(9, 0)$ ______________

From each equation of a circle, give the center and the radius.

13. $(x - 11)^2 + (y - 14)^2 = 196$ ______________
14. $(x - 9)^2 + (y + 9)^2 = 9$ ______________
15. $(x + 3)^2 + (y - 6)^2 = 576$ ______________
16. $(x + 2)^2 + (y + 12)^2 = 256$ ______________
17. $(x - 4.7)^2 + (y + 6)^2 = 1$ ______________
18. $(x + 2.5)^2 + (y + 1.6)^2 = 4$ ______________

Write an equation for each circle. Point *C* represents each center.

19.

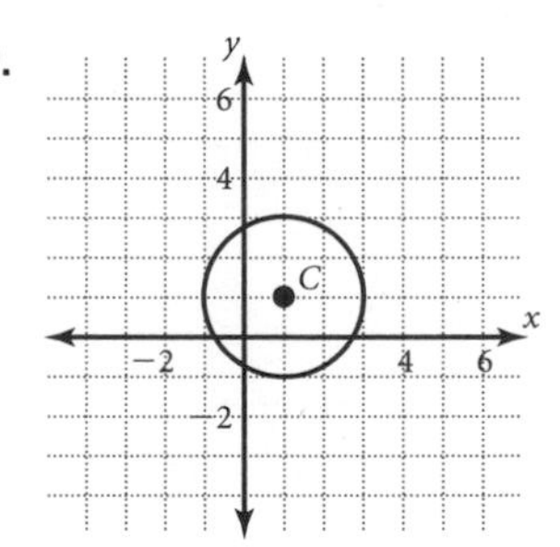

20.

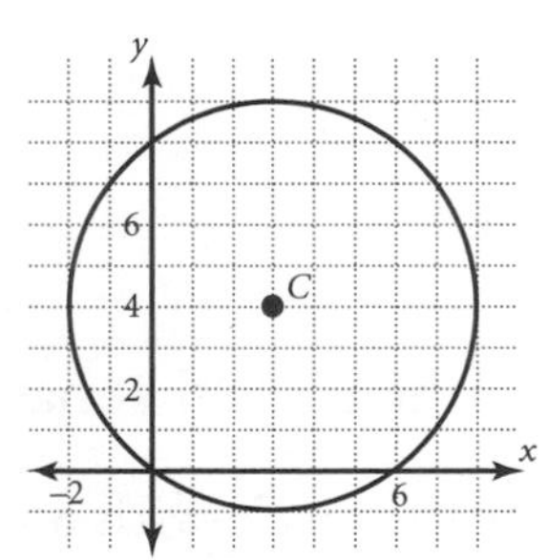

______________ ______________

21. Test the Triangle Midsegment Theorem for the triangle shown.

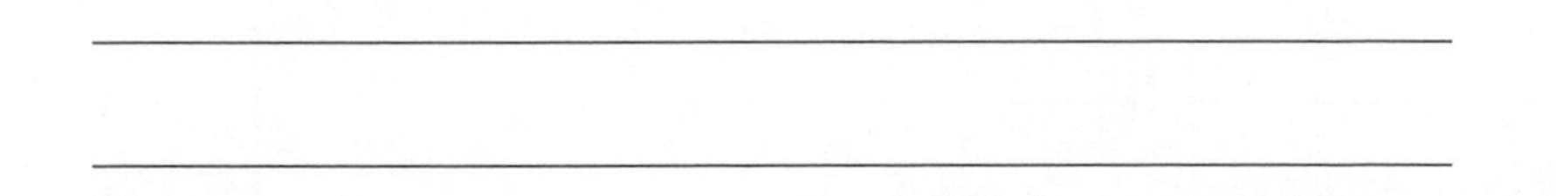

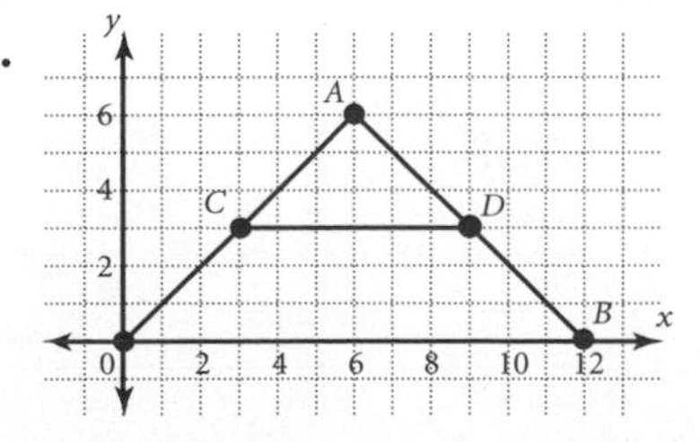

NAME ______________________ CLASS ____________ DATE ____________

Practice Masters Level C

12.5 Geometric Properties

From each equation of a circle, give the center and the radius.

1. $(x + 16)^2 + (y + 32)^2 = 1296$ ________

2. $(x + 17)^2 + (y - 9)^2 = 64$ ________

3. $(x - 21)^2 + (y - 12)^2 = 100$ ________

4. $(x + 10)^2 + (y + 30)^2 = 400$ ________

5. $(x - 3)^2 + (y + 7)^2 = \dfrac{9}{16}$ ________

6. $(x - 12)^2 + (y - 6)^2 = \dfrac{225}{729}$ ________

7. $(x + 4)^2 + (y + 1)^2 = \dfrac{27}{3}$ ________

8. $(x + 3)^2 + (y - 7)^2 = \dfrac{128}{2}$ ________

Write an equation for each circle.

9.

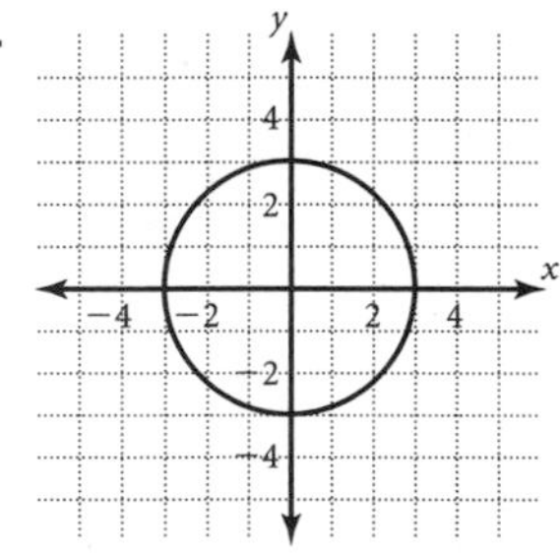

____

10.

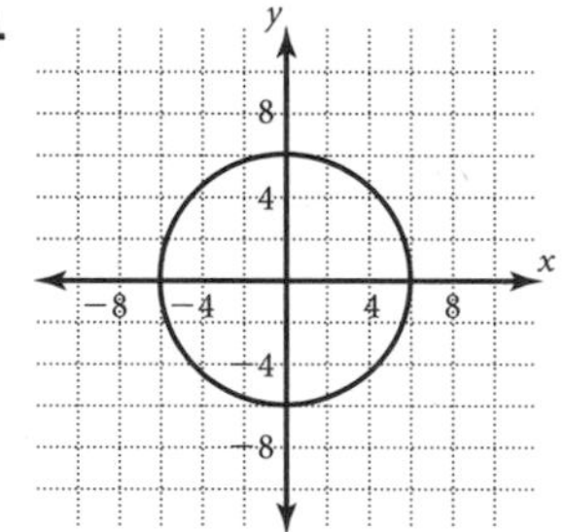

____

11.

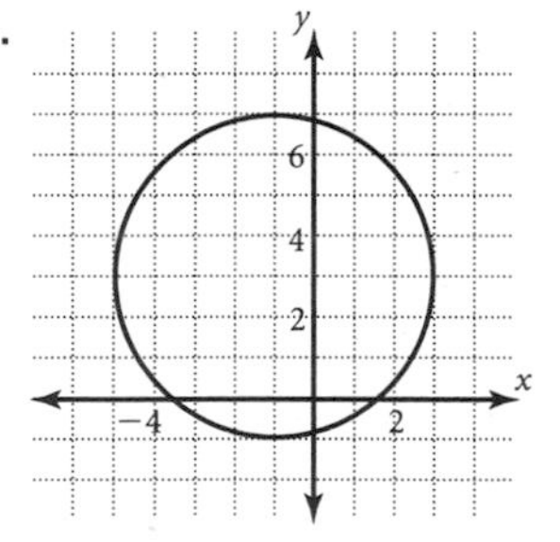

____

12.

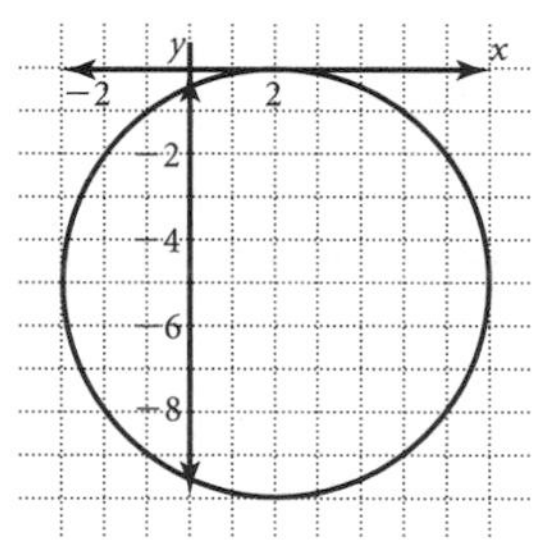

____

13. Write an equation for the circle that has endpoints at (1, 5) and (7, 11). ________

14. Write an equation for the circle with an area of 32π and a center at (6, 3). ________

15. Test the Triangle Midsegment Theorem for the triangle shown.

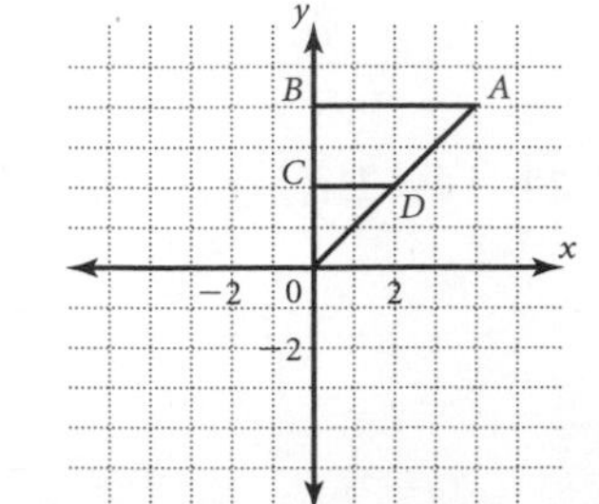

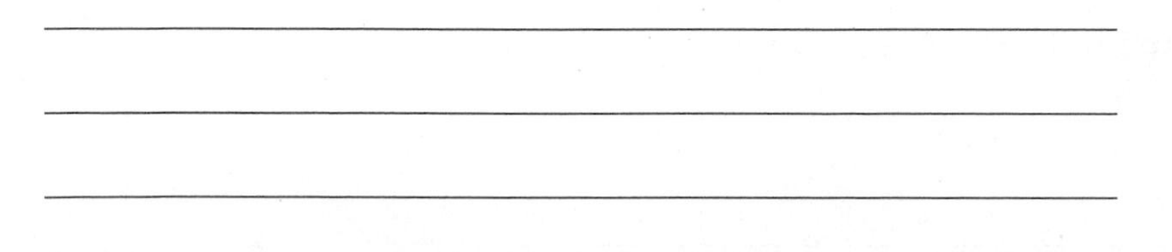

NAME ______________________ CLASS ____________ DATE ____________

Practice Masters Level A

12.6 The Tangent Function

Use a calculator to find the tangent of each angle to the nearest ten-thousandth.

1. 15° ______________________
2. 36° ______________________
3. 50° ______________________
4. 40° ______________________

Find the measure of the angle whose tangent is given. Give each angle measure to the nearest whole number of degrees.

5. 0.5000 ______________________
6. 0.1800 ______________________
7. 0.9876 ______________________
8. 1.2345 ______________________

Find the tangents of the acute angles in each right triangle below. Then find the measures of the acute angles by using the [TAN^{-1}] key of your calculator. Round answers to the nearest degree.

9.

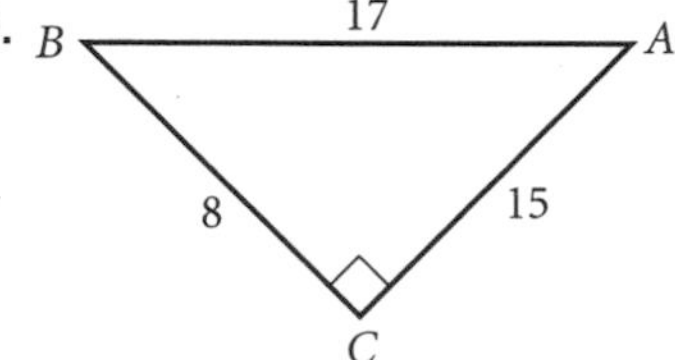

____

10.

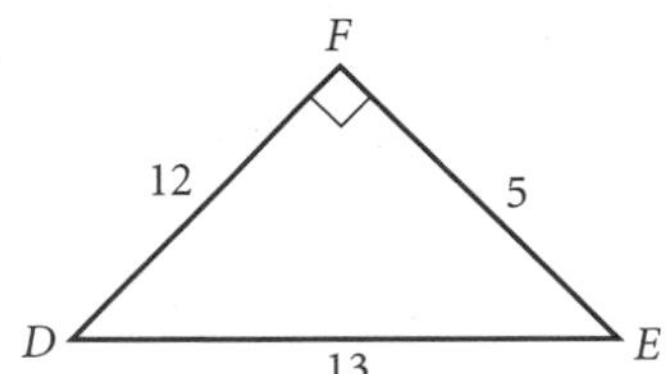

____

11.

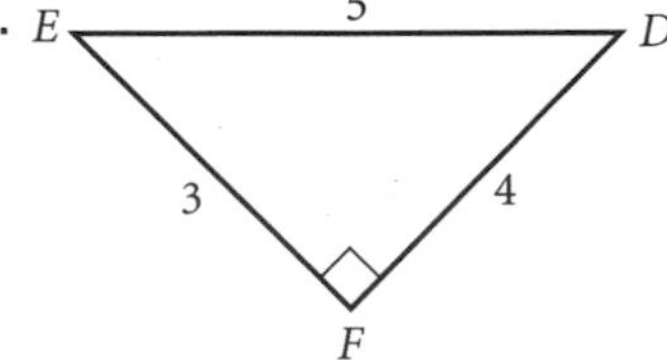

____

12.

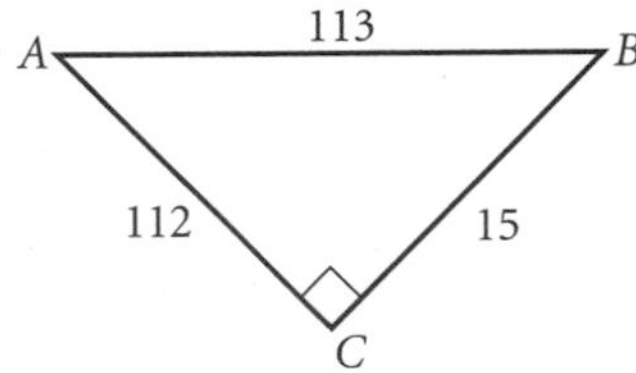

____

13.

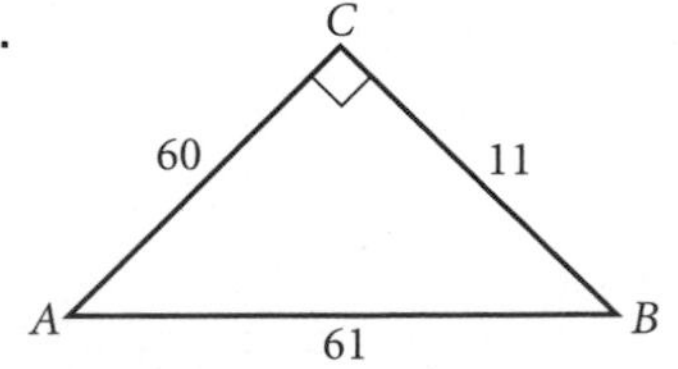

____

14.

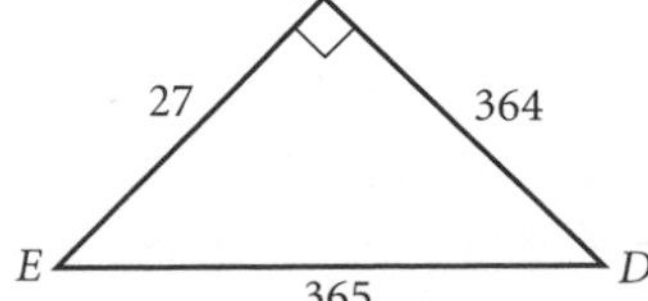

____

NAME ______________________ CLASS ______________ DATE ______________

Practice Masters Level B

12.6 The Tangent Function

Use a calculator to find the tangent of each angle to the nearest ten-thousandth.

1. 5.6° ______________
2. 63.9° ______________
3. 87.4° ______________
4. 31.2° ______________

Find the measure of the angle whose tangent is given. Give each angle measure to the nearest whole number of degrees.

5. 0.7652 ______________
6. 0.4831 ______________
7. 0.6250 ______________
8. 5.0102 ______________

Find the tangents of the acute angles in each right triangle below. Then find the measures of the acute angles by using the [TAN^{-1}] key of your calculator. Round answers to the nearest degree.

9.

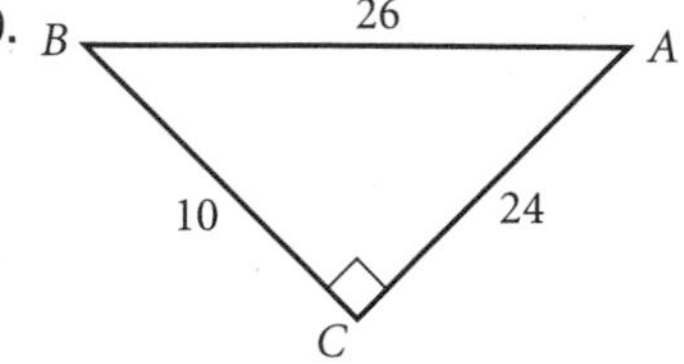

____

10.

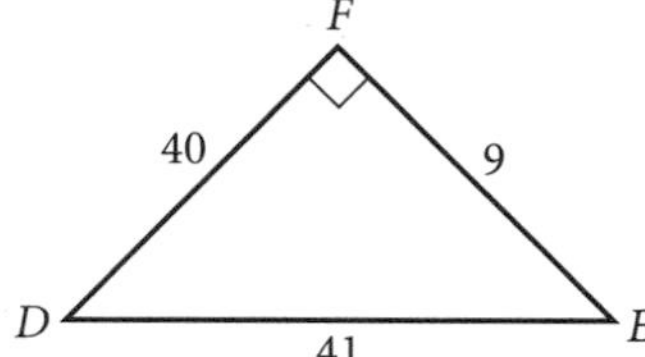

____

11.

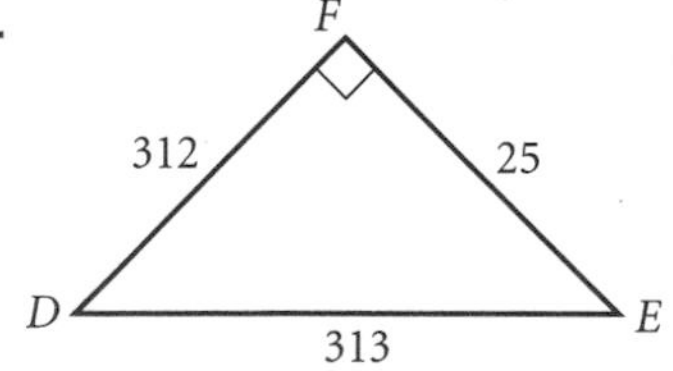

____

12.

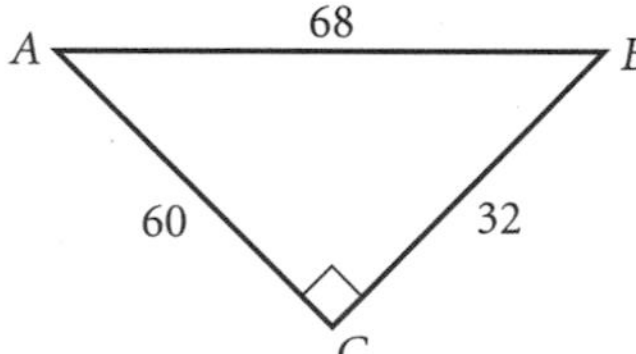

____

13.

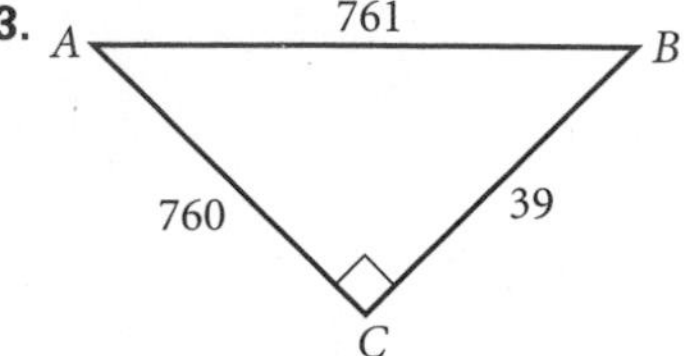

____

14.

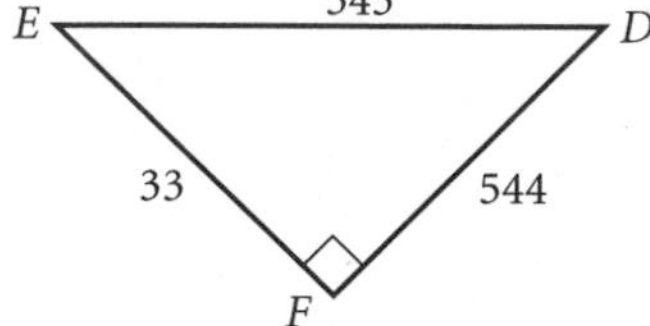

____

NAME ______________________ CLASS __________ DATE __________

Practice Masters Level C

12.6 The Tangent Function

Use a calculator to find the tangent of each angle to the nearest ten-thousandth.

1. 88° ______________________
2. 56° ______________________
3. 70.8° ______________________
4. 73.625° ______________________

Find the measure of the angle whose tangent is given. Give each angle measure to the nearest whole number of degrees.

5. 2.3517 ______________________
6. 1.1999 ______________________
7. 0.5625 ______________________
8. 10 ______________________

Use the diagram below for Exercises 9–18. Round your answer to the nearest hundredth.

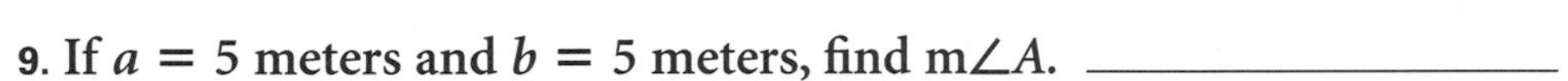

9. If $a = 5$ meters and $b = 5$ meters, find m$\angle A$. ______________________
10. If m$\angle B = 60°$ and $b = 4$ inches, find a. ______________________
11. If m$\angle B = 17°$ and $a = 2$ feet, find b. ______________________
12. If $a = 3.07$ centimeters and $b = 8$ centimeters, find m$\angle A$. ______________________

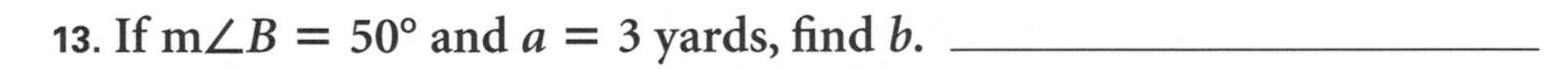

13. If m$\angle B = 50°$ and $a = 3$ yards, find b. ______________________
14. If m$\angle A = 63°$ and $b = 2.7$ meters, find a. ______________________
15. If $a = 2.10$ feet and $b = 20$ feet, find m$\angle A$. ______________________
16. If m$\angle B = 46°$ and $a = 12$ inches, find b. ______________________
17. If m$\angle A = 59°$ and $b = 10$ yards, find a. ______________________
18. If m$\angle B = 78°$ and $b = 500$ feet, find a. ______________________
19. As the measure of an acute angle in a right triangle increases, what happens to the tangent ratio of that angle? ______________________
20. The base of a loading dock is 5 feet from the ground. How long should a ramp be so the incline from the ground up to the dock is 19°? ______________________
21. The sun casts a shadow of a 36-foot tree that is 22 feet long. What angle does the sun's rays make with the ground? ______________________

NAME ______________________ CLASS __________ DATE __________

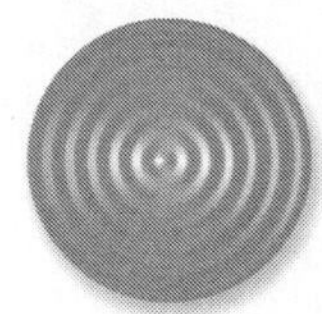

Practice Masters Level A

12.7 The Sine and Cosine Functions

Approximate each sine and cosine. Give your answers to the nearest ten-thousandth.

1. $\cos 35°$ ______________
2. $\sin 47°$ ______________
3. $\sin 69°$ ______________
4. $\cos 73°$ ______________
5. $\cos 56°$ ______________
6. $\sin 31°$ ______________

Find the acute angle measure for each approximate sine value below. Give your answer to the nearest tenth of a degree.

7. 0.95 ______________
8. 0.23 ______________
9. 0.49 ______________
10. 0.81 ______________

Find the acute angle measure for each approximate cosine value below. Give your answer to the nearest tenth of a degree.

11. 0.85 ______________
12. 0.47 ______________
13. 0.53 ______________
14. 0.21 ______________

For Exercises 15–22, use right triangle *ABC* below. Find the indicated side length or angle measure to the nearest tenth.

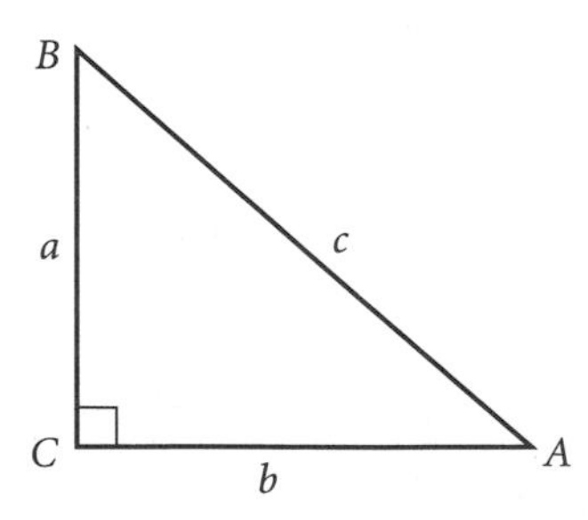

15. $\sin A =$ ______________
16. $\cos B =$ ______________
17. $b = 4, c = 5, m\angle B = ?$ ______________
18. $m\angle A = 26°, c = 9, a = ?$ ______________
19. $m\angle B = 60°, c = 12, a = ?$ ______________
20. $a = 5, c = 11, m\angle A = ?$ ______________
21. $a = 13, c = 25, m\angle B = ?$ ______________
22. $m\angle B = 21°, c = 14, b = ?$ ______________

NAME ______________________ CLASS __________ DATE __________

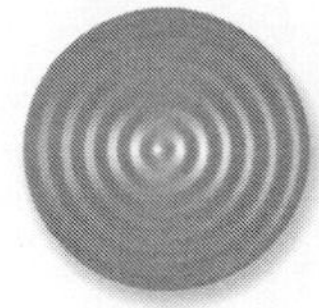

Practice Masters Level B

12.7 The Sine and Cosine Functions

Approximate each sine and cosine. Give your answer to the nearest ten-thousandth.

1. sin 29° ____________
2. cos 28° ____________
3. sin 80° ____________
4. sin 7° ____________
5. cos 11° ____________
6. cos 3° ____________

Find the acute angle measure for each approximate sine value below. Give your answer to the nearest tenth of a degree.

7. 0.607 ____________
8. 0.196 ____________
9. 0.325 ____________
10. 0.008 ____________

Find the acute angle measure for each approximate cosine value below. Give your answer to the nearest tenth of a degree.

11. 0.909 ____________
12. 0.524 ____________
13. 0.314 ____________
14. 0.001 ____________

For Exercises 15–22, use right triangle *ABC* below. Find the indicated side length or angle measure to the nearest tenth.

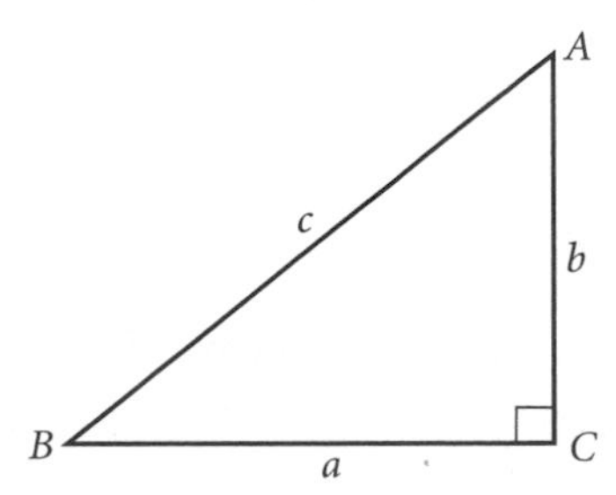

15. The ratio $\frac{a}{c}$ can be used to find the measure of which angle(s)? ____________
16. $b = 10, c = 20, m\angle A = ?$ ____________
17. $m\angle A = 45°, b = 11, a = ?$ ____________
18. $m\angle A = 13°, c = 19, b = ?$ ____________
19. $a = 20, c = 28, m\angle A = ?$ ____________
20. $a = 23, c = 32, m\angle A = ?$ ____________
21. $m\angle B = 39°, c = 16, b = ?$ ____________
22. If $m\angle A = 60°$ and $c = 100$ yards, what is the area of $\triangle ABC$? ____________
23. A 10-foot ladder is leaning against a house. The ladder makes a 72° angle with the ground. How far is the base of the ladder from the building? ____________

NAME ______________________ CLASS ____________ DATE ____________

Practice Masters Level C

12.7 The Sine and Cosine Functions

Approximate each sine and cosine. Give your answer to the nearest ten-thousandth.

1. $\cos 35°$ ______________
2. $\sin 35°$ ______________
3. $\sin 72°$ ______________
4. $\cos 72°$ ______________

Find the length of side *s* in the diagram below.

5. 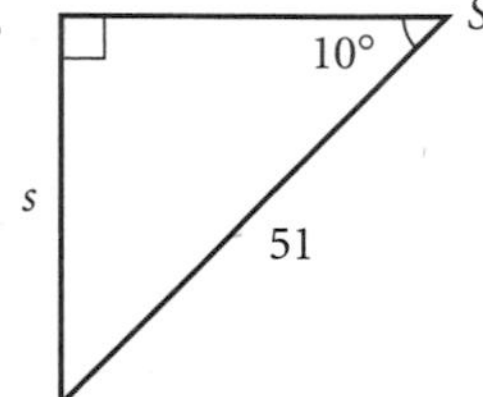

6. 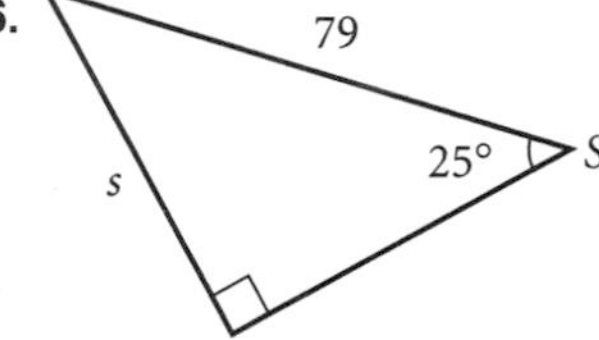

______________ ______________

Find the measure of angle *X* in the diagram below.

7. 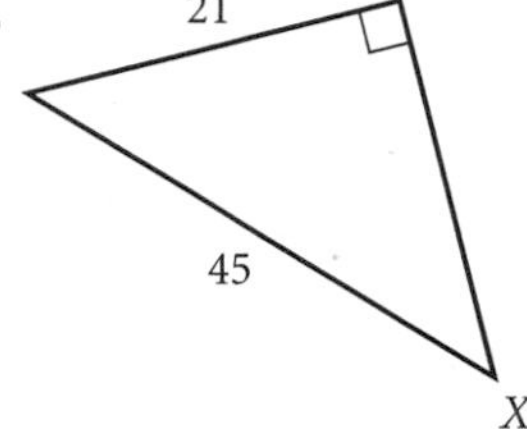

8. 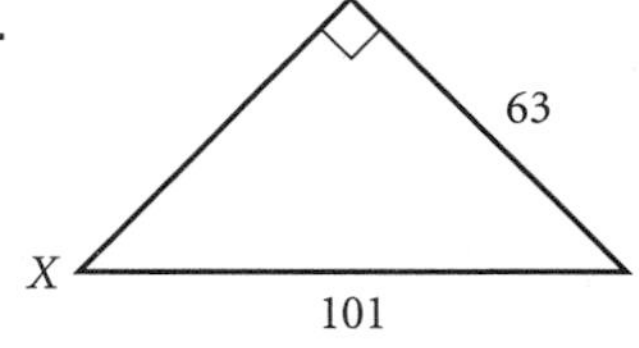

______________ ______________

For Exercises 9–12, use right triangle *ABC* below. Find the indicated side length or angle measure to the nearest tenth.

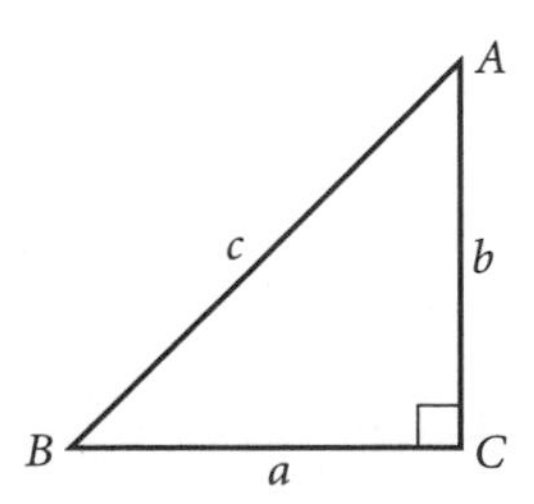

9. $a = 32, c = 48, m\angle A = ?$ ______________
10. $m\angle A = 24°, c = 10, b = ?$ ______________
11. $m\angle B = 25°, b = 14, c = ?$ ______________
12. $a = 99, c = 101, m\angle B = ?$ ______________
13. Write a paragraph proof showing that $\tan x = \dfrac{\sin x}{\cos x}$. ______________

____

____

NAME ______________________ CLASS ____________ DATE ____________

Practice Masters Level A

12.8 Introduction to Matrices

Determine which two of the three matrices are equal.

1. $H = \begin{bmatrix} 4 & 2(-1) \\ \frac{1}{3}(9) & -6 \end{bmatrix}$ $I = \begin{bmatrix} 2(2) & 1 \\ 3 & -2(3) \end{bmatrix}$ $J = \begin{bmatrix} 6-2 & -2 \\ -5+8 & -3(2) \end{bmatrix}$ ____________

2. $H = \begin{bmatrix} -3(3) & (5-2) \\ (1+3) & 2(5) \end{bmatrix}$ $I = \begin{bmatrix} (1-10) & (1+2) \\ (2+2) & 5(2) \end{bmatrix}$ $J = \begin{bmatrix} -1(9) & (8-5) \\ 2(2) & (8-2) \end{bmatrix}$ ____________

Add or subtract as indicated.

3. $\begin{bmatrix} 7 & -4 \\ -5 & 3 \end{bmatrix} + \begin{bmatrix} 3 & -2 \\ 5 & -7 \end{bmatrix}$ ____________

4. $\begin{bmatrix} 7 & -4 \\ -5 & 3 \end{bmatrix} - \begin{bmatrix} 3 & -2 \\ 5 & -7 \end{bmatrix}$ ____________

5. $\begin{bmatrix} -3 & 9 \\ 5 & -1 \end{bmatrix} + \begin{bmatrix} 4 & -6 \\ 1 & -2 \end{bmatrix}$ ____________

6. $\begin{bmatrix} -3 & 9 \\ 5 & -1 \end{bmatrix} - \begin{bmatrix} 4 & -6 \\ 1 & -2 \end{bmatrix}$ ____________

Perform each scalar multiplication.

7. $7\begin{bmatrix} 3 & 4 & -5 \end{bmatrix}$ ____________

8. $6\begin{bmatrix} -1 & 3 & -2 & 0 \end{bmatrix}$ ____________

9. $3\begin{bmatrix} -4 & 6 \\ 2 & -8 \end{bmatrix}$ ____________

10. $-2\begin{bmatrix} -1 & -9 \\ -5 & 6 \end{bmatrix}$ ____________

11. $-5\begin{bmatrix} 3 & -5 \\ -2 & 4 \end{bmatrix}$ ____________

12. $4\begin{bmatrix} -3 & 5 \\ -6 & 7 \end{bmatrix}$ ____________

Find the identity matrix for each matrix below.

13. $\begin{bmatrix} 0 & 0 & 0 & 1 \\ 0 & 0 & 1 & 0 \\ 0 & 1 & 0 & 0 \\ 1 & 0 & 0 & 0 \end{bmatrix}$ ____________

14. $\begin{bmatrix} 2 & 0 & 0 & 0 \\ 0 & 2 & 0 & 0 \\ 0 & 0 & 2 & 0 \\ 0 & 0 & 0 & 2 \end{bmatrix}$ ____________

NAME ______________________ CLASS __________ DATE __________

Practice Masters Level B

12.8 Introduction to Matrices

Determine which two of the three matrices below are equal.

1. $H = \begin{bmatrix} 3(12) & (20-3) & 8(9) \\ -5^2 & 2(3+6) & (20+3) \end{bmatrix}$ $I = \begin{bmatrix} 36 & 17 & 72 \\ -25 & 18 & 23 \end{bmatrix}$ $J = \begin{bmatrix} 36 & -25 \\ 17 & 18 \\ 72 & 23 \end{bmatrix}$ ____________

Add or subtract as indicated.

2. $\begin{bmatrix} -2 & 4 & -8 \\ 3 & -6 & 9 \\ -1 & 1 & -1 \end{bmatrix} - \begin{bmatrix} 5 & -2 & 1 \\ 3 & 2 & 4 \\ 1 & 1 & 5 \end{bmatrix}$ ____________

3. $\begin{bmatrix} -2 & 4 & -8 \\ 3 & -6 & 9 \\ -1 & 1 & -1 \end{bmatrix} + \begin{bmatrix} 5 & -2 & 1 \\ 3 & 2 & 4 \\ 1 & 1 & 5 \end{bmatrix}$ ____________

Perform each scalar multiplication.

4. $-3\begin{bmatrix} 5 & -2 & 0 \\ -7 & 4 & 2 \\ -1 & -3 & 8 \end{bmatrix}$ ____________

5. $5\begin{bmatrix} -7 & 1 & -4 \\ 5 & 6 & 0 \\ 2 & -3 & 9 \end{bmatrix}$ ____________

Find each product matrix, if it exists.

6. $\begin{bmatrix} 5 & -3 \\ 1 & 7 \end{bmatrix}\begin{bmatrix} -2 & 9 \\ 6 & -4 \end{bmatrix}$ ____________

7. $\begin{bmatrix} 4 & 8 \\ -6 & -2 \\ 7 & 9 \end{bmatrix}\begin{bmatrix} 5 & -1 & 0 \\ -3 & 5 & -3 \end{bmatrix}$ ____________

Find the identity matrix for each matrix below.

8. $\begin{bmatrix} 1 & 0 & 0 \\ 0 & 1 & 0 \\ 0 & 0 & 1 \end{bmatrix}$ ____________

9. $\begin{bmatrix} 5 & -6 & 3 & 1 \\ -7 & 2 & 8 & 4 \\ 9 & 3 & -2 & 6 \\ 1 & 0 & 7 & -5 \end{bmatrix}$ ____________

Practice Masters Level C

12.8 Introduction to Matrices

Matrices *M* and *N* are equal. Find the values of *a, b, c,* and *d*.

1. $M = \begin{bmatrix} 2^2 & (5 - a) & 3(1 + 2) \\ 3(-2) & 0(4 + 2) & -4(1 - 6) \\ d & -3(3) & 1^3 \end{bmatrix}$ $N = \begin{bmatrix} 2^3 - c & (2 - 6) & 3^2 \\ -2(3) & 5(7 - 7) & 10(2) \\ (4 - 7) & -(4 + 5) & b^2 \end{bmatrix}$ ____________

Perform the indicated matrix operations. If a solution is not possible, explain why.

2. $\begin{bmatrix} -10 & -4 & 8 \\ -12 & 6 & 5 \end{bmatrix} + \begin{bmatrix} 8 & -1 & -7 \\ 9 & 4 & -2 \end{bmatrix}$ ____________

3. $\begin{bmatrix} -9 & 4 \\ 11 & 1 \\ 6 & -5 \end{bmatrix} - \begin{bmatrix} 2 & -6 & 8 \\ 12 & 1 & -10 \end{bmatrix}$ ____________

4. $\begin{bmatrix} 10 & 15 \\ 17 & -12 \end{bmatrix}\begin{bmatrix} -5 & -2 \\ -3 & 7 \end{bmatrix}$ ____________

For the last two years, youth groups in Cleveland, Chicago, and Tulsa have had a candy sale. The tables below summarize the case sales of milk chocolate, dark chocolate, and white chocolate bars that each group sold. Use this information for Exercises 5–6.

This Year

	Milk	Dark	White
Cleveland	712	401	386
Chicago	935	342	358
Tulsa	463	240	112

Last Year

	Milk	Dark	White
Cleveland	710	475	417
Chicago	899	286	364
Tulsa	503	183	99

5. Create matrices to show the total candy sold by each group for both years.

____

6. How much total candy was sold by the groups in both years? ____________

NAME ______________________ CLASS ____________ DATE ____________

Practice Masters Level A

13.1 Theoretical Probability

Find the sample space for each experiment.

1. rolling a number cube ______________________
2. tossing a coin twice ______________________
3. tossing a coin three times ______________________
4. rolling a number cube twice ______________________

Find the number of favorable outcomes in the sample space for each experiment.

5. A number cube is rolled twice. The sum of the rolls is 4. ____________
6. A number cube is rolled twice. The sum of the rolls is 8. ____________
7. A coin is tossed three times. All three tosses are tails. ____________
8. A coin is tossed three times. The result is exactly two heads. ____________

Find the probability of each outcome.

9. Two rolls of a number cube will have a sum of 4. ____________
10. Two rolls of a number cube will have a sum of 8. ____________
11. Three tosses of a coin will be three tails. ____________
12. Three tosses of a coin will be exactly two heads. ____________

Which has the greater probability?

13. 1 chance out of 5 or 7 chances out of 10 ____________
14. 12 chances out of 20 or 21 chances out of 30 ____________
15. 15 chances out of 16 or 99 chances out of 100 ____________
16. 23 chances out of 40 or 34 chances out of 60 ____________

An integer between 1 and 25, inclusive, is drawn at random. Find each probability.

17. The integer is a multiple of 4. ________
18. The integer is a multiple of 9. ________
19. The integer is a multiple of 6. ________
20. The integer is a multiple of 7. ________
21. The integer is a multiple of 8. ________
22. The integer is a multiple of 10. ________

NAME ______________________ CLASS __________ DATE __________

Practice Masters Level B

13.1 Theoretical Probability

Find the sample space for each experiment.

1. tossing a coin five times ______________________
2. rolling a number cube and then tossing a coin ______________________

Find the number of favorable outcomes in the sample space for each experiment.

3. A coin is tossed three times. The result is exactly one head. ______________________
4. A number cube is rolled twice. The sum of the rolls is 6. ______________________
5. A number cube is rolled twice. The sum of the rolls is 11. ______________________
6. A coin is tossed four times. All four tosses are heads. ______________________

Suppose that you select a letter of the English alphabet at random. Find the probability of each event.

7. The letter is in the word *Mississippi.* ______________________
8. The letter is in the word *Ohio.* ______________________
9. The letter is in the word *probability.* ______________________
10. The letter is in the word *algebra.* ______________________
11. The letter is in the word *geometry.* ______________________
12. The letter is in the word *calculus.* ______________________
13. The letter is in the word *math.* ______________________

Find the probability of each outcome.

14. Two rolls of a number cube will have a sum of 7. ______________________
15. Two tosses of a coin will be two heads. ______________________
16. Two rolls of a number cube will have an odd sum. ______________________
17. Three tosses of a coin will be at least two heads. ______________________
18. Three tosses of a coin will be exactly two tails. ______________________
19. Two rolls of a number cube will contain at least one 5. ______________________

NAME ______________________ CLASS ____________ DATE ____________

Practice Masters Level C

13.1 Theoretical Probability

Find the sample space for each experiment.

1. tossing a coin and then rolling a number cube ____________
2. rolling a number cube and then tossing a coin twice ____________

Suppose that you select a letter in the English alphabet at random. Find the probability of each event.

3. The letter is in the word *coefficient.* ____________
4. The letter is in the word *coordinate.* ____________
5. The letter is in the word *perpendicular.* ____________
6. The letter is in the word *discriminant.* ____________
7. The letter is in the word *Lincoln.* ____________
8. The letter is in the word *Washington.* ____________
9. The letter is in the word *Jefferson.* ____________

Find probability of each outcome.

10. Three tosses of a coin will be at most two tails. ____________
11. Four tosses of a coin will be two heads and two tails. ____________
12. Four tosses of a coin will be more heads than tails. ____________
13. Three rolls of a number cube will have exactly two fives. ____________
14. Two rolls of a number cube will have an even sum. ____________
15. Four rolls of a number cube will have exactly one four. ____________

An integer between 51 and 99, inclusive, is drawn at random. Find the probability of each event.

16. The integer is even. ____________
17. The integer is greater than 5. ____________
18. The integer is less than 71. ____________
19. The integer is a multiple of 10. ____________
20. The integer is less than 5. ____________
21. The integer is greater than 57. ____________
22. The integer has at least one digit that is a 6. ____________

NAME ______________________ CLASS __________ DATE __________

Practice Masters Level A

13.2 Counting the Elements of Sets

In a deck of 52 playing cards, how many are there of each type listed below?

1. an ace OR a spade ______
2. a club OR black ______
3. a two OR a three ______
4. a heart AND red ______
5. a diamond AND a five ______
6. an eight AND black ______
7. a diamond OR a five ______
8. a face card AND red ______

List the integers from 1 to 10 inclusive that are

9. odd. ______
10. multiples of 5. ______
11. odd AND multiples of 5. ______
12. odd OR multiples of 5. ______
13. multiples of 4. ______
14. multiples of 6. ______
15. multiples of 4 OR multiples of 6. ______
16. multiples of 4 AND multiples of 6. ______

If you draw a card at random from a complete deck of 52 playing cards, what is the probability that you will draw the card below?

17. a four OR a heart ______
18. a spade OR black ______
19. a four AND a heart ______
20. a spade AND black ______
21. a face card OR red ______
22. a face card AND red ______

At the local state university a certain business class is made up of both Asian and American students. Use the table below for Exercises 23–27.

	Asian	American	Total
Male	9	1	19
Female	4	12	16
Total	13	22	35

23. How many in this class are female? ______
24. How many in this class are male AND Asian? ______
25. How many in this class are male OR Asian? ______
26. How many in this class are female AND American? ______
27. How many in this class are female OR American? ______

NAME ______________________ CLASS ____________ DATE ____________

Practice Masters Level B

13.2 Counting the Elements of Sets

List the integers from 1 to 20 inclusive that are multiples of the following numbers:

1. 6 ______________________
2. 9 ______________________
3. 6 AND 9 ______________________
4. 6 OR 9 ______________________
5. 2 ______________________
6. 5 ______________________
7. 2 OR 5 ______________________
8. 2 AND 5 ______________________

If you draw a card at random from a complete deck of 52 playing cards, what is the probability that you will draw the cards below?

9. an odd numbered card AND a red card ______________________
10. an odd numbered card OR a red card ______________________
11. a face card OR a black card ______________________
12. a face card AND a black card ______________________
13. a diamond OR a ten ______________________

Sly took a survey to get student opinions about a new graduation requirement. Use the table below for Exercises 14–20.

	Favor requirement	Oppose requirement	Total
Freshman	12	4	16
Sophomore	6	8	14
Junior	9	15	24
Senior	2	21	23
Total	29	48	77

14. How many of those surveyed are freshmen AND oppose the requirement? ____________
15. How many of those surveyed are freshmen OR oppose the requirement? ____________
16. How many of those surveyed are juniors AND favor the requirement? ____________
17. How many of those surveyed are juniors OR favor the requirement? ____________
18. How many of those surveyed are seniors OR oppose the requirement? ____________
19. How many of those surveyed are seniors AND oppose the requirement? ____________
20. How many of those surveyed are sophomores OR favor the requirement? ____________

NAME ________________ CLASS ________ DATE ________

Practice Masters Level C

13.2 Counting the Elements of Sets

List the integers from 1 to 25 inclusive that are multiples of the following numbers:

1. 3 ________________
2. 4 ________________
3. 5 ________________
4. 10 ________________
5. 3 AND 4 ________________
6. 3 AND 5 ________________
7. 5 OR 10 ________________
8. 5 AND 10 ________________

List the integers from 100 to 150 inclusive that are

9. multiples of 10. ________________
10. multiples of 25. ________________
11. multiples of 10 AND multiples of 25. ________________
12. multiples of 10 OR multiples of 25. ________________

If you draw a card at random from 2 complete decks of 52 playing cards, what is the probability that you will draw the cards below?

13. a two OR a club ________________
14. an eight AND a black card ________________
15. a numbered card OR a red card ________________

A survey of student participation in extra curricular activities has the results shown in the following table. Use the table below for Exercises 16–20.

	Participated	Did not participate
Boys	135	189
Girls	98	213

16. How many students are girls AND did not participate? ________________
17. How many students are girls OR did not participate? ________________
18. How many students are boys OR participated? ________________
19. How many students are boys AND participated? ________________
20. How many students are girls OR participated? ________________

NAME ______________________ CLASS __________ DATE __________

Practice Masters Level A

13.3 The Fundamental Counting Principle

Use a tree diagram to find the combined number of choices that are possible in each situation.

1. choosing a cheeseburger from a menu with 3 types of cheese and 2 types of bun ____________
2. flipping a dime and then flipping a quarter ____________
3. rolling a number cube and then flipping a coin ____________
4. choosing a rental car from 5 types of cars and 3 colors ____________
5. choosing a computer from 6 brands and 5 sizes of hard drive ____________
6. choosing dance partners from 8 boys and 8 girls ____________
7. choosing a lunch from a menu of 4 entrees and 6 desserts ____________

Use the Fundamental Counting Principle to find the number of choices that are possible for each situation.

8. A one-topping pizza is offered with 4 choices of toppings and 2 choices of crust. ____________
9. A business woman has 5 skirts that go with 7 blouses. ____________
10. A painter has 10 choices of color and 5 choices of brushes. ____________
11. A secretary has 3 choices of envelopes and 4 choices of stamps. ____________
12. A baseball manager has 3 choices of catchers and 9 choices of pitchers. ____________
13. A local delivery service can choose from 15 trucks and 17 drivers. ____________
14. A family has 45 choices of TV channels and 3 TV sets. ____________

Use the Fundamental Counting Principle to answer each question.

15. A car dealer has the following options on a new car: 3 different models, 7 different colors, and 4 different engine models. How many choices are there? ____________
16. A grocery store stocks 11 types of canned beans. Each type has 3 sizes of cans and 3 different brands. How many cans of beans must the store carry to have at least one can of each type, size, and brand? ____________

NAME ______________________ CLASS ______________ DATE ______________

Practice Masters Level B

13.3 The Fundamental Counting Principle

Use a tree diagram to find the combined number of choices that are possible in each situation.

1. A college has 9 choices for a college algebra class and 6 choices of teachers. ______________
2. Tom has 15 choices for lunch and 27 choices for dinner. ______________
3. Lupa is choosing between 4 colleges and 21 majors. ______________
4. Lyle has 33 choices of pens and 14 choices of notebooks. ______________
5. Sara has 2 choices of lunch entrees and 6 choices of beverage. ______________

Use the Fundamental Counting Principle to find the number of possibilities for each situation.

6. In how many ways can Tina choose a mystery and a drama from a collection of 6 mysteries and 5 dramas? ______________
7. A family is planning on driving across country. They can choose from 5 cars and 4 drivers. ______________

Use the Fundamental Counting Principle to answer each question.

8. George is landscaping around his home. If he can choose from 8 types of trees and 6 types of shrubs, how many ways are there to pick one of each? ______________
9. Sally has 7 flight choices for the first leg of her trip and 9 choices for the second. How many different combinations of flights can she make to get to her final destination? ______________
10. Jean is planning a cookout. She will give her guests 5 choices of sandwiches and 8 choices of salads. If each person chooses a sandwich and a salad, how many different types of plates can be made? ______________
11. Kelley has a video collection of 25 mysteries, 4 dramas, and 19 cartoons. If she wants to put one of each type on a videotape how many different tapes can she make? ______________
12. A store stocks 19 choices of pants, 25 choices of shirts, and 11 choices of ties. How many different outfits can be created? ______________
13. Simon is planning to purchase a motorcycle. He is considering 4 different models, 7 different colors, and 28 different helmets. How many different packages are possible? ______________

NAME ______________________ CLASS ____________ DATE ____________

Practice Masters Level C

13.3 The Fundamental Counting Principle

Use a tree diagram to find the combined number of choices that are possible in each situation.

1. A menu has 12 choices of sandwiches and 25 choices of side dishes. ____________

2. flipping a coin, then rolling a number cube, and then flipping another coin ____________

Use the Fundamental Counting Principle to answer each question. Two decks of playing cards are shuffled separately. A card is drawn from each deck.

3. How many ways are there of drawing a heart from the first deck AND a black card from the second deck? ____________

4. How many ways are there of drawing a number card from the first deck AND a number card from the second deck? ____________

5. How many ways are there of drawing a black card from the first deck AND a red card from the second deck? ____________

Use the Fundamental Counting Principle to find the number of possibilities for each situation.

6. If you choose one boy out of 9 boys and one girl out of 16 girls, how many ways can this be done? ____________

7. A department store stocks 13 dress shirts. Each shirt is available in 6 sizes. How many shirts must the store carry in order to have one shirt of each size on hand? ____________

8. In how many ways can Sean choose a CD and a tape from a collection of 51 CDs and 27 tapes? ____________

9. How many different 7-digit phone numbers are possible if the first 3 digits are 286? Assume that zero is not allowed. ____________

10. How many different 7-digit phone numbers are possible if the first digit cannot be a 0, 1 or 9? ____________

11. How many different 10-digit phone numbers are possible if the first digit cannot be a 0, 1 or 9? ____________

12. A menu contains 6 appetizers, 5 salads, 7 main courses, and 3 desserts. How many ways are there to choose one of each? ____________

NAME ______________________ CLASS __________ DATE __________

Practice Masters Level A

13.4 Independent Events

A red number cube and a green number cube are rolled together. Find the probability of each event.

1. Both numbers are odd. __________
2. Both numbers are greater than 4. __________
3. Both numbers are less than 4. __________
4. Both numbers are greater than 2. __________
5. The green cube is less than 3 AND the red cube is greater than 2. __________
6. The green cube is less than 4 AND the red cube is less than 4. __________

Use area models to find each probability below.

7. Brianna is flying from Tulsa to Cleveland for her vacation. The flight from Tulsa to St. Louis is on time 9% of the time and the flight from St. Louis to Cleveland is also on time 9% of the time. What is the probability that both flights will be on time? __________
8. Susannah is planning a business trip. If her flight is on time 5% of the time and the train is on time 92% of the time, what is the probability that both will be on time? __________
9. Tsu has an 85% chance of getting Question 1 correct on his Algebra exam and a 7% chance of getting Question 2 right. What is the probability that he will get both questions correct? __________

Suppose that you draw two cards with replacement from a regular deck of 52 playing cards. Find the probability of each event.

10. Both cards are jacks. __________
11. Both cards are diamonds. __________
12. Both cards are red. __________
13. Both cards are black queens. __________

Two cards are drawn with replacement from an ordinary deck of 52 cards. Find the probability of each event.

14. At least one card is *not* a jack. __________
15. At least one card is *not* a diamond. __________
16. At least one card is *not* red. __________
17. At least one card is *not* a black queen. __________

NAME ______________________ CLASS __________ DATE __________

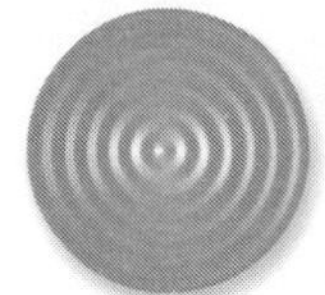

Practice Masters Level B

13.4 Independent Events

A red number cube and a green number cube are rolled together. Find the probability of each event.

1. The green cube is odd and the red cube is even. ______________
2. The green cube is less than 5 and the red cube is less than 6. ______________
3. The green cube is greater than 2 and the red cube is greater than 5. ______________
4. The green cube is less than 6 and the red cube is greater than 1. ______________

Use area models to find each probability below.

5. If the school bus is on time 8% of the time and you are on time 7% of the time, what is the probability that both you and the bus will be on time? ______________
6. Tim is taking the subway to the airport. If the subway is on time 65% of the time and his flight is on time 78% of the time, what is the probability that both will be on time? ______________
7. If Perry is late for work 4% of the time and his boss is late 8% of the time, what is the probability that both will be late? ______________
8. Gina is on time for class 85% of the time and her teacher is on time 98% of the time. What is the probability that both will be on time? ______________
9. If the mail is on time 98% of the time and Anthony pays his bills on time, 95% of the time, what is the probability that his bill payment will be received on time? ______________

Suppose that you draw two cards with replacement from a regular deck of 52 playing cards. Find the probability of each event.

10. Both cards are face cards. ______________
11. Both cards are number cards. ______________
12. Both cards are either an ace or two. ______________

Two cards are drawn with replacement from an ordinary deck of 52 cards. Find the probability of each event.

13. At least one card is not a face card. ______________
14. At least one card is not a number card. ______________
15. At least one card is neither an ace nor a two. ______________

NAME ______________________ CLASS ______________ DATE ______________

Practice Masters Level C

13.4 Independent Events

Use area models to find each probability below.

1. When commuting by bus, Stephan notices that his first bus is on time 53% of the time and his connecting bus is on time 67% of the time. What is the probability that both will be on time? ______________

2. Lee is ordering food for a party. If the pizza delivery is on time 85% of the time and the Chinese delivery is on time 76% of the time, what is the probability that both deliveries will be on time? ______________

3. To maximum her lunch hour, Kelli called in her order to the sub shop. If Kelli is on time to pick up her sandwich 83% of the time and the sandwich shop has their subs available on time 59% of the time, what is the probability that the subs will be available when Kelli arrives? ______________

One number is selected from the list {2, 3, 4, 6}. Another is selected from this list {1, 2, 3, 5, 7}. Find the probability of each event.

4. The two numbers selected are even. ______________

5. The two numbers selected are the same. ______________

6. The sum of the two numbers selected is even. ______________

7. The number selected from the second list is greater than the number selected from the first list. ______________

Ten chips numbered 1 through 10 are placed in a bag. A chip is drawn and replaced. Then a second chip is drawn. Find the probability of each event.

8. Both chips are odd. ______________

9. Both chips are even. ______________

10. The first chip is even AND the second chip is odd. ______________

11. One chip is even AND the other is odd. ______________

Seven chips numbered 1 through 7 are placed in a bag. A chip is drawn and replaced. Then a second chip is drawn. Find the probability of each event.

12. Both chips are odd. ______________

13. Both chips are even. ______________

14. The first chip is even AND the second chip is odd. ______________

NAME ______________________ CLASS ____________ DATE ____________

Practice Masters Level A

13.5 Simulations

Use coin tosses to simulate a 50% chance of rain for a two-day weekend. Perform 20 trials and combine your results with those of your classmates to answer the questions below.

	Trial 1	Trial 2	Trial 3	Trial 4	Trial 5	Trial 6	Trial 7	Trial 8	Trial 9	Trial 10
Day 1										
Day 2										

	Trial 11	Trial 12	Trial 13	Trial 14	Trial 15	Trial 16	Trial 17	Trial 18	Trial 19	Trial 20
Day 1										
Day 2										

1. For the entire class, what was the percent of trials that had rain on both days? ____________
2. For the entire class, what was the percent of trials that had no rain? ____________
3. For the entire class, what was the percent of trials that had rain on one day? ____________
4. For the entire class, what was the percent of trials that had rain on at least one day? ____________

Describe one way to simulate the random selection of the following:

5. a day of the month ____________
6. an hour of the day ____________
7. a month of the year ____________
8. a season of the year ____________
9. 1 student out of 25 students ____________
10. 1 teacher out of 6 teachers ____________

NAME ______________________ CLASS __________ DATE __________

Practice Masters Level B

13.5 Simulations

Use coin tosses to simulate a 50% chance of rain for a four-day holiday weekend. Perform 20 trials and combine your results with those of your classmates to answer the questions below.

	Trial 1	Trial 2	Trial 3	Trial 4	Trial 5	Trial 6	Trial 7	Trial 8	Trial 9	Trial 10
Day 1										
Day 2										
Day 3										
Day 4										

	Trial 11	Trial 12	Trial 13	Trial 14	Trial 15	Trial 16	Trial 17	Trial 18	Trial 19	Trial 20
Day 1										
Day 2										
Day 3										
Day 4										

1. For the entire class, what was the percent of trials that had rain on all four days? ______________
2. For the entire class, what was the percent of trials that had no rain? ______________
3. For the entire class, what was the percent of trials that had rain on two days? ______________
4. For the entire class, what was the percent of trials that had rain on at least two days? ______________

Suppose that you want to simulate a 50% chance of snow on each of two consecutive days in order to find the probability of there being snow at least one of the two days. Explain how you would perform the simulation by using the following methods:

5. generating random integers from 1 to 10 inclusive ______________
6. using a numbered cube ______________
7. using slips of paper ______________

NAME ______________________________ CLASS ______________ DATE ______________

Practice Masters Level C

13.5 Simulations

Use a number cube to simulate a $66\frac{2}{3}\%$ chance of rain for a three-day weekend. Perform 2 trials and combine your results with those of your classmates to answer the questions below.

	Trial 1	Trial 2	Trial 3	Trial 4	Trial 5	Trial 6	Trial 7	Trial 8	Trial 9	Trial 10
Day 1										
Day 2										
Day 3										

	Trial 11	Trial 12	Trial 13	Trial 14	Trial 15	Trial 16	Trial 17	Trial 18	Trial 19	Trial 20
Day 1										
Day 2										
Day 3										

1. For the entire class, what was the percent of trials that had rain on all three days? ______________
2. For the entire class, what was the percent of trials that had no rain? ______________
3. For the entire class, what was the percent of trials that had rain on at most two days? ______________

Design a simulation for each situation in Exercises 4–6, and give the results of your simulation.

4. Pick two integers from 1 to 5 inclusive. Give the experimental probability that both numbers are less than or equal to 2. ______________

____

5. A person has a 2% chance of correctly guessing a number. What is the experimental probability that a person will correctly guess the number 4 out of 7 tries? ______________

____

6. Design a simulation and then give the experimental probability that a family with 3 children has 2 boys and 1 girl, assuming that the births of boys and girls are equally likely. ______________

____

NAME ______________________ CLASS ______________ DATE ______________

Practice Masters Level A

14.1 Graphing Functions and Relations

Tell whether the relation is a function. Give a reason for each answer.

1. $\{(3, -1), (5, 4), (2, -1), (5, 0)\}$ ______________________

2. $\{(-3, 5), (0, 2), (-1, 0), (2, 4)\}$ ______________________

3. $\{(0, -5,), (-6, 4), (3, 1), (0, 0)\}$ ______________________

4. $\{(1, 1,), (2, 1), (3, 1), (4, 1)\}$ ______________________

Find $f(-1)$, $f(5)$, and $f(-3)$ for each function.

5. $f(x) = 2x + 3$ __________

6. $f(x) = |x|$ __________

7. $f(x) = 2x^2$ __________

8. $f(x) = 1 - x$ __________

9. $f(x) = |x - 1|$ __________

10. $f(x) = -3x^2$ __________

Name the parent function for each graph.

11. ______________________

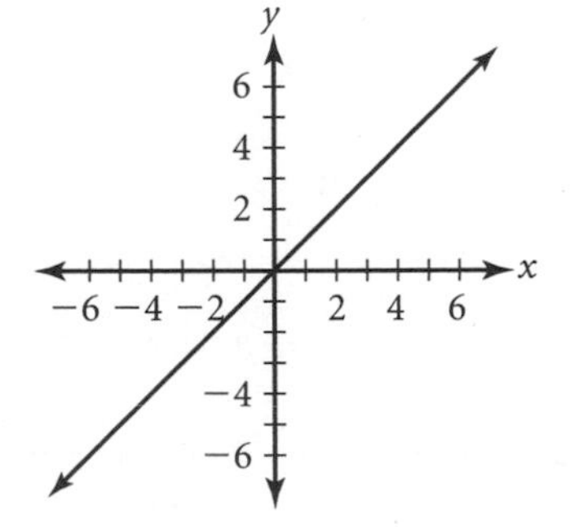

12. ______________________

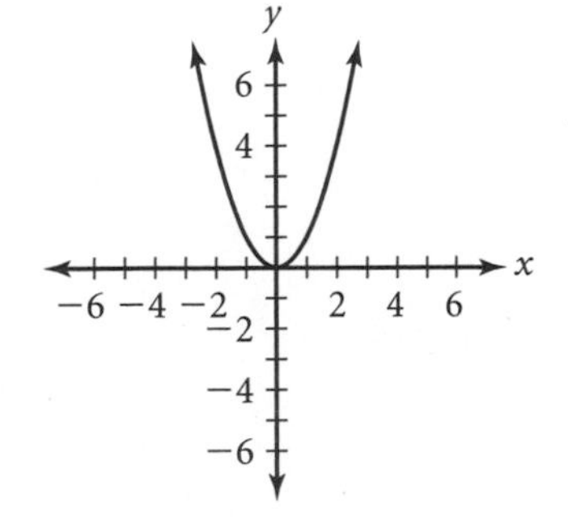

13. ______________________

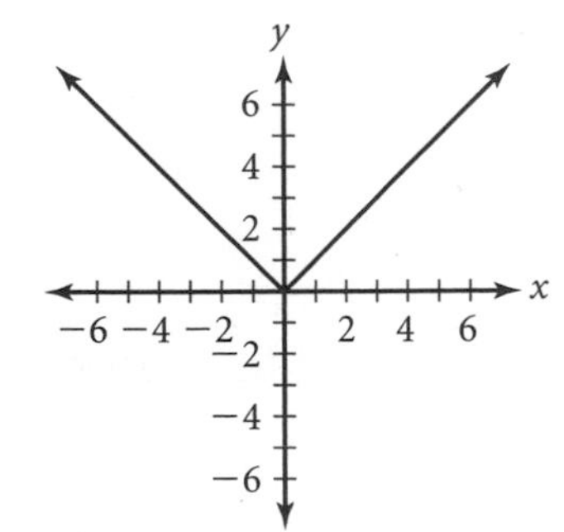

14. ______________________

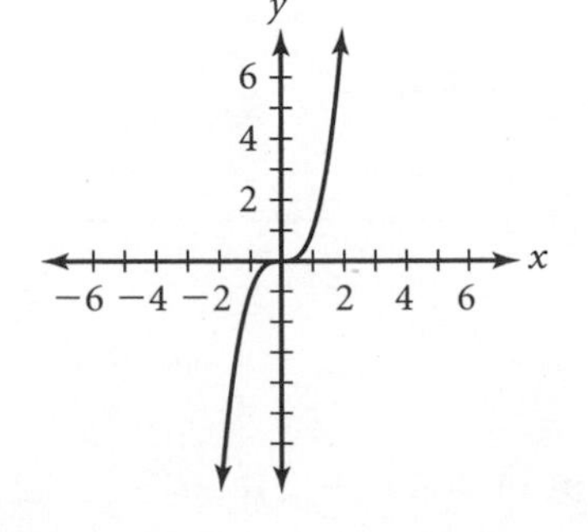

15. ______________________

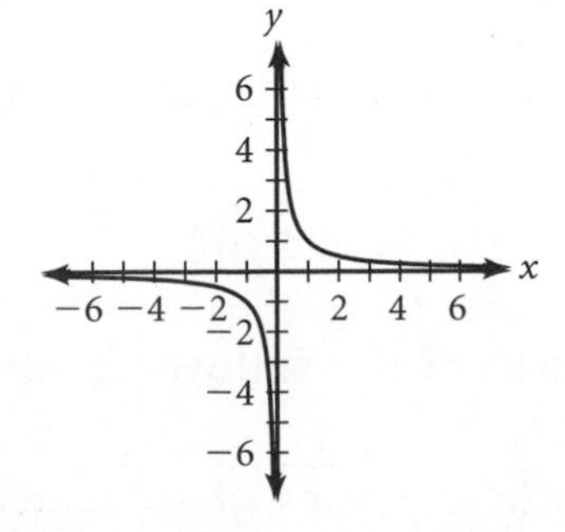

16. ______________________

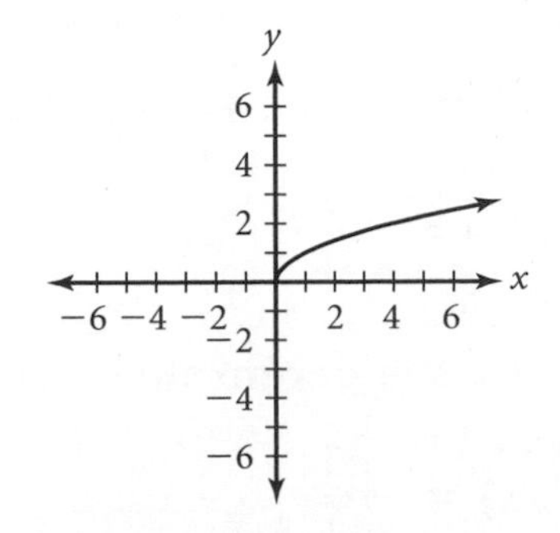

NAME ______________________ CLASS ____________ DATE ____________

Practice Masters Level B

14.1 Graphing Functions and Relations

Tell whether the relation is a function. Give a reason for each answer.

1. $\{(1, 2), (-1, 2), (5, 2)\}$ ______________________

2. $\{(3, 8)\}$ ______________________

3. $\{(0, -5,), (1, -6), (1, 2), (3, -8)\}$ ______________________

4. $\{(4, -1,), (-2, 3), (-2, 8)\}$ ______________________

Which of the following graphs represent functions? Give two reasons for each answer.

5. ______________ 6. ______________ 7. ______________

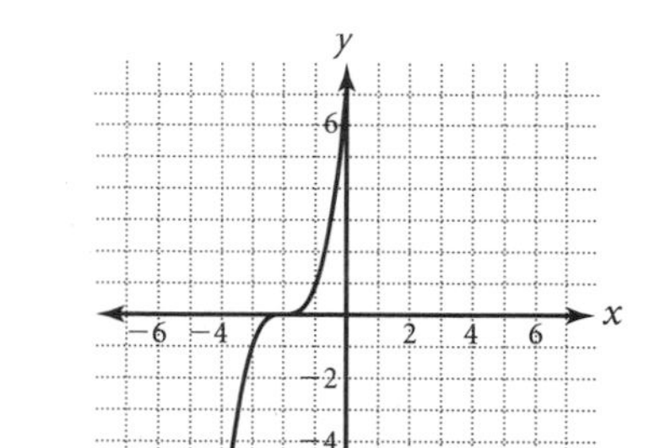

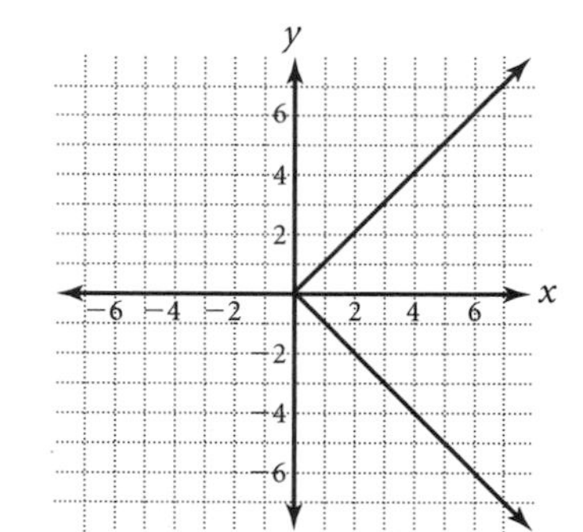

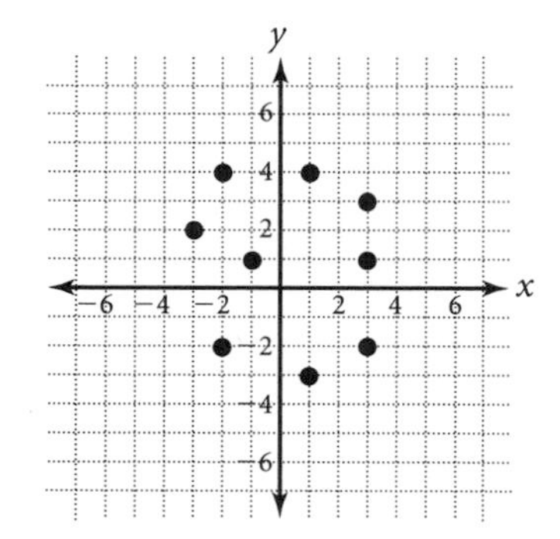

For $g(x) = x^2 - 5$, evaluate the following:

8. $g(0)$ ________ 9. $g(-1)$ ________ 10. $g\left(\frac{1}{2}\right)$ ________

11. $g(-2)$ ________ 12. $g(1)$ ________ 13. $g(5)$ ________

Evaluate each function for $x = -5$.

14. $f(x) = x^2 - 2x + 5$ ________ 15. $g(x) = |2x - 8|$ ________

16. $m(x) = \dfrac{1}{x + 2}$ ________ 17. $p(x) = \sqrt{x + 30}$ ________

Identify the parent function for each of the following functions.

18. $h(x) = \dfrac{1}{x + 5}$ ________ 19. $m(x) = \sqrt{x - 8}$ ________ 20. $p(x) = (x + 2)^2 + 5$ ________

NAME ______________________ CLASS __________ DATE __________

Practice Masters Level C

14.1 Graphing Functions and Relations

Give an example of each.

1. a set of ordered pairs that is a relation but is not a function ______________________

2. a rule for $f(x)$ that is a function ______________________

3. a model of a relation that is a function

____

4. a graph of a relation that is not a function

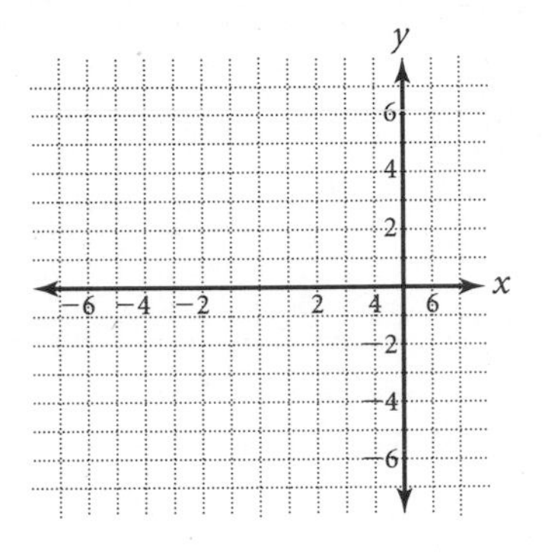

For $f(x) = 2(x - 1)^2 + 1$, evaluate the following:

5. $f(-3)$ ______________ 6. $f(2b)$ ______________ 7. $f(5)$ ______________

8. $f(0)$ ______________ 9. $f(a)$ ______________ 10. $f(100)$ ______________

Identify the parent function for each of the following functions.

11. $h(x) = \dfrac{-2}{x - 5}$ ______________ 12. $m(x) = -3\sqrt{2x + 1}$ ______________

13. $f(x) = -2(3 - x)^2 - 3$ ______________ 14. $k(x) = 2|x - 5| - 4$ ______________

Name the parent function for each graph.

15. ______________ 16. ______________ 17. ______________

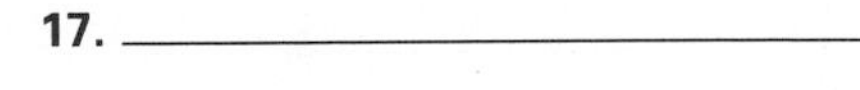

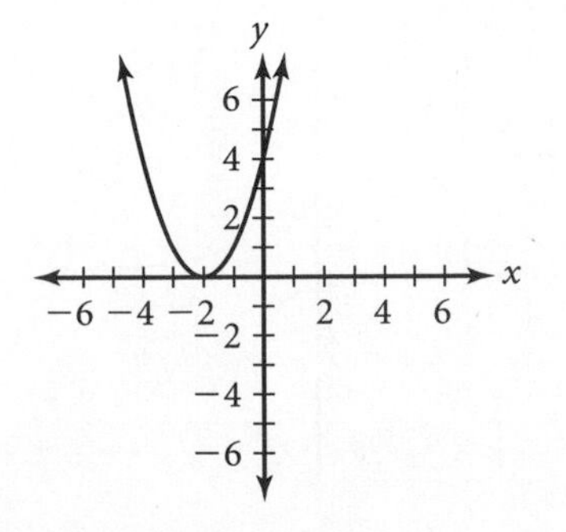

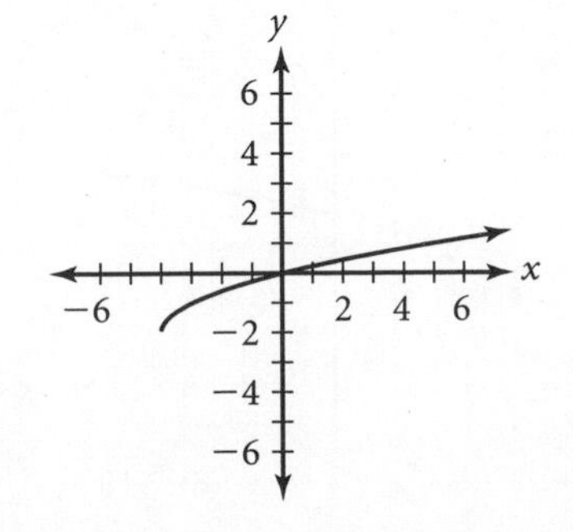

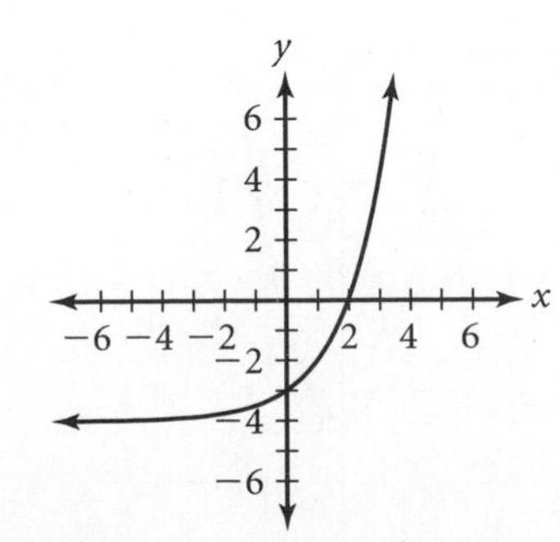

NAME ______________________ CLASS ____________ DATE ____________

Practice Masters Level A

14.2 Translations

Describe how to obtain the graph of each function from the graph of its parent function.

1. $y = x + 2$ ______________________

2. $y = x^2 - 6$ ______________________

3. $y = (x - 3)^2$ ______________________

4. $y = \frac{1}{x + 2}$ ______________________

5. $y = (x + 2)^2$ ______________________

6. $y = 10^x + 5$ ______________________

7. $y = \frac{1}{x} - 8$ ______________________

8. $y = x + 6$ ______________________

9. $y = \sqrt{x - 5}$ ______________________

10. $y = \sqrt{x} + 6$ ______________________

The point (3, 4) is on the graph of a function. Give the coordinates of the corresponding point on the new graph when the following translations are applied:

11. horizontal translation of 2 ____________

12. vertical translation of 5 ____________

13. vertical translation of -6 ____________

14. horizontal translation of 0 ____________

For each graph, determine the parent function and the translation.

15. ____________

16. ____________

17. ____________

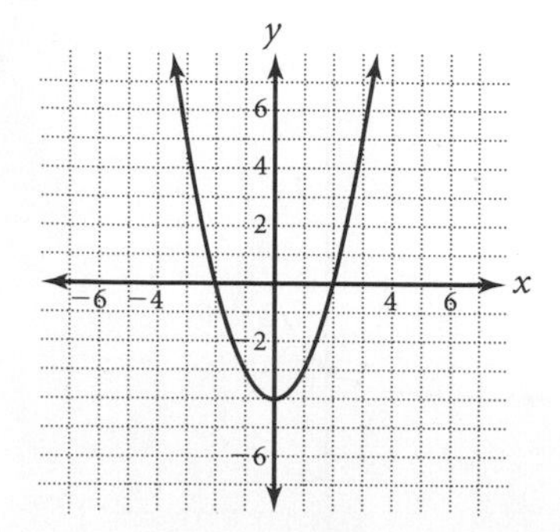

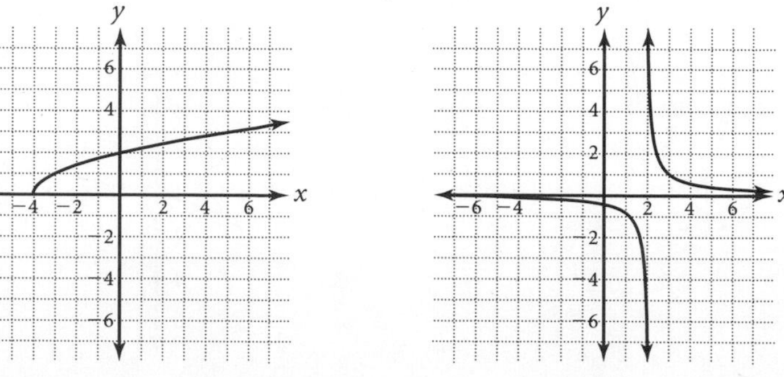

NAME ______________________ CLASS ______________ DATE ______________

Practice Masters Level B

14.2 Translations

Describe how to obtain the graph of each function from the graph of its parent function.

1. $y = |x + 1| - 3$ ______________
2. $y = (x + 3)^2 + 4$ ______________
3. $y = (x - 1)^2 + 3$ ______________
4. $y = |x - 2| + 5$ ______________
5. $y = \sqrt{x + 1} - 5$ ______________
6. $y = \dfrac{1}{x - 3} + 6$ ______________
7. $y = x^2 - 5$ ______________
8. $y = |x - 1| + 2$ ______________
9. $y = (x + 2)^2 - 5$ ______________
10. $y = \dfrac{1}{x + 1} - 3$ ______________

The point (3, 4) is on the graph of a function. Give the coordinates of the corresponding point on the new graph when the following translations are applied:

11. horizontal translation of 2, vertical translation of 1 ______________
12. horizontal translation of −1, vertical translation of −2 ______________
13. horizontal translation of 5, vertical translation of 0 ______________
14. horizontal translation of 3, vertical translation of −1 ______________
15. horizontal translation of −4, vertical translation of 5 ______________

For each graph, determine the parent function and the translation.

16. ______________ 17. ______________ 18. ______________

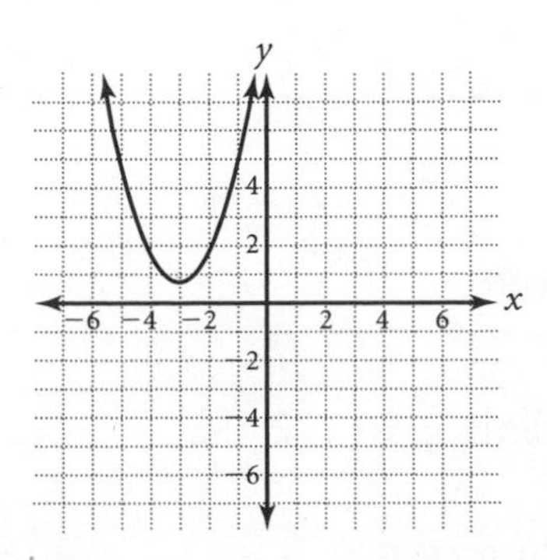

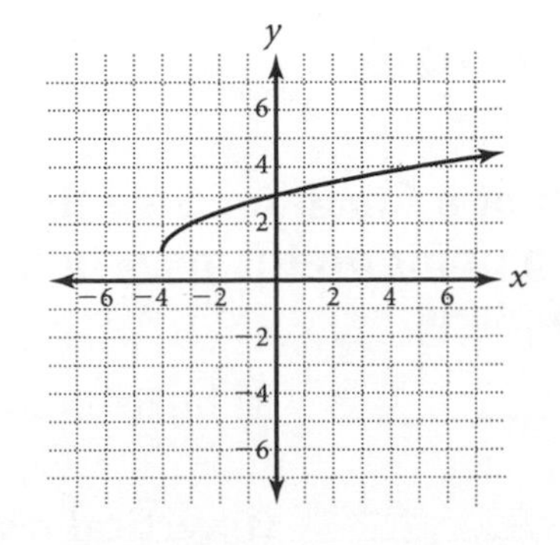

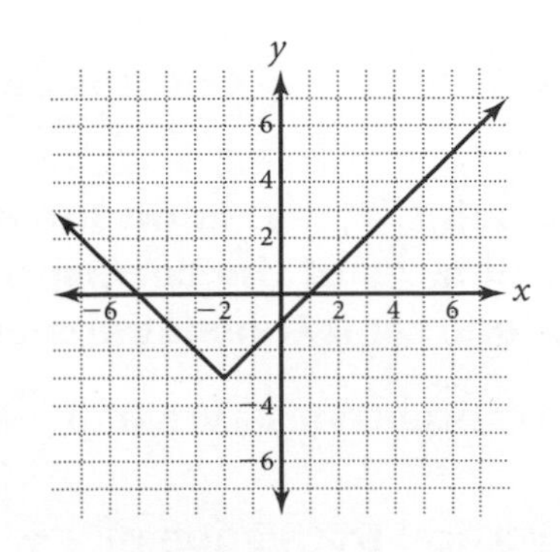

NAME ______________________ CLASS ____________ DATE ____________

Practice Masters Level C

14.2 Translations

For Exercises 1–3:

a. Identify the parent function.

b. Describe how to obtain the graph of each function from the graph of the parent function.

c. Graph and label the parent function and the transformed function.

1. $y = (x + 5)^2 + 1$

a. ______________

b. ______________

c.

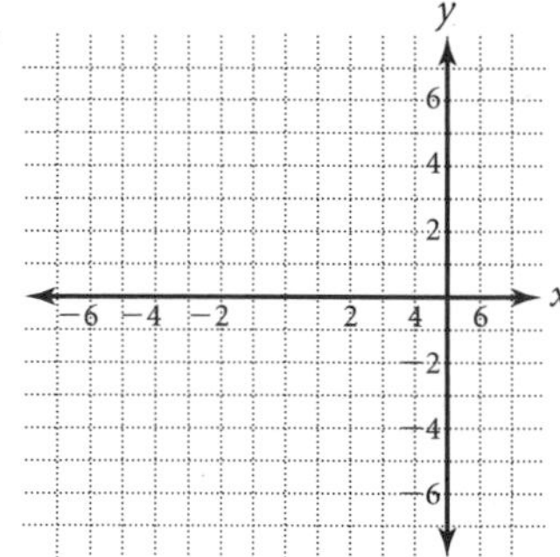

2. $y = |x - 4| - 2$

a. ______________

b. ______________

c.

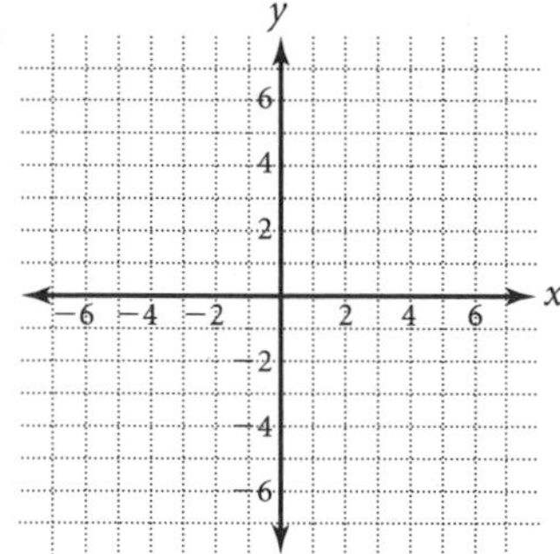

3. $y = \sqrt{x + 3} - 4$

a. ______________

b. ______________

c.

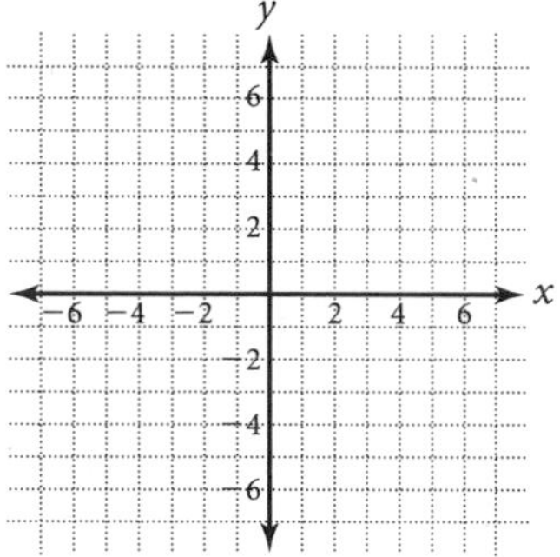

The point (−2, 2) lies on the graph of the parent function *f*(*x*) = |*x*|. Use this fact to describe how *f*(*x*) is translated to obtain each function.

4. $g(x) = |x + 1|$ contains the point $(-3, 2)$ ______________

5. $g(x) = |x - 1|$ contains the point $(-1, 2)$ ______________

6. $g(x) = |x| - 4$ contains the point $(-2, -2)$ ______________

7. $g(x) = |x| + 4$ contains the point $(-2, 6)$ ______________

The point (4, −7) is on the graph of a function. Give the coordinates of the corresponding point on the new graph when the following translations are applied:

8. vertical translation of 3 ______________

9. horizontal translation of -2 ______________

10. vertical translation of $y + 2$ ______________

11. vertical translation of $d + b$ ______________

NAME ______________________ CLASS ______________ DATE ______________

Practice Masters Level A

14.3 Stretches and Compressions

Determine which functions are vertical or horizontal stretches or compressions of their parent graph.

1. $y = 2x^2$ ______________
2. $y = 10^{3x}$ ______________
3. $y = |3x|$ ______________
4. $y = \left(\frac{1}{3}x\right)^2$ ______________
5. $y = \frac{5}{x}$ ______________
6. $y = \frac{1}{6x}$ ______________
7. $y = \frac{\sqrt{x}}{3}$ ______________
8. $y = \frac{1}{5}|x|$ ______________
9. $y = \sqrt{\frac{1}{2}x}$ ______________
10. $y = 2 \cdot 10^x$ ______________

Determine the parent function of each function, and use it to sketch the graph of each function.

11. $y = 4|x|$ ______________

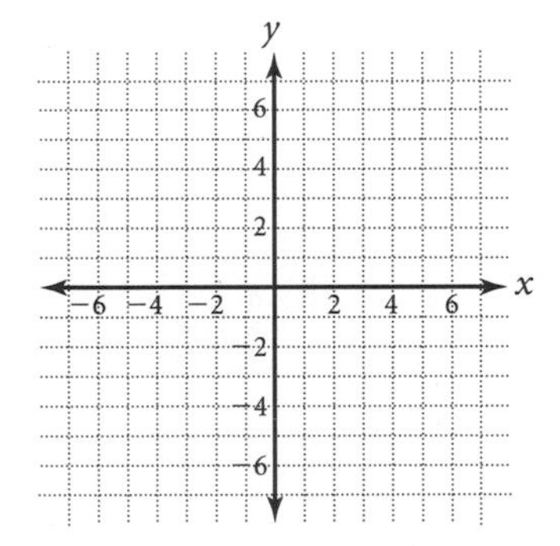

12. $y = \frac{\sqrt{x}}{4}$ ______________

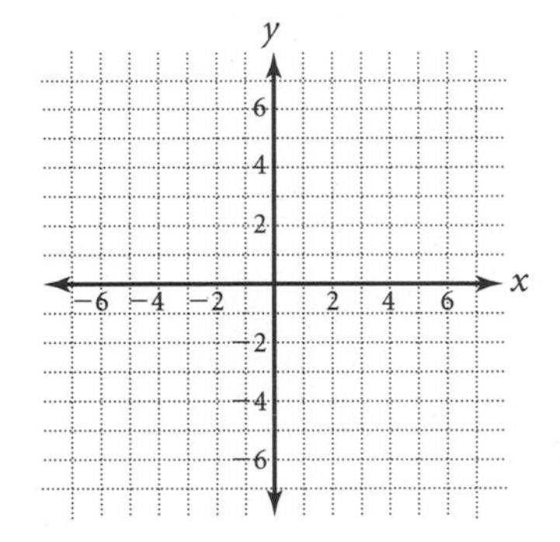

13. $y = \frac{1}{2}x^2$ ______________

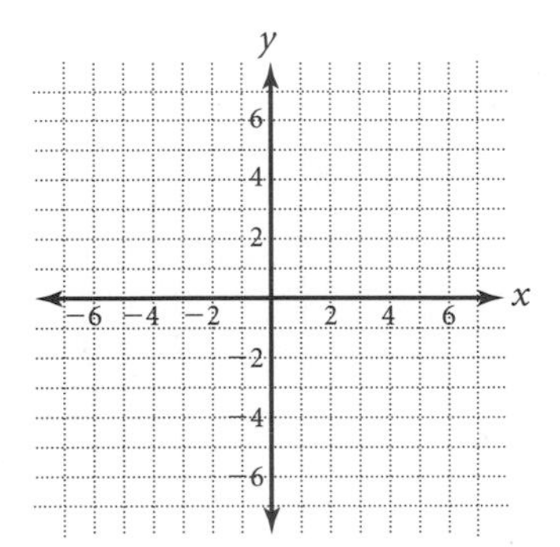

14. $y = |3x|$ ______________

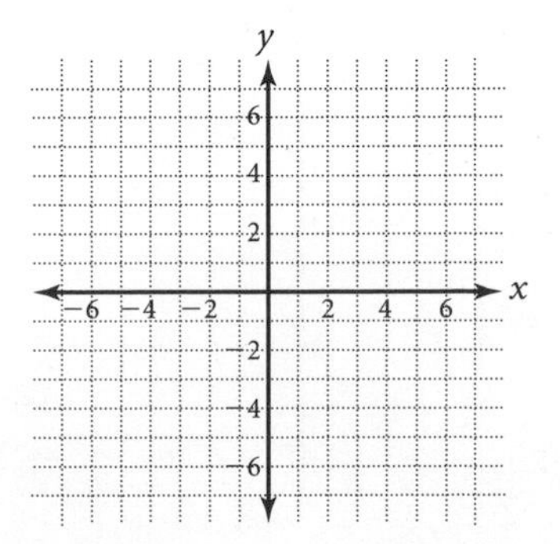

15. $y = (2x)^2$ ______________

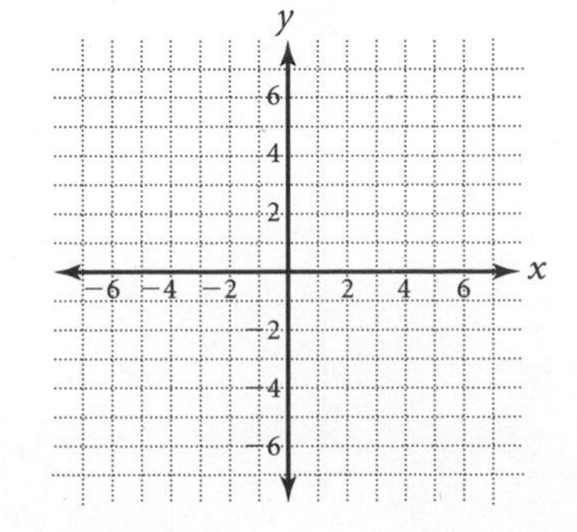

16. $y = \frac{1}{3x}$ ______________

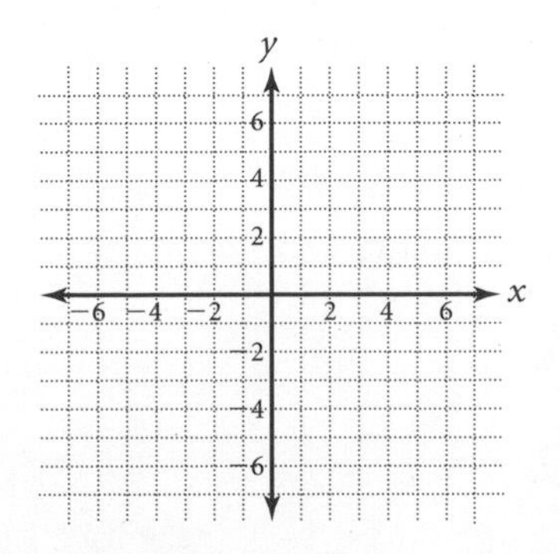

NAME ______________________ CLASS ______________ DATE ______________

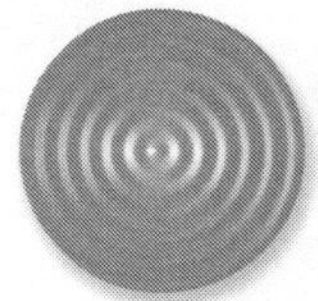

Practice Masters Level B

14.3 Stretches and Compressions

Determine whether each function is a vertical or horizontal stretch or compression, or neither, of their parent graph.

1. $y = 3x$ ______________
2. $y = \frac{1}{3}x^2$ ______________
3. $y = \frac{1}{x+2}$ ______________
4. $y = \frac{2}{x}$ ______________
5. $y = \frac{1}{3x}$ ______________
6. $y = |2x|$ ______________
7. $y = (4x)^2$ ______________
8. $y = 5\sqrt{x}$ ______________
9. $y = \sqrt{x-1}$ ______________
10. $y = |x| - 5$ ______________

Determine the parent function of each function, and use it to sketch the graph of each function.

11. $y = 0.25|x|$ ______________

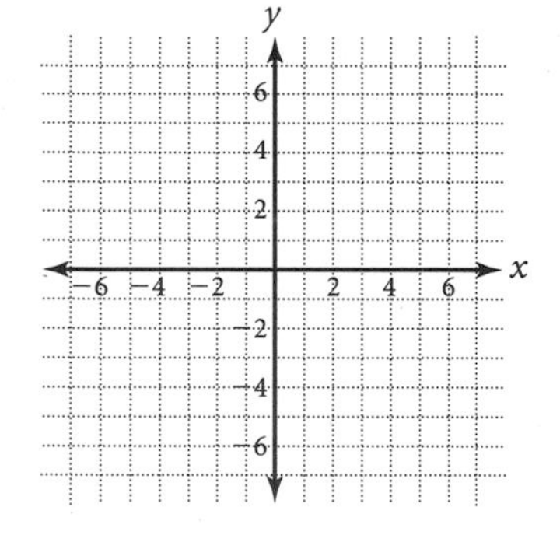

12. $y = 4\sqrt{2x}$ ______________

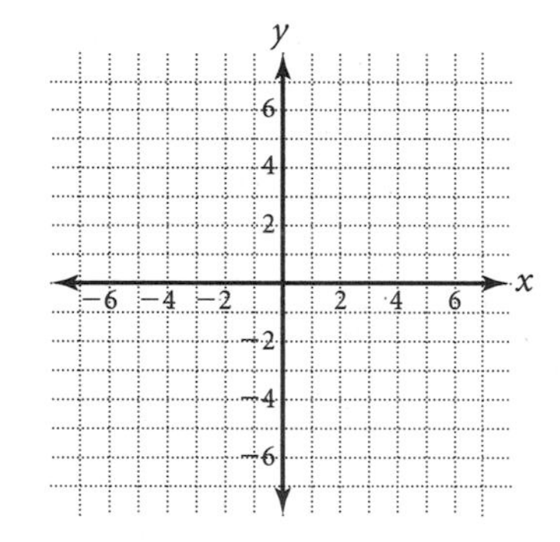

13. $y = (1.5x)^2$ ______________

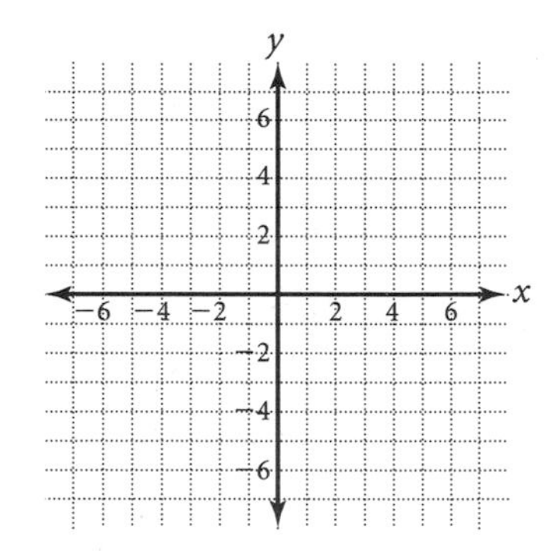

14. $y = \frac{1}{2}|3x|$ ______________

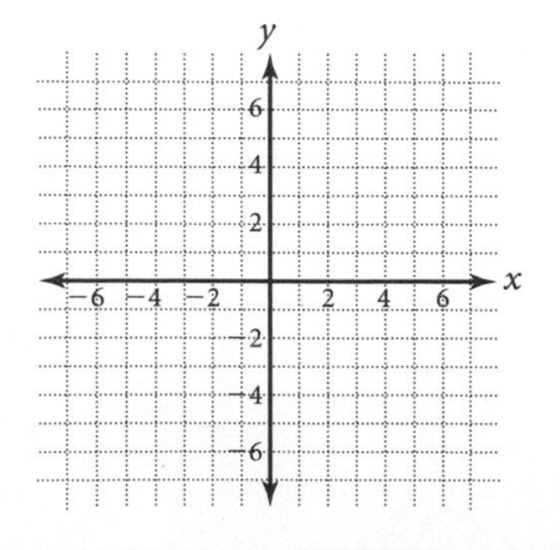

15. $y = 2(x + 1)^2$ ______________

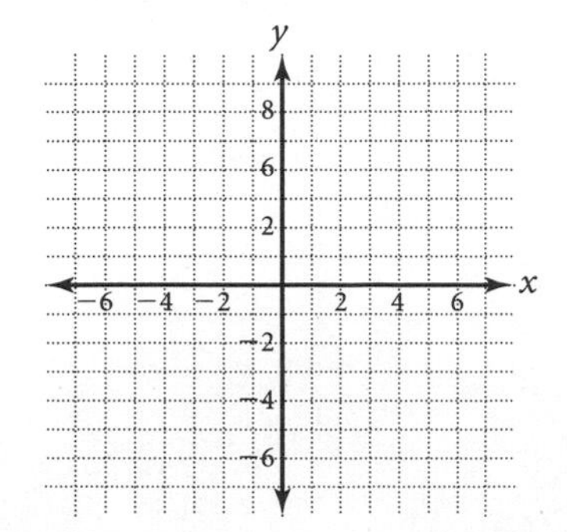

16. $y = \frac{2}{3x}$ ______________

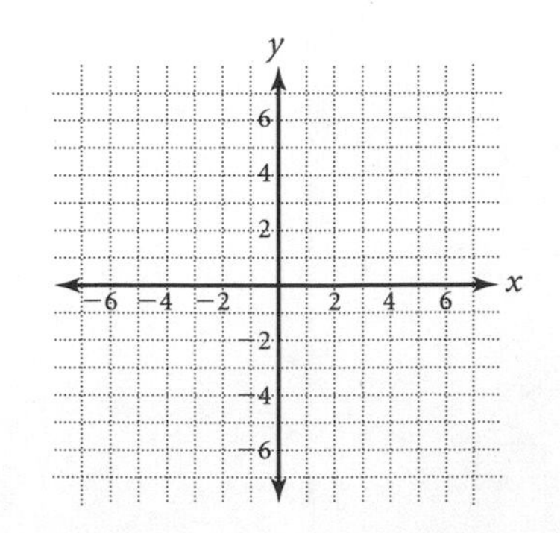

Practice Masters Level C

14.3 Stretches and Compressions

For Exercises 1–3 identify the parent function. Then sketch the graphs of both functions on the same coordinate plane. Each point (*x, y*) on the graph of parent function is transformed to what point on the new graph?

1. $y = \frac{1}{3}x^2$ ______________

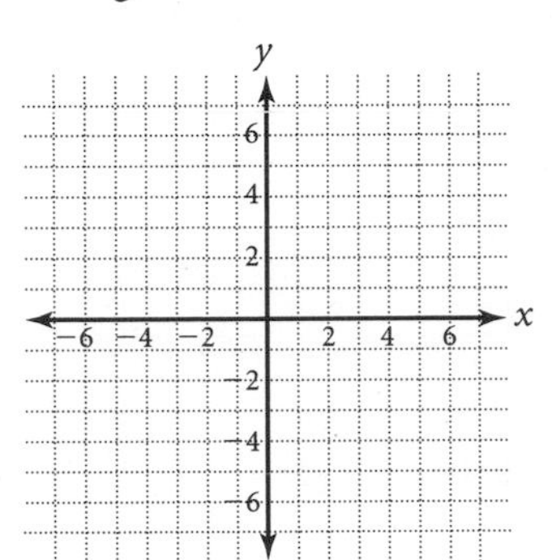

2. $y = |2x|$ ______________

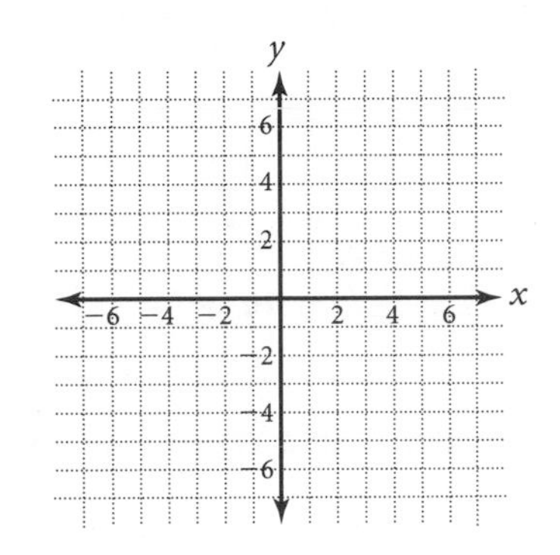

3. $y = 3\sqrt{x}$ ______________

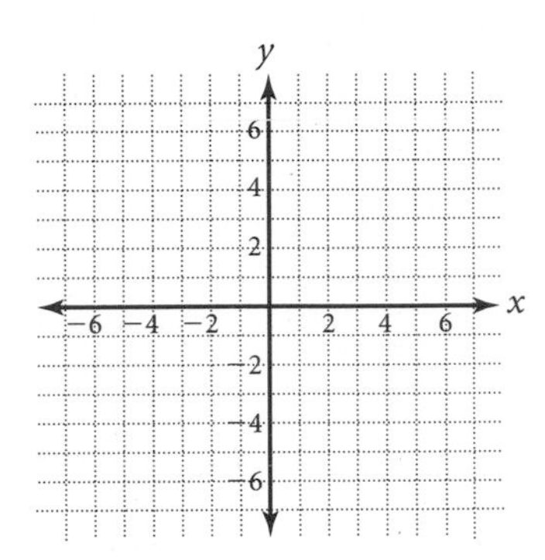

Write the equation for each graph.

4. ______________

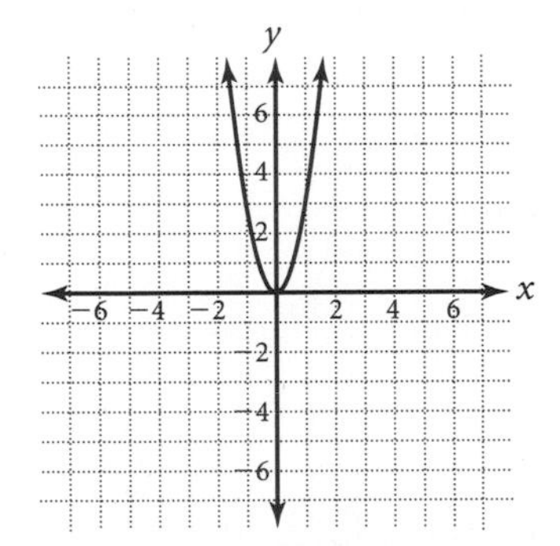

5. ______________

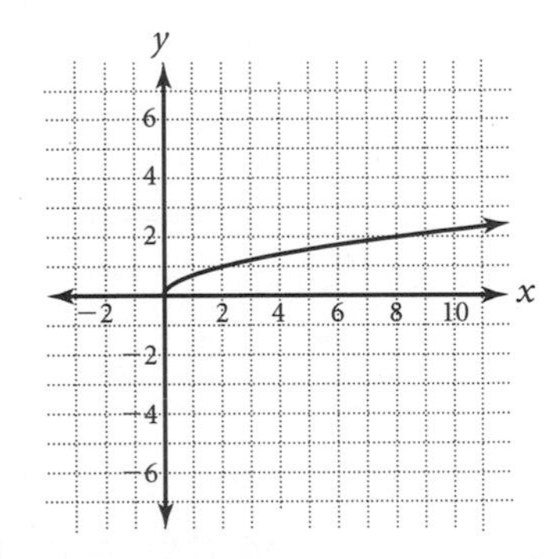

6. ______________

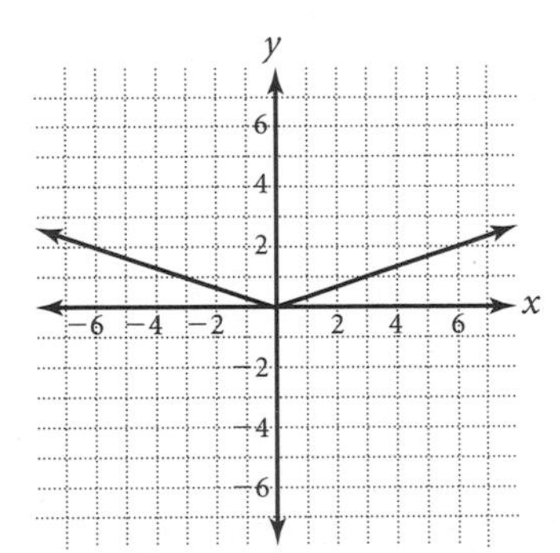

The point (−4, 16) lies on the graph of the parent function $f(x) = x^2$. Use this fact to describe how $f(x)$ is stretched or compressed to obtain each function.

7. $h(x) = \frac{1}{2}x^2$ contains the point $(-4, 8)$ ______________

8. $h(x) = 2x^2$ contains the point $(-4, 32)$ ______________

9. $h(x) = \left(\frac{1}{2}x\right)^2$ contains the point $(-8, 16)$ ______________

10. $h(x) = (2x)^2$ contains the point $(-2, 16)$ ______________

NAME ______________________ CLASS ______________ DATE ______________

Practice Masters Level A

14.4 Reflections

Determine whether the function is reflected over the *x*-axis or the *y*-axis.

1. $y = -x$ ______________

2. $y = -|x|$ ______________

3. $y = |-x|$ ______________

4. $y = \sqrt{-x}$ ______________

5. $y = -x^2$ ______________

6. $y = (-x)^2$ ______________

7. $y = \dfrac{1}{-x}$ ______________

8. $y = 10^{-x}$ ______________

9. $y = -\dfrac{1}{x}$ ______________

10. $y = -10^x$ ______________

Graph each pair of functions, and identify the transformations from *f* to *g*.

11. $f(x) = \sqrt{x - 1}, g(x) = -\sqrt{x - 1}$

____

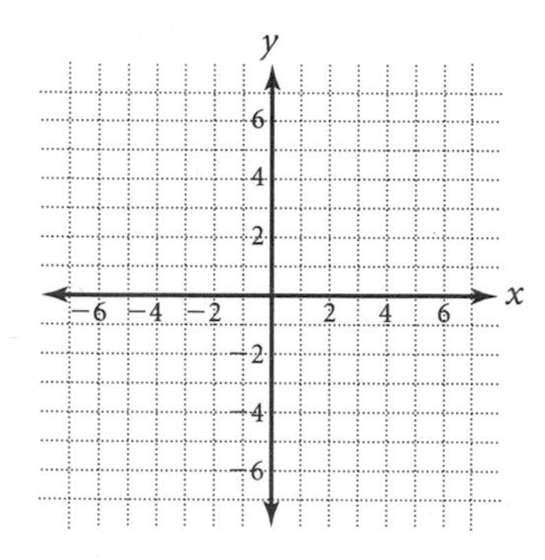

12. $f(x) = \dfrac{1}{x + 3}, g(x) = \dfrac{-1}{x + 3}$

____

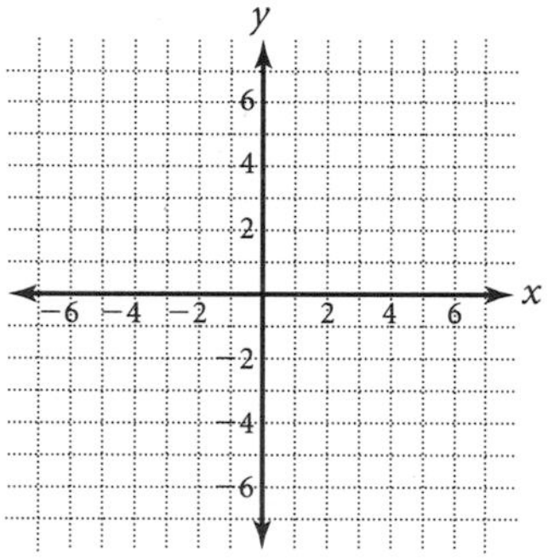

13. $f(x) = (x - 2)^2, g(x) = (-x - 2)^2$

____

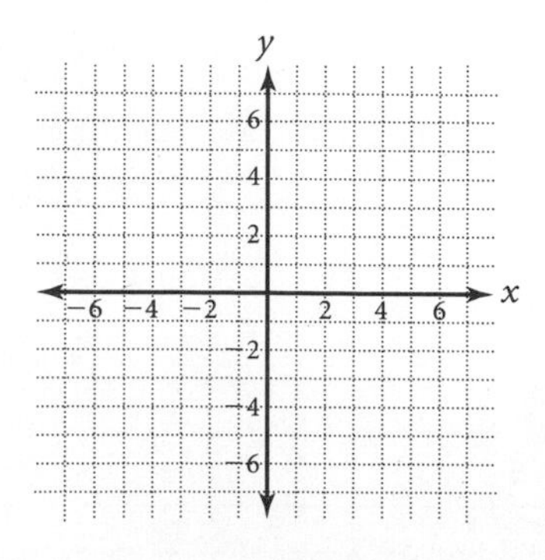

14. $f(x) = |x + 2|, g(x) = -|x + 2|$

____

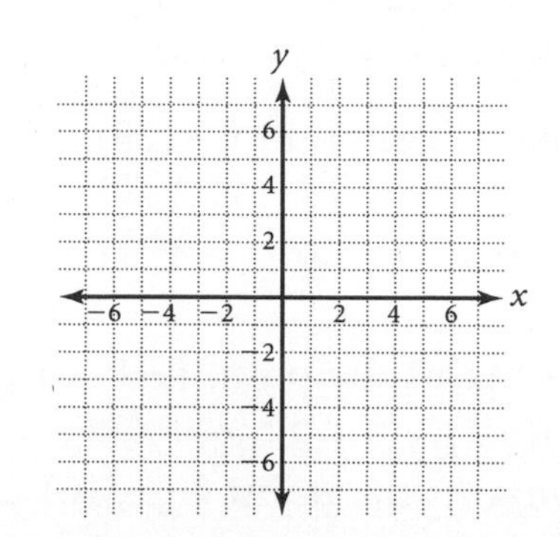

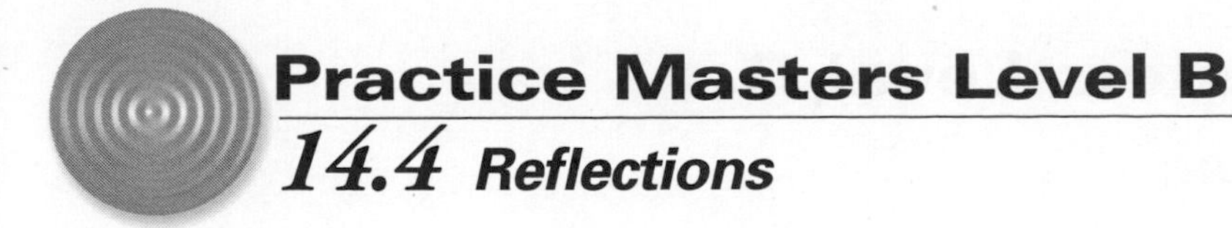

Practice Masters Level B

14.4 Reflections

Graph each function with its parent function on the same coordinate plane. Describe the transformation of the parent function.

1. $f(x) = -\sqrt{x + 1}$ ______________________

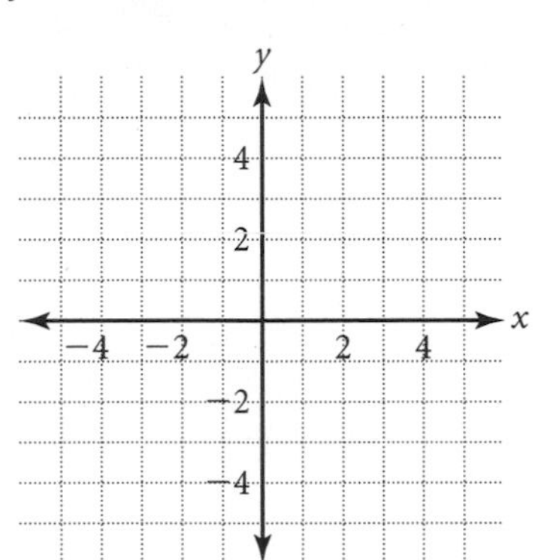

2. $g(x) = \dfrac{-1}{x - 2}$ ______________________

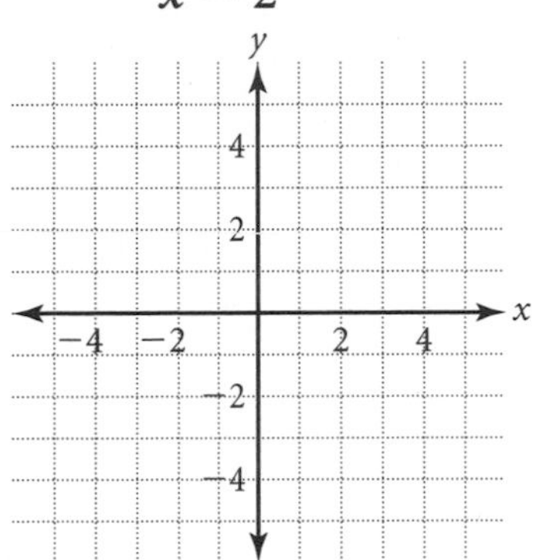

3. $f(x) = -(x + 2)^2$ ______________________

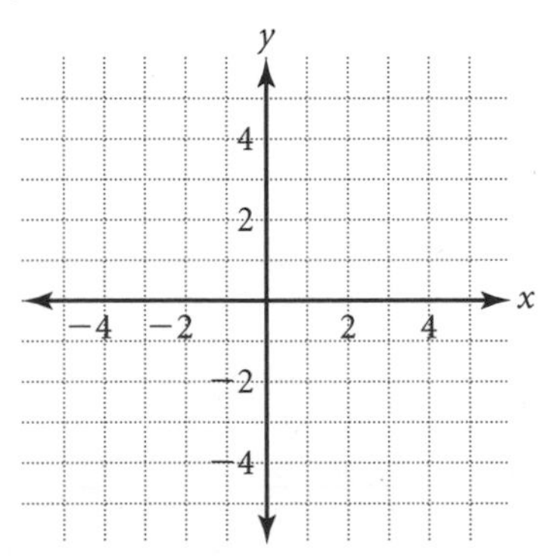

4. $f(x) = -|x - 3|$ ______________________

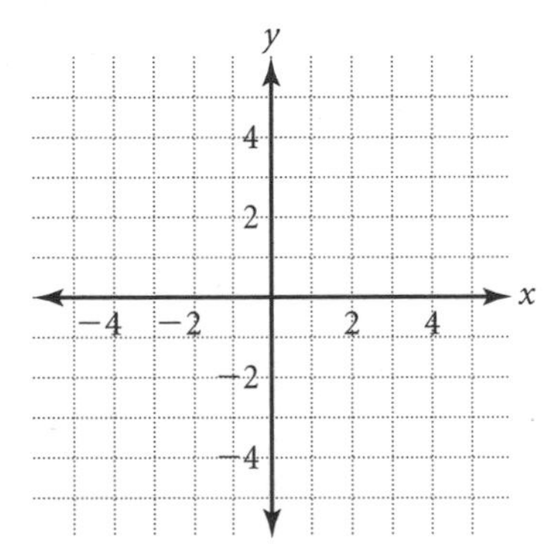

Tell which kind of reflection, horizontal or vertical, is applied to the function $y = 2x - 5$.

5. $f(x) = 2(-x) - 5$ ______________________

6. $f(x) = -(2x - 5)$ ______________________

Write the functions for the graph of each reflected across the *y*-axis and the *x*-axis.

7. $f(x) = 2x - 1$ ______________________

8. $f(x) = \sqrt{x - 1}$ ______________________

9. $f(x) = (x + 5)^2$ ______________________

10. $f(x) = |2x|$ ______________________

11. $f(x) = \dfrac{1}{x - 1}$ ______________________

12. $f(x) = \dfrac{3}{x}$ ______________________

NAME ______________________ CLASS ____________ DATE ____________

Practice Masters Level C

14.4 Reflections

Write the functions for the graph of each reflected across the *x*-axis and the *y*-axis.

1. $f(x) = (x + 2)^2$ ______________ ______________

2. $g(x) = \sqrt{x} - 2$ ______________ ______________

3. $m(x) = 2x - 1$ ______________ ______________

4. $n(x) = \frac{1}{2}x^2$ ______________ ______________

5. $r(x) = x + 5$ ______________ ______________

6. $d(x) = \frac{1}{2x}$ ______________ ______________

Write the equation for each graph.

7. ______________

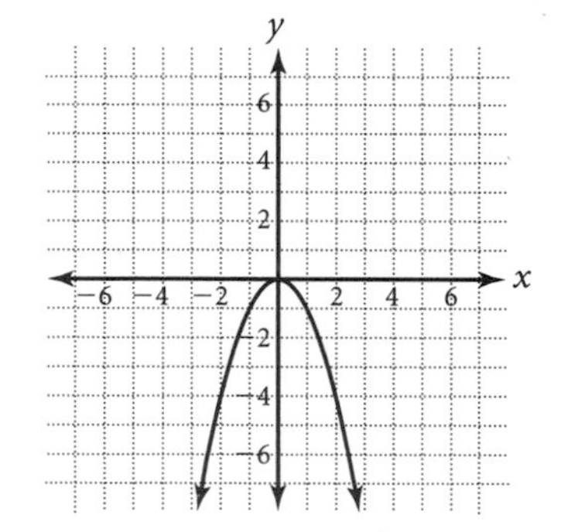

8. ______________

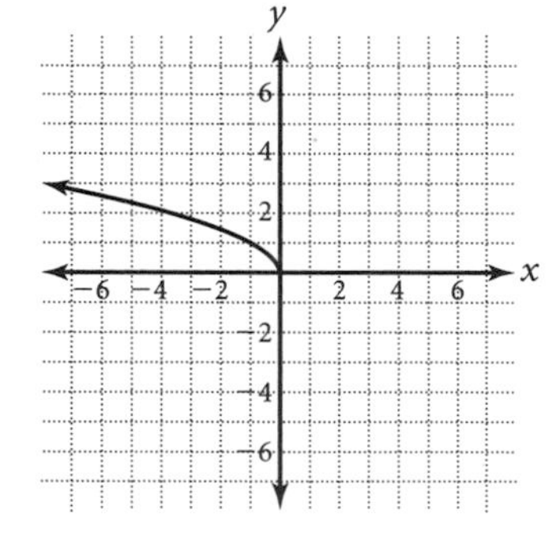

9. ______________

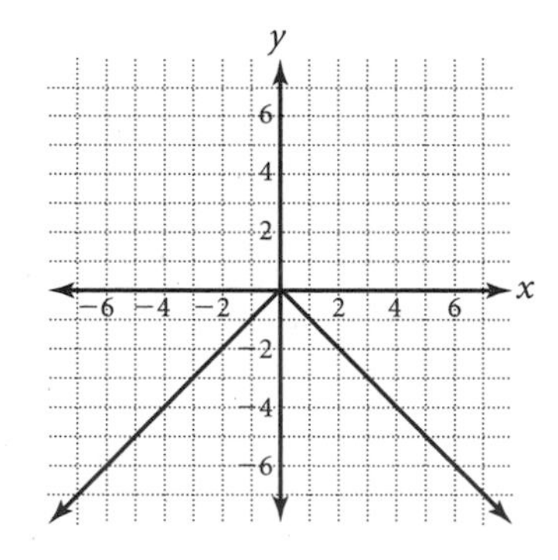

The point $\left(\frac{1}{4}, 4\right)$ lies on the graph of the parent function $f(x) = \frac{1}{x}$. Determine how this point is changed under the following:

10. reflected across the *x*-axis ______________

11. reflected across the *y*-axis ______________

12. reflected across the *x*-axis and then the *y*-axis ______________

13. reflected across the *y*-axis and then the *x*-axis ______________

Practice Masters Level A

14.5 Combining Transformations

Sketch the graph of each function.

1. $g(x) = \sqrt{x - 1} + 2$

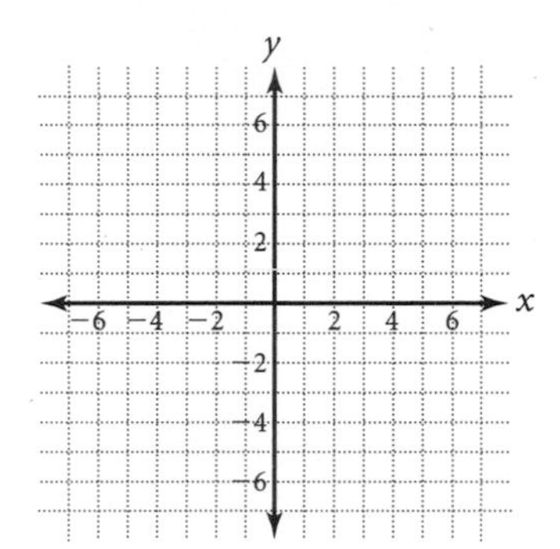

2. $h(x) = -\dfrac{1}{x + 3}$

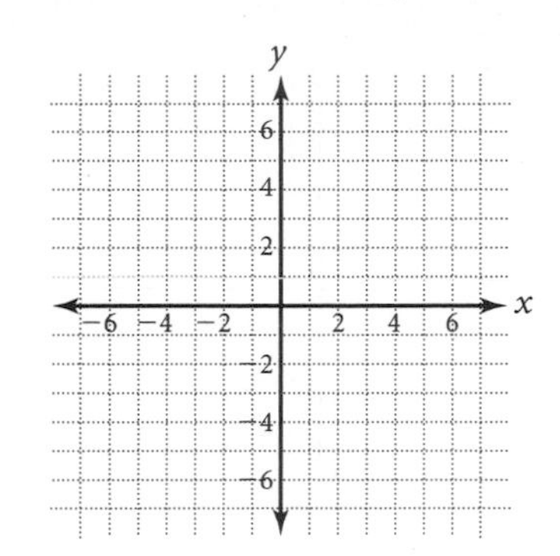

3. $f(x) = 2(x + 1)^2$

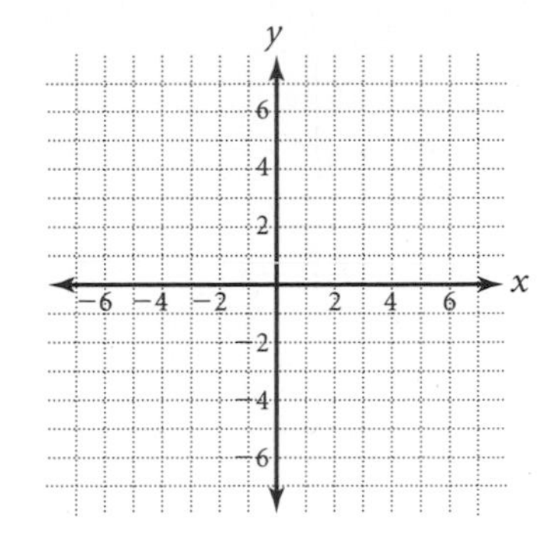

4. $p(x) = \dfrac{1}{2}|x - 1|$

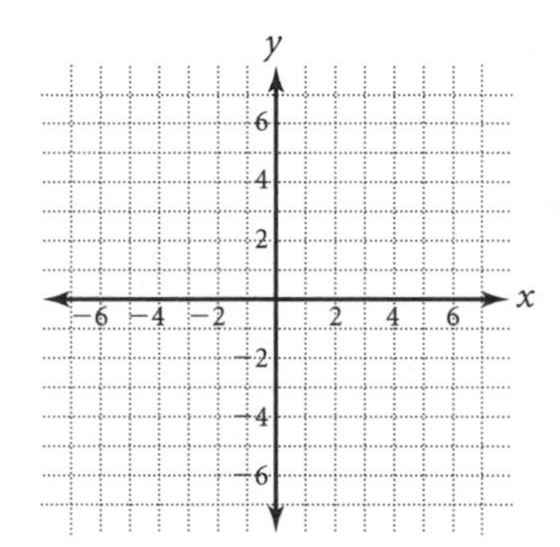

5. $q(x) = -x^2 - 2$

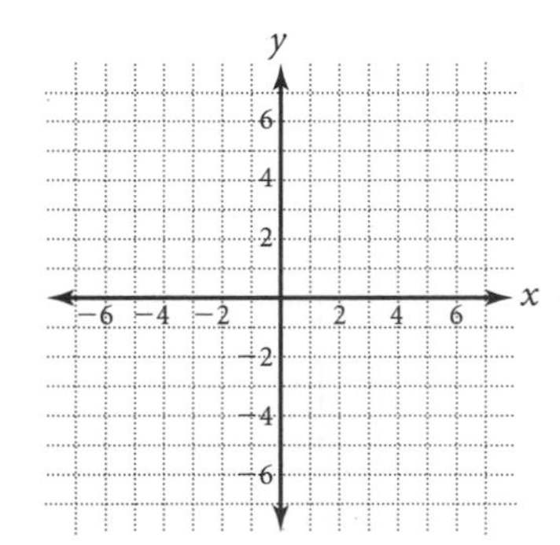

6. $f(x) = \sqrt{2x} + 5$

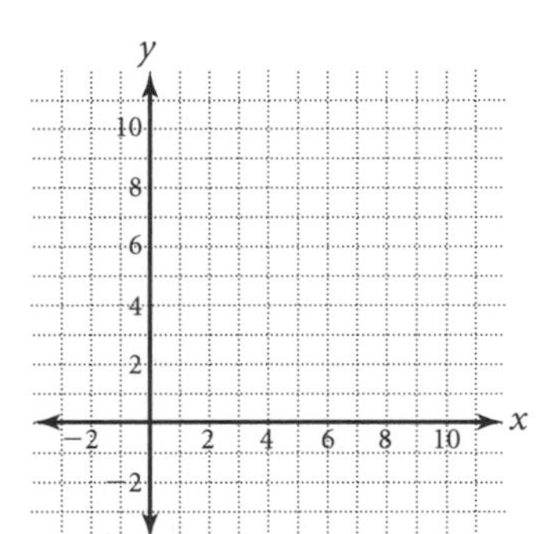

Name the point that corresponds to the point (6, −2) after applying the given transformations to the graph of $y = x - 8$.

7. a reflection across the x-axis, followed by a horizontal translation of −3 ______________

8. a reflection across the y-axis, followed by a vertical stretch of 2 ______________

A family is planning a vacation that will take them 2000 miles from home. They allow themselves 10 hours for resting.

9. Write the function that describes the rate, r, that they must travel in order to arrive in a certain number of hours, h. ______________

10. Graph the function from Exercise 9 and the parent function on the same coordinate grid.

Practice Masters Level B

14.5 Combining Transformations

Sketch the graph of each function.

1. $g(x) = \dfrac{-2}{x - 3}$

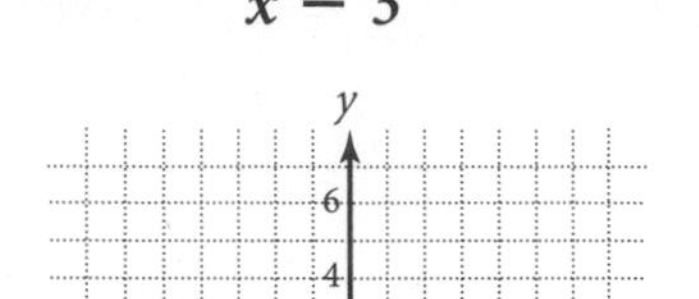

2. $h(x) = -\sqrt{x - 1} + 2$

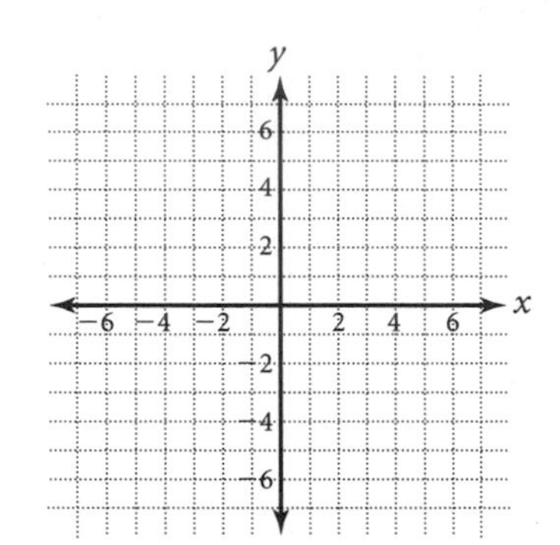

3. $f(x) = \dfrac{1}{2}(x - 1)^2 + 3$

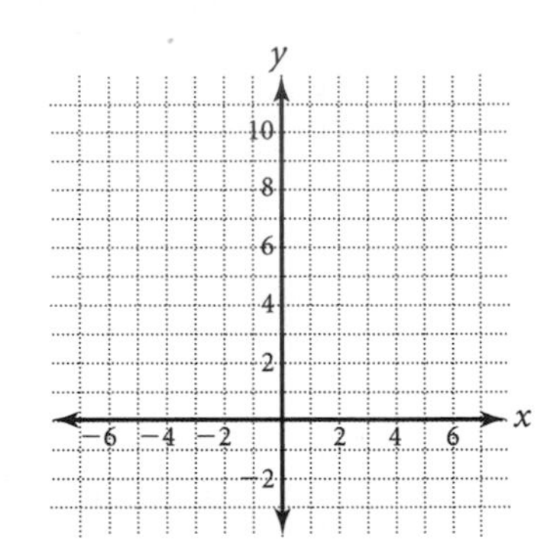

4. $p(x) = -2|x + 2| - 3$

5. $q(x) = -\dfrac{1}{2}x^2 + 1$

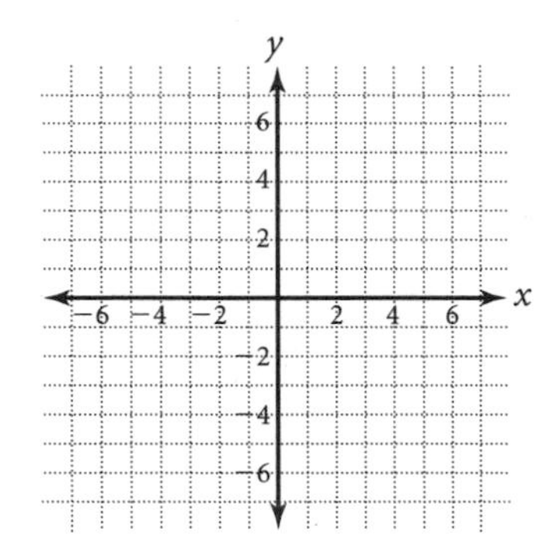

6. $f(x) = \dfrac{1}{4}\sqrt{3x} - 2$

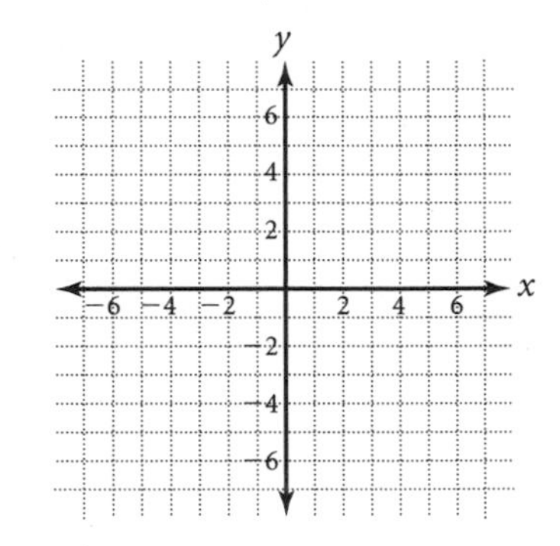

Name the point that corresponds to the point (2, 4) after applying the given transformations to the graph of $y = x^2$.

7. a horizontal stretch by a factor of 3, then a horizontal translation of 1, followed by a vertical translation of 2 ______________________

8. a reflection across the x-axis, then a reflection across the y-axis, followed by a vertical translation of -2 ______________________

9. a reflection across the y-axis, then a vertical stretch of 2, followed by a vertical translation of 2 ______________________

NAME ______________________ CLASS ____________ DATE ____________

Practice Masters Level C

14.5 Combining Transformations

Sketch the graph of each function.

1. $g(x) = \dfrac{3}{-x + 1} + 5$

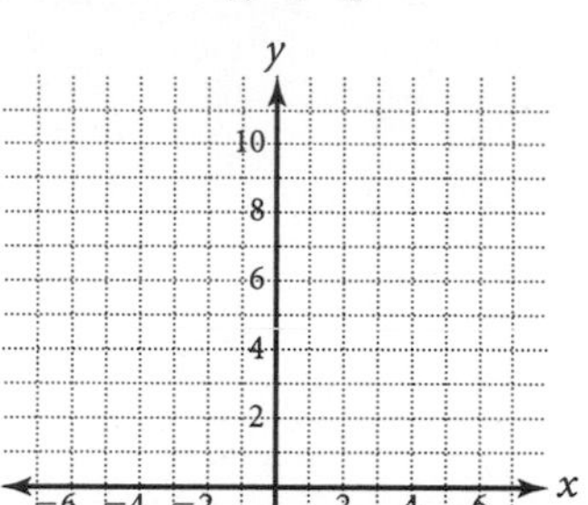

2. $h(x) = -2\sqrt{x + 2} + 1$

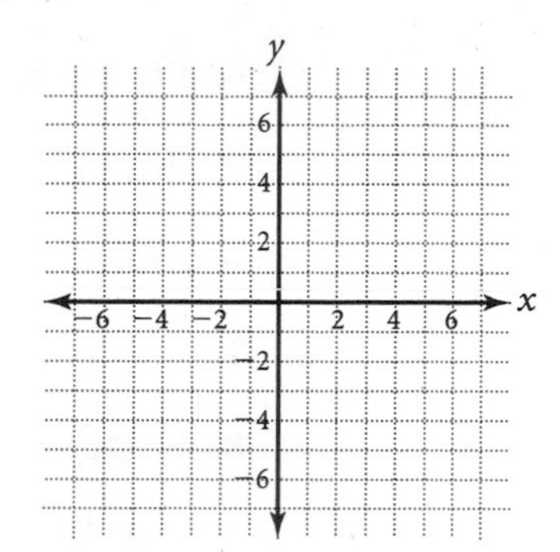

3. $f(x) = 3(x + 4)^2 - 2$

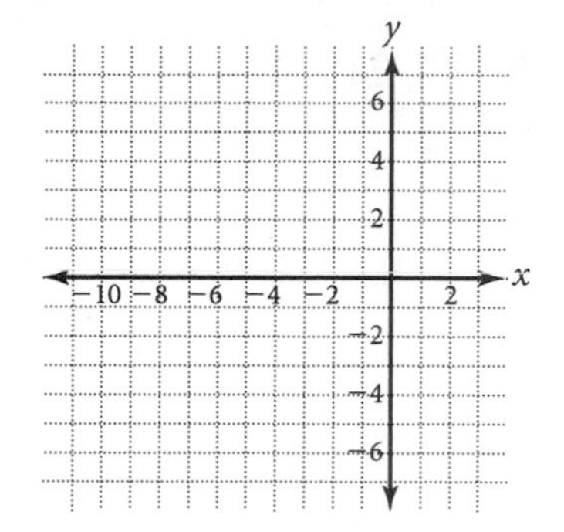

4. $p(x) = -\dfrac{1}{3}|x - 3| + 2$

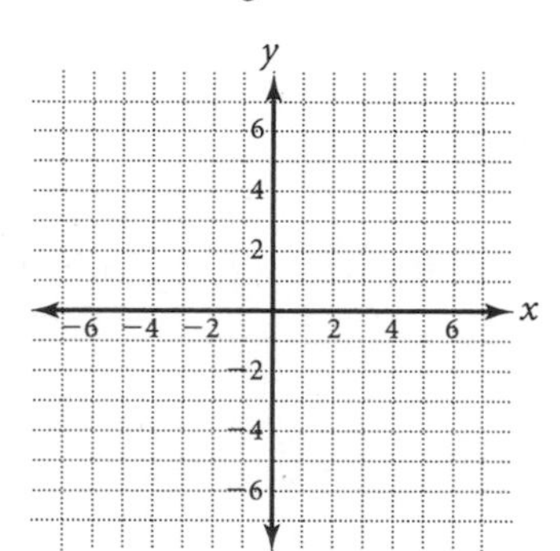

5. $q(x) = -2x^2 + 2$

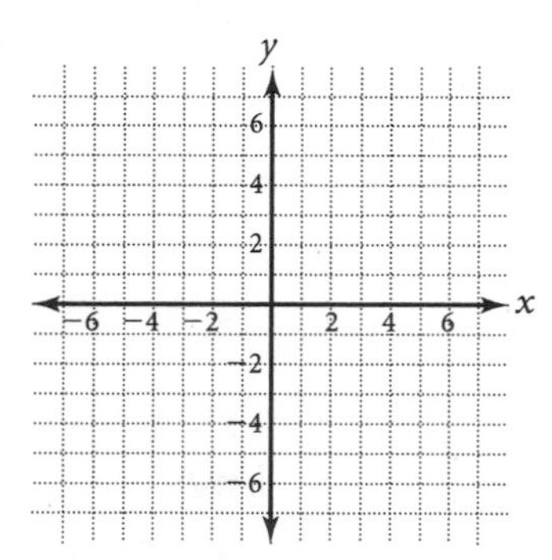

6. $f(x) = \dfrac{1}{2}\sqrt{2x} + 1$

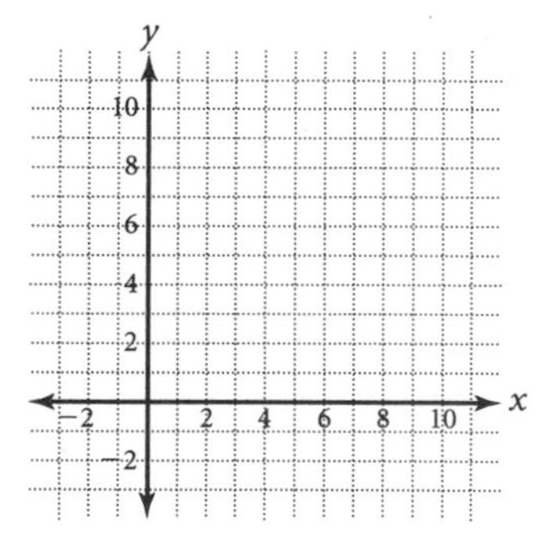

Name the point that corresponds to the point (4, 6) after applying the specified transformations to the graph of $y = \frac{3}{2}x$.

7. a horizontal compression by a factor of $\frac{1}{2}$, then a horizontal translation of 2, followed by a vertical translation of -3 ____________

8. a reflection across the y-axis, then a reflection across the x-axis, followed by a vertical translation of 2 ____________

9. Use the information in the table to find the transformed function that fits the data.

x	-3	-2	-1	0	1	2	3	4	5	6	7
y	19	9	3	1	3	9	19	33	51	73	99

____

Answers

Lesson 1.1
Level A

1. 5; 29, 34, 39
2. 18; 107, 125, 143
3. 19; 104, 123, 142,
4. 32; 187, 219, 251
5. 0.6; 3.5, 4.1, 4.7
6. 0.9; 13.4, 14.3, 15.2
7. −3; 64, 61, 58
8. −8; 57, 49, 41
9. −16; 34, 18, 2
10. −21; 56, 35, 14
11. −0.7; 3.0, 2.3, 1.6
12. −2.3; 5.5, 3.2, 0.9
13. 20, 26, 33
14. 31, 38, 46
15. 64, 77, 91
16. 161, 221, 291
17. 41, 35, 28
18. 57, 45, 31
19. 40, 23, 4
20. 152, 92, 22
21. 32, 37, 41
22. 79, 88, 96
23. 85, 95, 103
24. 402, 452, 492
25. 62, 59, 57
26. 450, 390, 340

Lesson 1.1
Level B

1. 63, 90, 122
2. 145, 119, 88
3. 4, 7, 5
4. 49, 47, 52
5. 19; 30; 37
6. 33; 21; 19
7. 2; 77; 107; 142
8. 4; 5; 3
9. 5, 8, 11, 14, 17
10. 29, 22, 15, 8, 1
11. 66, 79, 92, 105, 118
12. 159, 130, 101, 72, 43
13. 11, 19, 29, 41, 55
14. 33, 40, 51, 66, 85

Lesson 1.1
Level C

1. 67; 78; 79
2. 14; 40; 59; 82
3. 42; 126; 186; 246
4. 730; 690; 650; 10
5. 35, 43, 56, 74, 97
6. 56, 60, 62, 62, 60
7. 142, 136, 127, 115, 100
8. 1275
9. 5050
10. 66
11. 36
12. 120

Answers

Lesson 1.2
Level A

1. 7; 14; 21; 28; 35; 42
2. 8; 16; 24; 32; 40; 48
3. 10; 11; 12; 13; 14; 15
4. 2; 3; 4; 5; 6; 7
5. 7; 11; 15; 19; 23; 27
6. 3; 8; 13; 18; 23; 28
7. 9; 12; 18; 30; 36; 42
8. 0; 9; 27; 63; 135; 279
9. $q = 6$
10. $q = 22$
11. $x = 35$
12. $x = 55$
13. $k = 8$
14. $j = 21$
15. $k = 27$
16. $j = 7$
17. $2n = 28$; 14
18. $13t = 91$; 7
19. $3c = 48$; 16
20. $12t = 108$; 9

Lesson 1.2
Level B

1. 5; 11; 19; 37; 163; 285
2. 36; 101; 179; 387; 1297; 2610
3. \$12
4. \$25
5. \$11
6. \$39
7. $\$3s$
8. $\$3s + \$5l$
9. $x = 5$
10. $y = 5$
11. $v = 2$
12. $b = 9$
13. $h = 6$
14. $r = 3$
15. $f = 11$
16. $h = 14$
17. $r = 8$
18. $r = 4$
19. $x = 7$
20. $u = 9$
21. $w = 7$
22. $q = 5$
23. $d = 11$
24. $y = 12$
25. $0.95k = 4.75$; 5
26. $2k = 27$; rounding down, 13
27. $6n = 133$; rounding down, 22
28. $1.10p = 10$; rounding down, 9
29. $1.50b = 70$; rounding up, 47

Answers

Lesson 1.2
Level C

1. \$1.35
2. \$3.30
3. \$11.64
4. \$27.48
5. $\$0.42m$
6. $\$0.51n$
7. $\$1.26 + \$0.51n$
8. $\$0.42\text{w} + \$0.51n$
9. $y = 4x$
10. $y = 11x$
11. $y = x + 7$
12. $y = 2x + 1$
13. $y = 5x - 3$
14. 44
15. 104
16. 155
17. 2776
18. 79
19. 47
20. 240
21. 59

Lesson 1.3
Level A

1. 86
2. 31
3. 38
4. 10
5. 7
6. 92
7. 12
8. 23
9. 9
10. 4
11. 324
12. 1296
13. 26
14. 116
15. 18
16. 2
17. $(5 + 6) \cdot 7 = 77$
18. $3 \cdot (11 - 1) = 30$
19. $14 \div (4 + 3) - 1 = 1$
20. $3 + (6 - 5) \cdot 7 = 10$
21. $(3 + 5) \cdot 6 \div 2 = 24$
22. 5
23. 19
24. 4
25. 44
26. 200
27. 4
28. 40
29. 2
30. 4
31. 32
32. 100
33. 400

Answers

Lesson 1.3
Level B

1. 27
2. 125
3. 17
4. 22
5. 45
6. 18
7. 38
8. 39
9. 2
10. 1
11. $3 + (6^2 \div 2) = 21$
12. $(3 + 6)^2 - 1 = 80$
13. $[(2 + 3) \cdot 4 - 4] \cdot 4 = 64$
14. $(4 + 6) \cdot (5 - 3)^2 = 40$
15. 89
16. 234
17. 76
18. 80
19. 22
20. 12
21. 156
22. 51
23. 56
24. 12
25. 3.438
26. 1.258
27. 4.351
28. 110.889

Lesson 1.3
Level C

1. 723
2. 362
3. 117
4. 183
5. 197
6. 0
7. 9
8. 1
9. 627
10. 333.5
11. 664
12. 218
13. 11
14. 125
15. 810
16. 323.315
17. 3.188
18. 6.998
19. a. 28 in.^2
    b. 17.5 in.^2
    c. 60 in.^2
20. $4 \cdot (3 - 1) \div [5 \cdot (6 - 4) \div 5] = 4$

Lesson 1.4
Level A

1. I, (2, 3)
2. I, (7, 7)
3. III, (−8, −5)

Answers

4. IV, (4, −3)

5. II, (−9, 7)

6. II, (−4, 3)

7. I, (8, 4)

8. IV, (8, −3)

9. II, (−8, 3)

10. III, (−5, −2)

11.

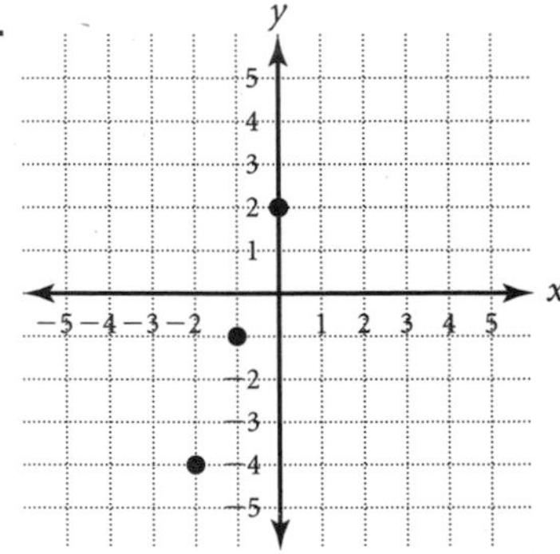

yes

12.

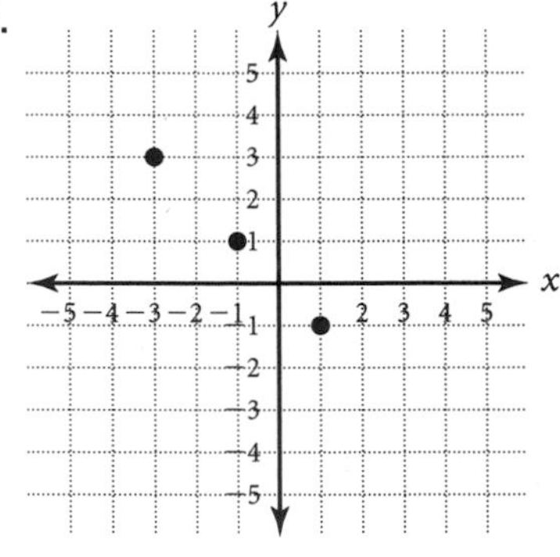

yes

13.

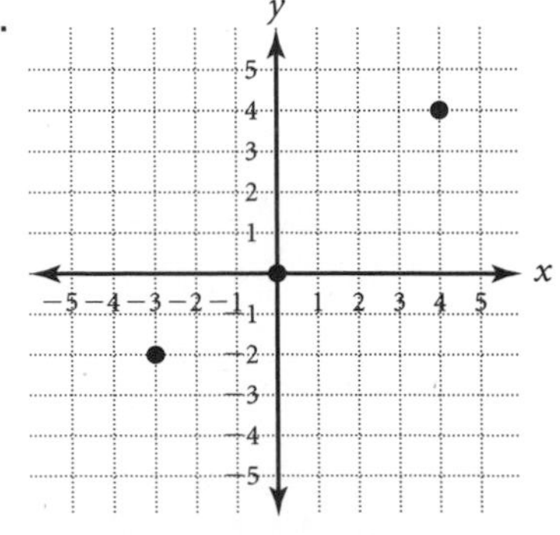

no

14.

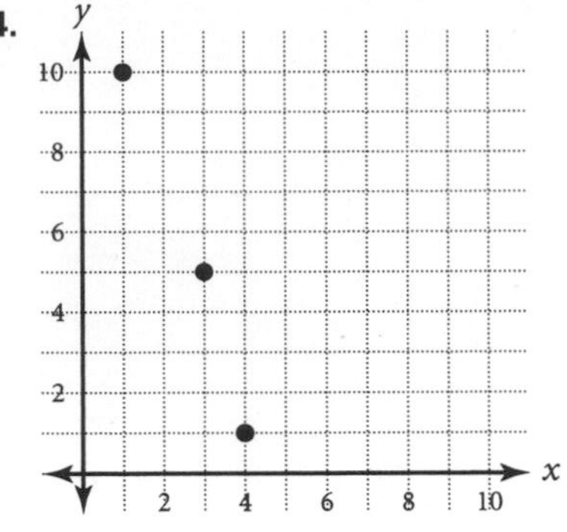

no

15.

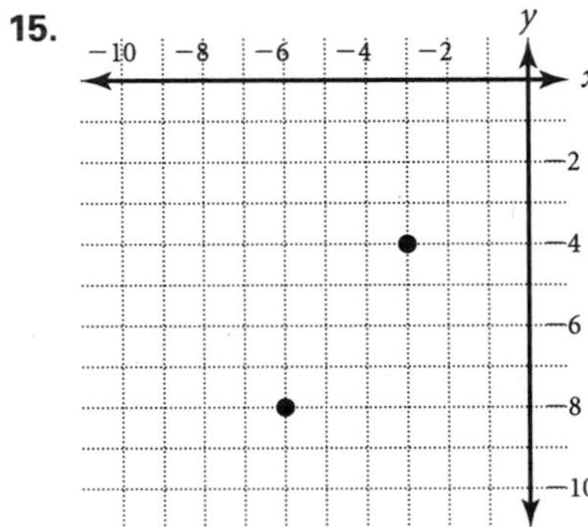

yes

16.

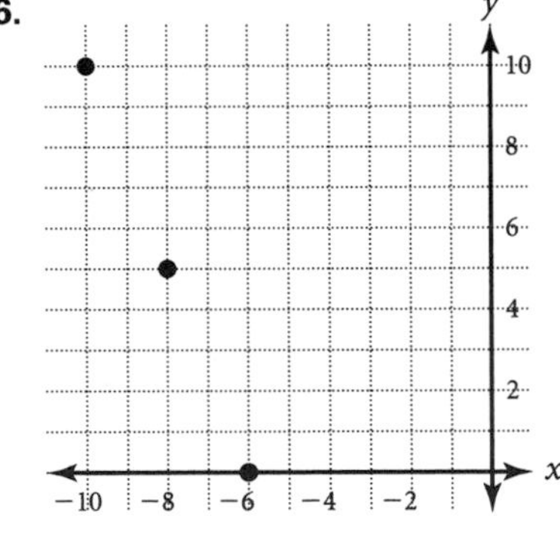

yes

Answers

Lesson 1.4
Level B

1. 1; 2; 3; 4; 5

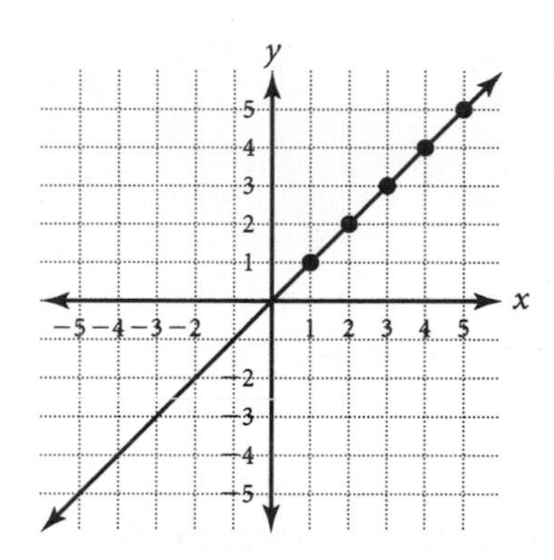

2. 3; 4; 5; 6; 7

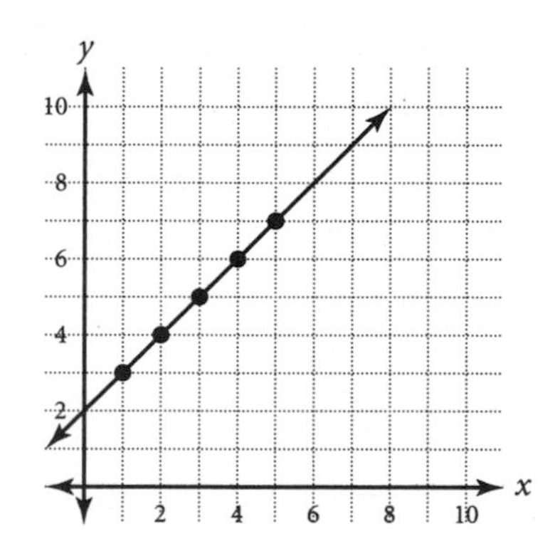

3. 5; 6; 7; 8; 9

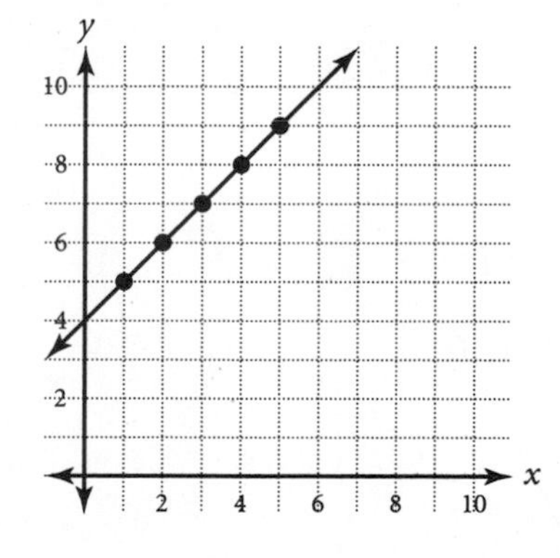

4. 0; 1; 2; 3; 4

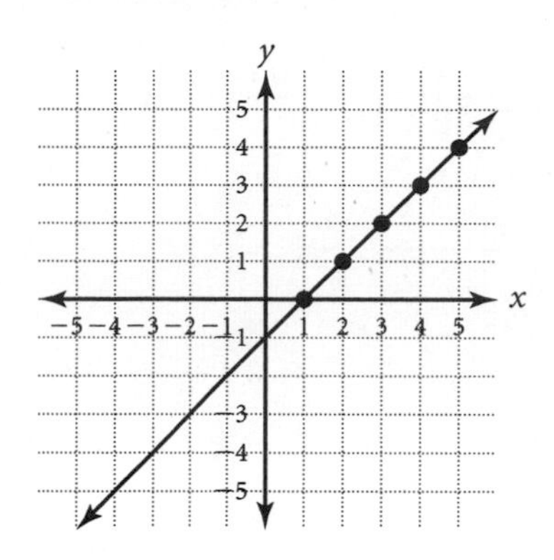

5. 1; 3; 5; 7; 9

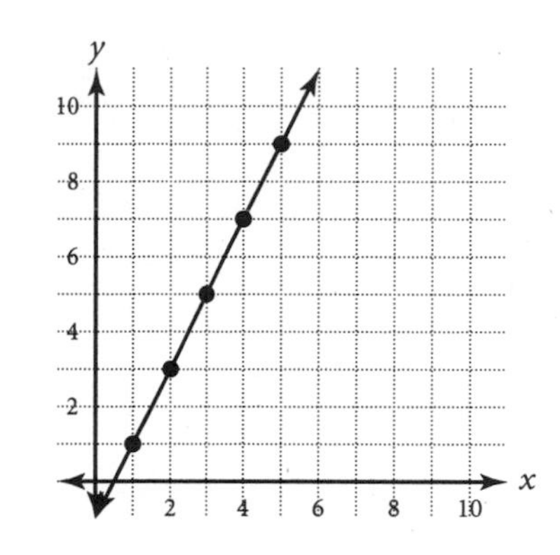

6. −1; −2; −3; −4; −5

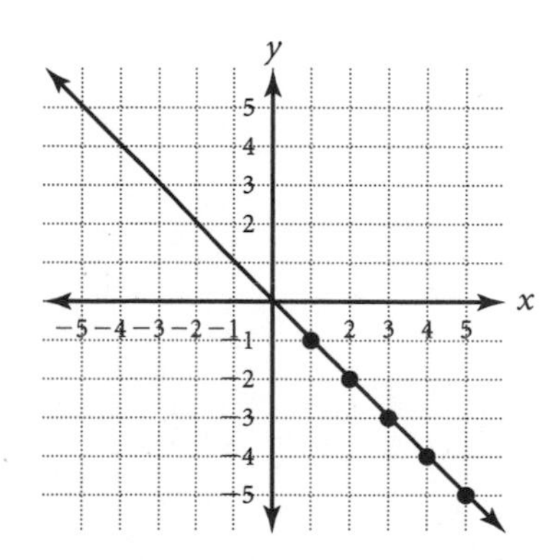

Answers

Lesson 1.4
Level C

1. $-2; -4; -6; -8; -10$

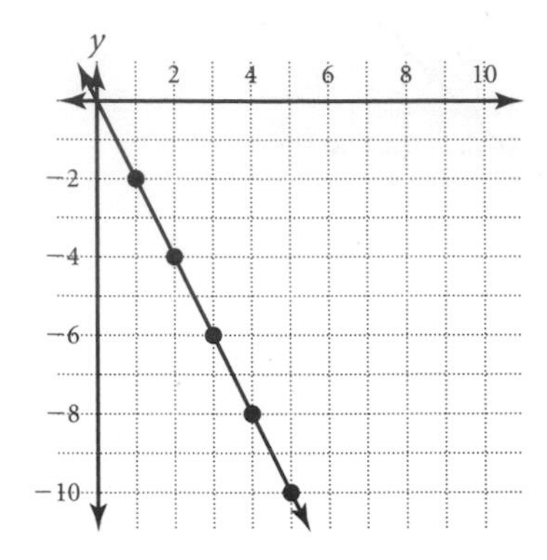

2. $-2; -3; -4; -5; -6$

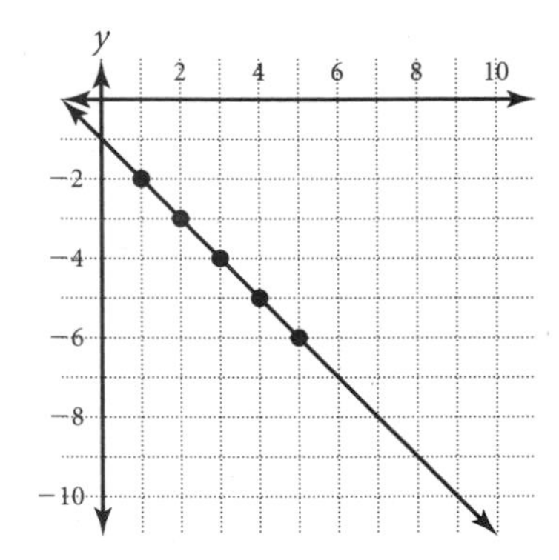

3. 5; 6; 7; 8; 9

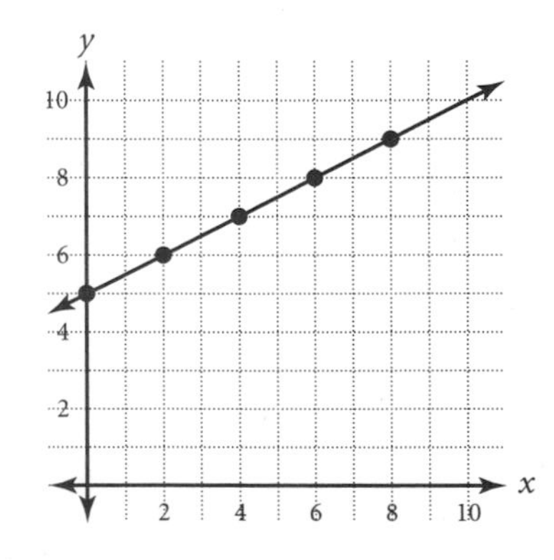

4. 0; 5; 10; 15; 20

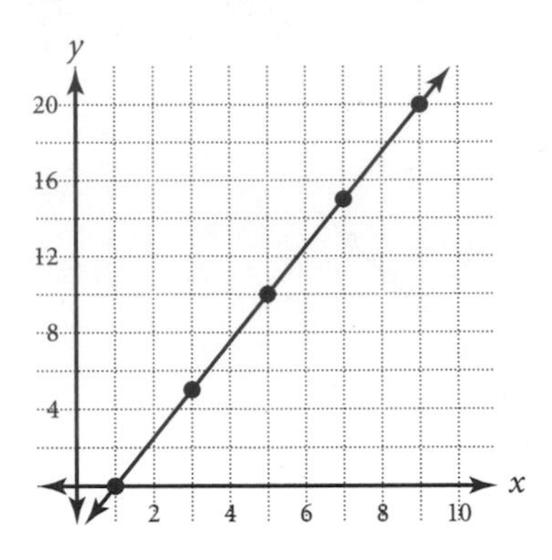

5. Ordered pairs may vary. Sample pairs are shown. Yes, they lie on a straight line.

(1, 4), (2, 5), (3, 6), (4, 7), (5, 8)

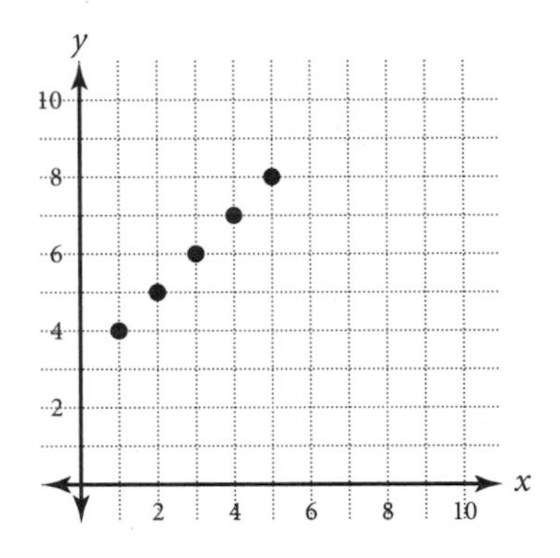

Lesson 1.5
Level A

1. 6; $y = 6x$
2. 7; $y = 7x + 5$
3. 11; $y = 11x + 19$
4. 20; 40; 60; 80; 100

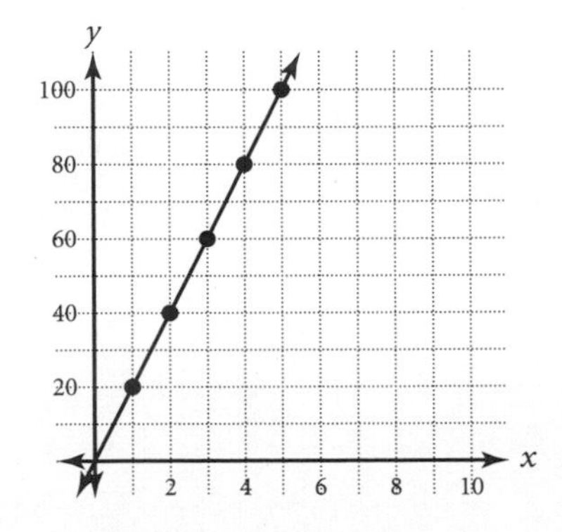

5. 20; 30; 40; 50; 60

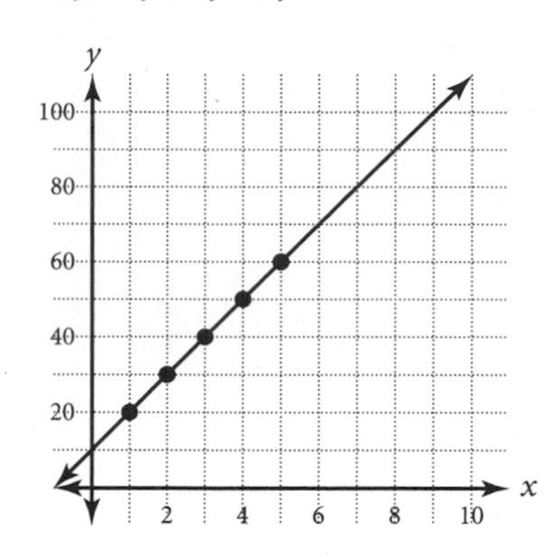

6. 19; 25; 31; 37; 43

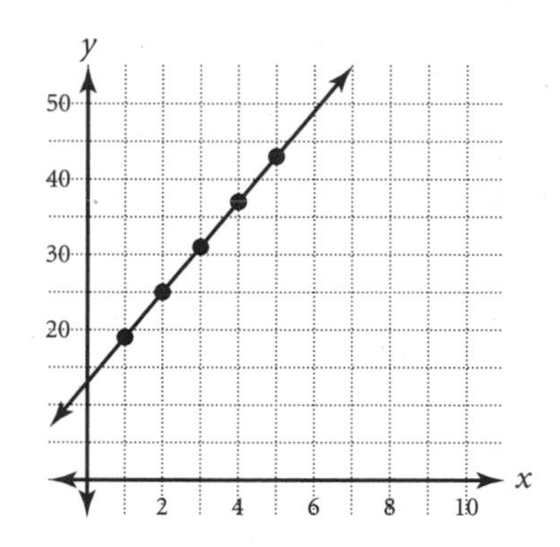

7. 3; 15; 27; 39; 51

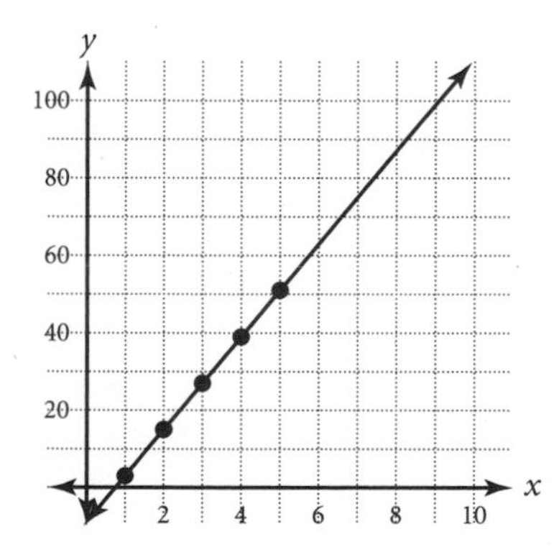

Lesson 1.5
Level B

1. 137; $y = 137x + 358$
2. 1.5; $y = 1.5x + 2$
3. $-\frac{3}{4}$; $y = 12\frac{3}{4} - \frac{3}{4}x$
4. 1.9; 3.9; 5.9; 7.9; 9.9

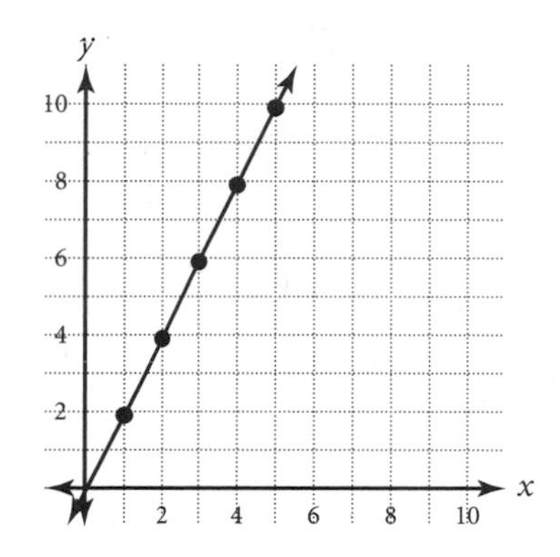

5. 7.5; 12.5; 17.5; 22.5; 27.5

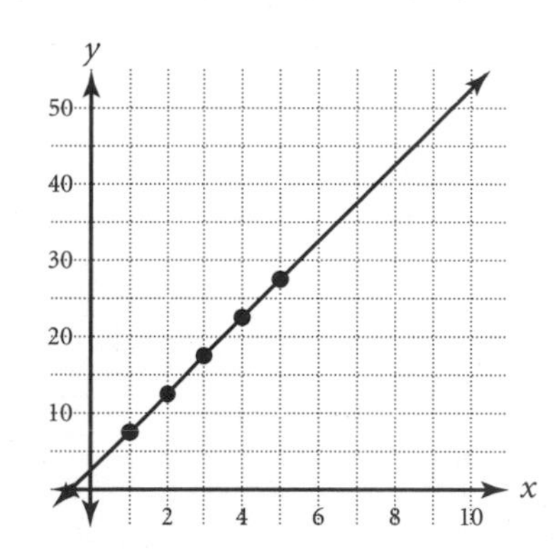

6. $4\frac{1}{2}$; 5; $5\frac{1}{2}$; 6; $6\frac{1}{2}$

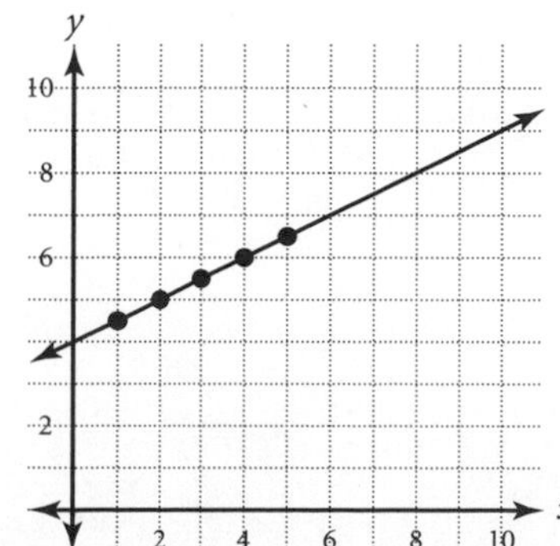

7. $2\frac{1}{2}$; 2; $1\frac{1}{2}$; 1; $\frac{1}{2}$

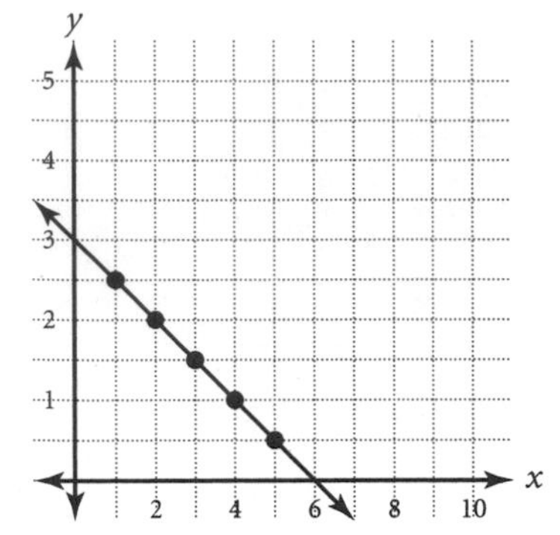

Lesson 1.5
Level C

1. -7.38; $y = 221.45 - 7.38x$

2. 1.6; $y = 1\frac{3}{5}x + \frac{1}{8}$

3. a. $w = 12000 - 1500h$
   b. \$12,000; \$10,500; \$9000; \$7500; \$6000; \$4500
   c. 8 hints

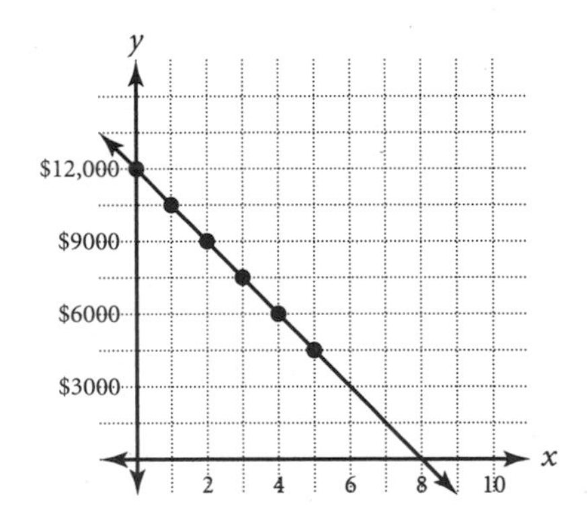

Lesson 1.6
Level A

1. little to none

2. strong negative

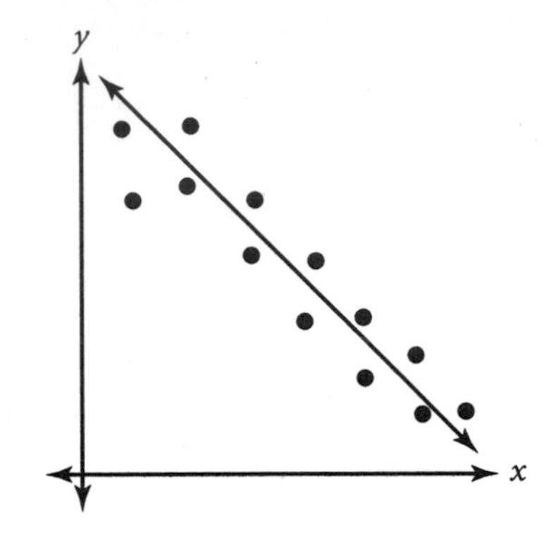

3. strong positive

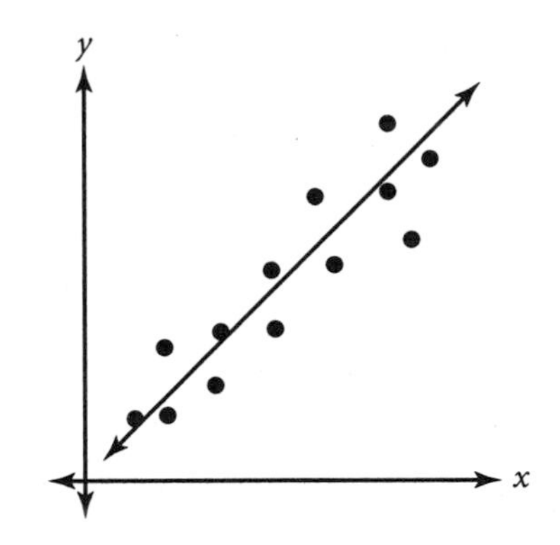

4. 3

5. 1

6. 2

7. little to none

8. negative

9. positive

10. positive

11. negative

12. little to none

13. negative

14. negative

15. little to none

16. positive

17. little to none

Lesson 1.6
Level B

1.

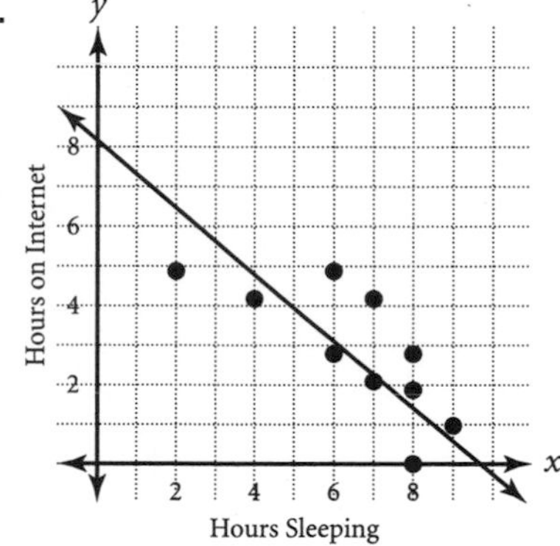

strong negative

2.

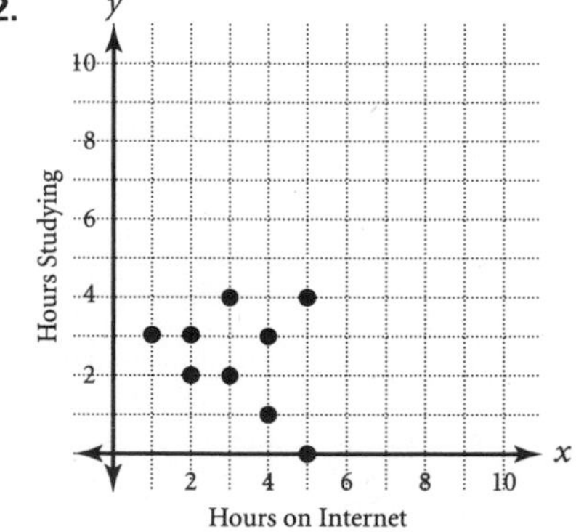

little to none

Lesson 1.6
Level C

1.

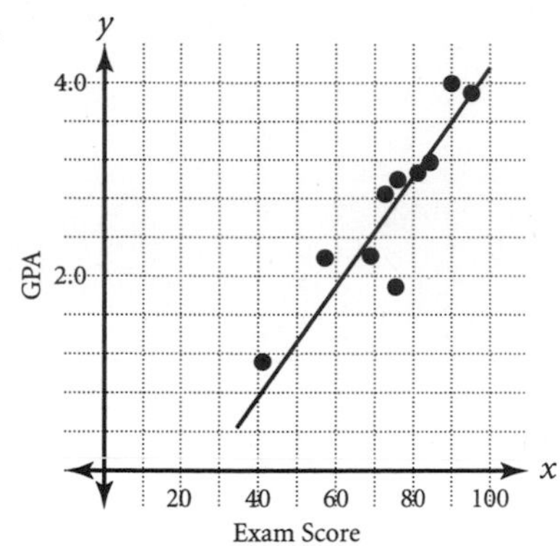

strong positive

2.

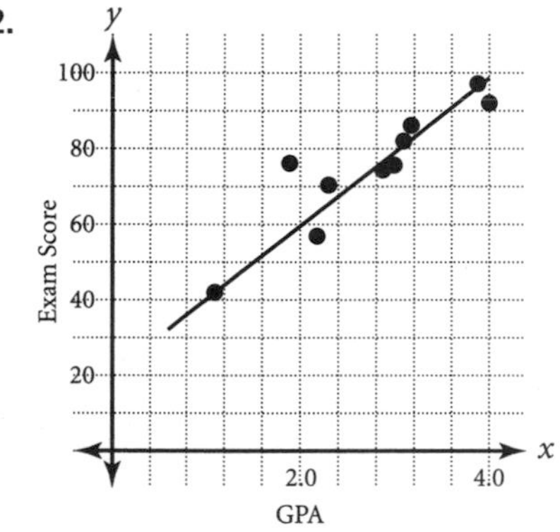

strong positive

3. No, the placement of the variables does not appear to make a difference in the correlation. The two variables have the same relationship regardless of their placement on the graph.

Answers

Lesson 2.1
Level A

1. $>$
2. $<$
3. $>$
4. $<$
5. $>$
6. $>$
7. $<$
8. $<$
9. $>$
10. $<$
11. $<$
12. $<$
13. $<$
14. $>$
15. $=$
16. $>$
17. $=$
18. $<$
19. $>$
20. $<$
21. $=$
22. -18
23. 5
24. 9
25. -6
26. 0
27. 1.4
28. $-4\frac{1}{8}$
29. $3\frac{2}{7}$
30. $-x$
31. $-m$
32. x
33. z
34. 18
35. 5
36. 9
37. 6
38. 1
39. 6
40. -12
41. -4
42. -32
43. -2
44. -7
45. 0
46. -0.5
47. 6
48. 101
49. 13
50. -12
51. 9
52. -45

Answers

Lesson 2.1
Level B

1. $>$
2. $>$
3. $<$
4. $=$
5. $>$
6. $=$
7. -13
8. -13
9. 16
10. -4
11. 24
12. -24
13. 24
14. 0
15. 144
16. 144
17. 2
18. 2
19. 4
20. -1.8
21. $1\frac{1}{4}$
22. $-\frac{5}{2}$
23. x
24. z
25. false
26. true
27. true
28. true
29. false

Lesson 2.1
Level C

1. -9
2. -3
3. -9
4. -3
5. 6
6. -2
7. -3
8. -2
9. -2
10. -2
11. 0
12. 18
13. 4
14. 5
15. 2
16. 18
17. true
18. true
19. false
20. true
21. false

Answers

22. $-5\frac{1}{2}$; $-8 + 5 - 2\frac{1}{2} = -5\frac{1}{2}$

23. 3, −2, −7

Lesson 2.2
Level A

1. 7
2. −7
3. 3
4. −3
5. 0
6. 0
7. 6
8. −5
9. −1
10. 7
11. 1
12. −9
13. 4
14. −3
15. −14
16. −4
17. −14
18. 4
19. 28
20. 12
21. 34
22. −80
23. −41
24. −29
25. −38
26. −21
27. 43
28. −71
29. 9
30. −90
31. 109
32. −109
33. −314
34. −2
35. −4.4
36. 1.7
37. −4.3
38. −2.9
39. −5.0
40. −2.9
41. 6.7
42. 4.8
43. −0.8
44. −13.5
45. −12.8
46. $\frac{1}{5}$
47. $-\frac{3}{5}$
48. $-\frac{1}{5}$
49. $\frac{1}{2}$

Answers

50. $-\frac{1}{2}$

51. -2

52. -5

53. 2

54. -2

Lesson 2.2
Level B

1. -21
2. 1
3. -1
4. -3
5. 3
6. 0
7. -891
8. -79
9. 544
10. -765
11. 178
12. 472
13. 553
14. -118
15. 136
16. 2.73
17. -1.48
18. -3.21
19. 5.14
20. -5.47
21. -1.51
22. -9.094
23. -3.107
24. -1.233
25. $\frac{13}{28}$
26. $-5\frac{11}{15}$
27. $7\frac{43}{45}$
28. $-\frac{7}{15}$
29. $-5\frac{3}{8}$
30. $-2\frac{4}{9}$
31. $5\frac{17}{25}$
32. $-21\frac{31}{56}$
33. $-2\frac{4}{5}$
34. 8
35. -4
36. 4
37. -8
38. -3
39. 15
40. -15
41. 3

42. 11

43. -7

44. 7

45. -11

46. -5

47. 13

48. -1

Lesson 2.2
Level C

1. 7

2. 3

3. -9

4. 6

5. -2

6. -7

7. 5

8. 7

9. 0

10. -17

11. 3.7

12. 7.5

13. 5.8

14. 2.4

15. $3\frac{1}{72}$

16. $-6\frac{3}{5}$

17. $2\frac{61}{105}$

18. $-18\frac{13}{72}$

19. 5

20. 4

21. -1.25

22. 24

23. 5

24. 22.5

25. $\frac{1}{4}$

26. $\frac{8}{3}$

27. yes; $8

28. 11°F

Lesson 2.3
Level A

1. 2

2. 6

3. -6

4. -2

5. -8

6. 2

7. -2

8. 8

9. 0

10. -12

11. 12

12. -7

13. -4

14. -8

15. 11

16. -10

17. -9

18. 8

19. -4

20. 3

21. -9

22. -3.1

23. -5.9

24. -2.3

25. -4

26. -31

27. 7

28. -9

29. 95

30. 52

31. 21

32. -21

33. -73

34. -101

35. -101

36. -3

37. -489

38. -1113

39. 489

40. -809

41. -435

42. 809

43. -1.8

44. -1.7

45. -3.5

46. -3.5

47. 3.5

48. 0.7

49. $-\frac{1}{5}$

50. $-\frac{2}{7}$

51. $-\frac{10}{7}$

52. $\frac{3}{4}$

53. $-\frac{1}{11}$

54. $-1\frac{1}{5}$

Lesson 2.3
Level B

1. -4

2. 20

3. 3

4. -1

5. 9

6. 9

7. -773

8. -1127

9. -112

10. -2.31

11. 2.31

12. -4.65

13. -7.38

14. -7.93

15. -10.37

16. 8.94

17. -8.301

18. -0.991

19. $-\frac{52}{55}$

20. $\frac{8}{55}$

21. $\frac{25}{56}$

22. $-\frac{1}{8}$

23. $-3\frac{7}{9}$

24. $3\frac{5}{72}$

25. $-3\frac{4}{11}$

26. $\frac{15}{4}$

27. $-14\frac{5}{7}$

28. 6

29. -2

30. 2

31. -6

32. 3

33. 11

34. -11

35. 9

36. -9

37. 8

38. 12

39. 9

40. 20

41. 11

42. 13

43. 29

44. 77

45. 322

Lesson 2.3
Level C

1. -2

2. -3

3. -2

4. 2

5. 21

6. 540

7. 210

8. -10.76

9. 0.15

10. $10\frac{19}{30}$

11. $-10\frac{79}{105}$

12. $6\frac{1}{9}$

13. $-21\frac{13}{14}$

14. $-\frac{3}{4}$

15. 4

16. 0

17. 3

18. $4\frac{3}{4}$

19. $7\frac{3}{4}$

20. $-3\frac{1}{4}$

21. -28

22. 11.6

23. 5

24. $8\frac{13}{35}$

25. $-13, -19, -24$

26. $-6\frac{2}{9}$ and $-4\frac{4}{9}$

Lesson 2.4
Level A

1. -48

2. -48

3. 48

4. -3

5. 3

6. -3

7. -6

8. -10

9. 0

10. 0

11. 39

12. -85

13. -84

14. -4

15. 3

16. 12

17. undefined

18. 90

19. -5

20. 192

21. 7

22. -3

23. -12

24. 37

25. -20

26. -3

27. 20

28. 0

29. -7.2

30. 20

31. $\frac{5}{2}$

Answers

32. $-\frac{3}{2}$

33. $-\frac{1}{7}$

34. -4

35. undefined

36. $\frac{2}{3}$

37. $-\frac{1}{2}$

38. $-\frac{2}{7}$

39. $\frac{3}{40}$

40. $-2\frac{1}{2}$

41. $\frac{3}{5}$

42. $-\frac{5}{6}$

43. -3

44. -10

45. 36

46. 6

47. -3

48. -16

Lesson 2.4
Level B

1. -52

2. 0

3. 3.36

4. -7

5. -8.5

6. -2

7. -3.11

8. -5

9. $\frac{2}{9}$

10. 0

11. -150

12. -6204

13. $5\frac{7}{10}$

14. $-7\frac{2}{3}$

15. 2

16. -5.332

17. $-5\frac{2}{3}$

18. -19

19. $12\frac{4}{5}$

20. -54

21. -9999

22. $-4\frac{4}{5}$

23. 0

24. -65

25. 45

26. $-4\frac{8}{21}$

27. undefined

28. -8

29. 8

30. -8

31. -7

32. 7

33. 7

34. -14

35. 14

36. 14

37. 14

38. 8

39. $10\frac{1}{4}$

40. 7

41. $1\frac{3}{4}$

42. $-\frac{7}{12}$

Lesson 2.4
Level C

1. 1

2. 6

3. -23.1

4. $-2\frac{1}{2}$

5. $4\frac{1}{2}$

6. 0

7. $16\frac{5}{14}$

8. 2.7632

9. -0.75

10. -0.75

11. -10

12. -3

13. -4.4375

14. -16.875

15. 37.125

16. -19.75

17. negative

18. positive

19. negative

20. zero

21. cannot be determined

22. positive

23. positive

24. undefined

25. positive

26. $3, -1, \frac{1}{3}$

27. $1973

Lesson 2.5
Level A

1. 18; 18; Associative; 100; 176

2. 40; Commutative; 40; Associative; 200; 1400

3. 80; 4; Distributive; 320; 28; 348

4. $20 \cdot (17 \cdot 5) = 20 \cdot (5 \cdot 17)$: Commutative Property of Multiplication; $= (20 \cdot 5) \cdot 17$: Associative Property of Multiplication; $= 100 \cdot 17 = 1700$

5. $(52 + 37) + 48 = (37 + 52) + 48$: Commutative Property of Addition; $= 37 + (52 + 48)$: Associative Property of Addition; $= 37 + 100 = 137$

6. 160

7. 10; 3; 90; 27; 117

Lesson 2.5
Level B

1. 2; 91; Commutative Property of Multiplication;
   2; 91; Associative Property of Multiplication; 10; 91; 910

2. 161; 39; Commutative Property of Addition;
   161; 39; Associative Property of Addition; 81; 200; 281

3. 5; 600; 5; 5; 3000; 25; 2975

4. 4; 30; 4; 1.5; 120; 6; 114

5. $8(x + y)$

6. $4(2p + 7q)$

7. $3(2a - 3b)$

8. $a(b + c)$

9. $5r(s + 5t)$

10. $g(3f - h)$

11. Commutative Property of Addition

12. Distributive Property

13. Associative Property of Multiplication

14. Associative Property of Addition

15. Commutative Property of Multiplication

16. Distributive Property

Lesson 2.5
Level C

1. 6; 5; 0.9; 6; 5; 6; 0.9; 30; 5.4; 35.4

2. 7; $\frac{1}{3}$; 9; 7; 9; $\frac{1}{3}$; 63; 3; 66

3. Commutative Property of Multiplication

4. Distributive Property

5. Commutative Property of Addition

6. $x = y$ (Given); $xa = xa$ (Reflexive Property of Equality); $xa = ya$ (Substitution Property of Equality)

7. $m = n$ (Given); $m + r = m + r$ (Reflexive Property of Equality); $m + r = n + r$ (Substitution Property of Equality)

8. $(x + y) + z = x + (y + z)$ (Associative Property of Addition); $= (y + z) + x$ (Commutative Property of Addition); $= (z + y) + x$ (Commutative Property of Addition)

9. $\frac{1}{x} \cdot (y \cdot x) = \frac{1}{x} \cdot (x \cdot y)$(Commutative Property of Multiplication); $= \left(\frac{1}{x} \cdot x\right) \cdot y$ (Associative Property of Multiplication); $= 1 \cdot y$ (Multiplicative Inverse Property); $= y$ (Multiplicative Identity Property)

Lesson 2.6
Level A

1. $5x + 3x = (5 + 3)x = 8x$

2. $6y + 7y = (6 + 7)y = 13y$

3. $6r - 3r = (6 - 3)r = 3r$

Answers

4. $14z - 5z = (14 - 5)z = 9z$
5. $-3q + 5q = (-3 + 5)q = 2q$
6. $8a + (-7a) = [8 + (-7)]a = a$
7. $2m - 9m = (2 - 9)m = -7m$
8. $-3p - 11p = (-3 - 11)p = -14p$
9. $-2x - 5$
10. $3z - 4$
11. $-8m + 7$
12. $4b + 6$
13. $-a - b + c$
14. $-2f + 3g - 1$
15. $2m - 5n + 9$
16. $3x - 4y + z$
17. $13x$
18. $16a$
19. $6p$
20. $4d$
21. y
22. $-2w$
23. $-6j$
24. $-17h$
25. $11t + 8$
26. $9n + 1$
27. $12b - 12$
28. $-2z - 3$
29. $2u - 4v$
30. $2q - 5r$
31. $6l + 3m$
32. 3a
33. $-2f - 2g$
34. $4c - 10d$
35. $8v - w$
36. $-11g - 3h$
37. $-2s + t$
38. $-4r + 3s$

Lesson 2.6
Level B

1. $8.5m + 1.3m = (8.5 + 1.3)m = 9.8m$
2. $\frac{11}{9}a - \frac{2}{9}a = \left(\frac{11}{9} - \frac{2}{9}\right)a = a$
3. $-6x - 6.1x = (-6 - 6.1)x = -12.1x$
4. $2q + \frac{1}{3}q = \left(2 + \frac{1}{3}\right)q = \frac{7}{3}q$
5. $1.7a + 4$
6. $-0.9g + 2$
7. $-0.9f - 3g$
8. $-9.7p + 0.4q$
9. $1.7x - 1.7y$
10. $13.1j - 2.2k$
11. $-1.78w$
12. $-1.23m - 0.3n$
13. $1\frac{4}{5}h$
14. $-9\frac{1}{3}y$
15. $1\frac{2}{3}k$

Answers

16. $\frac{3}{7}b$

17. $2\frac{1}{2}c + 3d$

18. $-\frac{1}{2}l - \frac{1}{2}m$

19. $10r + 12s$

20. $18m - 10n$

21. $5x + 4y$

22. $4a + 3b + 2c$

Lesson 2.6
Level C

1. $13x = (-14.1 + 27.1)x$
   $= -14.1x + 27.1x$

2. $-2.05z = (1 - 3.05)z = z - 3.05z$

3. $-1\frac{1}{8}a - 2\frac{1}{4}a = \left(-1\frac{1}{8} - 2\frac{1}{4}\right)a = -3\frac{3}{8}a$

4. $-\frac{5}{9}b + 2\frac{1}{3}b = \left(-\frac{5}{9} + 2\frac{1}{3}\right)b = 1\frac{7}{9}b$

5. $a + 6b - 5c$

6. $-10x + 3y - 2z$

7. $2.8j - 0.5k + 0.1l$

8. $-0.2m + 0.5n - 0.6p$

9. $3\frac{1}{3}r - 2s + 9\frac{3}{4}t$

10. $\frac{m}{8} - 2$

11. $10q - 4r + 9$

12. $x + 2y$

13. $5a + 10b - 2$

14. $3f + g + 5h - 1$

15. $2a + 3b$

Lesson 2.7
Level A

1. $12y$
2. $-16a$
3. $30r$
4. $54b$
5. $-28q$
6. $18c$
7. $4n$
8. $3h$
9. $-2q$
10. $-2f$
11. $7p$
12. z
13. $6k^2$
14. $-54j^2$
15. $-32w^2$
16. $24u^2$
17. $7h^2$
18. v^2
19. $6x + 3$
20. $20l + 15$
21. $12d - 48$
22. $49g - 7$
23. $-6m - 15$
24. $-11m + 22$

Answers

25. $10b^2 + 4b$

26. $28c^2 + 56c$

27. $8s^2 - 4s$

28. $-6y^2 - 15y$

29. $-60f^2 + 42f$

30. $-3g^2 + g$

31. $27d$

32. $-9l$

33. $7j + 2$

34. $-3h + 1$

35. $-18t - 36$

36. $13m + 16$

37. $3 + 2a$

38. $2x + 3$

39. $2 - y$

40. $2m - 1$

Lesson 2.7
Level B

1. $5y^2 - 3$
2. $9k^2 - 12$
3. $-3r$
4. $7.2m^2$
5. $6p$
6. $-12x$
7. $-4q$
8. $10t$
9. $10y - 15$
10. $-10.2a + 20.4$
11. $-30b + 20$
12. $5m + 10$
13. $-2s - 1$
14. $-2r + 3$
15. $3p + 9$
16. $-26c^2 + 14c$
17. $150 + 5k$
18. $4d^2 + 3d + 2$
19. 15 in.^{2}
20. 18 ft^2
21. $12a^2$
22. $52x^2$

Lesson 2.7
Level C

1. $8a + 5b^2$
2. $-8m + 0.5n$
3. $-2p^2 - p + 3$
4. $19r^2 - 10r + 4$
5. $15a - 9b + 5c$
6. $-3x^2 + 2x - 10y + 4$
7. 120 cm^3
8. 30 yd^3
9. $2a^2b$
10. $3mnp$
11. \$64
12. \$$(40h + 48)$

Answers

Lesson 3.1
Level A

1. Subtraction Property of Equality
2. Addition Property of Equality
3. Addition Property of Equality
4. Addition Property of Equality
5. Addition Property of Equality
6. Subtraction Property of Equality
7. Addition Property of Equality
8. Subtraction Property of Equality
9. Subtraction Property of Equality
10. Addition Property of Equality
11. Subtraction Property of Equality
12. Addition Property of Equality
13. Subtraction Property of Equality
14. Subtraction Property of Equality
15. Subtraction Property of Equality
16. $y = 16$
17. $h = 19$
18. $z = -13$
19. $t = -6$
20. $y = -9$
21. $d = 2$
22. $m = -1$
23. $k = -34$
24. $x = -23$
25. $x = -2$
26. $z = -20$
27. $j = 34$
28. $y = 21$
29. $h = -48$
30. $z = -40$

Lesson 3.1
Level B

1. Subtraction Property of Equality
2. Addition Property of Equality
3. Addition Property of Equality
4. Subtraction Property of Equality
5. $x = \frac{1}{2}$
6. $k = -2.3$
7. $z = \frac{2}{3}$
8. $h = 9.6$
9. $x = -6.4$
10. $a = \frac{5}{6}$
11. $y = 1\frac{1}{10}$
12. $b = 3\frac{1}{6}$
13. $m = -11.4$
14. $d = 1$
15. $n = -20.7$
16. $f = 32.2$
17. $j = -\frac{13}{21}$
18. $t = -27.21$

19. $p = 1\frac{2}{15}$

20. $q = 8.12$

21. $w = 11.1$

22. $y = -1\frac{31}{40}$

23. $m = 3\frac{17}{20}$

24. $x = 28.45$

25. $x = 2\frac{3}{4}$

26. $z = -12.4$

27. $b = 7.35$

28. $k = -2.3$

Lesson 3.1
Level C

1. $x = -2.9$
2. $k = -6$
3. $h = 1\frac{8}{15}$
4. $m = -6.675$
5. $g = 21$
6. $f = 0$
7. $p = 1.7$
8. $n = -2.03$
9. $k = 7$
10. $z = -3$
11. $18 + 7.5 + x = 27.57$; \$2.25
12. $30 + 68 + x = 180$; 82°
13. $x + 1375 = 2150$; 775 calories
14. $x + 295{,}000 = 315{,}000$; 20,000 people
15. $x + \frac{3}{4} = 1\frac{2}{3}, \frac{11}{12}$ cups
16. $350.28 + x = 652.75$; \$302.47
17. $x - 15 = 62$; $x = 77$

Lesson 3.2
Level A

1. Division Property of Equality
2. Multiplication Property of Equality or Division Property of Equality
3. Multiplication Property of Equality
4. Division Property of Equality
5. Multiplication Property of Equality
6. Multiplication Property of Equality
7. Division Property of Equality
8. Division Property of Equality
9. Multiplication Property of Equality
10. Division Property of Equality
11. $y = -4$
12. $h = 2$
13. $q = -100$
14. $t = 4$
15. $g = -3$
16. $d = -42$
17. $k = -48$
18. $h = -3$
19. $j = -26$
20. $p = -16$
21. $z = 16$

22. $s = 63$

23. $y = -2$

24. $f = -7$

25. $z = 6$

Lesson 3.2
Level B

1. Multiplication Property of Equality
2. Multiplication Property of Equality
3. Division Property of Equality
4. Multiplication Property of Equality
5. $x = \frac{-1}{5}$
6. $k = -3$
7. $z = 6$
8. $m = -5.5$
9. $n = 28$
10. $a = -7$
11. $b = 5$
12. $f = -10\frac{1}{2}$
13. $m = 6$
14. $d = -1\frac{1}{4}$
15. $n = -\frac{1}{3}$
16. $g = 2$
17. $v = -21\frac{1}{3}$
18. $m = \frac{2}{7}$
19. $n = -40$
20. $x = 3$
21. $k = 105$
22. $p = 1\frac{1}{5}$

Lesson 3.2
Level C

1. $m = 521$
2. $n = 3.75$
3. $h = -1\frac{23}{25}$
4. $k = 1$
5. $m = 0$
6. $f = -1$
7. $h = \frac{V}{lw}$
8. $d_2 = \frac{w_1 d_1}{w_2}$
9. $m = wr$
10. $p = \frac{mq}{n}$
11. $h = \frac{2A}{b_1 + b_2}$
12. $m = \frac{E}{c^2}$
13. $705 = 6x$; $117.5°$
14. $\frac{4}{8} = \frac{x}{6}$; 3 cups
15. $108x = 16.20$; \$0.15
16. $16x = 64$; 4 bulbs

17. $\frac{x}{12} = 8.5$; \$102

18. $\frac{3.99}{12} = \frac{x}{1}$; \$0.33

19. $\frac{1}{3}x = 20$; 60 mph

Lesson 3.3
Level A

1. $x = 7$
2. $m = -11$
3. $x = -15$
4. $k = 18$
5. $x = 7$
6. $n = 2$
7. $b = 63$
8. $f = 10$
9. $y = 2$
10. $z = 28$
11. $p = 11$
12. $g = 50$
13. $h = 56$
14. $q = -2$
15. $2x + 3 = 25$; 11 years old
16. $8 + 3x = 107$; 33 CDs
17. $6 + 8 + x = 28$; 14 inches
18. $7 + \frac{x}{4} = 52$; 180
19. $\frac{x}{3} - 11 = 18$; 87
20. $32 = 8 + 2x$; 12 inches
21. $5000 + 3x = 71000$; \$22,000
22. $9 + \frac{x}{5} = 21$; 60

Lesson 3.3
Level B

1. $d = -1$
2. $f = 3$
3. $b = -3.5$
4. $j = -8$
5. $h = -2$
6. $n = -16$
7. $b = 62$
8. $f = -10$
9. $y = -16$
10. $z = 7.5$
11. $p = 35$
12. $r = -54$
13. $h = 54$
14. $q = -12$
15. $z = -132$
16. $q = -16.68$
17. $k = -3$
18. $m = -27.6$
19. $b = -15$
20. $q = 3$
21. $0.25x + 5.15 = 7.90$; 11 quarters

22. $12 - \frac{x}{-2} = 15$; 6

23. $29.95 + 15.75x = 187.45$; 10 months

24. $\frac{x}{5} - 11.2 = 27.75$; 194.75

Lesson 3.3
Level C

1. $x = -2.2$
2. $m = -6$
3. $n = 24\frac{3}{4}$
4. $p = 37$
5. $q = 10\frac{4}{5}$
6. $r = 17\frac{1}{2}$
7. $s = 0$
8. $t = -4$
9. $w = -57$
10. $x = 11$
11. $y = 6$
12. $z = -6$
13. $a = -27$
14. $b = -2$
15. $c = 5\frac{2}{3}$
16. $d = -15$
17. $f = -20$
18. $g = -14\frac{1}{3}$
19. $h = 99$
20. $k = 36$
21. $0.1d + 15.75 = 27.95$; 122 dimes
22. $\frac{1}{2}p - 5.50 = 6.00$; \$23.00
23. $6c + 21.20 = 52.70$; \$5.25 per craft

Lesson 3.4
Level A

1. $y = 2$
2. $n = -4$
3. $y = 3$
4. $y = -1$
5. $x = 3$
6. $x = -3$
7. $m = -3$
8. $z = 5$
9. $y = 2$
10. $m = 2$
11. $h = 2$
12. $y = -1$
13. $x = -5$
14. $z = 4$
15. $h = 5$
16. $a = 3$
17. $x = 11$
18. $y = 3$
19. $h = -3$
20. $x = 2$
21. $a = -2$

22. $x = 2$

23. $y = 7$

24. $m = -2$

25. $y = -4$

26. $x = 3$

27. $x = 2$

28. $m = 4$

29. $a = -4$

30. $x = 2$

31. $8x + 5 = 4x - 3; -2$

32. $25 + 45x = 10 + 60x$; 1 hour

33. $8 - x = 3 + 5x; \frac{5}{6}$

Lesson 3.4
Level B

1. $x = 3$

2. $h = -3$

3. $y = 1.5$

4. $a = 3$

5. $x = -2.6$

6. $z = 2$

7. $y = 4$

8. $y = \frac{-1}{3}$

9. $a = -6$

10. $x = -4$

11. $m = -4$

12. $x = \frac{6}{7}$

13. $y = 3$

14. $h = -4$

15. $m = 2$

16. $a = -8$

17. $x = 3$

18. $a = -5$

19. $y = 5$

20. $h = -1.875$

21. $h = 5.8$

22. $x = \frac{3}{5}$

23. $z = 8$

24. $n = 4.5$

25. $f = \frac{10}{13}$

26. $k = 0.5625$

27. $x = \frac{-2}{3}$

28. $p = 0$

29. $b = 30$

30. $x = 1$

31. $0.15x + 3.4 + 1.19 = 0.1x + 4.99$; 8 pencils and 8 apples

32. $-3x - 1 = x + 11; -3$

33. $x - 4 = -5 + 2x; 1$

Answers

Lesson 3.4
Level C

1. $a = 24$
2. $b = -6\frac{2}{3}$
3. $c = -4$
4. $d = -\frac{1}{13}$
5. $f = -6$
6. $g = 22$
7. $h = -16$
8. $f = \frac{2}{11}$
9. $k = 3$
10. $m - 128$
11. $n = 100$
12. $p = 1$
13. $q = 6$
14. $r = 50$
15. $s = 140$
16. $t = \frac{1}{2}$
17. $w = \frac{4}{3}$
18. $x = -60$
19. $y = -\frac{3}{2}$
20. $z = -2$
21. $\frac{314 + x}{5} = 80; 86$
22. $6 - \frac{x}{2} = 8 + \frac{x}{4}; \frac{-8}{3}$

Lesson 3.5
Level A

1. $s = -4$
2. $c = 4$
3. $a = 3$
4. $y = 4$
5. $x = 1$
6. $m = -5$
7. $y = -\frac{6}{11}$
8. $r = -\frac{1}{3}$
9. $c = -13$
10. $d = 2$
11. $a = 3$
12. $x = 1$
13. $h = 1$
14. $h = 5$
15. $b = 1$
16. $d = 3$
17. $a = 7$
18. $x = 1$
19. $g = -\frac{1}{2}$

Answers

20. $x = 7$

21. $a = 1$

22. $g = 4$

23. $y = 2$

24. $x = 1$

25. $x = \frac{6}{7}$

26. $h = -12$

27. $y = -1$

28. $l = 0$

29. $f = 0$

30. $d = 7$

31. $g = 6$

32. $x = -3$

33. $16 + 8x = 12x$

34. $3(x + 5) = -2(3 - x)$

Lesson 3.5
Level B

1. $d = -12.5$

2. $a = 2\frac{5}{6}$

3. $x = \frac{2}{3}$

4. $a = \frac{8}{13}$

5. $x = -2$

6. $a = 4$

7. $x = 4$

8. $y = -8$

9. $a = -4$

10. $y = -5$

11. $y = 2$

12. $x = -4$

13. $z = 2$

14. $d = 7$

15. $a = 15$

16. $y = 8$

17. $x = -1$

18. $z = -\frac{1}{14}$

19. $y = -1\frac{2}{7}$

20. $m = -4$

21. $g = 3\frac{1}{3}$

22. $x = -3$

23. $z = 9$

24. $2(2 + x) + 2(2x - 1) = 4(x + 2); 3$

25. $3(x + 5) + 6 = -2(3 - x) - 5; -32$

Lesson 3.5
Level C

1. $a = 1$

2. $h = 3$

3. no solution

4. $g = 3.5$

5. $k = -1$

6. $m = \frac{3}{2}$

Answers

7. $x = 0.9$

8. $z =$ all real numbers

9. no solution

10. $b = -12.25$

11. $f = 1$

12. $a =$ all real numbers

13. $c = 1\frac{5}{6}$

14. $d = -2$

15. $x = \frac{3}{2}$

16. $j = 12.5$

17. no solution

18. $2(2x - 6) + 2(x - 10) = 2(x + 2) + 2(x + 4)$; 22; 100

19. $2x + 17 + 4x - 5 + 2x + 6 = 180$; $x = 20.25$; $m\angle A = 57°$; $m\angle B = 76°$; $m\angle C = 46.5°$

Lesson 3.6
Level A

1. 251.6°

2. 147.92°

3. 41°

4. 32°

5. 50°

6. 77°

7. $b = 3$

8. $b = 7$

9. $b = 14$

10. $b = 4$

11. $b = 10$

12. $b = 9$

13. $L = Q - S$

14. $L = S + Q$

15. $Q = L - S$

16. $f = g - d$

17. $f = g - d$

18. $d = -g - f$

19. $d = g + f$

20. $n = \frac{m}{r}$

21. $z = \frac{y}{5}$

22. $j = \frac{k}{h}$

23. $m = sr$

24. $d = -5k$

25. $R_1 = R_T - R_2$

26. $K = L + J$

27. $b = \frac{c}{a}$

28. $g = \frac{-h}{3}$

29. $n = m - p$

30. $b = -dc$

31. $t + w = s$

32. $g = h - k$

33. $t = \frac{-w}{s}$

Answers

34. $d = \frac{f}{c}$

35. $d - b = c$

36. $a = \frac{b}{2}$

Lesson 3.6
Level B

1. $b = 3$
2. $b = 17$
3. $b = 8$
4. $b = 24$
5. $b = 12$
6. $b = 18$
7. $b = 28$
8. $b = 5$
9. $b = 12$
10. $b = 4$
11. $m = \frac{y + 5}{x}$
12. $b = \frac{y - 2x}{2}$
13. $r = \frac{m - 2q}{3}$
14. $a = \frac{4 + 5b}{3}$
15. $m = \frac{p - 7n}{7}$
16. $x = \frac{d + 3}{q}$
17. $x = \frac{6 - c}{b}$
18. $q = 5m - p$
19. $x = \frac{1 - 3y}{2}$
20. $h = \frac{m - 5}{g}$
21. $y = \frac{3 - 5x}{2}$
22. $x = \frac{5 + 3y}{2}$
23. $b = j(t - h)$
24. $n = 2m$
25. $k = tj - b$
26. $k = \frac{q + p}{b}$
27. $n = -k + 3m$
28. $b = -c(d - 5)$
29. $s = \frac{w + t}{4}$
30. $x = \frac{3z - 6y}{4}$
31. $w = 4$ inches, $l = 8$ inches

Lesson 3.6
Level C

1. $b_2 = 28$
2. $b_2 = 23$
3. $b_2 = 2$
4. $b_2 = 11$
5. $b_2 = 10$

Answers

6. $b_2 = 20$

7. $b_2 = 12$

8. $b_2 = 10$

9. $b_2 = 55$

10. $b_2 = 14$

11. $g = \frac{-4}{3} - 2h$

12. $h = \frac{-4k}{3} - 2m$

13. $r = \frac{p - 2.5b}{3.1}$

14. $h = \frac{4}{3}\left(m - \frac{1}{2}g\right)$

15. $m = \frac{p + 7.2}{7.5}$

16. $x = \frac{1.1d - 3.2}{0.3}$

17. $g = 4(l - 5) - h$

18. $q = \frac{1.5m + 3.2}{5}$

19. $x = \frac{0.1 - 0.3y}{0.2}$

20. $k = 2b + m + 4$

21. 9.55 centimeters

22. 8 inches

Answers

Lesson 4.1
Level A

1. means: 7, 48 extremes: 12, 28
2. means: 32, 54 extremes: 9, 192
3. means: 5, 98 extremes: 14, 35
4. means: 45, 126 extremes: 18, 315
5. means: 50, 12 extremes: 10, 60
6. means: 73, 144 extremes: 36, 292
7. yes
8. no
9. yes
10. yes
11. no
12. no
13. yes
14. no
15. yes
16. $n = 10$
17. $x = 91$
18. $m = 528$
19. $y = 6$
20. $p = 2$
21. $z = 8.75$
22. $x = 10$
23. $y = 80$
24. $x = 270$
25. $x = 8\frac{2}{3}$
26. $x = 2250$
27. $x = 1.375$

Lesson 4.1
Level B

1. no
2. yes
3. no
4. yes
5. yes
6. no
7. $x = 10.5$
8. $y = 2.5$
9. $z = 1134$
10. $n = 1634$
11. $m = 385.2$
12. $p = 17.5$
13. $q = 1.72$
14. $x = 68.4$
15. $r = 2448$
16. $w = 800$
17. $112.00
18. 216 miles
19. $n = 7.2$

Lesson 4.1
Level C

1. yes
2. yes
3. no
4. no
5. no
6. no

Answers

7. $x = 83.72$
8. $y = 17$
9. $n = 128.4$
10. $m = 52$
11. $p = 25.5$
12. $q = 800$
13. $r = 82.1$
14. $w = 106$
15. 70 minutes
16. $7.20
17. $y = 4.8$
18. 15 bananas
19. 15 feet

Lesson 4.2
Level A

1. 0.04
2. 0.28
3. 0.67
4. 1.50
5. 0.72
6. 2.30
7. 0.55
8. 0.09
9. 0.32
10. 0.01
11. 0.85
12. 1
13. 12%
14. 8%
15. 185%
16. 200%
17. 57%
18. 360%
19. 70%
20. 43%
21. 3%
22. 26%
23. 120%
24. 98%
25. $\frac{23}{100}$
26. $\frac{1}{2}$
27. $\frac{4}{5}$
28. $\frac{9}{20}$
29. $\frac{16}{25}$
30. $\frac{9}{50}$
31. $\frac{73}{100}$
32. $\frac{9}{25}$
33. $\frac{7}{50}$

34. $\frac{7}{20}$

35. $\frac{81}{100}$

36. $\frac{99}{100}$

37. 16

38. 72

39. 16

40. 111

41. 350

42. 84

43. 352

44. 20

45. 43%

46. 60%

47. 20%

48. 80%

Lesson 4.2
Level B

1. 0.14
2. 0.60
3. 1.50
4. 0.165
5. 2.45
6. 0.0025
7. 7.81
8. 0.36
9. 0.89
10. 50%
11. 120%
12. 1%
13. 100%
14. 458%
15. 68%
16. 0.5%
17. 89%
18. 1.2%
19. $\frac{17}{20}$
20. $\frac{43}{100}$
21. $\frac{31}{25}$
22. $\frac{17}{200}$
23. $\frac{6}{25}$
24. $\frac{13}{5}$
25. 330
26. 15%
27. 160%
28. 33.6
29. 39
30. 24%
31. 3.08
32. 125

Answers

33. 22.4%

34. 19.76%

35. 1620

36. $23.32

Lesson 4.2
Level C

1. 0.005
2. 25.0
3. 0.08756
4. 268.7%
5. 1.5%
6. 64%
7. $3\frac{1}{2}$
8. $\frac{21}{25}$
9. $\frac{21}{200}$
10. $\frac{11}{25}$
11. $\frac{61}{250}$
12. $\frac{167}{100} = 1\frac{67}{100}$
13. 812.5
14. 15.03
15. 25%
16. 200
17. 400%
18. 124
19. $27.00
20. $28.22
21. $1.76
22. $17,520.00
23. a. 25.72%
    b. 40.16%

Lesson 4.3
Level A

1. 0.10 or 10%
2. 0.40 or 40%
3. 0.30 or 30%
4. 0.20 or 20%
5. 0.73 or 73%
6. 0.12 or 12%
7. 0.79 or 79%
8. 0.80 or 80%
9. 0.40 or 40%
10. 0.20 or 20%
11. 0.27 or 27%
12. 0.365 or 36.5%

Lesson 4.3
Level B

1. 0.812 or 81.2%
2. 0.268 or 26.8%
3. 0.52 or 52%
4. 0.792 or 79.2%
5. 0.2 or 20%
6. 0.9 or 90%

Answers

7. 0.5 or 50%
8. 0.1 or 10%
9. 0.3 or 30%
10. 0.4 or 40%
11. 0.2 or 20%
12. 1 or 100%
13. 0 or 0%
14. 0 or 0%
15. 0.6 or 60%
16. 0.6 or 60%
17. 0.8 or 80%

Lesson 4.3
Level C

1. 0.275 or 27.5%
2. 0.05 or 5%
3. 0.475 or 47.5%
4. 0.32 or 32%
5. 0 or 0%
6. 0.48 or 48%
7. 0.80 or 80%
8. 0.52 or 52%
9. 0.24 or 24%
10. 0 or 0%
11. 0.80 or 80%
12. 0.20 or 20%
13. 0.20 or 20%
14. 8

Lesson 4.4
Level A

1. mean: 11
   median: 10.5
   mode: 11
   range: 16
2. mean: 27
   median: 27
   mode: none
   range: 13
3. mean: 12.5
   median: 12
   modes: 10, 19
   range: 16
4. mean: 117
   median: 115.5
   mode: none
   range: 27
5. mean: 63
   median: 60
   mode: 60
   range: 17
6. mean: 12.6
   median: 15
   mode: none
   range: 19
7.

1	2	3	4	5	6
//	//	///	/	/	/

8. mean: 3
   median: 3
   mode: 3
   range: 5

Lesson 4.4
Level B

1. mean: 24.5
   median: 23
   mode: none
   range: 24

2. mean: 53.75
   median: 54.95
   mode: none
   range: 21

3. mean: 115
   median: 117.3
   mode: 117.3
   range: 21

4. mean: 222.456
   median: 221.6
   mode: none
   range: 60

5. mean: 0.828
   median: 0.865
   modes: 0.25, 1.3
   range: 1.05

6.

20	21	22	23	24	25	26
𝍸	𝍸	𝍸 //	///	///	////	///

7. mean: 22.6
   median: 22
   mode: 22
   range: 6

8. 99

Lesson 4.4
Level C

1. median

2. mode

3. the mode since this number occurs the most often

4. the median which is $833.50; the mean is approximately $1350 which is higher than all but one score

5.

78	79	80	81	82	83
//	///	𝍸 /	/	//	/

6.

79	80	81	82	83	84	85
𝍸	///	//	/	///		/

7. mean: 80.07
   median: 80
   mode: 80
   range: 5

8. mean: 80.87
   median: 80
   mode: 79
   range: 6

9. a. mean
   b. mode
   c. It's the lower score.

Lesson 4.5
Level A

1. a. comedy
   b. 70

2. a. biography
   b. 20

3. 200

4. 45

5. about 38

6. chocolate

7.

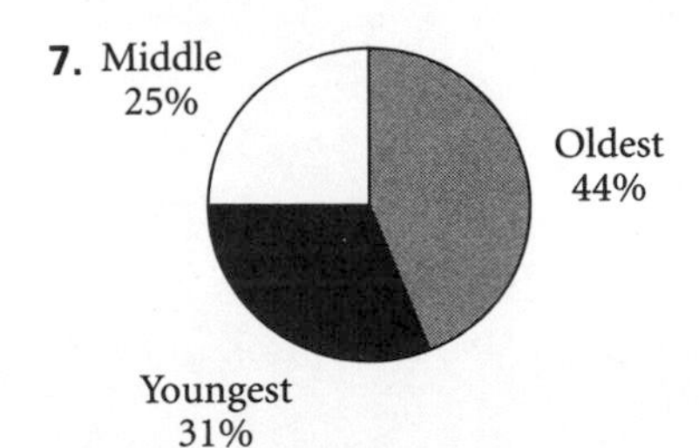

Answers

Lesson 4.5
Level B

1. 3 inches
2. Tropicville; March
3. Springtown; July
4. because the scales of the 2 graphs are different
5. 6 hours
6. People were watching less TV in 2000.
7. 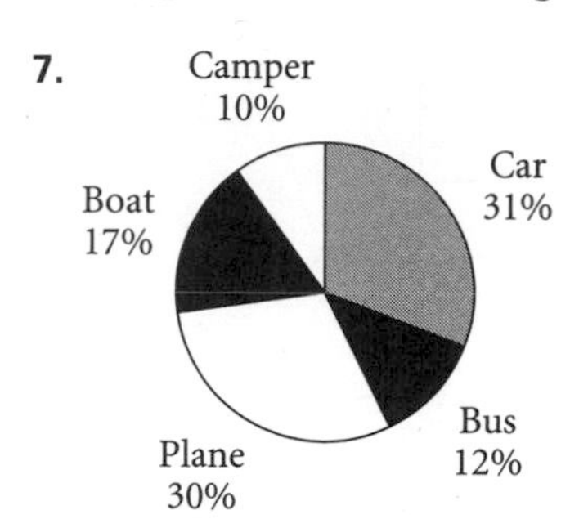

5. 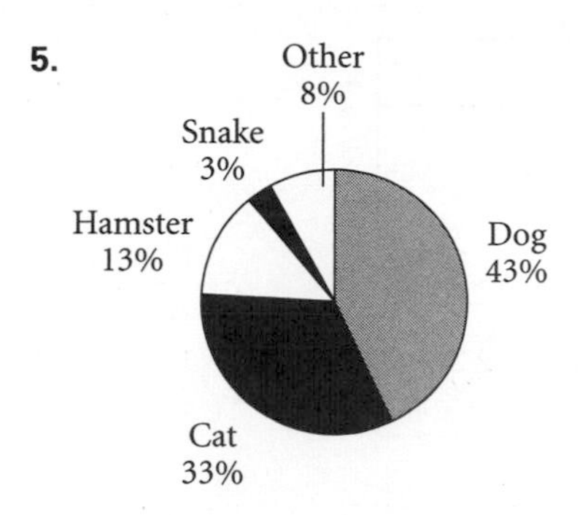

6. actual number of owners in each category
7. distribution of different pets among owners

Lesson 4.5
Level C

1. a. January
   b. $55
2. about $45
3. cost of heat in winter months & air conditioning in summer months
4. 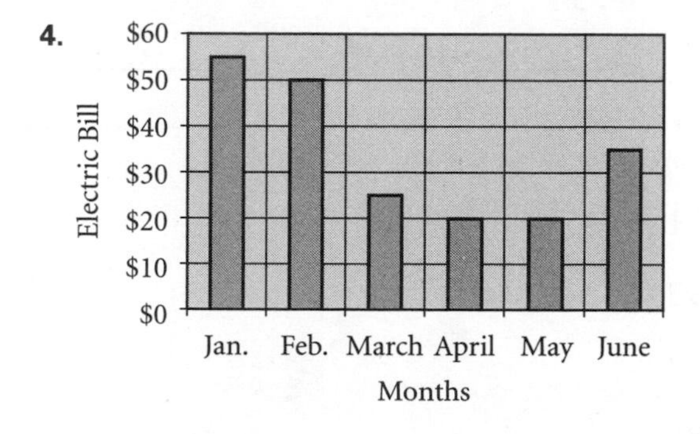

Lesson 4.6
Level A

1.

Stems	Leaves
1	0, 1, 2, 2, 5, 8, 8
2	0, 4, 5, 6, 7
3	1

2.

Stems	Leaves
4	2, 2, 5, 5, 8
5	5, 5, 6, 7
6	0, 1, 3, 6
7	
8	0, 3

3.

Stems	Leaves
1	1, 5, 6, 9
2	0, 1, 3, 6
3	3, 3, 5, 5, 7, 8

4.

Stems	Leaves
6	9
7	5, 6, 7, 7
8	0, 1, 2, 3, 7, 9, 9
9	
10	1, 3

5.

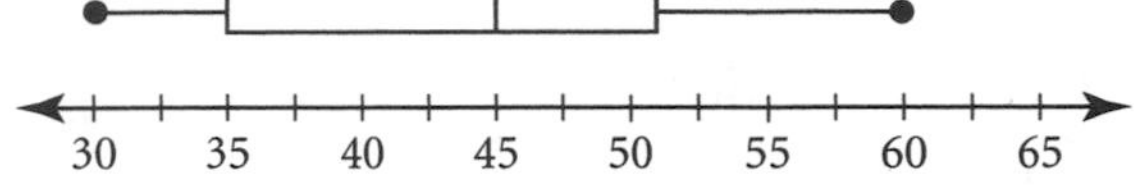

6.

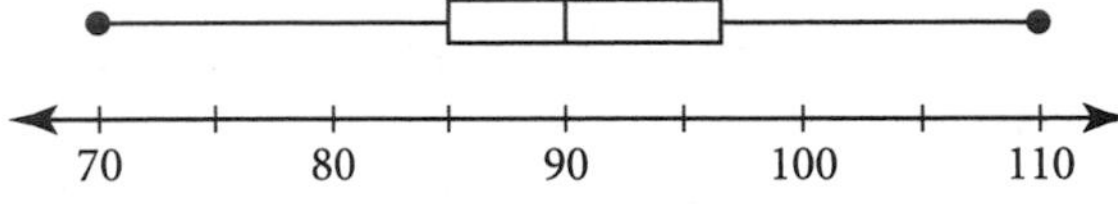

7.

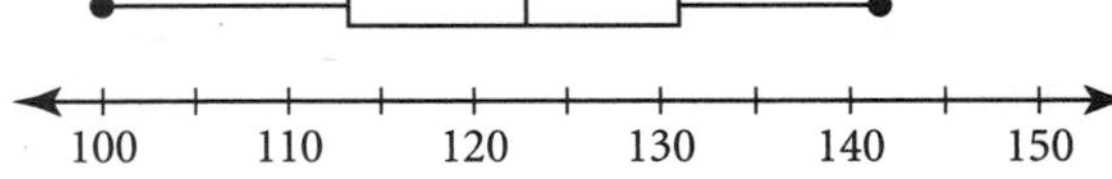

8.

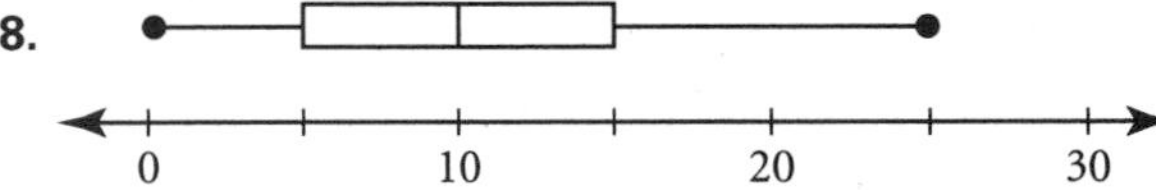

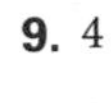

9.

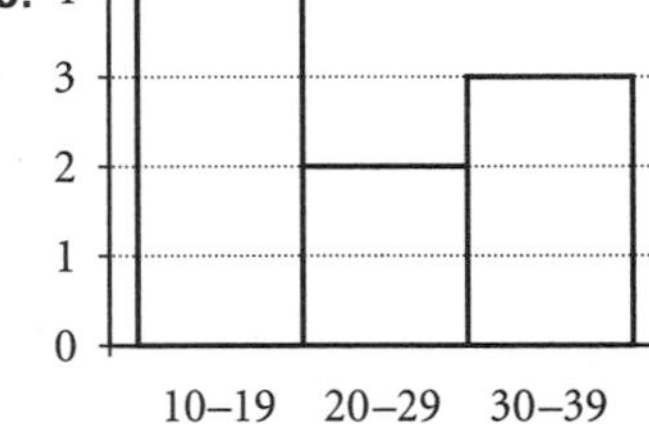

10.

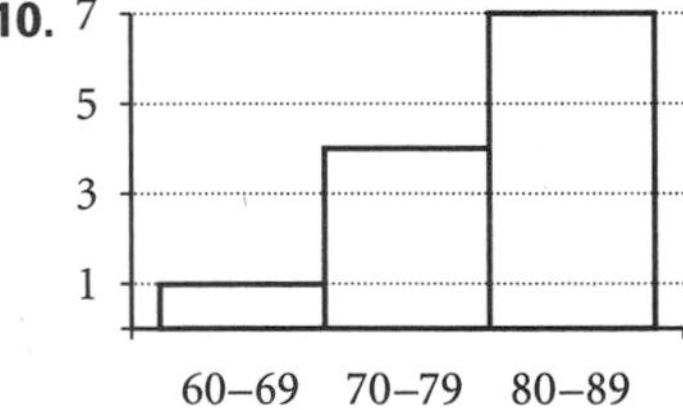

11. 22

12. 27

13. 35

14. 22, 35, 49

Lesson 4.6
Level B

1. 9

2. 20

3. 12

4. 15

5. No; The mode is not shown in a box-and-whisker plot.

6.

Stems	Leaves
3	5
4	2, 5, 5, 7, 8
5	0, 0, 4, 4, 6, 8
6	0, 2, 4, 7
7	2, 5

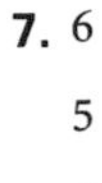

7.

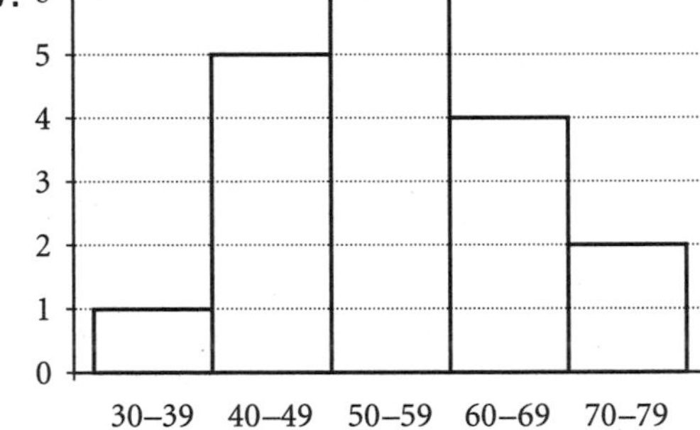

8. 54

9. 45, 50, 54

10.

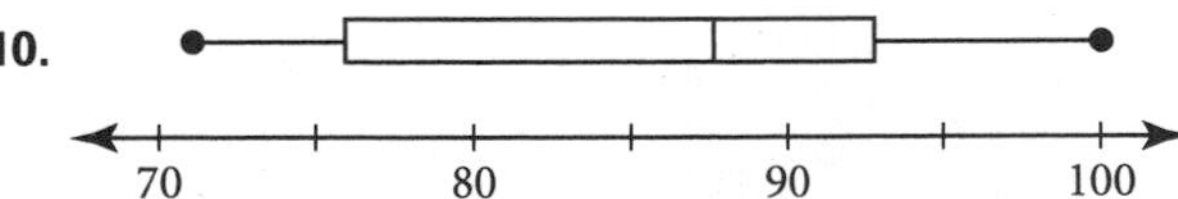

11. 71

12. 93

13. 28

14. 88

15. no

Answers

Lesson 4.6
Level C

1. 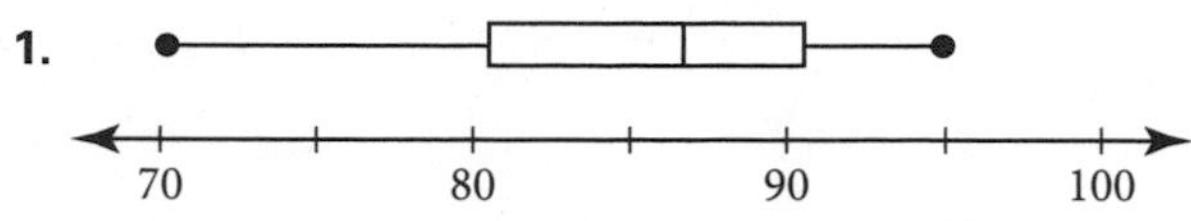

2. 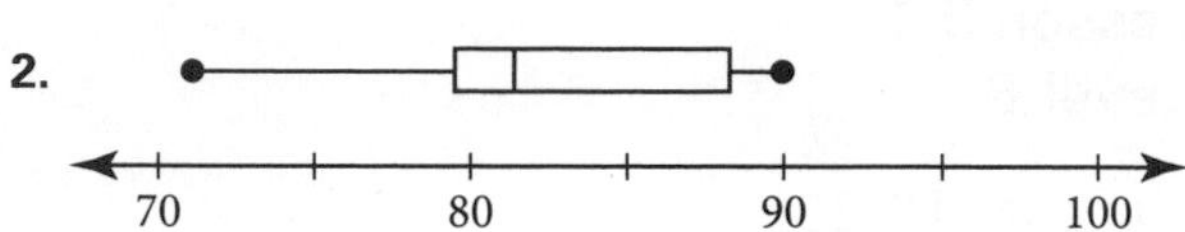

3. 70

4. 87

5. upper quartile for morning class, 91

6. lowest score for morning class, 70

7. Based on the middle 50%, the morning class did better.

8.

Stems	Leaves
7	0, 3, 8, 8
8	1, 2, 2, 5, 5, 6, 6, 7, 7, 8
9	1, 1, 2, 2, 3, 5

9. 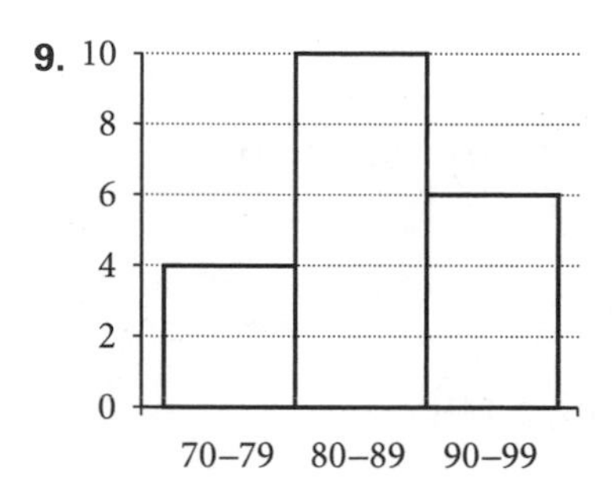

10. the mode

Answers

Lesson 5.1
Level A

1. a. {3, 5, 9, 13}
   b. {8, 10, 14, 18}
   c. function

2. a. {0, 1, 2, 3, 4}
   b. {4}
   c. function

3. a. {2, 10, 16}
   b. {0, 4, 6, 13}
   c. not a function

4. a. {8, 10, 15, 22, 31}
   b. {3, 6, 8, 20}
   c. function

5. a. {5, 14}
   b. {0, 1, 2, 3}
   c. not a function

6. a. {8, 10, 30, 34}
   b. {3, 7, 9}
   c. function

7. a. {16}
   b. {2, 10, 16, 25}
   c. not a function

8. a. {12, 20, 45}
   b. {12, 20, 45}
   c. function

9. 14

10. 4

11. 5

12. 8

13. 1

14. 15

15. 56

16. 11

17. 4

18. 8

19. 10

20. 88

Lesson 5.1
Level B

1. a. {8, 10, 15}
   b. {12.1, 12.3, 12.5}
   c. function

2. a. {22}
   b. {18.3, 25.5, 29, 31.4}
   c. not a function

3. −6

4. 9

5. −12

6. −1

7. 2

8. 7

9. $y = 2 + 0.10x$

10. {(12, 3.20), (15, 3.50), (24, 4.40), (36, 5.60)}

11. {12, 15, 24, 36}

12. {3.20, 3.50, 4.40, 5.60}

13. $y = 2x + 1$;

x	y
0	1
1	3
2	5
3	7

Answers

Lesson 5.1
Level C

1. time in hours
2. distance in miles
3. $y = 60x$
4. $\{0 \leq x \leq 7\}$
5. Each x value is paired with exactly one y value.
6. The 60 from the equation would increase to match his new speed.
7. Sample answer:

x	5	7	12	18	21
y	1	10	14	25	26

8. Sample answer:

x	5	5	12	5	21
y	1	10	14	25	26

9. \$250
10. 10 days
11. $\{2, 3, 4\}$
12. $\{21, 28, 35, 48\}$
13. No, it is not a function. There are 2 different prices for 2 pairs of jeans.

Lesson 5.2
Level A

1. 3
2. 1
3. 2
4. $\frac{1}{2}$
5. 3
6. $\frac{4}{3}$
7. $\frac{1}{3}$
8. 2
9. $\frac{1}{2}$
10. 4
11. $\frac{1}{7}$
12. $\frac{5}{3}$
13. 2
14. $\frac{2}{3}$
15. 0
16. 6
17. 1
18. undefined
19. 0
20. undefined

Lesson 5.2
Level B

1. -3
2. 1
3. $\frac{1}{2}$
4. $-\frac{1}{4}$

5. 2

6. $-\frac{2}{3}$

7. $-\frac{1}{3}$

8. $\frac{4}{3}$

9. $-\frac{3}{2}$

10. 0

11. $\frac{4}{3}$

12. undefined

13. undefined

14. -2

15. $\frac{2}{5}$

16. $\frac{14}{5}$

Lesson 5.2
Level C

1. $-\frac{4}{3}$

2. $\frac{2}{5}$

3. 5

4. b

5. 3

6. a

7. -2

8. $-\frac{68}{77}$

9. $-\frac{3}{4}$

10. -1

11. $\frac{1}{8}$

12. $\frac{2}{5}$

13. less than

Lesson 5.3
Level A

1. 4; $y = 4x$

2. 5; $y = 5x$

3. 9; $y = 9x$

4. 6; $y = 6x$

5. 16; $y = 16x$

6. 3; $y = 3x$

7. 8; $y = 8x$

8. $\frac{1}{2}$; $y = \frac{1}{2}x$

9.

x	0	1	2	3
y	0	3	6	9

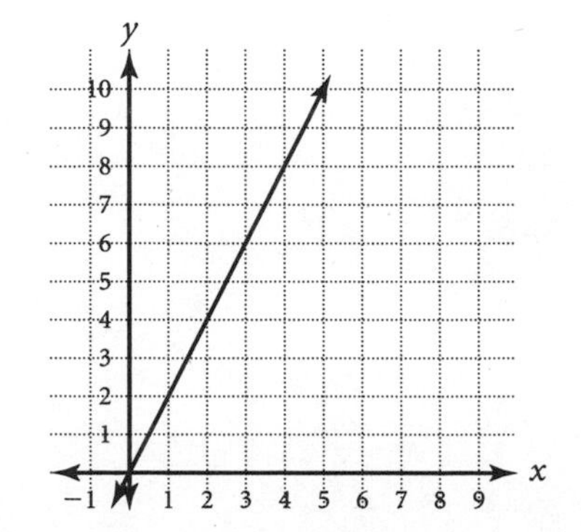

Answers

10.

x	0	1	2	3
y	0	2	4	6

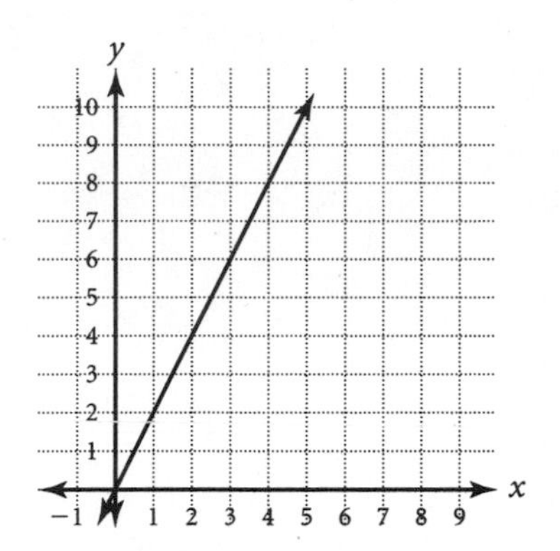

11. Danielle did not sort any boxes from 10 to 30 minutes.

12. 5 boxes per minute; 1 box per minute

Lesson 5.3
Level B

1. \$4 per class
2. \$2 per class
3. 7; $y = 7x$
4. 3; $y = 3x$
5. 5; $y = 5x$
6. 3.3; $y = 3.3x$
7. 20
8. 77
9. 98.4
10. 12
11. $y = 6x$

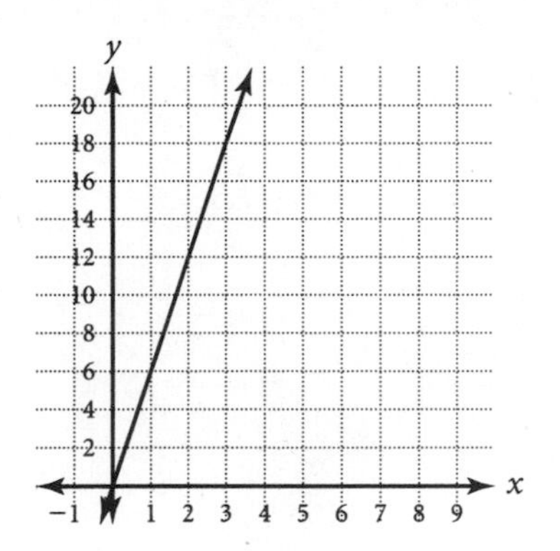

Lesson 5.3
Level C

1. 4 feet per second
2. Aaron stood still from 3 to 7 seconds.
3. 2 feet per second
4. 12 feet; 18 feet
5. Aaron's speed
6. $y = 60x$

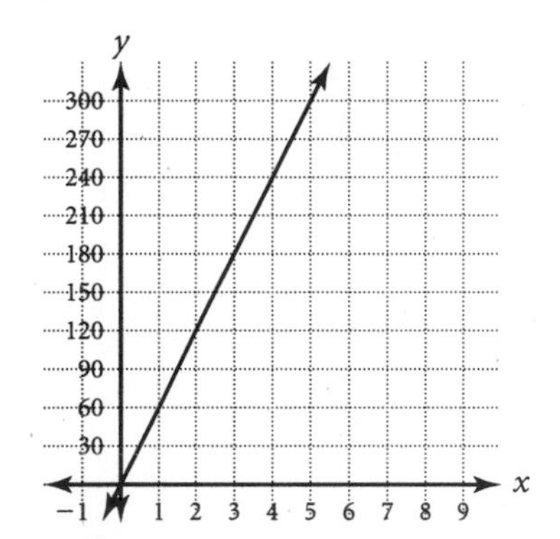

7. 25; $y = 25x$
8. 9 hours
9. 0.8; $y = 0.8x$
10. greater than

Answers

Lesson 5.4
Level A

1. $-2, 12$
2. $-2, 10$
3. $7, -7$
4. $2, 8$
5. $-5, 50$
6. $\frac{1}{2}, 4$
7. 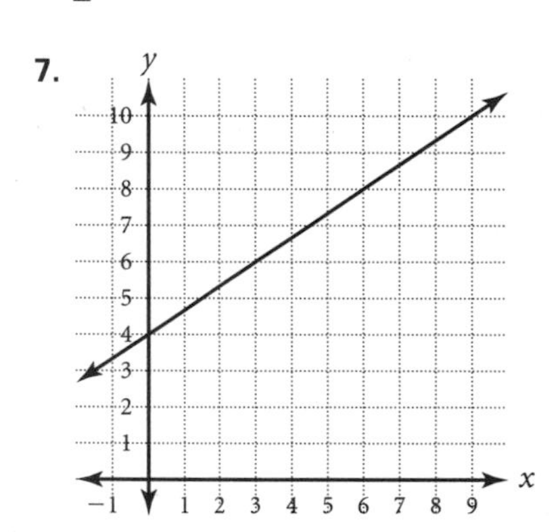
8. 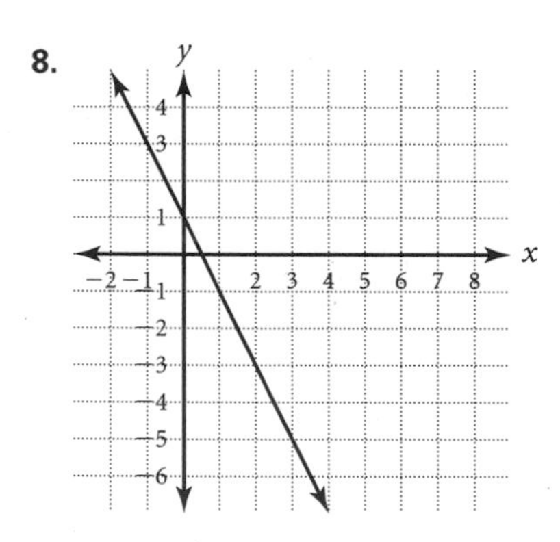
9. $y = 5x - 3$
10. $y = \frac{3}{4}x + 8$
11. $y = 2x + 10$
12. $y = 3x - 5$
13. $y = \frac{1}{2}x + 3$

Lesson 5.4
Level B

1. $-\frac{1}{3}, 1$
2. $-9, 6$
3. $4, -3$
4. 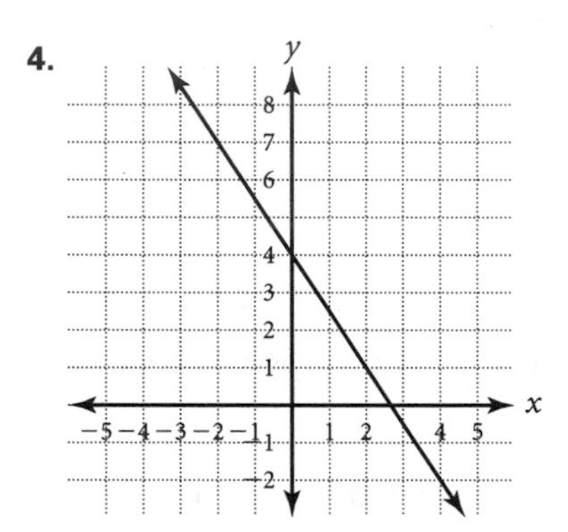
5. 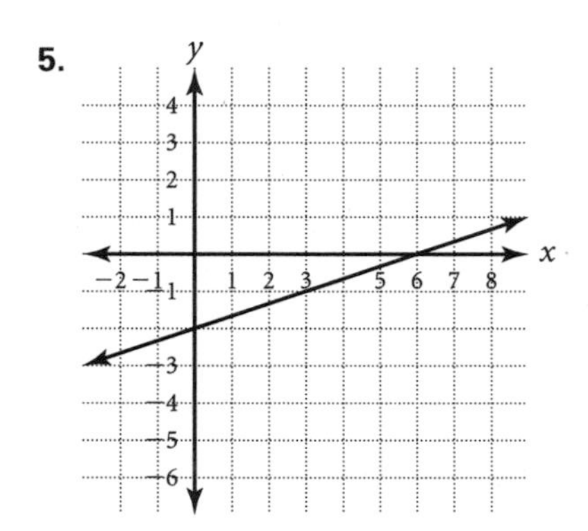
6. $y = -\frac{1}{2}x + 4$
7. $y = -\frac{4}{3}x - 1$
8. $y = \frac{2}{3}x + 5$
9. $y = -\frac{1}{3}x + 3$
10. $y = 7x$
11. $y = 6x - 3$

Answers

12. crosses the y-axis at 7 and has a slope of -4

13. crosses the y-axis at -3 and has a slope of $\frac{4}{5}$

Lesson 5.4
Level C

1. c
2. d
3. a
4. e
5. b
6. 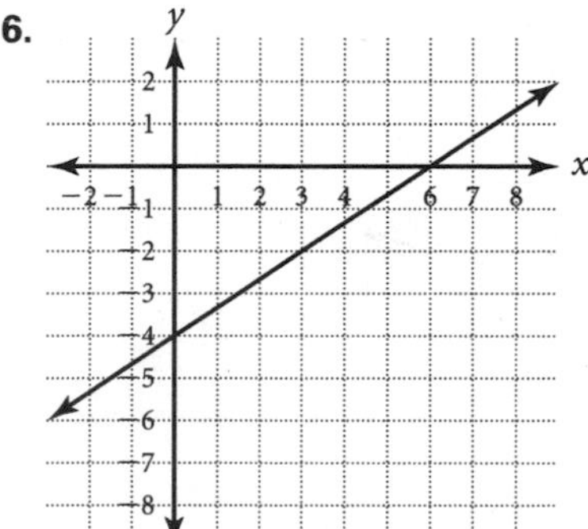

7. $y = -\frac{1}{2}x + 3$
8. a. $y = 1.5x + 2$
   b. 4 weeks
9. a. $y = -3x + 80$
   b. $35
10. $y = 1.75x$

Lesson 5.5
Level A

1. $6x - 4y = 20$
2. $-2x - 5y = 16$
3. $5x + 10y = 3$
4. $8x - 21y = -13$
5. $19x + y = 2$
6. $-2x + 14y = 18$
7. $12x - 9y = 35$
8. $12x - 34y = 25$
9. $3x + 5y = 1$
10. $10x + 2y = 8$
11. $y - 8 = 7(x - 1)$
12. $y = 2(x - 4)$
13. $y - 2 = 4(x - 7)$
14. $y - 3 = 5(x - 6)$
15. $y - 4 = 3(x - 8)$
16. $y - 1 = 10(x - 5)$
17. 2; 5
18. 4; 6
19. 5; 3
20. 7; 14
21. 15; 3
22. 3; 7
23. 3; -10
24. 4; 18
25. $y - 4 = 4(x - 5)$
26. $y - 3 = 5(x - 1)$
27. $y - 14 = 4(x - 7)$
28. $y - 4 = 4(x - 6)$
29. $6x - y = 3$
30. $5x - y = 3$
31. $x - 5y = -31$
32. $2x - y = 2$

Lesson 5.5
Level B

1. $8x - 3y = -12$

2. $3x - 9y = 15$

3. $-14x + 22y = 17$

4. $-2x + 5y = -15$

5. $y - 6 = \frac{2}{3}(x - 5)$

6. $y - (-5) = -\frac{1}{4}(x - 7)$

7. $y - 8 = 5(x - (-2))$

8. $y - 1 = -2(x - 3)$

9. $4; -12$

10. $32; -4$

11. $-3; 45$

12. $\frac{8}{5}; -12$

13.

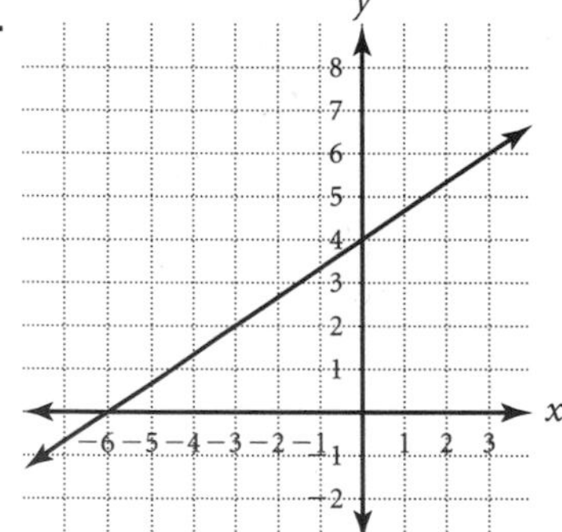

14.

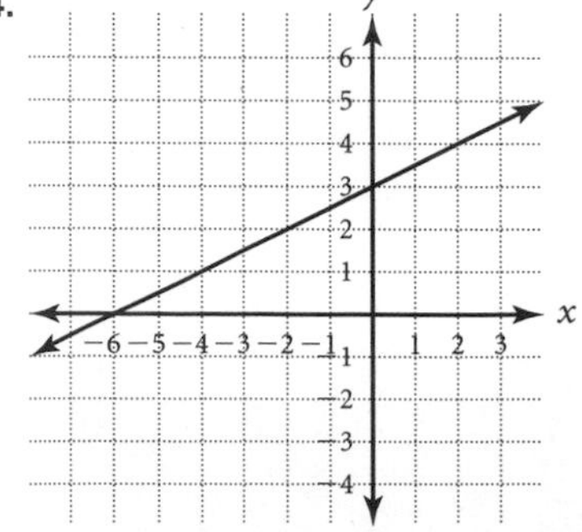

15. $4x + 3y = 36$

16. $3x - 4y = -2$

17. $y - 60 = 5(x - 4)$

Lesson 5.5
Level C

1. c

2. b

3. e

4. a

5. d

6.

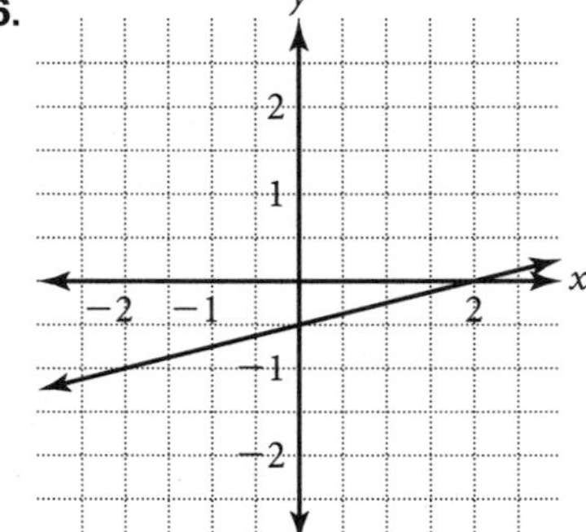

7.

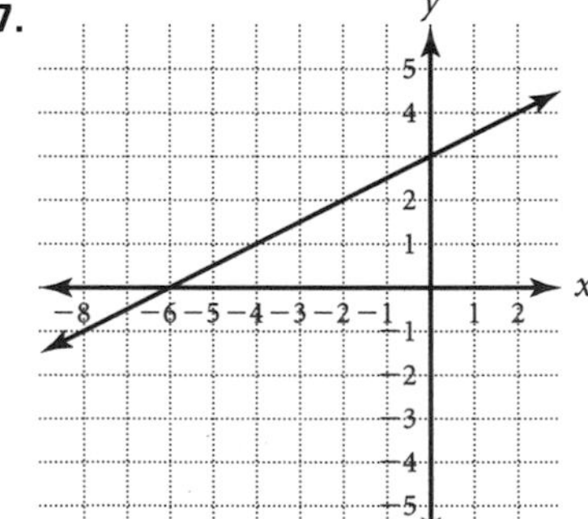

8. $y - 190 = -2(x - 4)$

9. **a.** $0.25x + 0.50y = 15$
   b. 20 cookies

10. **a.** $5x + 7y = 700$
    b. 70 games

Answers

Lesson 5.6
Level A

1. 2
2. -3
3. $\frac{1}{2}$
4. -2
5. 3
6. -3
7. $-\frac{1}{5}$
8. $\frac{3}{2}$
9. $-\frac{3}{4}$
10. $\frac{1}{2}$
11. $\frac{1}{4}$
12. -2
13. True
14. True
15. False
16. True
17. True
18. False
19. $y - 8 = 5(x - 5)$
20. $y - 1 = \frac{-1}{3}(x - 6)$
21. $y - 5 = 8(x - 2)$

Lesson 5.6
Level B

1. 2
2. $-\frac{2}{3}$
3. -2
4. $\frac{1}{20}$
5. $-\frac{1}{2}$
6. $-\frac{5}{2}$
7. $-\frac{1}{3}$
8. 16
9. e
10. a
11. d
12. b
13. c
14. $y - 5 = 3(x - 2)$
15. $y - 10 = -2(x - 4)$
16. $x = 1$
17. $y - (-8) = \frac{4}{3}(x - (-3))$

Answers

Lesson 5.6
Level C

1. $\frac{1}{4}$
2. $\frac{3}{2}$
3. $\frac{5}{2}$
4. $-\frac{1}{4}$
5. b
6. a
7. d
8. e
9. c
10. $y = -3x - 1$
11. $y = -\frac{4}{3}x$
12. $y = x + 2$
13. $y = -x + 3$
14. yes
15. The line will be perpendicular to both parallel lines.

Answers

Lesson 6.1
Level A

1. true
2. false
3. true
4. false
5. false
6. false
7. false
8. true
9. false
10. false
11. false
12. false
13. true
14. true
15. false
16. $x > 5$

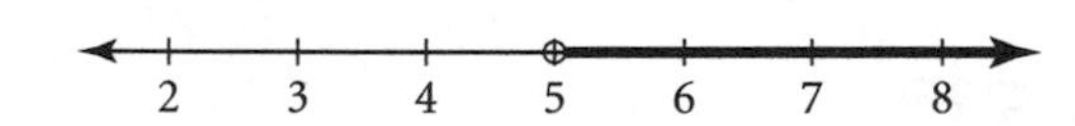

17. $c < -2$

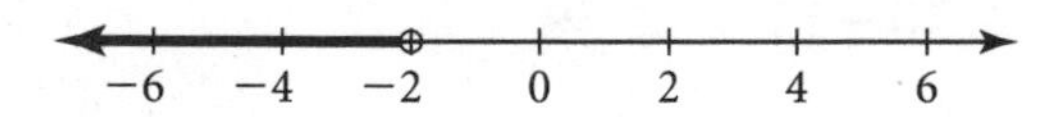

18. $d \leq 6$

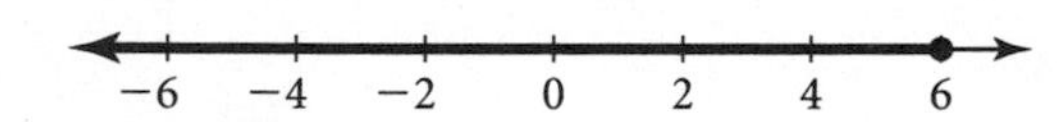

19. $k < 3$

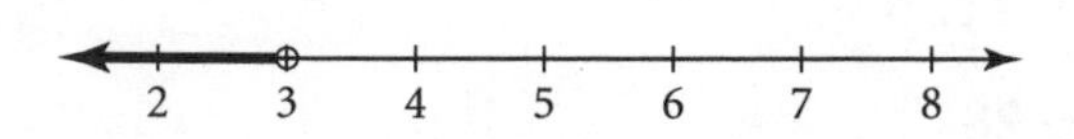

20. $n \geq 12$

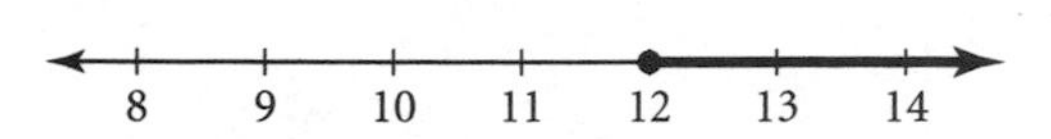

21. $p > -3$

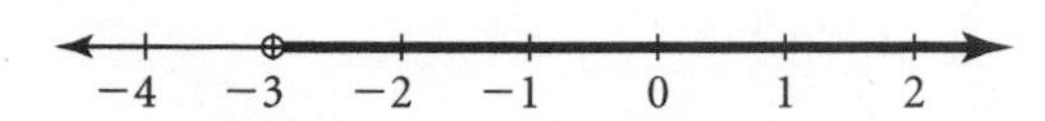

22. $w < -2$

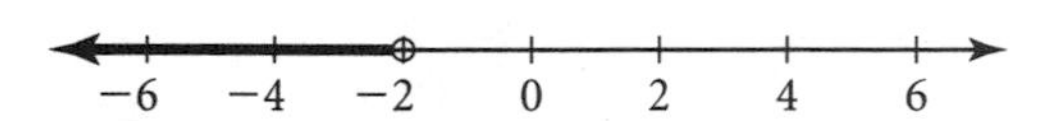

23. $t \leq 6$

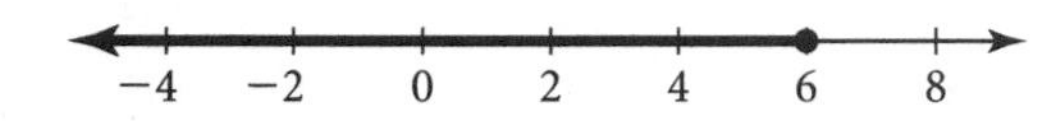

24. $h \geq -12$

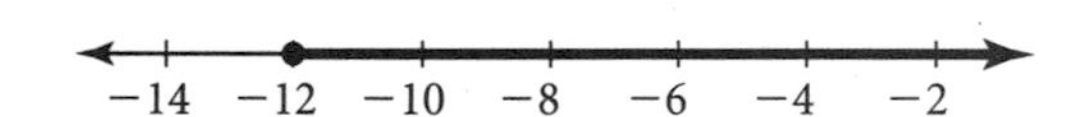

25. $v > -5$

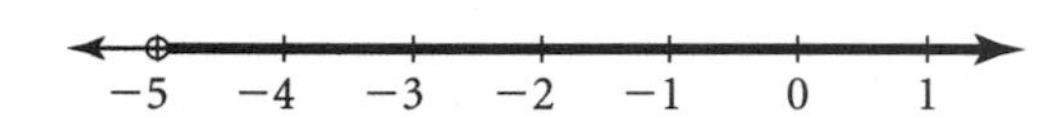

26. $x < 7$

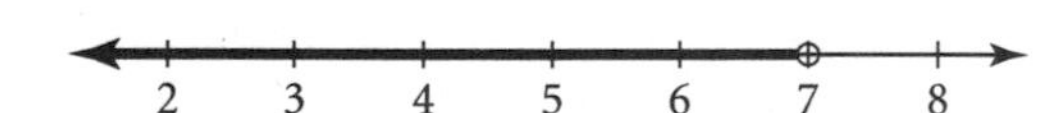

27. $y < 0$

28. $x > -1$
29. $y < 13$
30. $b \leq 1$
31. $y \geq 21$
32. $g \leq 13$
33. $f < -21$
34. $h < 0$
35. $t \leq 29$

Answers

36. $y < -20$
37. $q > 9$
38. $t > -1$
39. $x \geq -1$
40. $h < -2$
41. $y \geq 16$
42. $m \leq -13$
43. $c > 8$
44. $a > -64$
45. $n < 8$

Lesson 6.1
Level B

1. $x > -2$
2. $y < 1.3$
3. $v \geq -6.6$
4. $t < 6$
5. $r \leq \frac{3}{8}$
6. $w > 1.16$
7. $p \leq \frac{26}{27}$
8. $g < 21.5$
9. $b > -1.1$
10. $d > -\frac{1}{3}$
11. $q < 5.57$
12. $t \geq \frac{69}{14}$
13. c
14. d
15. g
16. a
17. f
18. e
19. b
20. $x > 3$
21. $x > -2$
22. $x > 6$
23. $x \leq 1.5$
24. $-2 < x \leq 3$

Lesson 6.1
Level C

1. $75 \leq x < 110$
2. $x \leq -14$
3. $x > 32$
4. $x < 55$
5. $20 < x \leq 41$
6. $x < 25$
7. $x \leq 42 - 29, x \leq 13$
8. $x \leq 70 - 11, x \leq 59$
9. $x \leq \$15.81$
10. $x < 1.5$
11. $\frac{1}{5}$
12. 5° drop
13. -28

Answers

Lesson 6.2
Level A

1. $y > x$
2. $c \leq -12$
3. $-h < 1.9$
4. $0 < z \leq 5$
5. $-18.9 \leq n \leq -14.3$
6. false
7. true
8. false
9. true
10. false
11. true
12. false
13. true
14. true
15. false
16. true
17. true
18. subtract 3 from each side
19. add 2 to each side
20. add 8 to each side
21. multiply each side by -3
22. divide each side by 3.9
23. multiply each side by 3
24. $y \leq 5$
25. $x > \frac{21}{5}$
26. $p > -2$
27. $r \geq 121$
28. $k > -6$
29. $t \geq 2$

Lesson 6.2
Level B

1. add seven to each side, then divide by 2
2. subtract 1 from each side, then multiply by 8
3. add $(-4 - p)$ to each side, then divide by -4
4. subtract 9 from each side, then multiply by -6
5. add 5 to each side, then multiply by $\frac{3}{2}$
6. distribute the -1, add 8 to each side, then divide by -1
7. $2x - 5 < 4$
8. $\frac{x}{3} + 8 \geq 10$
9. $\frac{x}{(-7)} \leq 3x$
10. $2(x + 6) > \frac{x}{10}$
11. $y > -2.5$
12. $z < 1$
13. $h \geq 1$
14. $n > \frac{-7}{2}$
15. $g < 6$
16. $p \leq -12.18$
17. $u \geq \frac{-1}{2}$

18. $x > 6$

19. $y > -1$

20. $h < 1$

21. $u \leq -5$

22. $x > \frac{8}{3}$

Lesson 6.2
Level C

1. $t < \frac{25}{9}$

2. $k \leq \frac{-11}{4}$

3. $m < \frac{-67}{3}$

4. $h \geq 1.5$

5. $u > \frac{1}{8}$

6. $q > \frac{37}{51}$

7. $n \leq \frac{-5}{6}$

8. $w \leq \frac{80}{9}$

9. $x > \frac{913}{430}$

10. $c \leq \frac{-8}{7}$

11. $b < \frac{33}{41}$

12. $x \leq \frac{5}{119}$

13. 22 guests

14. $\frac{-22}{3}$

15. $\frac{9}{5}$

16. 340 miles

17. 8 shares

18. 41 cars

19. 216 centimeters^2

Lesson 6.3
Level A

1. $4 < h < 10$

2. $-7 < c \leq 2$

3. $d \leq 0$ OR $d > \frac{2}{3}$

4. $k \leq 4$ OR $k \geq 5$

5. $-12 < y < 9$

6. $0.25 \leq x \leq 1$

7. $-7 < x < 2$

8. $m \leq -7$ OR $m > -4$

9. $13 < w \leq 15$

10. $x < -3$ OR $x \geq 1$

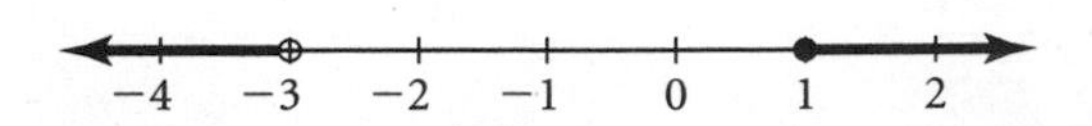

Answers

11. $14 < u \leq 17$

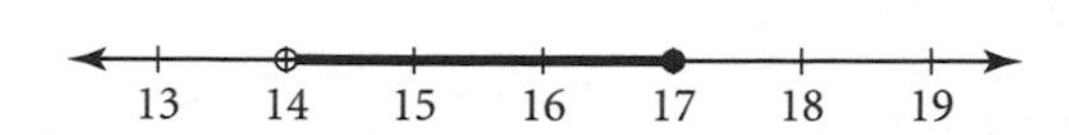

12. $13 < x \leq 14$

13. $1 \leq t \leq 3$

14. $g \leq 3$ OR $g < 38$

15. $2 < m < 3$

16. $6 < w \leq 8$

17. $\frac{-25}{4} \leq h < 0$

18. $u \leq 2$ OR $u > 4$

19. $1 \leq k < 27$

20. $-3.7 < y < 0.2$

21. $r < 2$ OR $r > 3$

22. $2 \leq p < \frac{5}{2}$

23. $p \leq 1.5$ OR $p > 2$

24. $0 \leq n < 3.2$

25. $j > 4$ OR $j \leq 3$

Lesson 6.3
Level B

1. add 4
2. add 1, then divide by 3
3. multiply by 4 on left side, add 2 on right side
4. subtract 6, then divide by -1
5. subtract 2, then divide by 3 on left side, multiply by 5 on right side
6. subtract 1, then divide by 2
7. $h \leq -1$ OR $h > \frac{-3}{5}$

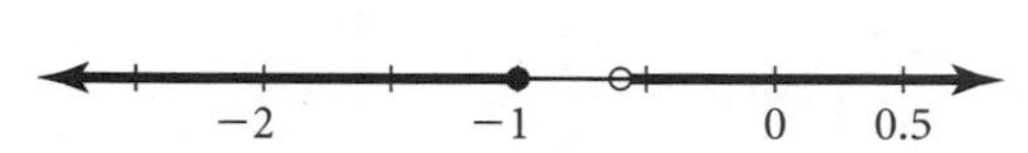

8. $1.7 \leq b \leq 3.2$

9. $0 \leq c < 7.5$

10. $f \geq -4.1$

11. $m < 9$ OR $m \geq 10$

12. $\frac{-2}{3} < d \leq 3$

13. $-2 \leq m \leq 3$

14. $v \leq -10$ OR $v > 1$

15. $w < -5$ OR $w > 1$

16. $-3 < y < -1.5$

17. $a \leq \frac{1}{2}$ OR $a > 6$

18. $3 \leq q < 4$

19. $-1 \leq x < 15$

20. $t < \frac{1}{2}$ OR $t \geq 35$

21. $-88 < x < -12$

Answers

Lesson 6.3
Level C

1. $v < \frac{-7}{2}$ OR $v > \frac{14}{15}$

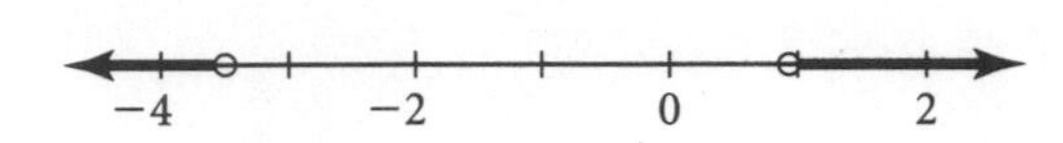

2. $2 \leq c \leq 5.5$

3. $\frac{40}{7} \leq x < \frac{64}{7}$

4. $p < 0$ OR $p > 18$

5. $u \leq 2$ OR $u > \frac{10}{3}$

6. $\frac{-11}{4} < g < \frac{-1}{2}$

7. 13 and 17 targets

8. $32 < T < 212$

9. 5.5 and 7

10. 8 point rise

11. 7 point drop

12. $-\frac{1}{7}$

Lesson 6.4
Level A

1. 32

2. 5.1

3. $\frac{4}{3}$

4. 0.35

5. 14

6. $\frac{11}{23}$

7. 19.21

8. 0.1

9. 5.4

10. 0

11. $\frac{1}{21}$

12. 100

13. 0.7

14. 7.03

15. $\frac{14}{3}$

16. $\frac{1}{10{,}000}$

17. 5

18. 3

19. 8

20. 11

21 1.9

22 -7

23. 13

24. 11

25. 0

26. 17

27. -16

28. 4

29. 39

30. 59

31. 4

32. 4.3

33. -5

34. 10

35. -0.1

36. 1.5

37. 0

38. 18

39. 18

40. $\frac{2}{15}$

41. domain: all real numbers, range: $y \geq 0$

42. domain: all real numbers, range: $y \geq 2$

43. domain: all real numbers, range: $y \geq 0$

44. domain: all real numbers, range: $y \geq 0$

45. domain: all real numbers, range: $y \geq 3$

Lesson 6.4
Level B

1. $\frac{1}{18}$

2. -1.4

3. 8.9

4. 6.11

5. $\frac{67}{56}$

6. $\frac{-29}{16}$

7. -41.4

8. 28

9. 5.8

10. 24.32

11. $\frac{253}{36}$

12. -5

13. -0.09

14. 62.62

15. $\frac{61}{56}$

16. 10

17. 56

18. 1

19. 120

20. 67

21. 0

22. -336

23. 17

24. 1

25. domain: all real numbers, range: $y \leq 7$

26. domain: all real numbers, range: $y \geq 0$

27. domain: all real numbers, range: $y \geq 0$

28. domain: all real numbers, range: $y \leq -9$

29. domain: all real numbers, range: $y \leq 0$

30. domain: all real numbers, range: $y \geq 2$

31. 1 unit to the right

32. 13 units to the left

33. 1 unit to the right and 9 units down

34. reflected across x-axis and translated 6 units down

Answers

Lesson 6.4
Level C

1. translated left 4 units and then translated down 1 unit
2. translated right 17 units and reflected across the x–axis
3. translated right 10 units and translated up 2 units
4. reflected across the x–axis and translated up $\frac{1}{3}$ unit
5. reflected across the x–axis, translated right 9 units, translated down 4 units, and vertically stretched by a factor of 2
6. translated left 3 units and vertically compressed by a factor of $\frac{1}{3}$
7. $y = |x - 11| + 4$
8. $y = -|x + 7|$
9. $y = -|x + 2| - 8$
10. $y = -|x - 6| - 2$
11. Tom: -155, 155 Kim: 262, 262
    Mike: -348, 348 Sue: 24, 24
12. If the students were permitted to go over the actual number, then Sue is the winner because she has the lowest absolute error. However, if they had to be closest without going over then Tom is the winner. Either answer is acceptable.
13. 2 hours

Lesson 6.5
Level A

1. -5, 13
2. -4, -2
3. -28, 2
4. 5
5. 4, 12
6. -7, 4
7. $\frac{-1}{4}, \frac{3}{4}$
8. 0.1, 14.1
9. -2.2, 0.8
10. -165, 35
11. 0, 16
12. 16.7, 26.5
13. 4, 4
14. -0.5, 0.5
15. $\frac{-2}{15}, \frac{2}{15}$
16. -5, 5
17. 0.9, 0.9
18. 3, 3
19. $\frac{7}{72}, \frac{7}{72}$
20. -0.5, 0.5
21. -8, 8
22. 0.02, 0.02
23. $-3 < x < 7$

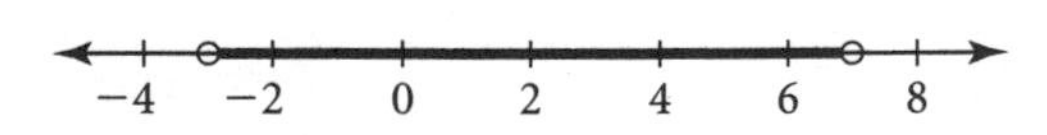

24. $x < -10$ OR $x > 4$

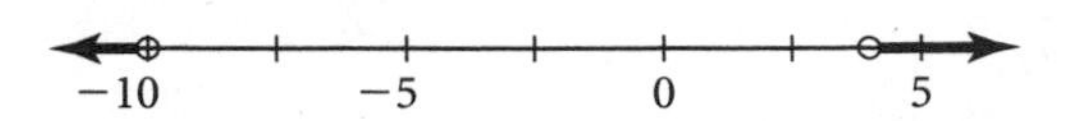

Answers

25. $6 < x < 12$

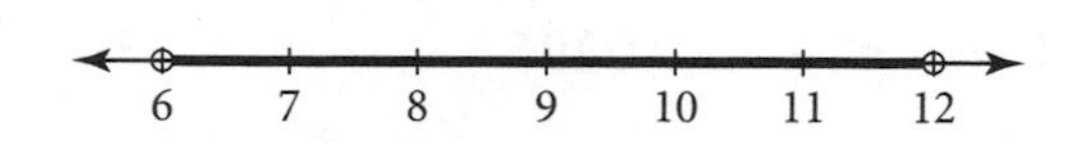

26. $-1.5 \leq x \leq 4.5$

27. $-5 \leq x \leq 3$

28. $x < -24$ OR $x > -4$

29. $x < -4$ OR $x > 20$

30. $-2.4 < x < 13.2$

31. $-24 \leq x \leq 0$

32. all real numbers

33. $x < 0$ OR $x > 18$

34. $-35 < x < 7$

Lesson 6.5
Level B

1. $|x - 8| \leq 4$

2. $|x - 2| < 15$

3. $|x - 5| \leq 3$

4. $|x - 14| \geq 17$

5. $|x - 22| < 28$

6. no solution

7. no solution

8. $\frac{31}{4}, \frac{25}{4}$

9. $-2, \frac{12}{11}$

10. $\frac{3}{5}$

11. $-77, 75$

12. $6, 10$

13. $\frac{-41}{15}, -2.6$

14. $x < -8$ OR $x > 1$

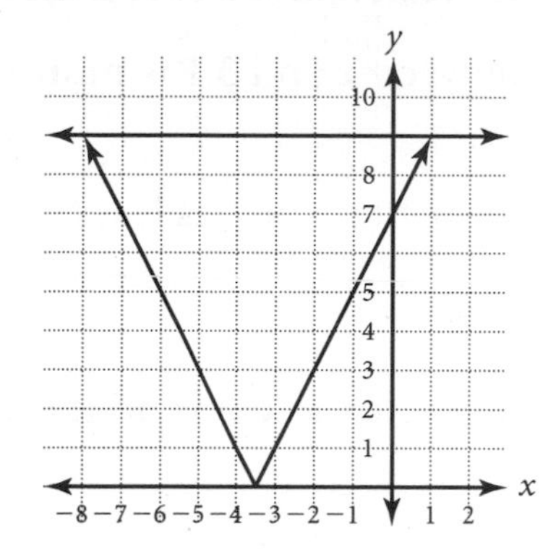

15. $2 \leq x \leq 8$

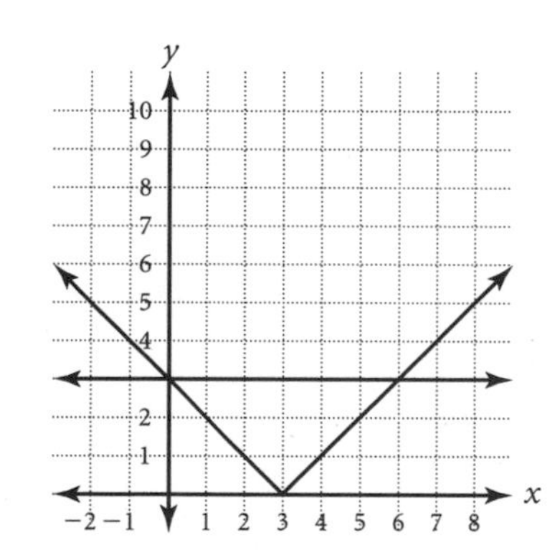

16. $\frac{-2}{11} < x < \frac{10}{11}$

17. $x < -16$ OR $x > 1$

18. $x \leq \frac{8}{9}$ OR $x \geq \frac{10}{9}$

19. $1 < x < \frac{5}{3}$

20. $\frac{-14}{11} \leq x \leq \frac{19}{11}$

21. $x < \frac{-20}{13}$ OR $x > \frac{30}{13}$

Answers

Lesson 6.5
Level C

1. The distance between x and -4 is less than 8.
2. The distance between x and 5 is at least 3.
3. The distance between x and 3.1 is more than 9.2.
4. The distance between x and $\frac{1}{4}$ is less than $\frac{2}{7}$.
5. The distance between x and 5 is no more than 10.
6. $-10 \leq x \leq 22$
7. $x < \frac{48}{91}$ OR $x > \frac{58}{91}$
8. $\frac{7}{2} < x < \frac{49}{2}$
9. $-1 \leq x \leq \frac{-1}{9}$
10. $x \leq -7.5$ OR $x \geq -5$
11. $-0.026 < x < 0.0296$
12. $\frac{3}{23} < x < \frac{183}{23}$
13. $\frac{-19}{8} \leq x \leq \frac{-1}{8}$
14. $|x - 21| \leq 0.75$, $20.25 \leq x \leq 21.75$, 21.75 ounces and 20.25 ounces
15. $|x - 1.1| \leq 0.001$, $1.099 \leq x \leq 1.101$

    1.101 cm and 1.099 cm

16. $|x - 120| \leq 2.5$, $117.5 \leq x \leq 122.5$

    122.5 degrees and 117.5 degrees

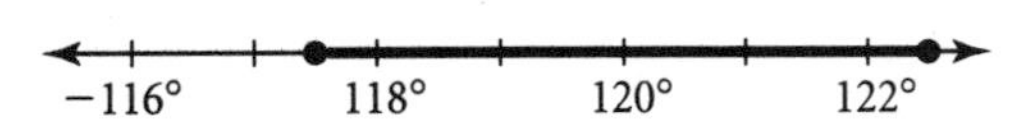

Answers

Lesson 7.1
Level A

1. yes
2. no
3. no
4. yes
5. yes
6. no
7. no
8. yes
9. $y = \left(-\frac{1}{3}\right)x - \frac{5}{6}$
10. $y = 2x + 3$
11. $y = 4x - \frac{3}{4}$
12. $y = -5x - 25$
13. $y = \left(-\frac{1}{2}\right)x + \frac{2}{3}$
14. $y = \left(-\frac{1}{2}\right)x + \frac{13}{4}$
15. $y = \left(-\frac{1}{3}\right)x + 2$
16. $y = -x + 2$
17. $(-1, -2)$

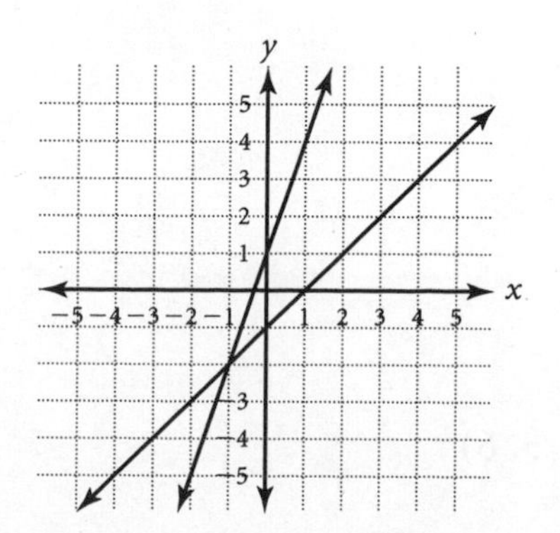

18. $(1, -4)$

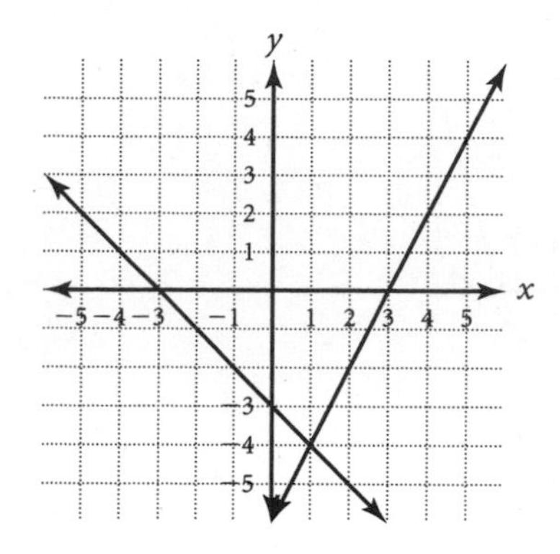

19. $(0, 5)$

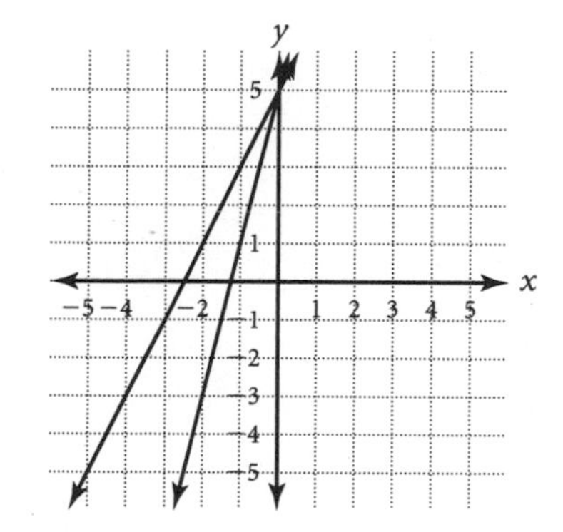

20. $(-1, 4)$

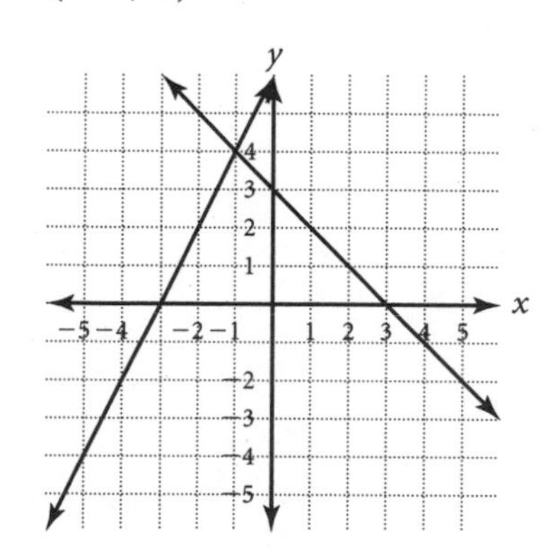

Answers

21. $(1, -1)$

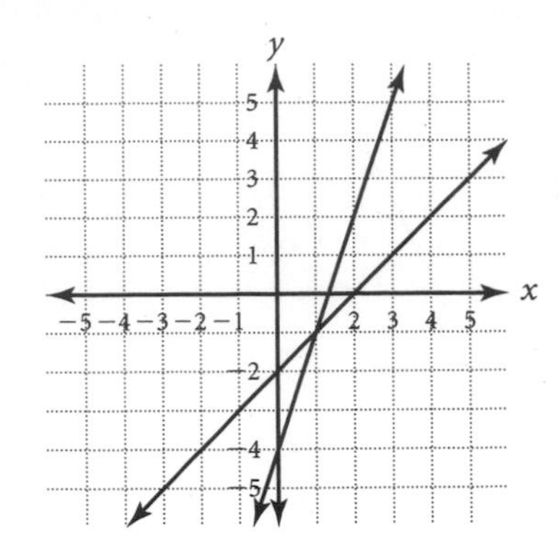

22. $(2, 2)$

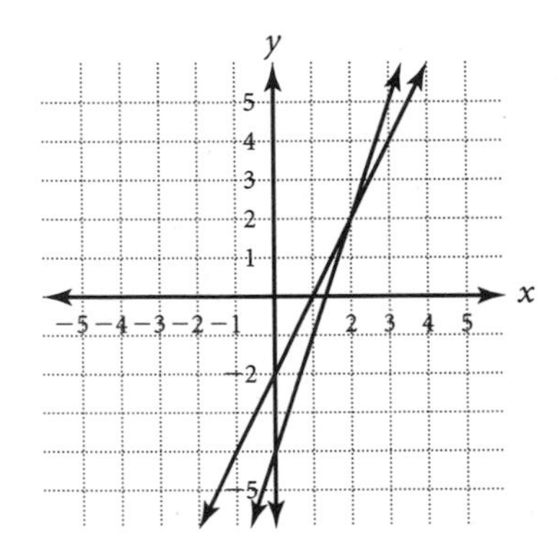

Lesson 7.1
Level B

1. yes
2. yes
3. no
4. yes
5. no
6. no
7. yes
8. no
9. 4
10. 7
11. 13
12. $(2, 3)$

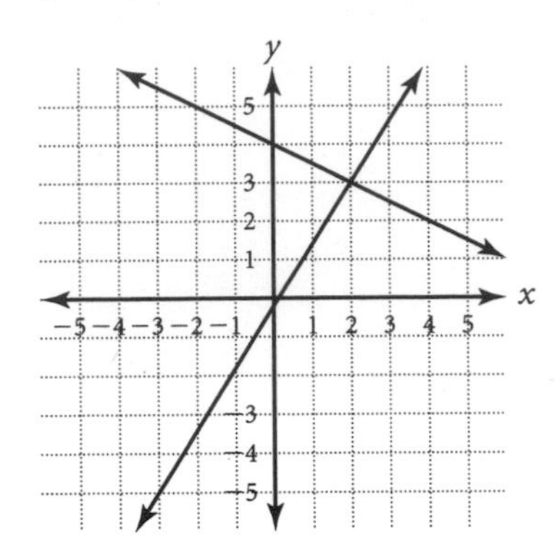

13. $(-2.3, -0.3)$

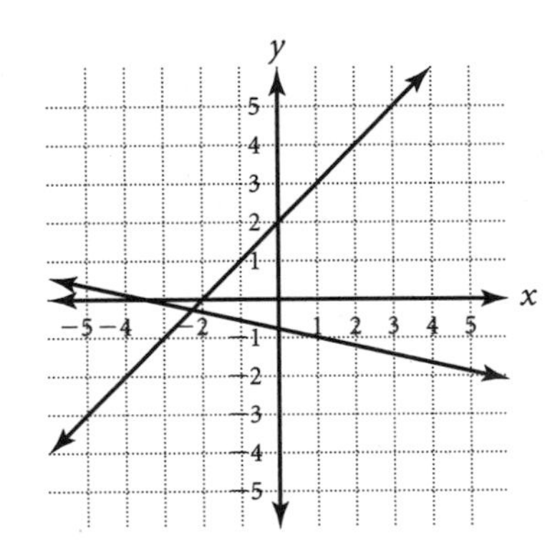

14. $(5, -4.8)$

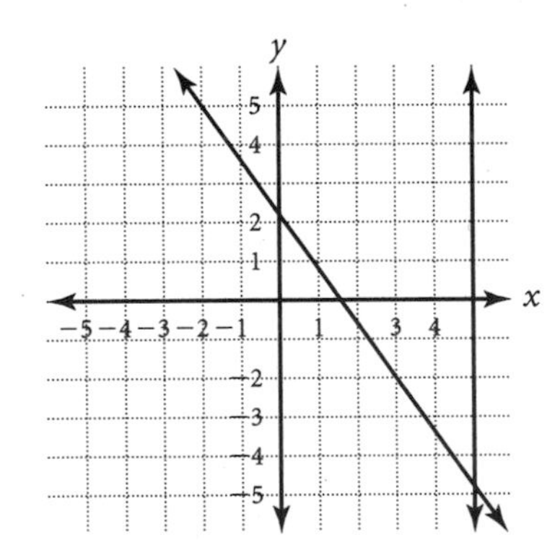

Lesson 7.1
Level C

1. $y = -3x - 3$
2. $y = 6$
3. $y = x + 1$
4. $(-1, 0); (-3, 6)$

Answers

5. 15 months
6. \$90
7. 4 minutes
8. 484 feet
9. 15.6 minutes

Lesson 7.2
Level A

1. yes
2. no
3. no
4. yes
5. no
6. yes
7. no
8. yes
9. (4, 10)
10. (−1, 9)
11. (1.8725, 12.125)
12. (0, 0)
13. (−4, 1)
14. (17, −53)
15. (10, 129)
16. (0, 3)
17. (−2, 0)
18. (0, −4)
19. (−1, 1)
20. (2, 0)
21. $\left(\frac{1}{2}, 2\right)$
22. (2, 8)
23. (−15, −22)
24. (1, 4)
25. (−1.5, −17.5)
26. (0, 2)
27. (−1, 4)
28. (10, 48)

Lesson 7.2
Level B

1. (−1.25, 3.75)

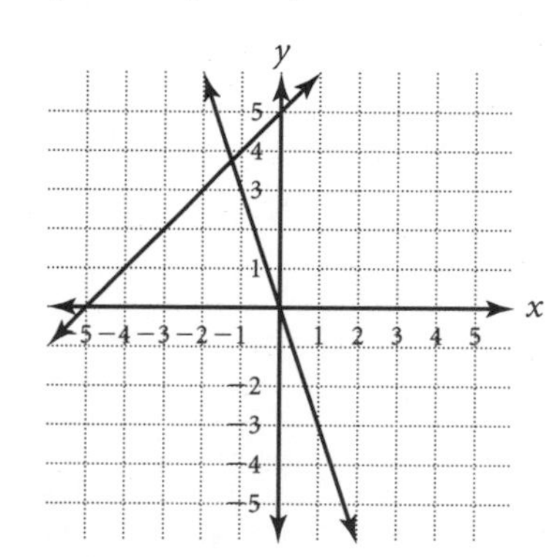

2. (0.75, −1.625)

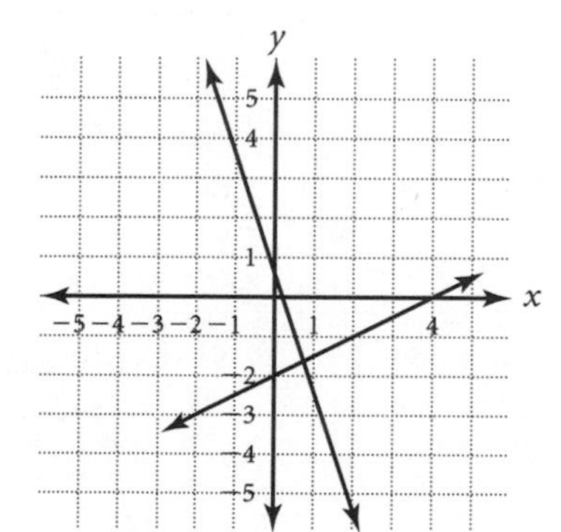

Answers

3. $(-2.75, -3.25)$

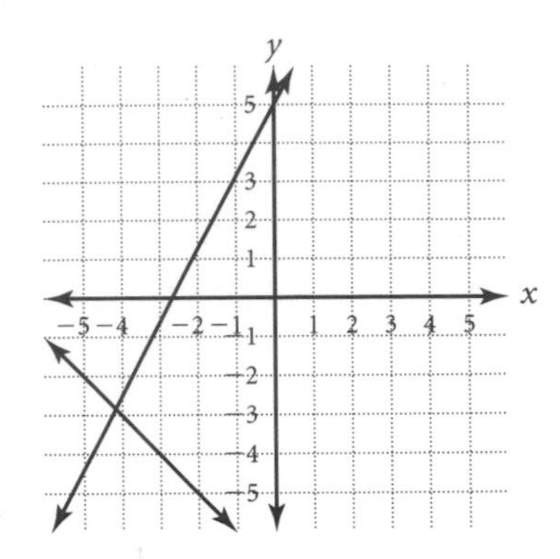

4. $\left(\frac{1}{3}, \frac{2}{3}\right)$

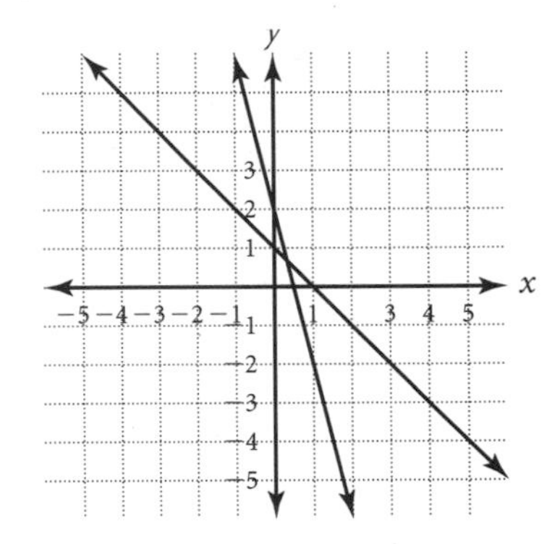

5. $\left(\frac{1}{6}, \frac{1}{12}\right)$

6. $(6, 4)$

7. $(12, 1)$

8. $(1, 1)$

9. $\left(-\frac{1}{2}, \frac{1}{3}\right)$

10. $(2, 4)$

11. $(2, -3)$

12. $(-3, 4)$

13. $(10, 2)$

14. $(-1.125, 2.125)$

15. $\left(\frac{5}{12}, 2\right)$

16. $(-1, 1)$

Lesson 7.2
Level C

1. $-4.6, -4.4$

2. $-12, -\frac{1}{2}$

3. 30 feet by 20 feet

4. 2.94 seconds

5. 140.6 feet

6. 53 adults, 41 students

7. 4 of last year's models and 7 new models

8. 35.9 seconds

9. 64.6 feet beneath the surface

Lesson 7.3
Level A

1. opposites: $2x$ and $-2x$. Add the equations together and solve for y.

2. opposites: $8d$ and $-8d$. Add the equations together and solve for c.

3. opposites: $1.5b$ and $-1.5b$. Add the equations together and solve for v.

4. opposites: n and $-n$. Add the equations together and solve for m.

5. opposites: $17x$ and $-17x$. Add the equations together and solve for y.

6. opposites: $\left(\frac{1}{5}\right)y$ and $-\left(\frac{1}{5}\right)y$. Add the equations together and solve for x.

7. $x = 1, y = 2$

8. $x = 3, y = 1$

9. $n = 6, m = -\frac{8}{7}$

10. $t = 1, p = -2$

11. $h = 15, s = \frac{77}{5}$

12. $j = 3, k = 7$

13. $x = 31, y = 4$

14. $m = 5, n = -\frac{23}{7}$

15. $u = 1, c = 2$

16. $h = 2, g = \frac{9}{2}$

17. $x = 2, y = 20$

18. $m = -1, n = -1$

19. $w = 4, p = 5.25$

20. $g = 11, b = -6$

Lesson 7.3
Level B

1. elimination; There is a pair of opposites.
2. substitution; There is an equation solved for y.
3. elimination; Multiply by -2 to obtain a pair of opposites.
4. graph; Estimate the solution.
5. $x = 2, y = -\frac{2}{3}$
6. $m = -3, n = 1$
7. $q = 7, b = -1$
8. $u = 1, v = 2$
9. $x = 8, y = 2$
10. $q = -1, w = 4$
11. $h = 7, g = 7$
12. $x = 5, y = 3$
13. $x = 2, y = 14$
14. $p = 9, q = 5$
15. $m = 1, n = 1$
16. $x = -1.1, y = 2.9$
17. $c = 3, d = 2$
18. $x = 2, y = 2$
19. $k = -3, p = 3$
20. $v = 0, r = 5$
21. $x = \frac{1}{2}, y = 1$
22. $m = 5, n = -2$

Lesson 7.3
Level C

1. $x = \frac{1}{2}, y = \frac{3}{7}$
2. $b = 26, c = 22$
3. $m = 10, n = 100$
4. $q = -0.8, w = -0.7$
5. $500 weekly salary, $16.25 per hour overtime
6. rose bush $15.50, and bed of petunias $9.75
7. 62.8 seconds

8. 803.6 feet
9. 63 hot dogs, and 114 hamburgers
10. \$6.50 for first hour, and \$0.08 per minute after first hour

Lesson 7.4
Level A

1. inconsistent
2. consistent
3. inconsistent
4. consistent
5. inconsistent
6. inconsistent
7. consistent
8. consistent
9. inconsistent
10. consistent
11. independent
12. independent
13. dependent
14. dependent
15. independent
16. independent
17. dependent
18. independent
19. no solution; inconsistent
20. $(-5.5, -34.5)$ consistent and independent
21. infinitely many solutions; consistent and dependent
22. no solution; inconsistent
23. no solution; inconsistent
24. infinitely many solutions; consistent and dependent
25. $(0, 7)$ consistent and independent
26. infinitely many solutions; consistent and dependent

Lesson 7.4
Level B

1. sample answer: $y = 3x + 1$
2. sample answer: $m = 4n + 6$
3. sample answer: $8y = 26t + 16$
4. sample answer: $4d + 6c = 5$
5. sample answer: $K = -4.5p + 3$
6. sample answer: $r = -2q - 4$
7. sample answer: $y = -x - 3$
8. sample answer: $c = \frac{10}{3}b + \frac{1}{3}$
9. sample answer: $y = 7x + 3$
10. sample answer: $c = 2d + 2$
11. sample answer: $r = 2t + 1$
12. sample answer: $n = m + 1$
13. sample answer: $-2y - 4t = -14$
14. sample answer: $x = y - 2$
15. sample answer: $3r + 7t = 2$
16. sample answer: $u = c - 7$
17. no solution; inconsistent
18. infinitely many solutions; consistent and dependent
19. $\left(\frac{116}{112}, \frac{23}{14}\right)$; consistent and independent

Answers

20. $(-1, -2)$; consistent and independent

21. $(-2, 2)$; consistent and independent

22. $\left(0, \frac{11}{7}\right)$; consistent and independent

23. infinitely many solutions; consistent and dependent

24. no solution; inconsistent

25. infinitely many solutions; consistent and dependent

26. no solution; inconsistent

27. $(1, 2)$; consistent and independent

28. $\left(\frac{2}{26}, \frac{6}{13}\right)$; consistent and independent

29. $(4, 4)$; consistent and independent

30. $(6, -17)$; consistent and independent

Lesson 7.4
Level C

1. $x = -2, y = 7$

2. consistent and independent

3. They have different slopes.

4. infinitely many solutions

5. consistent and dependent

6. They have the same slope.

7. AB: $y = 3x - 3$, CD: $y = 3x + 2$; no solution

8. inconsistent

9. Answers will vary. Sample answer: $y = -3x - 3$

10. 285 pounds, 220 pounds

11. 51, 12

12. $p = -8$

13. Juan $= 135 + 2.1t$, Tom $= 2.1t$. These form an inconsistent system and have no solution. Since the event of Tom catching Juan is represented by the intersection of this system, it can never happen.

Lesson 7.5
Level A

1. yes

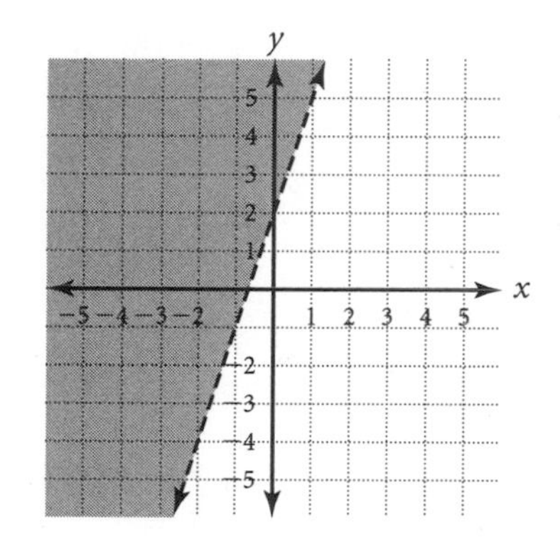

2. yes

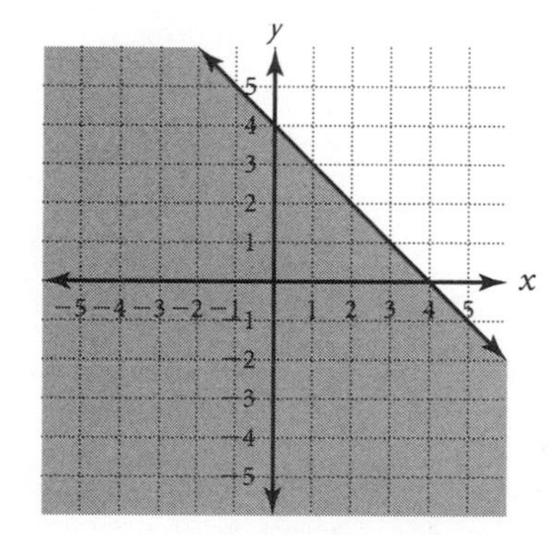

3. yes

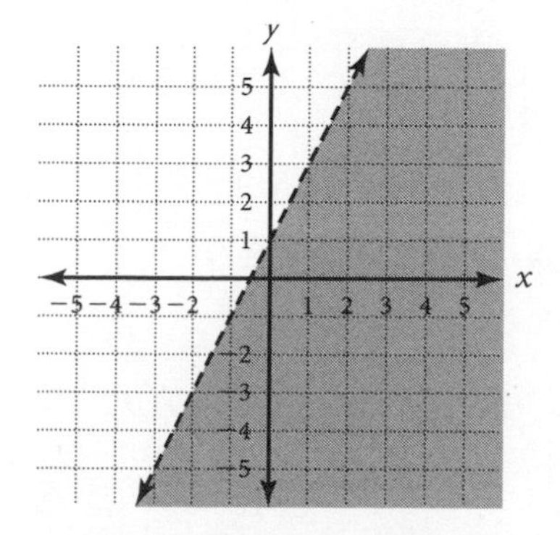

4. yes

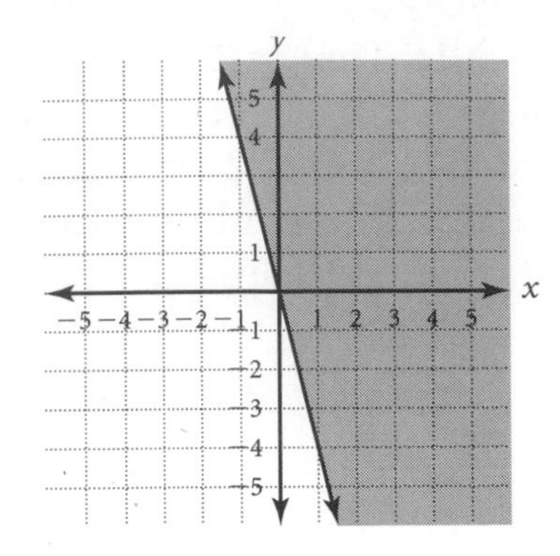

5. yes

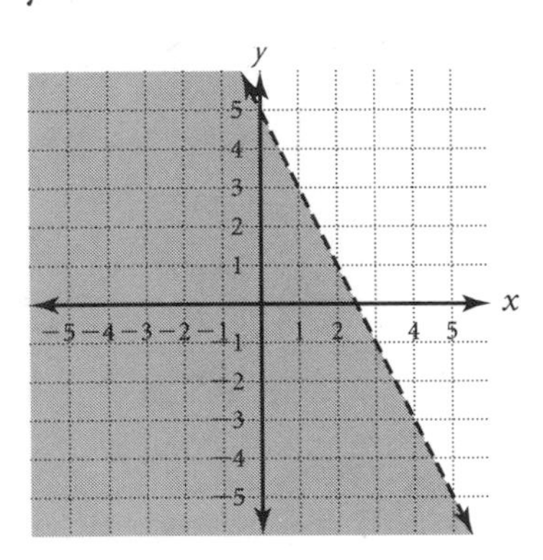

6. yes

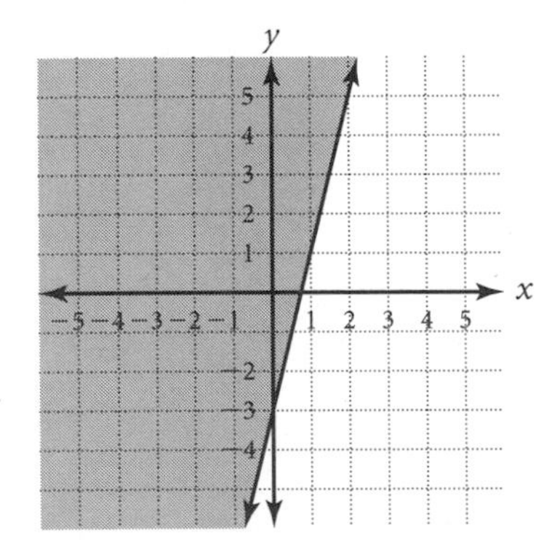

7.

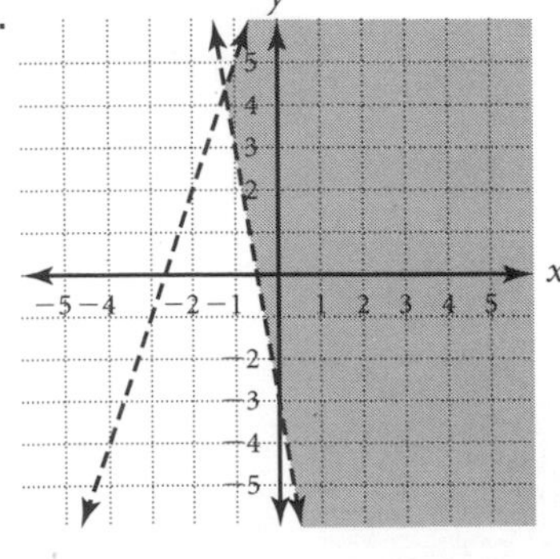

8.

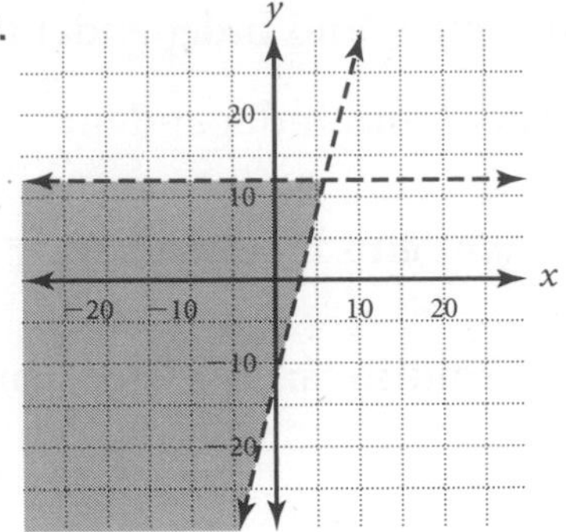

9.

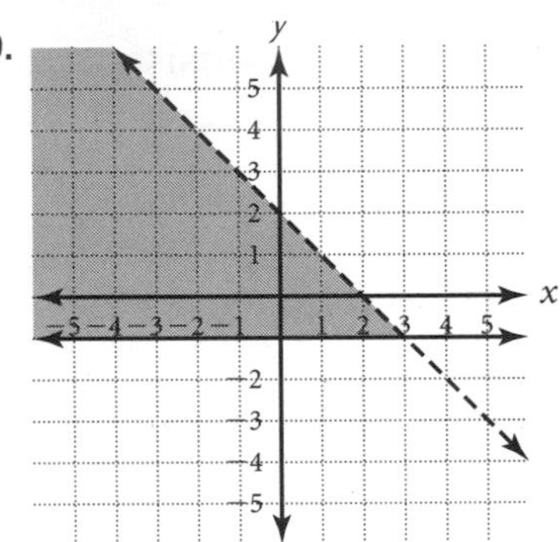

Lesson 7.5
Level B

1.

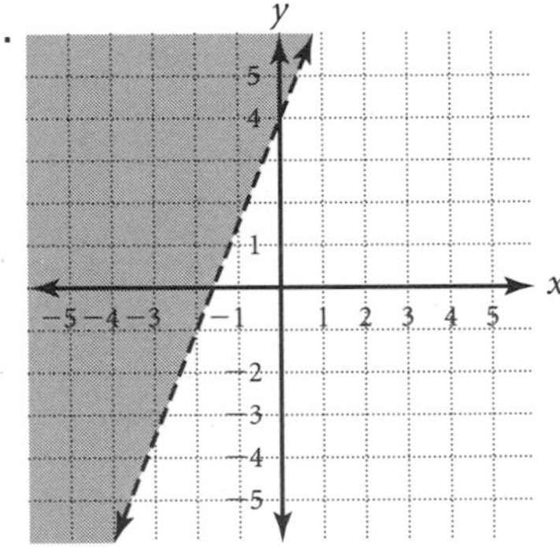

2.

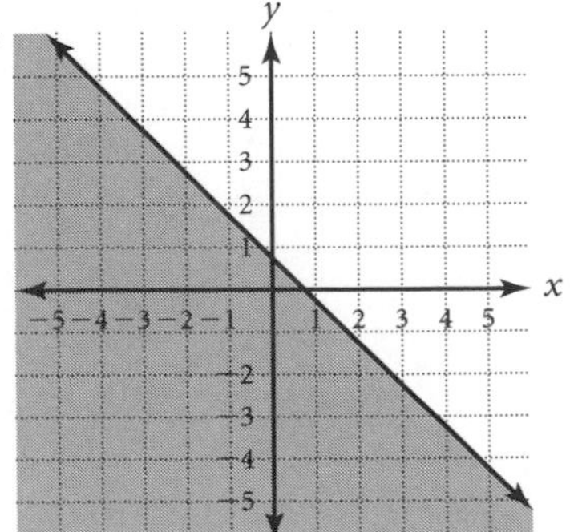

Answers

3.

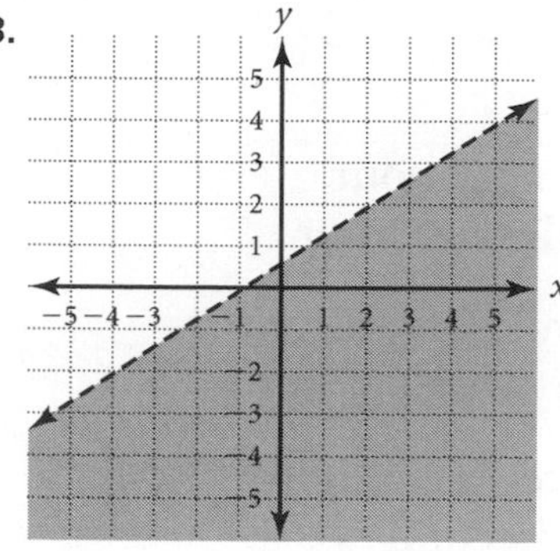

4. $y < 5$

5. $y \geq x - 5$

6. $y \leq x + 4$

7. $x > 7$

8. $y < -x + 1$

9. $x < 7$

10. $y < x - 10$

11. $y > 3x$

12. $x > 8$

13. $y \leq 5$

14.

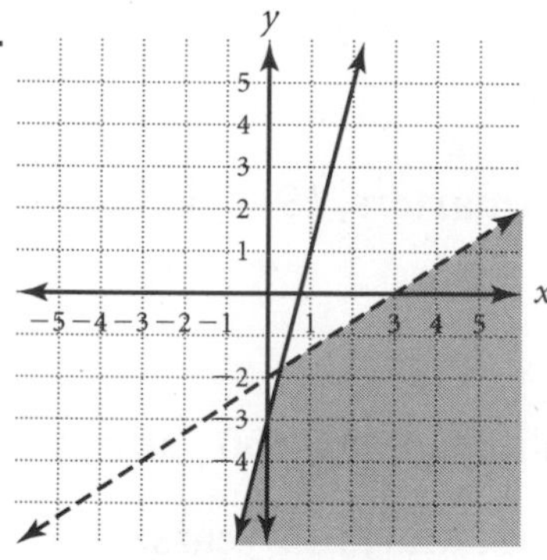

15.

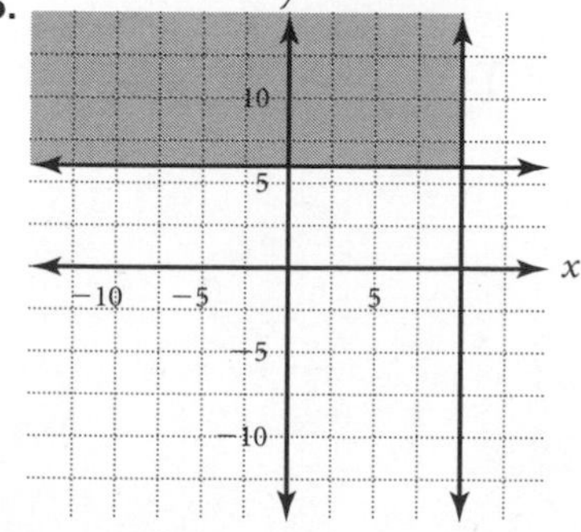

16.

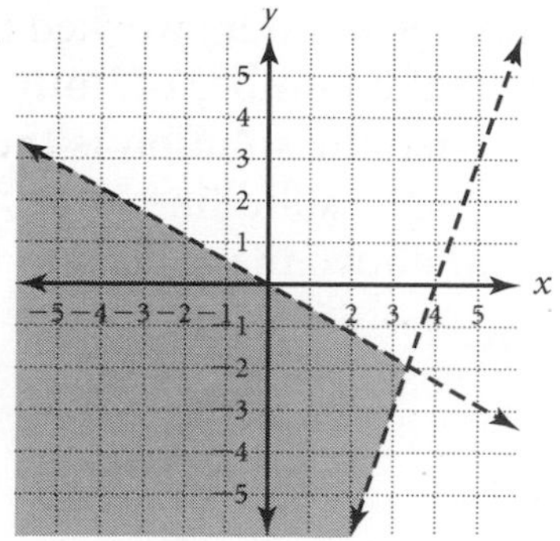

Lesson 7.5
Level C

1. $y \geq 0, x \geq 0$

2. 3

3. 4

4. $x \leq 4, y \geq -5, y \leq 2x - 1$

5. $x < 5, y \leq 13, x > -8, y \geq -19$

6. 16 trucks

7. Let y represent number of goldfish and x represent number of coy. Then the solution is the shaded area above the solid line $y = -x + 15$ and below the solid line $y = -1.42x + 30.4$.

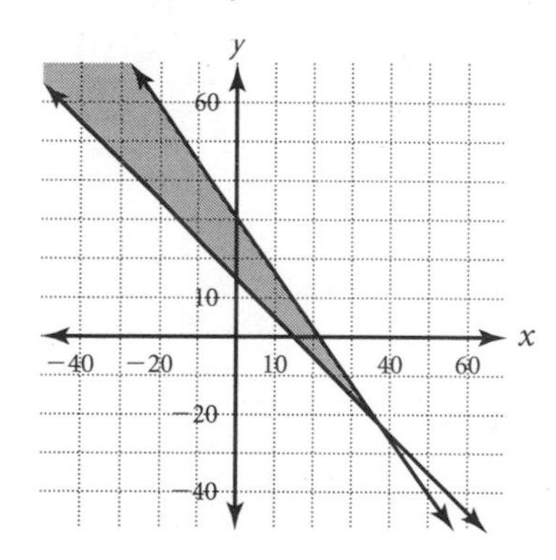

8. yes; The point (10, 10) lies in the solution area.

9. Let x be the number of hours worked in the park and y be the number of hours in the restaurant. Then the solution is the shaded area below the solid line $y = -x + 40$ and above the solid line $y = \left(-\frac{5}{6}\right)x + 30$.

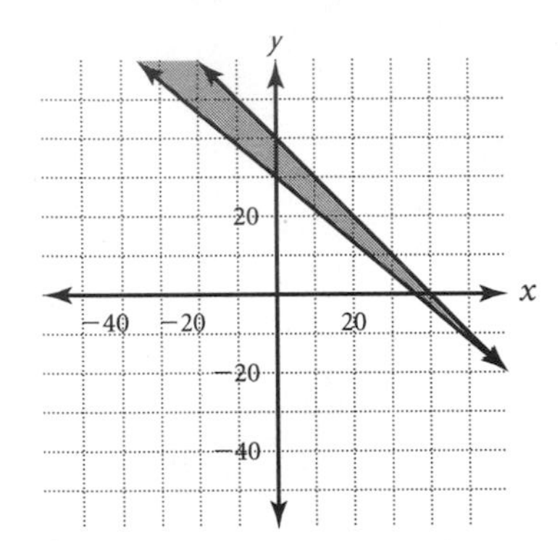

10. no; It is not possible to work a negative number of hours.

Lesson 7.6
Level A

1. 20, 9
2. 66, 32
3. 78, 50
4. 12 quarters, 6 dimes
5. 4 \$20s, 19 \$5s
6. 32
7. 733
8. 39 years, 16 years
9. 13 years, 8 years
10. 63 years, 19 years
11. 99
12. Vicky: 36 years, Pam: 31 years
13. Leo: 49 years, Adam: 11 years

Lesson 7.6
Level B

1. wind: 25 miles per hour, plane: 275 miles per hour
2. current: 4.4 miles per hour, swimmer: 5.6 miles per hour
3. current: 5.4 feet/second, diver: 9.5 feet/second
4. 12 \$1s, 37 quarters
5. 135 pennies, 61 nickels
6. runner: 9.26 meters/second, wind: 0.78 meters/second
7. 54
8. Harrison: 33 years, twins: 6 years
9. 573

Lesson 7.6
Level C

1. A: 36 ounces, B: 50 ounces
2. 3%: 80 gallons, 5%: 120 gallons
3. 1937
4. 11 dollars, 15 quarters, 19 nickels, 6 pennies
5. 75,613
6. 50%: 1600 gallons, 80%: 400 gallons
7. Nancy: 35 years, Mark: 19 years, Claire: 11 years
8. Answers will vary.

Answers

Lesson 8.1
Level A

1. base of 6, exponent 3
2. base of 4, exponent 2
3. base of 8, exponent 5
4. base of 10, exponent 5
5. base of 5, exponent 5
6. base of 2, exponent 7
7. base of 12, exponent 2
8. base of 20, exponent 1
9. 243
10. 15
11. 729
12. 64
13. 49
14. 1,000
15. 24
16. 64
17. 5^5
18. 8^{11}
19. 9^6
20. 10^7
21. 7^7
22. 10^{12}
23. 2^9
24. 4^8
25. 6^5
26. 10^4
27. 3^8
28. 5^8
29. 10^9
30. 6^9
31. 7^8
32. 4^{10}
33. $27x^5$
34. $10x^6y^5$
35. $12x^5y^9$
36. $-42a^9b^3$
37. $-10c^9d^3$
38. $36t^7v^8$
39. $-4a^7$
40. $-24x^7y^5z^6$
41. $-7m^8n^9$
42. $30f^4g^6$
43. $-24r^7s^4$
44. $15a^3b^2c^8$
45. $-4x^4y^5$
46. $-36h^5i^6$
47. $30u^6v^9w^2$
48. $-24a^5b^6$
49. $-30s^7t^4$
50. $24n^7p^9w^5$
51. $15d^6e^5$
52. $-14x^7y^7$
53. $56a^2b^6c^5e^7$
54. $-20r^7s^9$

Answers

Lesson 8.1
Level B

1. base of 5, exponent 7
2. base of 6, exponent 6
3. base of 2, exponent 12
4. base of 7, exponent 4
5. 1,000,000,000
6. 16,384
7. 20,736
8. 59,049
9. 9^{14}
10. 6^{14}
11. 10^{18}
12. 5^{15}
13. 4^{13}
14. 8^{12}
15. 10^{20}
16. 12^{9}
17. 2^{11}
18. 3^{11}
19. 10^{x+y}
20. 6^{a+b}
21. $-7x^9$
22. $-45x^{10}y^7$
23. $-60a^7b^{12}$
24. $-56x^{10}y^5z^9$
25. $-35r^9s^{12}$
26. $72a^4b^3c^{13}$
27. $-56x^8w^{10}$
28. $-48a^4b^7$
29. $81x^8$
30. $-40e^7f^{11}$
31. $8a^2b^7$
32. $-60x^9y^{10}$
33. $140f^3g^9h^2$
34. $96a^5b^6c^6$
35. $-72r^2s^8t^8$
36. $90x^9y^7z^5$
37. $-24a^{10}c^9$
38. $-160x^5y^9z^2$
39. $12xyz$
40. $20x^2y$
41. $16a^2b^2c^2$
42. $36f^2g^2h^3$
43. $45x^3y^3z^2$
44. $84ab^3c$

Lesson 8.1
Level C

1. $-120a^9$
2. $-70x^7y^3z^5$
3. $24a^9b^8$
4. $140c^7d^{10}e^8$
5. $-2a^{14}b^{12}$
6. $-120r^{12}s^{12}t^8$
7. $16x^{a+c}y^b$
8. $-84a^{x+y}b^{x+z}$
9. $-100r^{2a+b}$

Answers

10. $336c^{f+g}d^{4g}$

11. $x = 4$

12. $x = 9$

13. $x = 1$

14. $x = 9$

15. $(a + b)^{3x}$

16. $x = 2$ and $y = 1$ or $x = 1$ and $y = 4$

17. $7x^2z^4$

18. 4,096 mg

19. $a = -3$

20. $x = 3$

21. $a = \frac{17}{7}$

22. $a = 7$ and $b = 3$

23. $m = 5$ and $n = 1$

24. any number between 0 and 1

Lesson 8.2
Level A

1. 4096
2. 15,625
3. 262,144
4. 1
5. y^3
6. 100,000,000
7. k^4
8. 0
9. b^8
10. 144
11. 729
12. 2401
13. z^{12}
14. t^{72}
15. a^6
16. b^7
17. w^{15}
18. k^{26}
19. x^4y^4
20. j^{4z}
21. h^{5w}
22. a^9b^9
23. m^4n^2
24. k^7p^{35}
25. c^8d^{12}
26. $s^{20}p^2$
27. h^2g^3
28. x^3y^{27}
29. $z^{2m}y^m$
30. $u^{25}v^{25}$
31. $m^{2p}n^{2q}$
32. $g^{7p}h^{14}$
33. x^8y^{12}
34. 225
35. 512
36. 3375
37. 512
38. 20,736

Answers

39. 100
40. 7776
41. 81
42. 10,000
43. 0
44. 18
45. 3969
46. 45
47. 32
48. -96
49. 200
50. 45
51. 36
52. 91,125
53. 900
54. 144
55. 8,503,056
56. 91,125
57. 36

Lesson 8.2
Level B

1. -9
2. r^2
3. -1
4. g^8
5. -9
6. 2401
7. $-(w^4)$
8. 4096
9. -4096
10. -100
11. $-16{,}777{,}216$
12. $-p^{15}$
13. 16
14. -8000
15. 2500
16. 8
17. 10,000
18. 108
19. 15,625
20. 225
21. 27,000
22. 1600
23. 512
24. $-32{,}768$
25. y^8x^6
26. g^{28}
27. $432b^2c^{10}$
28. $1296x^{10}y^{22}$
29. $169h^{10}p^{16}$
30. $-125t^{12}v^{27}$
31. x^6y^6
32. $-1944m^{20}n^5$
33. $1600c^7d^{11}$
34. $12g^7h^8$
35. $-19683a^9b^7$

36. $-1000q^9$

37. $256a^{15}b^{22}$

38. $x^{24}y^{42}$

39. $-9b^9$

40. $-15u^8t^{24}$

41. $-256g^4h^{15}k^{35}$

42. $-243n^{12}p^{17}$

43. a^8

44. $-c^{13}d^{19}$

45. $-a^4b^2c^{19}d^9$

46. $-243m^{10}n^{72}p^{18}$

47. $-a^{200}b^{201}$

48. $v^8w^{22}u^{10}$

Lesson 8.2
Level C

1. $\frac{1}{32x^{15}y^{20}}$

2. $\frac{-27}{64a^9b^{18}c^6}$

3. $\frac{1}{2401m^{20}n^{12}p^4}$

4. $\frac{1}{1024q^{25}w^{55}}$

5. $\frac{-1}{128u^{14}v^{28}}$

6. $\frac{-a^{21}b^9d^{30}}{27}$

7. $-27u^{46}v^{44}w^{33}$

8. $-4b^{52}c^{141}d^{70}$

9. $-1000x^{89}y^{119}z^{56}$

10. $-36m^{18}n^{25}p^{51}$

11. $72c^{13}b^{10}$

12. 72 centimeters^3

13. $\frac{324m^9}{n^{14}}$ meters^2

14. $3xy^3z^3$ meters

15. $27.50

Lesson 8.3
Level A

1. 49

2. 144

3. $\frac{1}{169}$

4. 625

5. 1000

6. $\frac{1}{14}$

7. 6561

8. $\frac{1}{64}$

9. $\frac{1}{900}$

10. 225

11. 125,000

12. 512

13. x^{m-n}

14. c^4

15. k^p

16. h^{13-d}

Answers

17. b^{2u}

18. n^{2-k}

19. y^{5-q}

20. mn

21. v

22. b^{c-d}

23. g^{6-h}

24. y^{2x}

25. f^{4-w}

26. c^3

27. n^9

28. ab^2

29. $2n$

30. $\frac{1}{2hg^2}$

31. $(xy)^{c-d}$

32. $\frac{u}{v^2}$

33. $\frac{kh}{g^2}$

34. $\frac{f^2p^3}{3}$

35. $6m^4$

36. 13

37. $\frac{1}{5x}$

38. $-2h^6$

39. $\frac{2f^3}{d^2}$

Lesson 8.3
Level B

1. $\frac{c^5}{d^5}$

2. $\frac{g^6}{f^{15}}$

3. $\frac{49y^2}{4x^6}$

4. $\frac{16a^{16}}{81b^{28}}$

5. $\frac{64}{u^{12}}$

6. $\frac{-h^{21}}{g^{14}}$

7. $\frac{81w^8}{q^{12}}$

8. $\frac{5^{2p}m^{8p}}{n^{14p}}$

9. $\frac{g^{12}f^{12}}{16}$

10. $\frac{8t^9}{125v^6}$

11. $\frac{1}{16x^6y^4}$

12. $\frac{3^{3v}z^{3v}}{x^{3v}}$

13. $\frac{-1}{27u^3v^6}$

14. x^7y^{14}

15. $\frac{k^{2fh}p^{10f}}{g^{4f}}$

Answers

16. -64

17. $\frac{1}{32}$

18. $\frac{-1}{144}$

19. -0.5

20. 1

21. -27

22. -4

23. -256

24. -8

25. $5b^2$

26. $\frac{-1}{13xy^5}$

27. $\frac{n^2}{3m^3}$

28. $\frac{2a}{b^7c^2}$

29. $\frac{-z^8}{5y}$

30. $\frac{x^3z^4}{y^6}$

31. $\frac{-9u}{w^{43}v^3}$

32. $\frac{-36m^2}{n^{16}p^9q}$

33. $\frac{-x^6z^{10}v}{4}$

34. $\frac{5h^2f^{18}p}{g^{23}}$

Lesson 8.3
Level C

1. $\frac{-u^{50}w^{115}}{32v^{40}}$

2. $\frac{81z^{28}}{16x^{52}y^{28}}$

3. $\frac{10000n^{32}}{m^{40}p^{108}}$

4. $\frac{1024q^{70}w^{150}}{v^{45}}$

5. $\frac{-1}{8u^3v^{33}w^6}$

6. $\frac{a^2}{9b^{122}d^{36}}$

7. $\frac{-1}{32a^{58}b^2c^{70}}$

8. $\frac{-27u^{42}w^{48}}{8v^{87}}$

9. 3.2 cars per family

10. $\frac{8a^6}{b^{18}c^3}$ centimeters^3

11. $\frac{24a^4}{b^{12}c^2}$ centimeters^2

12. $\frac{\pi z^{18}}{x^{10}y^2}$ meters^2

Lesson 8.4
Level A

1. $\frac{1}{16}$

2. $\frac{1}{81}$

Answers

3. -8
4. $\frac{1}{49}$
5. $\frac{1}{125}$
6. 1
7. 8
8. $\frac{1}{10{,}000}$
9. -64
10. $\frac{1}{15}$
11. $\frac{1}{36}$
12. $\frac{1}{13}$
13. 1
14. 1
15. 144
16. $\frac{1}{x^3}$
17. $\frac{1}{p^5}$
18. 1
19. 1
20. $\frac{5}{w^2}$
21. 13
22. $\frac{8}{b^{12}}$
23. 1
24. $\frac{1}{c^6}$
25. $\frac{1}{2y^5}$
26. $\frac{5}{k^7}$
27. $\frac{1}{78125k^7}$
28. $\frac{14c}{d^2}$
29. 1
30. $\frac{d^5}{w^6}$
31. $\frac{1}{a^4}$
32. $\frac{1}{xy^5}$
33. h^9
34. $\frac{1}{64}$
35. $\frac{1}{9}$
36. $\frac{1}{128}$
37. $\frac{1}{16}$
38. $\frac{1}{4}$
39. 16

Answers

40. $\frac{1}{2}$

41. $\frac{9}{32}$

42. $\frac{1}{576}$

43. -27

44. 2

45. $-\frac{8}{3}$

46. $\frac{1}{8}$

47. 1

48. $-\frac{16}{27}$

Lesson 8.4
Level B

1. 1
2. $\frac{-m^3}{n^4}$
3. $\frac{-1}{27b^6c^3}$
4. $\frac{-3w^3}{q^8}$
5. $\frac{q^{28}}{16p^8}$
6. $\frac{4s^5}{3t}$
7. $5h^7$
8. $x^{12}y^4$
9. 1
10. $\frac{-v^{15}}{27}$
11. $\frac{27}{8a^9b^6}$
12. w
13. $\frac{3z^2}{y^{12}}$
14. $\frac{-m^6}{8}$
15. 3
16. $\frac{2}{9}$
17. $\frac{27}{8}$
18. $-\frac{1}{216}$
19. $\frac{2}{27}$
20. 24
21. $\frac{8}{19{,}683}$
22. $\frac{2}{9}$
23. $\frac{64}{9}$
24. -108
25. $\frac{d^{13}}{2c^5f^2}$
26. $\frac{q^{14}}{169w^{10}}$

Answers

27. $\frac{1}{25x^2}$

28. $\frac{-p^5}{2}$

29. 1

30. $\frac{w^5}{8u^6v^2}$

31. $\frac{8a^5}{c^9}$

32. $\frac{2x^2z^5}{y^{10}}$

33. $\frac{-g^{12}}{f^{24}h^{36}}$

34. $\frac{-9b}{ac^3}$

35. $\frac{t^5}{4r^5s^7}$

36. $\frac{w^{20}}{16p^{10}q^2}$

Lesson 8.4
Level C

1. $\frac{125y^9z^6}{x^{27}}$

2. $\frac{b^{16}c^3}{4a^{16}}$

3. 1

4. $\frac{x^{18}y^{12}}{z^{36}}$

5. $\frac{-2u^7w}{v^9}$

6. $16m^{104}n^{68}p^{32}$

7. $\frac{-g^{71}}{25f^{20}h^{50}}$

8. 1

9. $\frac{q^{80}}{m^{30}n^6}$

10. $\frac{-x^{141}}{9z^{34}}$

11. $\frac{m^4p\pi}{432n}$ meters^3

12. 87.3 meters^3

13. For any nonzero number x and any integer d, the rational expression $\frac{x^d}{x^d} = x^0$ is a nonzero number divided by itself, which is always 1.

Lesson 8.5
Level A

1. 5×10^5

2. 4×10^7

3. 1×10^{14}

4. 2.35×10^8

5. 1.7×10^{11}

6. 6×10^{-7}

7. 7.7×10^{-5}

8. 1×10^{-11}

9. 3×10^7

10. 8.5×10^{-7}

11. 9.78×10^{-3}

12. 4.12×10^{-2}

13. 9,000,000,000

14. 60,000

15. 180,000
16. 20,710,000
17. 0.00000002
18. 4,900,000
19. 300,100,000
20. 0.4
21. 274
22. 0.0086
23. 2,070,000,000
24. 0.003
25. 8×10^{11}
26. 9×10^{7}
27. 1×10^{10}
28. 2.8×10^{15}
29. 3.5×10^{4}
30. 1.15×10^{15}
31. Answers may vary: It is a shorter simpler method to represent large numbers.

Lesson 8.5
Level B

1. 8×10^{12}
2. 2.2×10^{1}
3. 3.85×10^{10}
4. 4.008×10^{7}
5. 9.5×10^{2}
6. 2.7×10^{0}
7. 3.04×10^{-4}
8. 6.819×10^{-7}
9. 5.12×10^{20}
10. 7×10^{1}
11. 1×10^{-5}
12. 3.7×10^{-13}
13. 2,000,000,000,000,000
14. 80,000
15. 76,000,000,000
16. 80,750,000,000,000,000
17. 0.00032
18. 0.000009
19. 0.00000003001
20. 0.00045
21. 719,000
22. 0.0000000106
23. 4,300,000,000
24. 0.2
25. 6×10^{7}
26. 2.4×10^{14}
27. 4×10^{22}
28. 3×10^{7}
29. 7.9996×10^{9}
30. 5.2×10^{8}

Lesson 8.5
Level C

1. 6×10^{27}
2. 4×10^{37}
3. 1.2×10^{21}
4. 6.468×10^{41}
5. 6×10^{65}

6. 3.87×10^{24}

7. 3×10^{0}

8. 5×10^{8}

9. 2×10^{-3}

10. 3.375×10^{2}

11. 2×10^{14}

12. 4×10^{-2}

13. 5×10^{15}

14. 2.2×10^{10}

15. 9.955×10^{10}

16. 3.988×10^{8}

17. 3.6916×10^{7}

18. 6.9998×10^{3}

19. 1.045×10^{21}

20. 9.48×10^{-7}

Lesson 8.6
Level A

1–4.

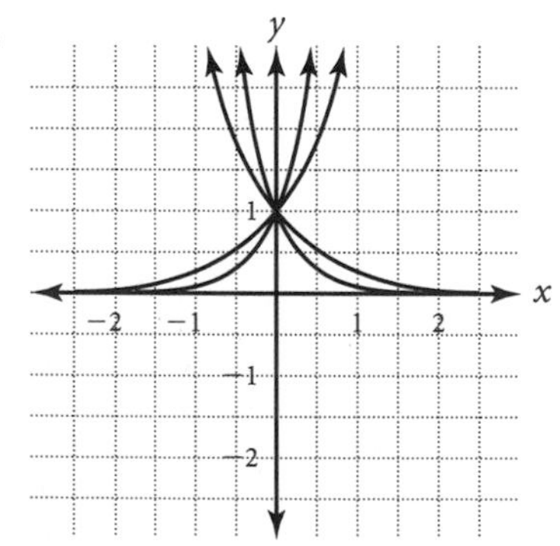

5. (0, 1)

6. right; left

7.

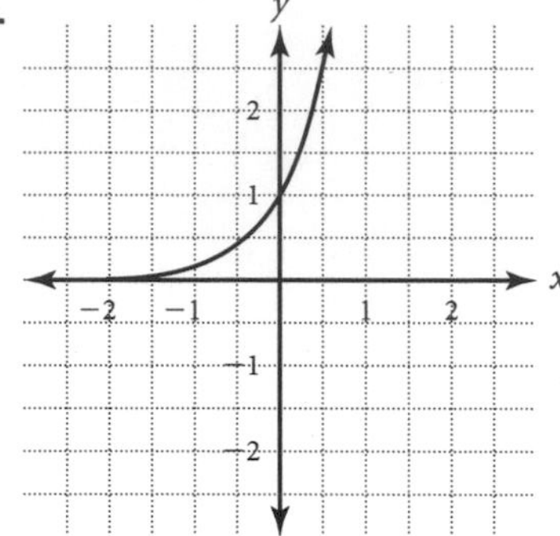

8.

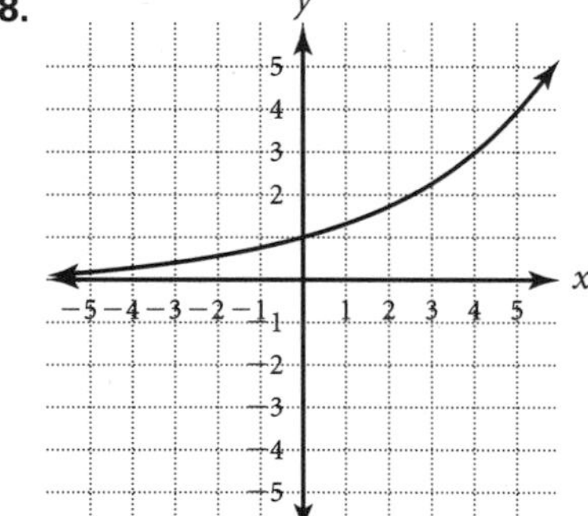

9.

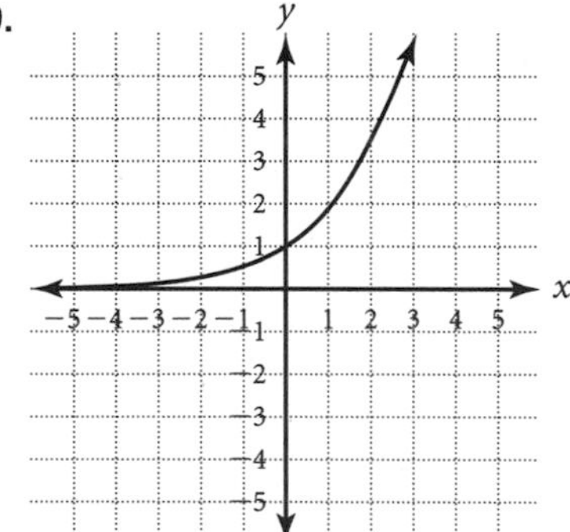

10. $y = 2^{x}$

Answers

Lesson 8.6
Level B

1–4.

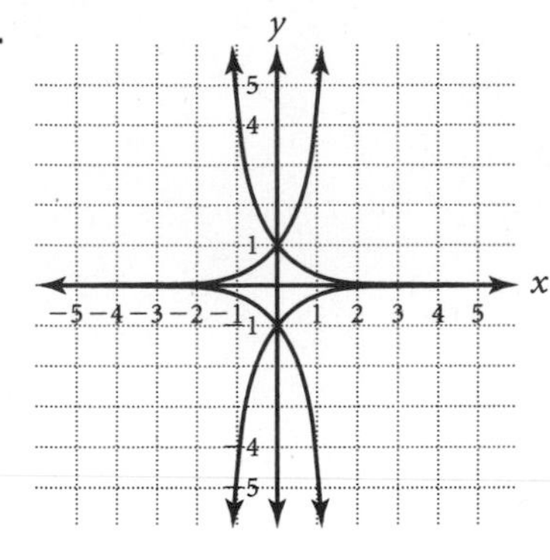

5. 0

6. I and II

7. III and IV

8. $y = -(4)^x$

9.

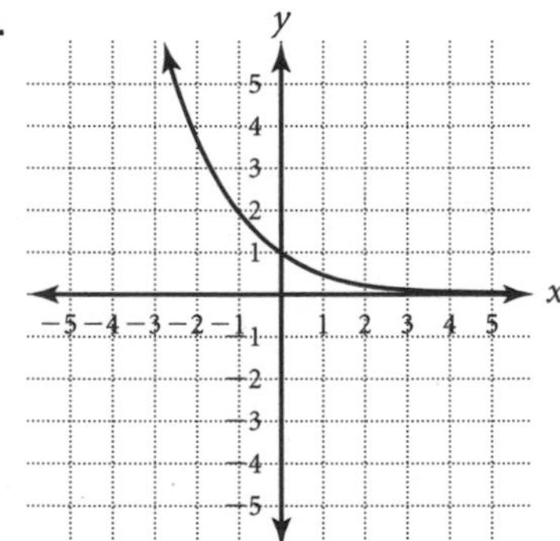

10.

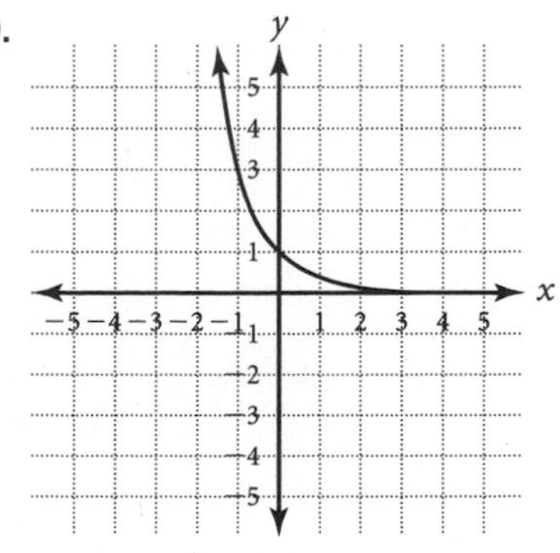

11.

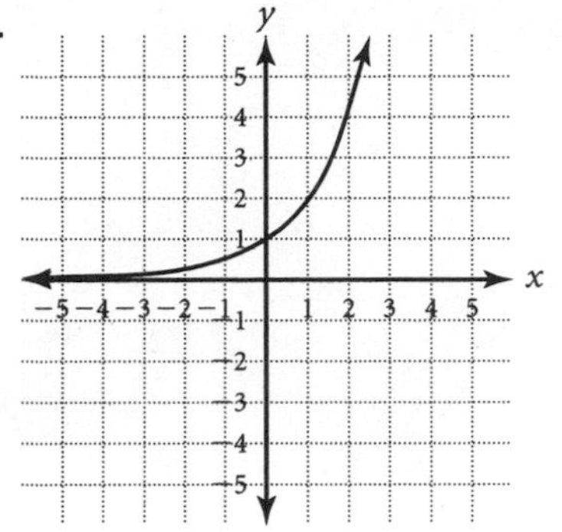

Lesson 8.6
Level C

1–6.

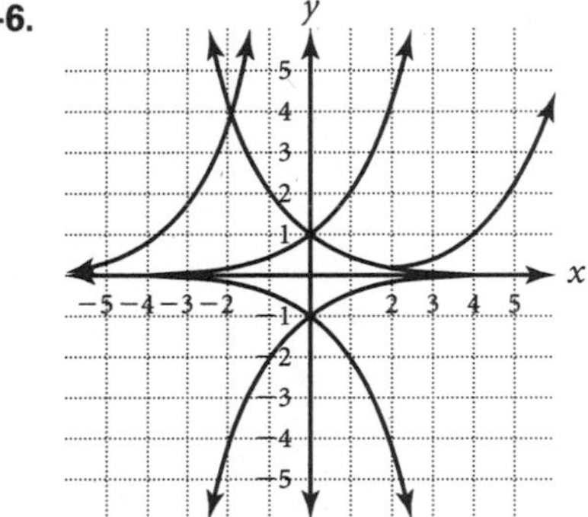

7. x

8. It is a reflection across the y-axis.

9. It slides the graph left or right.

10. 518,321

11. 718,320

Lesson 8.7
Level A

1. 5700 and 11,400

2. estimate: about 8000 years

3. 0 and 5700

4. estimate: about 2800 years

5. 11,400 and 17,100

6. estimate: about 13,700 years

Answers

7. 28,500 and 34,200
8. estimate: about 32,000 years
9. $1,777.99
10. $110,800
11. $140,400

Lesson 8.7
Level B

1. about 21%
2. about $\frac{2}{3}$
3. 0.1%
4. about 10 days
5. about 1 day
6. $2333.78
7. less; $22.39
8. $125,600
9. $186,700
10. 25%
11. $12,182
12. 8 years old

Lesson 8.7
Level C

1. 10 years
2. about 23 years
3. 0.1%
4. $11,314.08
5. $12,762.82
6. $7500
7. 2029
8. 2015
9. $2802.30
10. $22,000

Answers

Lesson 9.1
Level A

1. $a^2 + 2a + 3$
2. $3n^2 + n + 3$
3. $x^2 - 2x + 3$
4. $y^2 + y - 2$
5. $z^2 - 2z + 1$
6. $-2z^2 + z + 5$
7. $5x + 3$
8. $11a$
9. $3a + 2$
10. 0
11. $x^2 + 7x + 1$
12. $3y^2 - y - 3$
13. $5b + 5$
14. $4m + 13$
15. $-7t - 5$
16. 5
17. $2r^2 + 7r + 4$
18. $-2c^2 + 2c + 12$
19. $5x + 3$
20. $-3t - 5$
21. $-2z + 14$
22. $-6d - 8$
23. $2n - 2$
24. $v - 7$
25. $-5z - 4$
26. $5t - 5$
27. $6m^2 + 3$
28. $2m^2 + 3m$

Lesson 9.1
Level B

1. $-3b^3 + b^2 - 2b + 5$
2. $-5z^5 + 3z^3 - 3z^2 + 7$
3. $r^5 - 2r^3 - 3r^2 + 8$
4. $-2w^6 - w^3 + w^2 - w$
5. $-s^7 + 5s^2 - 3s + 3$
6. $-2x^7 - 2x^3 + x - 5$
7. $6v^2 - v$
8. $-10z - 4$
9. $4b^5 + 4b^4 - 9b^2 - 4b + 2$
10. $-7c^5 - 6c - 5$
11. $-7u^2 + 4u + 6$
12. $-14h^2 - 4$
13. $6k^5 - 5k^4 - 2k^2 - 2k$
14. $-3y^2 + 3y - 1$
15. $6a + 6$
16. $12d + 16$
17. $12z + 12$
18. $24n - 12$
19. $14n + 10$
20. $19a - 5$

Lesson 9.1
Level C

1. $-9a^2 + 14a - 2$
2. $9n^2 + 17n + 18$
3. $12z^2 - 18z - 17$
4. $-2x^3 + 17x^2 + x + 1$
5. $-3y^3 + 7y^2 - 2y - 9$

Answers

6. $5m^2 + 9$

7. $\frac{-2m - 5}{-2m - 5} = 1$

8. $\frac{3n^2 - 1}{3n^2 - 1} = 1$

9. $\frac{3a^2 - a}{-3a^2 + a} = -1$

10. $22a + 2b$

11. $\frac{53}{4}n$

12. $\frac{3}{2}$ units

13. $x = 0$

14. $y = -\frac{8}{3}$

15. no solution

Lesson 9.2
Level A

1. $(x + 1)(x + 3); x^2 + 4x + 3$

2. $(x + 2)(x - 1); x^2 + x - 2$

3. $x^2 + 5x + 6$

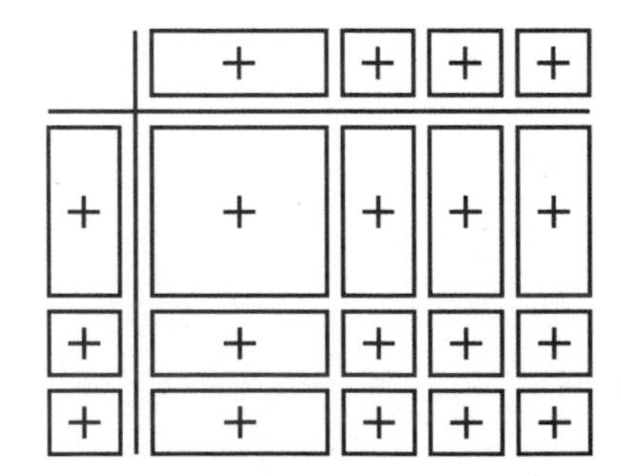

4. $x^2 + 2x - 3$

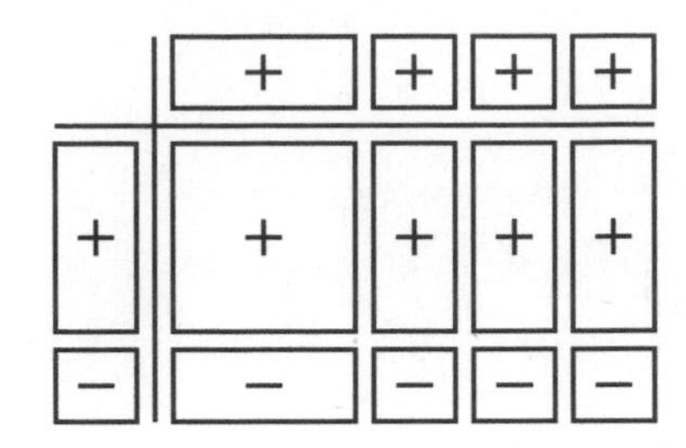

5. $x^2 - 2x - 3$

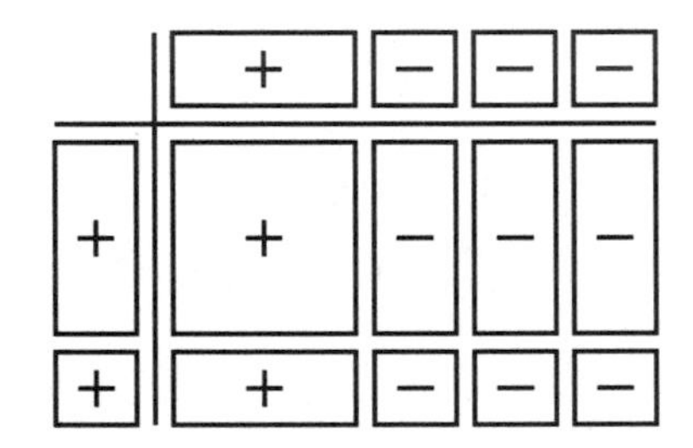

6. $x^2 - 4x + 4$

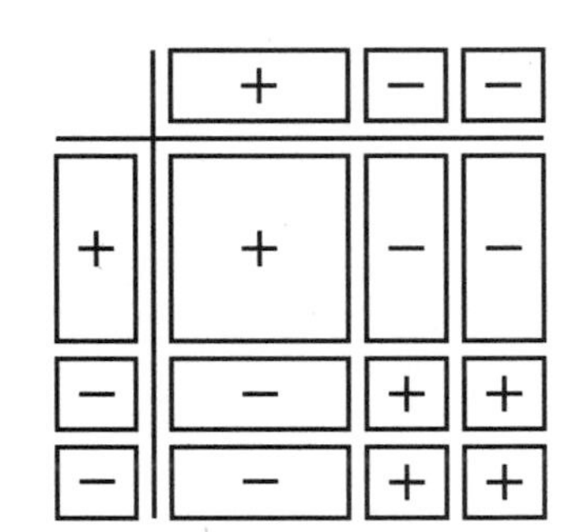

7. $x^2 - 9$

8. $x^2 + 2x + 1$

9. $x^2 - 25$

10. $x^2 - 6x + 9$

11. $3x + 15$

12. $7x - 21$

13. $16x - 40$

14. $27x - 12$

15. $2x^2 + 6$

16. $-35x^2 + 10x$

Answers

Lesson 9.2
Level B

1. $2x^2 + 4x - 6$

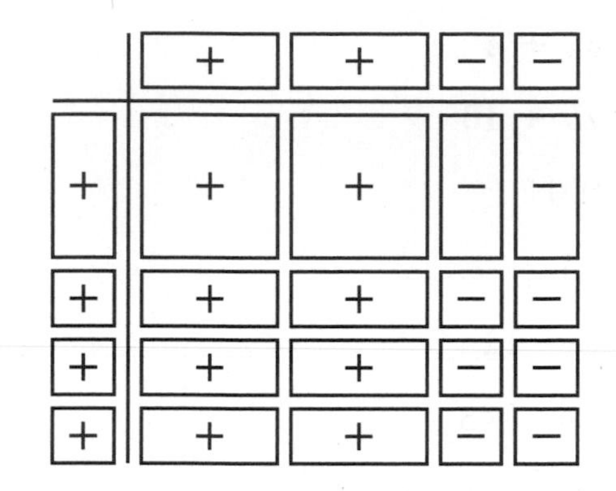

2. $4x^2 + 8x + 3$

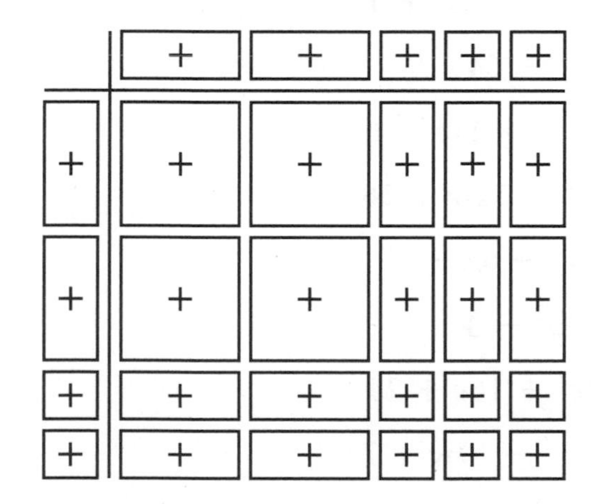

3. $4x^2 - 4x - 3$

	+	+	−	−	−
+	+	+	−	−	−
+	+	+	−	−	−
+	+	+	−	−	−

4. $2x^2 - 7x + 3$

	+	+	−
+	+	+	−
−	−	−	+
−	−	−	+
−	−	−	+

5. $9x^2 - 25$

6. $x^2 - 64$

7. $16x^2 - 56x + 49$

8. $81x^2 + 36x + 4$

9. $4x^2 + 20x + 25$

10. $9x^2 - 24x + 16$

11. $9x^2 + 24x + 16$

12. $\frac{1}{4}x^2 - 7x + 49$

13. $33x - 77$

14. $-2x^2 + 12x$

15. $18x^2 - 45x$

16. $27x^3 - 12x^2$

17. $-14x + 2x^2$

18. $5x - x^3$

19. $9a^2 - 49b^2$

20. $81x^2 + 198xy + 121y^2$

Answers

Lesson 9.2
Level C

1. $10x^5 - 14x^3 + 10x^2$
2. $x^2 - 64x^6$
3. $16x^4 - 40x^3 + 25x^2$
4. $121x^2 - 44x^3 + 4x^4$
5. $9x^6 + 24x^3 + 16$
6. $9x^2 - 25x^4$
7. $(x + 1)(x + 3)$

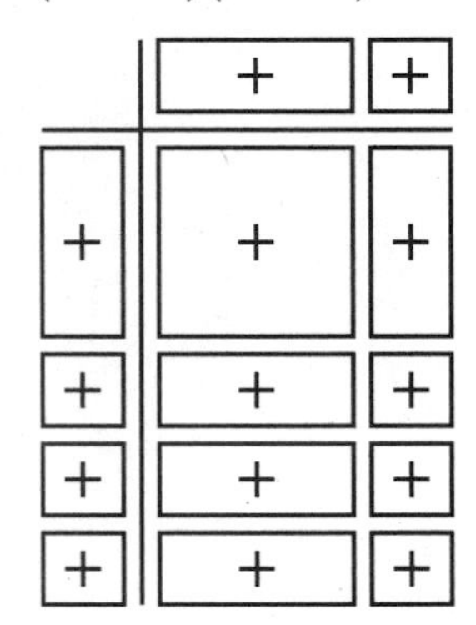

8. $(2x - 1)(x + 2)$

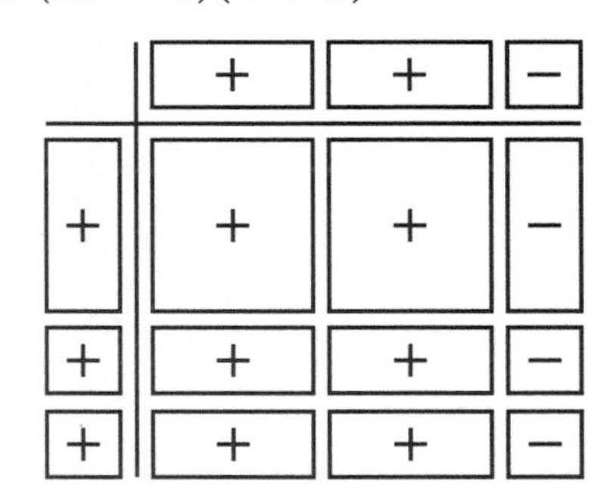

9. $2(6a - 5b) = 12a - 10b$
10. $n = 4$
11. $(x - 2)^2 = x^2 - 4x + 4$
12. $2(x - 2) = 2x - 4$
13. $2(x - 2) = 2x - 4$
14. $(x + 9)^2 = x^2 + 18x + 81$
15. $(x + 9)^2 - [(x - 2)^2 + 2(x - 2) + 2(x - 2)] = 18x + 85$

Lesson 9.3
Level A

1. $9s^2 + 54s$
2. $x^2 + 5x + 4$
3. $b^2 - 7b + 10$
4. $c^2 - 100$
5. $n^2 - 16$
6. $y^2 - y - 12$
7. $p^2 - 4p - 45$
8. $w^2 - 4w - 21$
9. $49f^2 + 28f + 4$
10. $9z^2 - 64$
11. $10z^2 + 11z - 6$
12. $w^2 - 12w + 36$
13. $9x^2 - 6x - 8$
14. $49q^2 - 4$
15. $2c^2 + 16c + 24$
16. $6w^2 - 17w + 7$
17. $10x^2 - x - 2$
18. $2x^2 + 13x - 45$
19. $6x^2 + 11x - 10$
20. $5x^2 + 8x - 21$
21. $b^2 - b - 6$
22. $8x^2 - 12x - 8$
23. $25d^2 - 20d + 4$

Answers

Lesson 9.3
Level B

1. $2x^2 + 3xy + y^2$
2. $7a^2 - 26ab + 15b^2$
3. $30d^2 - cd - c^2$
4. $16n^2 - 9m^2$
5. $s^6 - 16t^2$
6. $w^6 - w^4$
7. $48e^2 + 10eg - 3g^2$
8. $2d^4 - 13d^3 + 20d^2$
9. $h^2 - \frac{1}{9}$
10. $k^2 + \frac{1}{2}k + \frac{1}{25}$
11. $q^5 - 7q^4 + 2q^3 - 14q^2$
12. $8r^2 - 8t - 5r^2t^2 + 5t^3$
13. $4.14k^4 + 8.8k^2 - 40$
14. $16d^2 - 48d + 36$
15. $16d^2 + 34d - 15$
16. $5a^2 + 39a - 38$

Lesson 9.3
Level C

1. $6a^3 - 29a^2 + 28a$
2. $3.64x^4 - 1.1x^2y - 1.14y^2$
3. $-66z^4 + 55z^3 + 12z^2 - 10z$
4. $\frac{4}{3}x^2 - \frac{3}{2}x - \frac{5}{24}$
5. $25y^6 + 20x^2y^3 + 4x^4$
6. $a = 4$
7. $(2w + 25)(2w + 10)$
   $= 4w^2 + 70w + 250$
8. $269.50
9. $x = -5.5$
10. $y = -6$

Lesson 9.4
Level A

1. True
2. False
3. True
4. False
5.

x	$(x-8)(x+4)$	$x^2 - 4x - 32$
−3	−11	−11
−2	−20	−20
−1	−35	−35
0	−32	−32
1	−35	−35
2	−36	−36
3	−35	−35

6.

x	$(x-3)^3$	$x^3 - 9x^2 + 27x - 27$
−3	−216	−216
−2	−125	−125
−1	−64	−64
0	−27	−27
1	−8	−8
2	−1	−1
3	0	0

Answers

7.

x	$x^3 + 64$	$(x + 4)(x^2 - 4x + 16)$
-3	37	37
-2	56	56
-1	63	63
0	64	64
1	65	65
2	72	72
3	91	91

8.

x	$x^2 + 12x + 36$	$(x + 6)^2$
-3	9	9
-2	16	16
-1	25	25
0	36	36
1	49	49
2	64	64
3	81	81

9. $V = x^3$

10. $V = (3x)^3 = 27x^3$

11. $V = 8x$

12. $V = 6\pi x^2$

Lesson 9.4
Level B

1.

x	-2	-1	0	1	2
$(2x + 1)^2$	9	1	1	9	25
$4x^2 + 4x + 1$	9	1	1	9	25

2.

x	-2	-1	0	1	2
$(3x - 2)(-2x + 5)$	-72	-25	-10	3	4
$-6x^2 + 19x - 10$	-72	-25	-10	3	4

3.

x	-2	-1	0	1	2
$(x - 5)(x^2 + 5x + 25)$	-133	-126	-125	-124	-117
$x^3 - 125$	-133	-126	-125	-124	-117

4.

x	-2	-1	0	1	2
$x^3 - 12x^2 + 48x - 64$	-216	-125	-64	-27	-8
$(x - 4)^3$	-216	-125	-64	-27	-8

5. $V = 216x^3$ meters^3

6. $S = 6x^2$ inches^2

7. $S = 28x + 90$ inches^2

8. $V = 108\pi x^2$ meters^3

9. $V = 18.515\pi x^2$ yards^3

10. Original Volume $\approx$ 283 cm^3

New Volume $\approx$ 354 cm^3

Height $\approx$ 13 cm

11. Original Volume $\approx$ 147 cm^3

New Volume $\approx$ 191 cm^3

Height $\approx$ 10 cm

Answers

Lesson 9.4
Level C

1.

x	-2	-1	0	1	2
$(2x+3)(-4x-5)$	-3	-1	-15	-45	-91
$-8x^2-22x-15$	-3	-1	-15	-45	-91

2.

x	-2	-1	0	1	2
$(3x-7)^2$	169	100	49	16	1
$9x^2-42x+49$	169	100	49	16	1

3.

x	-2	-1	0	1	2
$(2x+3)(x^2-2x+1)$	-9	4	3	0	7
$2x^3-x^2-4x+3$	-9	4	3	0	7

4.

x	-2	-1	0	1	2
$(x-6)(x^2+6x+36)$	-224	-217	-216	-215	-208
x^3-216	-224	-217	-216	-215	-208

5. $S = 132.54x^2$ meters^2
6. $S = 116x + 240$ inches^2
7. $V = 27\pi x^2$ meters^3
8. $V \approx 91.89\pi x^3$ yards^3
9. $V = 4x^3 - 72x^2 + 288x$ inches^3
10. Original Volume ≈ 283 cm^3

    New Volume ≈ 240 cm^3

    Height ≈ 9 cm
11. 11% larger

Lesson 9.5
Level A

1. $3(x - 4)$
2. $4z(2z - 1)$
3. $5(x^2 - x - 4)$
4. $q^3(q^3 - 1)$
5. $3(3x^2 + 12x + 5)$
6. $2(6s^2 - 3s + 4)$
7. $10(10 - 2d^3 + d)$
8. $7b^2(b^2 + 1)$
9. $16t(t + 2)$
10. $15c(4c^2 - 3c + 1)$
11. $z^2(2z^2 - z + 5)$
12. $3s^2(8s^2 - 5s + 3)$
13. $(y + 3)(y + 2)$
14. $(w + 6)(7 - x)$
15. $(r + 15)(t + 3)$
16. $(k + 5)(k - 8)$
17. $(x - 3)(2x - 5)$
18. $(s - 9)(11 - 3t)$
19. $(6 + d)(5 - e)$
20. $(4 - t^2)(p + 9)$
21. $(b + 2)(a - c)$
22. $(m + 3)(4 - n)$
23. $(3 - g)(2f - 5)$
24. $(x + 7)(4x - 5)$
25. $(a + b)(3 + x)$
26. $(x - 1)(y + 3)$
27. $(c + 2)(d - 3)$
28. $(x + 5)(x + 3)$
29. $(a + 3)(x + y)$
30. $(x - 2)(x - 5)$

Answers

31. $(y + 5)(y + 5)$

32. $(2x - 3)(4x - 3)$

33. $(t - 3)(7t + 8r)$

34. $(m + 3)(4 + 3) = 7(m + 3)$

Lesson 9.5
Level B

1. $3(s^2 + 3s - 1)$
2. $5v^2(v^2 - 3 + 25v)$
3. $4gh(3g - h + 6gh)$
4. $6x^2y^4(2x - 6y - 9x^2)$
5. $3d^3e^2(5d^2e^4 - 9d + 14e^3)$
6. $5(25f^6 + 5f^3 - 3)$
7. $8s^5t^3(8s^3t^3 - 6 + 9s)$
8. $6q^2(4q^3 - 6)$
9. $(w + 5)(9 - r)$
10. $(2w + r)(r^2 - 5)$
11. $(r + 8)(t - 3s)$
12. $(s - 5)(11 - 4a)$
13. $(3 - y)(x + 1)$
14. $(p + 9)(q - 1)$
15. $(6 - c)(3ab - 7)$
16. $(k - 5)(2k + 7)$
17. $(4a + 1)(2a + b)$
18. $(z + 5)(2y + x)$
19. $(5x + 1)(a - 3b)$
20. $(2y - 3)(4x - 3)$
21. $(3 - a^2)(4 - 5a)$
22. $(d^2 + 3)(c + 4)$
23. $(x^2 + y^2)(y^3 + x^3)$
24. $(2 + 5r)(3 + \pi)$
25. $(ay - 2bx)(6ay + 10bx)$
26. $(a + 8), (a + 3)$
27. $(3r + 5), (3r + 5)$
28. It is a square.

Lesson 9.5
Level C

1. $x^5 - x$
2. $3 + 8x^{2n+2}$
3. $e^{n+3} - 4e^{4n-2} + 2 - 3e^{3n-3}$
4. $x^n(x - 1)$
5. $y^n(y^5 - y^2 - 1)$
6. $(x^n + y^n)(y - 1)$
7. $(x^a + 2)(3a^x - 6)$
8. $(x - 2)(a + b - c)$
9. $(2x - 3)(x - y - z)$
10. $(z - 5)(x + 3y + 2)$
11. $(x - y - z)(x^2 + y^2 + z^2)$
12. $4a - 4$
13. $(7 - 5x)(7 + 11x)$
14. $-15x^2y, (2x - 3y)(2y^3 + 5x^2)$

Lesson 9.6
Level A

1. $(x + y)^2$
2. $(c - d)^2$
3. $(x - 3)^2$
4. $(a + 8)^2$
5. $(x - 7)^2$
6. $(r + 3)^2$

Answers

7. $(s + 9)^2$
8. $(5 + t)^2$
9. $(w - x)(w + x)$
10. $(3d - c)(3d + c)$
11. $(5 - k)(5 + k)$
12. $(2f - 7g)(2f + 7g)$
13. $(4x - 9z)(4x + 9z)$
14. $(8s - 5)(8s + 5)$
15. $(5 - 6d)(5 + 6d)$
16. $(10a - 3)(10a + 3)$
17. $(x - 4)^2$
18. $(x - 2)(x + 2)$
19. $(6e - 5)(6e + 5)$
20. $(5c - 1)^2$
21. $(10s - 3)^2$
22. $(c - 12)^2$
23. $(8q - 7r)(8q + 7r)$
24. $(10 - 3d)(10 + 3d)$
25. $(2 - 3s)^2$
26. $(9q - 12p)(9q + 12p)$
27. $(6q - 1)^2$
28. $(3y - 2)^2$
29. $(2q + 5)^2$
30. $(11x - 9y)(11x + 9y)$
31. $(8y - 5x)(8y + 5x)$
32. $(7y + 4)^2$
33. $(5s - 3)^2$
34. $(xy + zw)(xy - zw)$

Lesson 9.6
Level B

1. 2475
2. 3591
3. 9964
4. 616
5. $(xy - 4z)^2$
6. $(x^2 - 3)(x^2 + 3)$
7. $(4r^2 - 7s^2)(4r^2 + 7s^2)$
8. $(x^2 + 3)^2$
9. $(2x^3y - 5z)(2x^3y + 5z)$
10. $(7a^2 + 9b)^2$
11. $(x - 3)(x + 3)(x^2 + 9)$
12. $(a - b)(a + b)(a^2 + b^2)(a^4 + b^4)$
13. 81
14. $126y$
15. $9y^2$
16. 16
17. $(4x - 5)^2$
18. $(12x - 11)(12x + 11)$
19. $4(x^2 - 2)(x^2 + 2)$
20. $(4x - 9)^2$
21. $(2x - 17)^2$
22. $3x + 4$

Answers

Lesson 9.6
Level C

1. $\left(x^2 + \frac{1}{9}\right)\left(x - \frac{1}{3}\right)\left(x + \frac{1}{3}\right)$
2. $4x(x^2 - 3)(x^2 + 3)$
3. $(y - 3)(y + 3)(x - 2)(x + 2)$
4. $(x - 2)^3$
5. $4x^2(4x^2 - 50)$
6. $4(y - 1)(y + 8)$
7. $(x^2 + y^2)(x - y)(x + y)(z - w)(z + w)$
8. $\left(\frac{1}{6}p - \frac{2}{5}q\right)\left(\frac{1}{6}p + \frac{2}{5}q\right)$
9. $\left(\frac{2}{9}s - \frac{3}{4}\right)^2$
10. $(x^n y^{2n} - 3)(x^n y^{2n} + 3)$
11. $a^4(a^{2x+1}b^x - 4b^2)(a^{2x+1}b^x + 4b^2)$
12. $-48y^3$
13. $256x^{n+2}$
14. $\pm 22x^n y^3$
15. $\pm 6a^{x+2}b^{2x+3}$

Lesson 9.7
Level A

1. $(x + 3)(x + 2)$

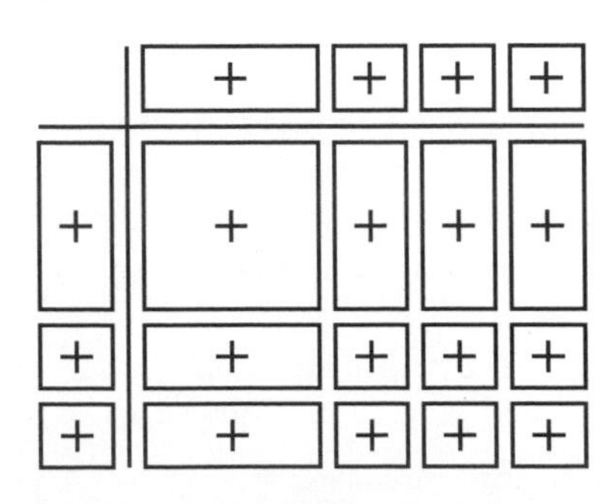

2. $(x - 3)(x + 1)$
3. $(x + 1)(x + 4)$
4. $(x + 5)(x + 1)$
5. $(c - 6)(c - 3)$
6. prime
7. $(x - 3)(x + 1)$
8. $(h + 3)(h - 1)$
9. $(r - 6)(r + 1)$
10. $(a + 7)(a - 4)$
11. prime
12. $(h - 6)(h - 4)$
13. $(z - 5)(z - 4)$
14. prime
15. $(q - 5)(q - 3)$
16. $(e + 16)(e - 2)$
17. $(t - 16)(t + 3)$
18. $(e + 8)(e + 4)$
19. $(s - 25)(s + 4)$
20. $(x + 15)(x - 5)$
21. $2(x + 11)(x + 2)$
22. $3(x + 4)(x + 3)$
23. $2(s - 7)(s + 3)$
24. $5(x + 10)(x - 1)$

Answers

25. $3(k - 3)(k - 1)$
26. $4(x + 8)(x - 5)$
27. $6s(s - 8)(s + 2)$
28. $3(x + 18)(x - 2)$
29. $x(x - 5)(x - 1)$
30. $xy(x - 5)(x + 4)$

Lesson 9.7
Level B

1. $(r + 18)(r - 2)$
2. $(3x - 1)(2x + 5)$
3. $(5c + 7)(c + 1)$
4. $(12x - 3)(x + 1)$
5. $(3h + 4)(h + 5)$
6. $(2d - 3)(d + 5)$
7. $(2g - 9)(g - 4)$
8. $(4t + 15)(t - 4)$
9. $(3w + 7)(2w - 5)$
10. prime
11. prime
12. $(2x - 15)(2x - 3)$
13. $(5 + 2x)(8 + 3x)$
14. $(6f - 1)\left(f - \frac{1}{12}\right)$
15. $(9x - 1)(5x - 12)$
16. $6s(s^2 - 7)(s + 2)(s - 2)$
17. $(3x + 5)(2x + 8)$
18. $4(4s^2 - 3)^2$
19. $4x, 2x - 3, 3x + 5$

Lesson 9.7
Level C

1. $(2x - 3)(x + 2)$

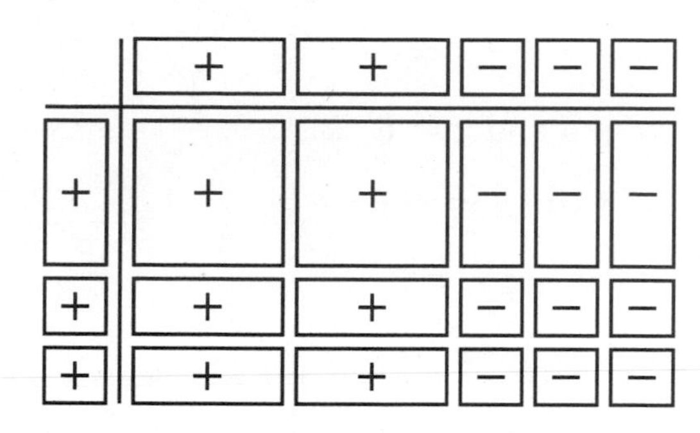

2. $(3x - 2)(x - 2)$

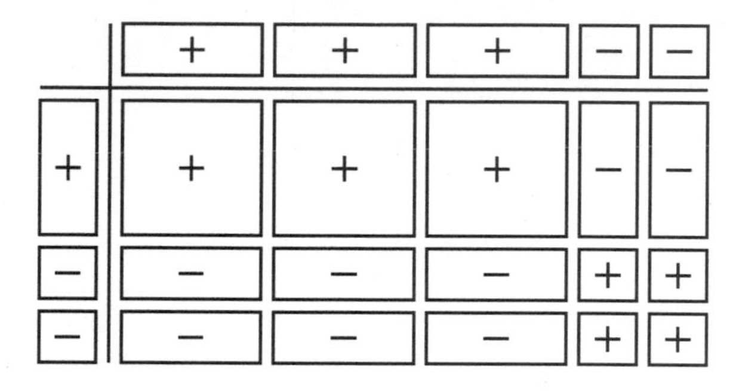

3. $(4x - 1)(2x - 3)$
4. $(2b - 1)(b + 3)$
5. $(2x + 3)(x - 5)$
6. $(2x^3y^3 + 1)(x^3y^3 - 4)$
7. $(2x + 3)(y - 9)(y + 5)$
8. $(5b - 2)(3a + 7)(2a - 3)$
9. $(3t + 4)(s - 5)(s + 4)$
10. $(2d - 3)(2d + 3)(3c - 4)^2$
11. $\pm 11, \pm 13, \pm 17, \pm 31$
12. $\pm 4, \pm 7, \pm 11, \pm 17, \pm 28, \pm 59$
13. $7x + 8, x - 1$
14. a) $2y, 3x + 8, 5x - 4$
    b) $6xy + 16y, 10xy - 8y,$
    $15x^2 + 28x - 32$
15. $x = 7,\ y = 5$

Answers

Lesson 9.8
Level A

1. $x = 4$ and $x = 3$
2. $x = -7$ and $x = -2$
3. $x = -4$ and $x = 6$
4. $x = 8$ and $x = -1$
5. $x = 10$ and $x = 7$
6. $x = 9$ and $x = -5$
7. $x = 0$ and $x = -8$
8. $x = -11$ and $x = 12$
9. $x = -2$ and $x = -2$
10. $x = 15$
11. $x = \frac{1}{2}$ and $x = -3$
12. $x = 0$ and $x = \frac{3}{4}$
13. $x = 6$ or $x = 0$
14. $x = 0$ or $x = -12$
15. $x = 0$ or $x = 4$
16. $x = 0$ or $x = -6$
17. $x = -5$ or $x = -4$
18. $x = 4$ or $x = 3$
19. $x = -4$
20. $x = 2$
21. $x = -1$ or $x = -2$
22. $x = 1$ or $x = 5$
23. $x = 7$ or $x = 4$
24. $x = 5$ or $x = -2$
25. $x = 9$ or $x = -4$
26. $x = 5$ or $x = -3$
27. $x = 6$ or $x = 3$
28. $x = 8$ or $x = -5$
29. $x = \frac{1}{2}$ or $x = \frac{-2}{3}$
30. $x = 1$ or $x = \frac{-1}{3}$
31. $x = -8$ or $x = \frac{-3}{2}$
32. $x = 2$ or $x = \frac{3}{5}$
33. $x = 15$ or $x = -5$
34. $x = -2$ or $x = \frac{3}{4}$
35. $x = 1$ or $x = \frac{1}{3}$
36. $x = -\frac{1}{2}$ or $x = -1$
37. $x = -\frac{1}{2}$ or $x = 1$
38. $x = \frac{4}{3}$ or $x = -3$

Lesson 9.8
Level B

1. $x = 3$
2. $x = \frac{2}{3}$ or $x = -3$
3. $x = \frac{3}{2}$ or $x = -5$
4. $x = \frac{15}{4}$ or $x = -4$

Answers

5. $x = \frac{1}{3}$ or $x = -5$

6. $x = 8$

7. $x = 5$ or $x = -5$

8. $x = \frac{6}{5}$ or $x = \frac{3}{2}$

9. $x = -\frac{3}{4}$ or $x = -4$

10. $x = -\frac{8}{3}$ or $x = \frac{5}{2}$

11. $x = \frac{2}{3}$ or $x = -\frac{5}{2}$

12. $x = \frac{3}{2}$

13. $x = \frac{4}{3}$ or $x = -\frac{5}{2}$

14. $x = \frac{3}{4}$ or $x = \frac{7}{3}$

15. $x = -\frac{11}{2}$ or $x = 8$

16. $x = 0$ or $x = 2$ or $x = -2$

17. $x = -\frac{2}{3}$ or $x = 18$

18. $x = \frac{1}{2}$ or $x = \frac{11}{2}$

19. $x = -\frac{5}{6}$ or $x = \frac{5}{6}$

20. $x = -\frac{1}{5}$ or $x = \frac{2}{5}$

21. $x = 2$ inches

22. $x = 4$ meters

23. length = 8 centimeters, width = 3 centimeters

24. length = 13 inches, width = 6 inches

Lesson 9.8
Level C

1. $x = 0$ or $x = \frac{1}{6}$

2. $x = -\frac{4}{3}$ or $x = \frac{10}{9}$

3. $x = \pm\frac{4}{3}$ or $x = -\frac{5}{2}$

4. $x = 6$ or $x = -\frac{2}{3}$

5. sample: $x^2 - 11x - 24 = 0$

6. sample: $x^2 + 7x + 10 = 0$

7. sample: $x^2 - 5x - 36 = 0$

8. sample: $x^2 - 7x = 0$

9. sample: $3x^2 + 29x - 10 = 0$

10. sample: $6x^2 - 31x + 35 = 0$

11. 24 yards, 8 yards

12. $70

13. 24 units, $\frac{5}{4}$ units

14. 1 second

Answers

Lesson 10.1
Level A

1. shifted 5 units up
2. shifted 3 units to the right
3. shifted 4 units right and 3 units up
4. shifted 1 unit to the left and 10 units down
5. shifted 12 units to the right and 7 units down
6. shifted 6 units to the left and 8 units up
7. $(1, 2), x = 1$

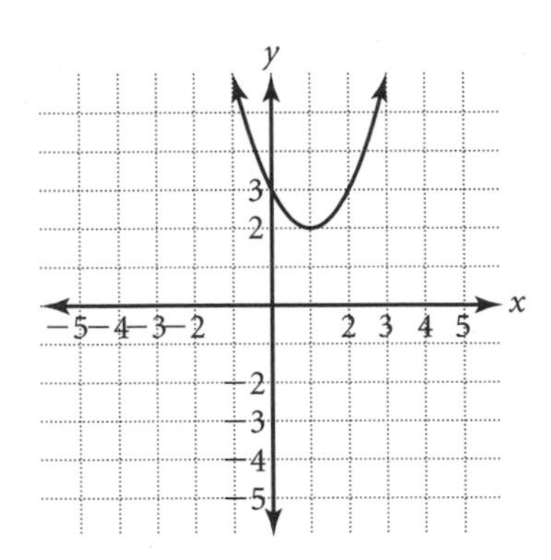

8. $(-2, -1), x = -2$

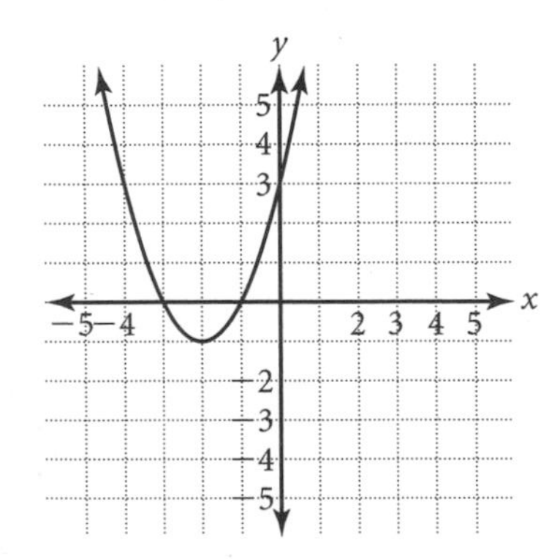

9. $x = 8$ and $x = -3$
10. $x = 9$ and $x = 4$
11. $x = -3$ and $x = -5$
12. $x = 3$ and $x = -6$
13. zeros: $x = 6$ and $x = 4$
    vertex: $(5, -1)$
14. zeros: $x = 5$ and $x = -1$
    vertex: $(2, -9)$

Lesson 10.1
Level B

1. shifted 2 units up and 8 units to the left
2. shifted 12 units right and 9 units down
3. vertically reflected, then shifted 3 units down
4. The graph is vertically reflected, then shifted 5 units to the left and 6 units down.
5. $(-2, 3); x = -2$

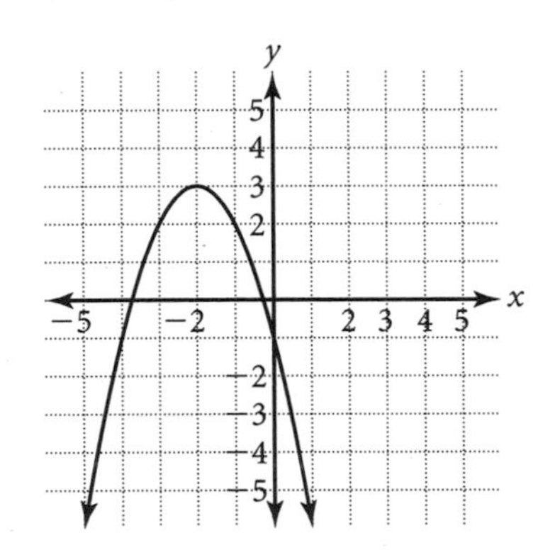

6. $(-1, -2); x = -1$

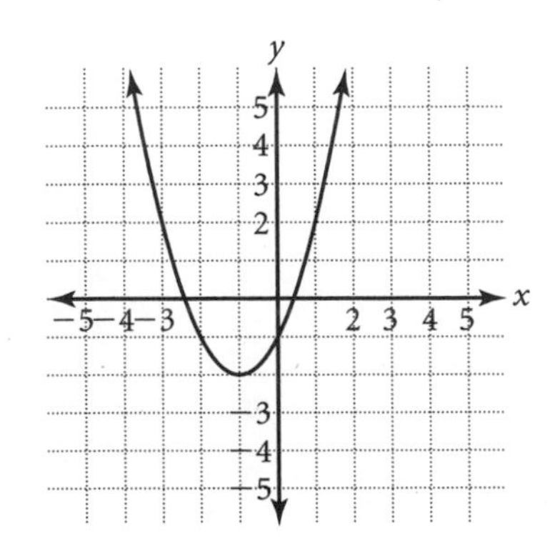

Answers

7. $(1, -2); x = 1$

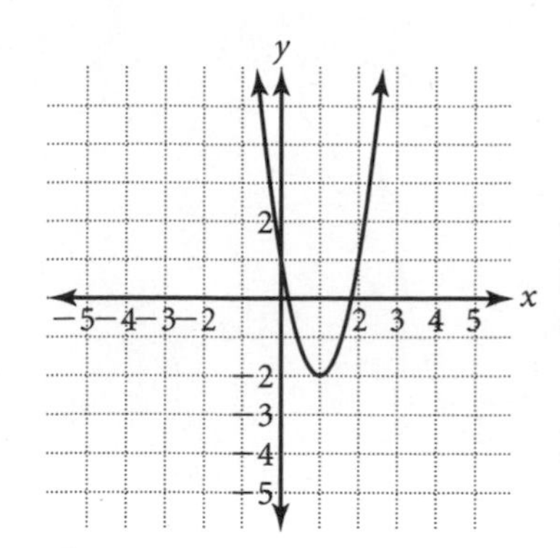

8. $(2, 0); x = 2$

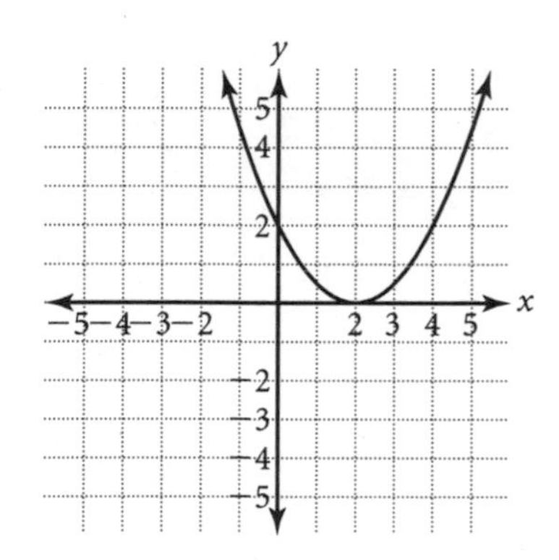

9. zeros: $x = 36$ and $x = -4$
   vertex: $(16, -400)$

10. zeros: $x = 12$ and $x = -11$
    vertex: $\left(\frac{1}{2}, -132\frac{1}{4}\right)$

6. zeros: -0.5 and 1.5
   vertex: $\left(\frac{1}{2}, -4\right)$

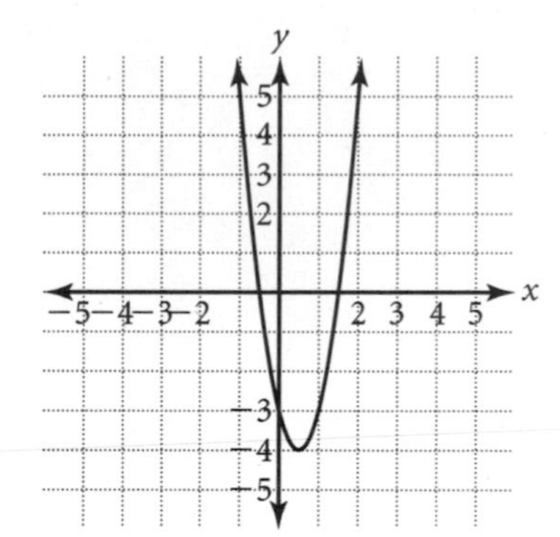

7. zeros: 3.5 and 2.5
   vertex: $(3, -1)$

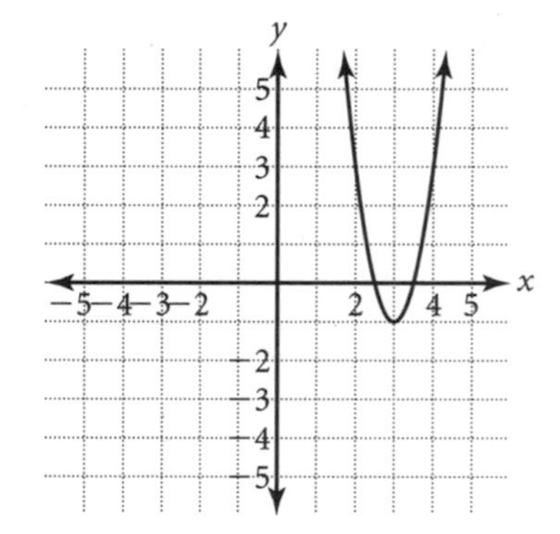

8. a. 9 feet
   b. 0.5 second

Lesson 10.1
Level C

1. shifted $\frac{5}{3}$ units to the right and 2 units down
2. reflected vertically, shifted $\frac{7}{2}$ units to the left, and shifted 1 unit up.
3. $y = -(5x - 2)^2 - \dfrac{3}{2}$
4. $y = -x^2 + 5$
5. $y = -(x + 4)^2 - 2$

Lesson 10.2
Level A

1. 14
2. 10
3. 15
4. 9.80
5. 25
6. 11.18

7. 6.71

8. 5.48

9. $x = \pm 6$

10. $x = \pm 7$

11. $x = \pm 12$

12. $x = \pm 17$

13. $x = \pm 4.47$

14. $x = \pm 20$

15. $x = \pm 24.49$

16. $x = \pm 5.66$

17. $x = \pm\frac{3}{4}$

18. $x = \pm\frac{1}{5}$

19. $x = \pm\frac{8}{11}$

20. $x = \pm\frac{9}{16}$

21. $x = \pm 4$

22. $x = \pm 30$

23. $x = \pm 8.66$

24. $x = \pm 7.07$

25. $x = 7$ or $x = -1$

26. $x = -5$ or $x = 1$

27. Zeros: 1 and 3
Vertex: $(2, -1)$
Axis of Symmetry: $x = 2$

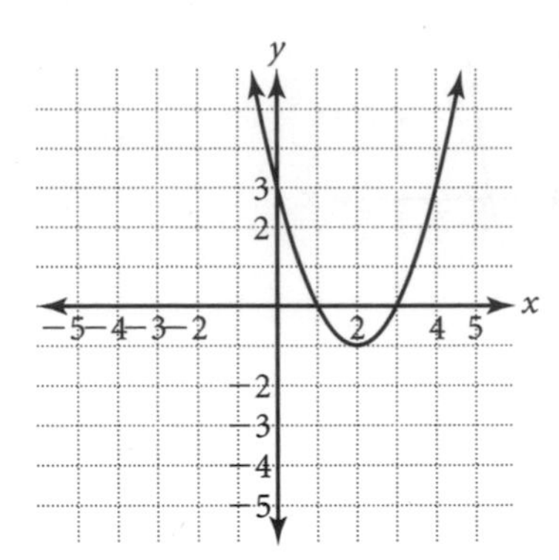

28. Zeros: 1 and -3
Vertex: $(-1, -4)$
Axis of Symmetry: $x = 2$

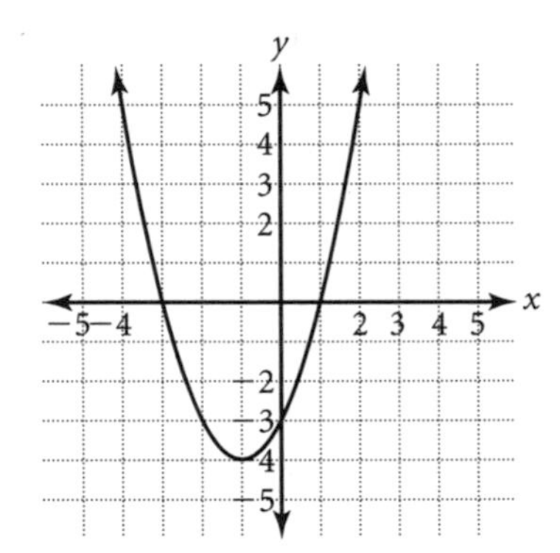

Lesson 10.2
Level B

1. $x = \pm 19$

2. $x = \pm 15$

3. $x = \pm 6.40$

4. $x = \pm 1.5$

5. $x = \pm 2$

6. $x = \pm 3.74$

7. $x = \pm 1.414$

8. $x = 7$ or $x = -1$

9. $x = -6.13$

10. $x = 1.71$ or $x = 12.29$

11. $x = 0$ or $x = 10$

12. $x = 8$ or $x = 0$

13. $x = 1$ or $x = \frac{1}{3}$

14. $x = 0$ or $x = -\frac{2}{3}$

15. Vertex: $(2, -8)$
Axis of Symmetry: $x = 2$
Zeros: 0 and 4

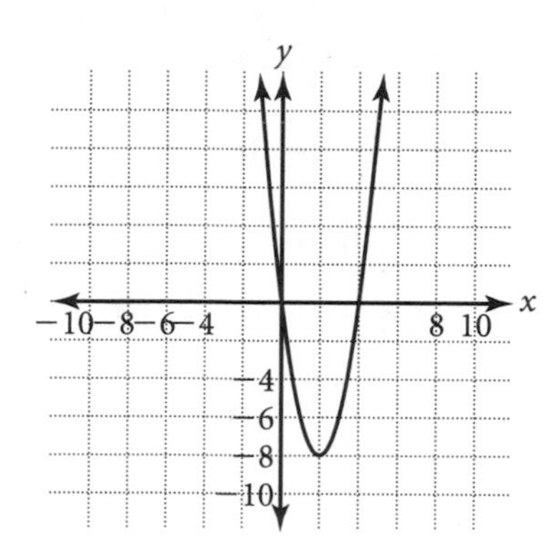

16. Vertex: $(-2, 8)$
Axis of Symmetry: $x = -2$
Zeros: 0 and -4

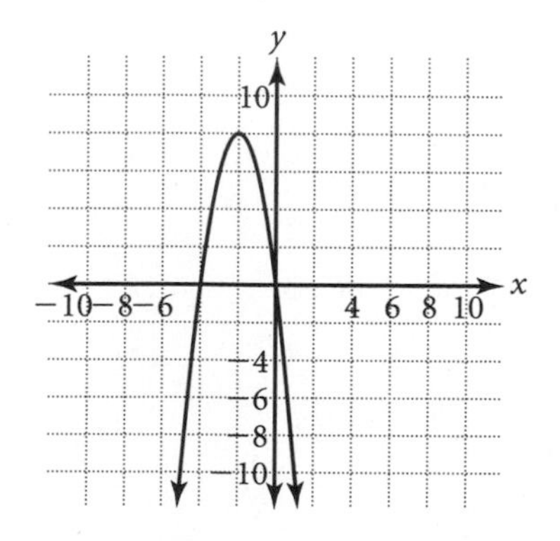

17. 7 seconds

18. 7.5 seconds

19. 2.82 inches

Lesson 10.2
Level C

1. samples: $x^2 = 196$; $2x^2 = 392$

2. samples: $(x + 3)^2 = 25$; $4(x + 3)^2 = 100$

3. samples: $x^3 = 21$; $2x^2 = 42$

4. samples: $(x - 8)^2 = 0$; $3(x - 8)^2 = 0$

5. $x = 9.91$ or $x = -15.91$

6. $x = -1.42$ or $x = -2.58$

7. $x = 1.75$ or $x = -0.15$

8. $x = 105.66$ or $x = 94.34$

9. Vertex: $(-2, 1)$
Axis of Symmetry: $x = -2$
Zeros: -2.5 and -3.5

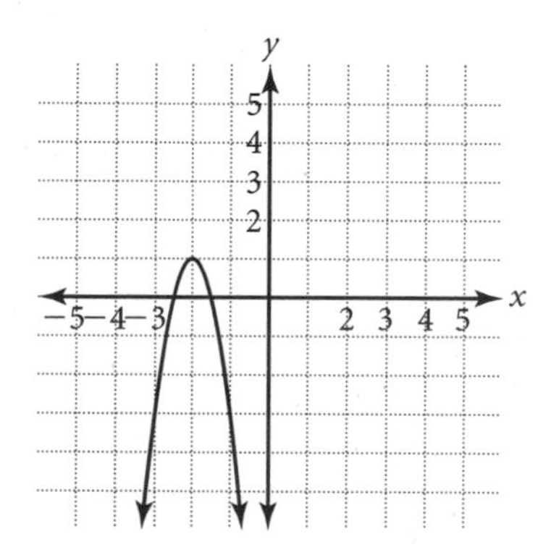

10. Vertex: $(2, -1)$
Axis of Symmetry: $x = 2$
Zeros: none

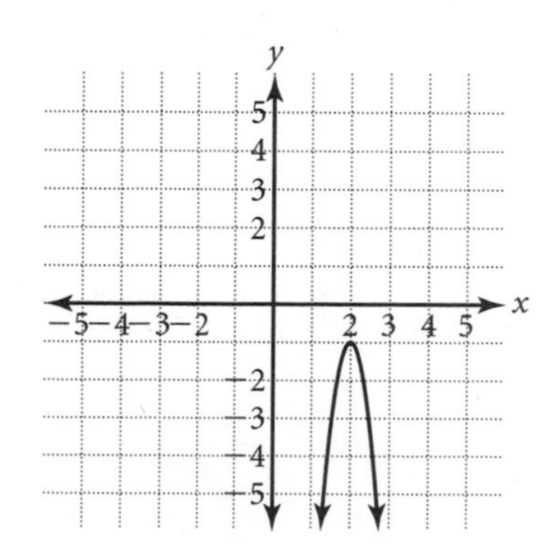

Answers

11. $t = \frac{\sqrt{s - h}}{4}$

12. $x = 4$ and $y = 6$
    $x = 0$ and $y = 0$

13. $x = \frac{2}{3}$ and $y = 1$

    $x = -\frac{1}{3}$ and $y = -3$

Lesson 10.3
Level A

1. $y = x^2 + 2x + 1$

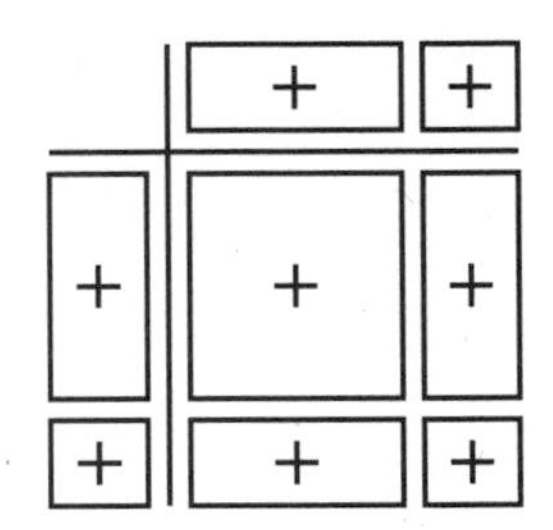

2. $y = x^2 - 4x + 4$

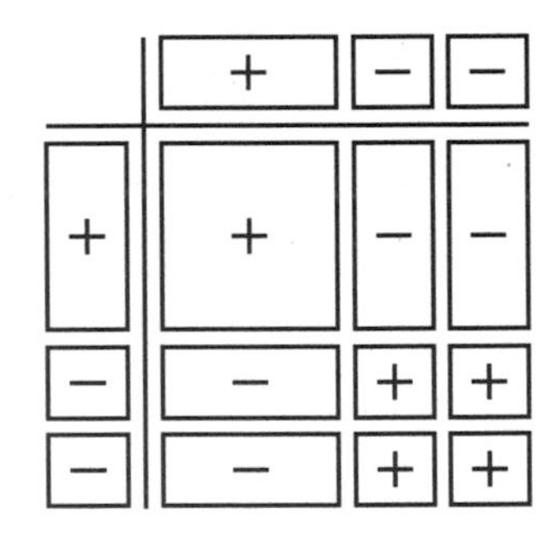

3. $y = (x - 3)^2 - 9$
4. $y = (x + 5)^2 - 25$
5. $y = (x - 7)^2 - 49$
6. $y = (x + 4)^2 - 16$
7. $(x + 2)^2$
8. $(x - 6)^2$
9. $(x - 9)^2$
10. $(x + 11)^2$
11. $(0, 0)$
12. $(0, 5)$
13. $(0, -3)$
14. $(0, -2)$
15. $y = (x - 0)^2 + 7; (0, 7)$
16. $y = (x - 0)^2 - 2; (0, -2)$
17. $y = (x - 2)^2 - 4; (2, -4)$
18. $y = (x + 3)^2 - 9; (-3, -9)$
19. $y = (x + 1)^2 - 1; (-1, -1)$
20. $y = (x - 6)^2 - 36; (6, -36)$
21. $y = (x - 10)^2 - 100; (10, -100)$
22. $y = (x + 12)^2 - 144; (-12, -144)$
23. $y = (x - 5)^2 - 24; (5, -24)$
24. $y = (x + 2)^2 - 2; (-2, -2)$

Lesson 10.3
Level B

1. $x^2 + 28x + 196$
2. $x^2 - 30x + 225$
3. $x^2 - x + \frac{1}{4}$
4. $x^2 + \frac{9}{2}x + \frac{81}{4}$
5. $x^2 - 25x + 625$
6. $x^2 + 15x + \frac{225}{4}$

7. $x^2 - \frac{1}{2}x + \frac{1}{16}$

8. $x^2 + \frac{2}{5}x + \frac{1}{25}$

9. $y = (x - 0)^2 - 8$

10. $y = (x - 3)^2 - 9$

11. $y = \left(x - \frac{3}{2}\right)^2 - \frac{9}{4}$

12. $y = \left(x + \frac{21}{2}\right)^2 - \frac{441}{4}$

13. $y = (x + 3)^2 - 5$

14. $y = (x - 5)^2 - 30$

15. $y = (x + 9)^2 - 31$

16. $y = \left(x + \frac{3}{2}\right)^2 - 2$

17. $y = \left(x - \frac{9}{2}\right)^2 - 17$

18. $y = \left(x - \frac{11}{2}\right)^2 - \frac{101}{4}$

19. $y = (x - 11)^2 - 117$

20. $y = \left(x - \frac{3}{2}\right)^2 - \frac{7}{4}$

21. $y = (x - 2)^2 - 3$; Vertex: $(2, -3)$

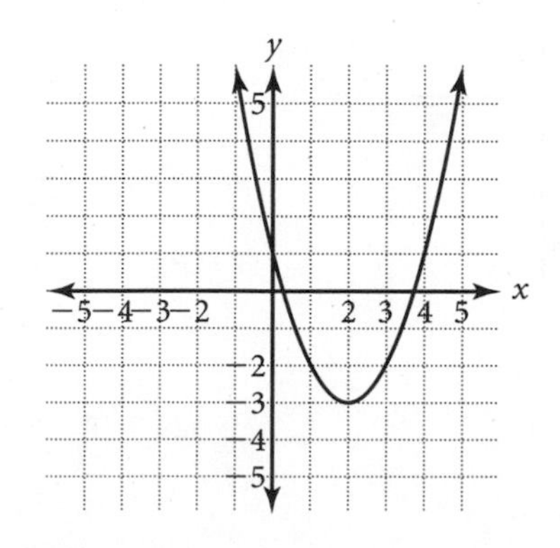

22. $y = (x + 3)^2 - 2$; Vertex: $(-3, -2)$

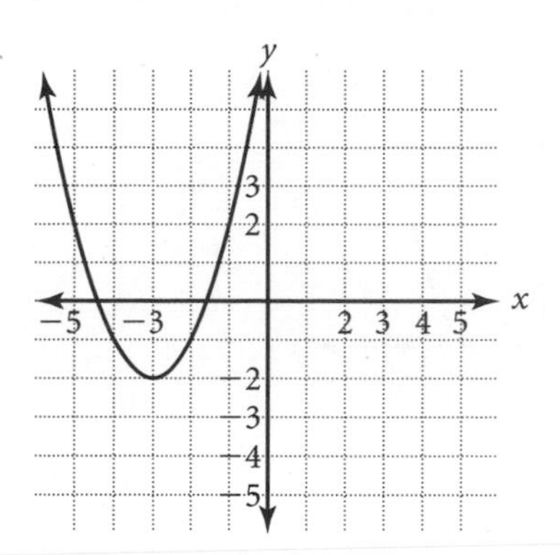

Lesson 10.3
Level C

1. $y = \left(x - \frac{19}{2}\right)^2 - \frac{361}{4}$

2. $y = \left(x + \frac{25}{2}\right)^2 - \frac{625}{4}$

3. $y = \left(x - \frac{1}{4}\right)^2 + \frac{31}{16}$

4. $y = \left(x + \frac{1}{6}\right)^2 + \frac{71}{36}$

5. $y = \left(x - \frac{3}{2}\right)^2 - \frac{35}{12}$

6. $y = \left(x + \frac{1}{6}\right)^2 + \frac{7}{36}$

7. $y = \left(x - \frac{5}{2}\right)^2 - \frac{37}{4}$

8. $y = \left(x + \frac{3}{2}\right)^2 + \frac{13}{4}$

9. $y = \left(x - \frac{7}{3}\right)^2 - \frac{115}{9}$

10. $y = (x + 6)^2 - \frac{93}{2}$

Answers

11. Minimum: $\left(\frac{3}{2}, -\frac{1}{4}\right)$

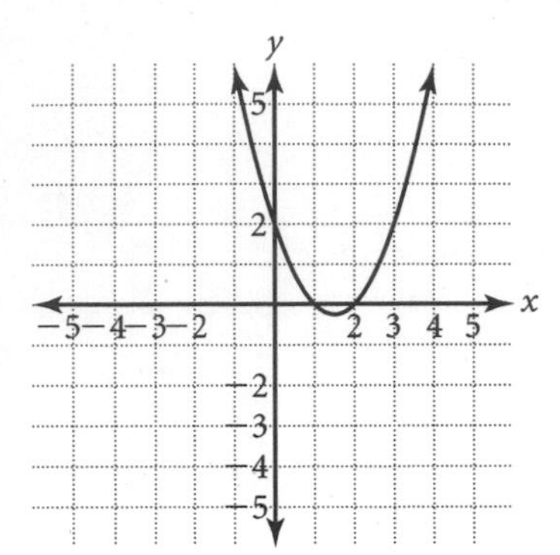

12. Minimum: $\left(\frac{1}{2}, \frac{7}{4}\right)$

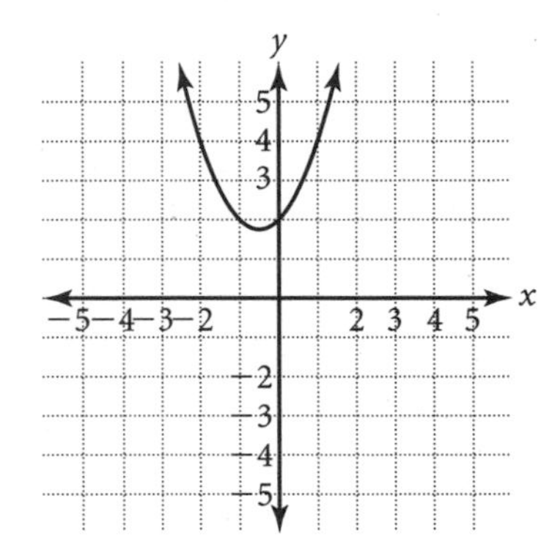

13. $y = 2(x + 2)^2 - 8$

14. $y = 3(x + 3)^2 - 27$

15. $h = -16\left(t - \frac{9}{2}\right)^2 + 324$

Lesson 10.4
Level A

1. 5 and -1
2. 2 and -6
3. $x = 15$ or $x = 3$
4. $x = -9$ or $x = 3$
5. $x = 25$ or $x = -3$
6. $x = -8$ or $x = 3$
7. $x = 5$
8. $x = -12$ or $x = 3$
9. $x = 7$ or $x = -7$
10. $x = 16$ or $x = -3$
11. $x = 1$ or $x = -3$
12. $x = 1$ or $x = -9$
13. $x = 16$ or $x = -4$
14. $x = -9$ or $x = -11$
15. $x = 1.46$ or $x = -5.46$
16. $x = 4.24$ or $x = -0.24$
17. $d = 6$ or $d = 0$
18. $r = 6.32$ or $r = -0.32$
19. $s = 3.73$ or $s = 0.27$
20. $q = 8.42$ or $q = -1.42$
21. $s = 12$ or $s = -5$
22. $s = 8$ or $s = -5$
23. (2, 4) and $(-2, 4)$

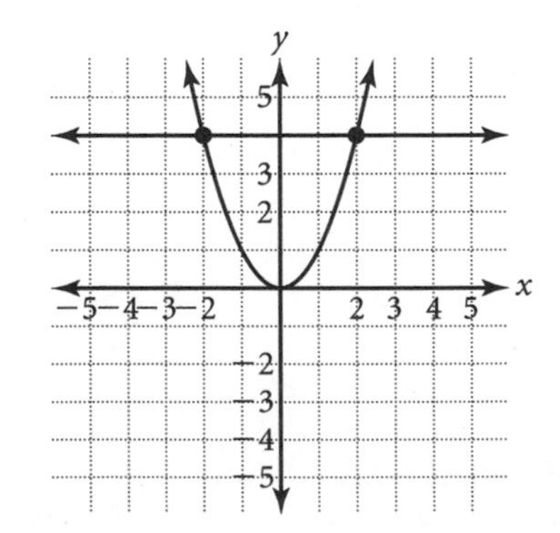

Answers

24. (1, 2) and (−1, 2)

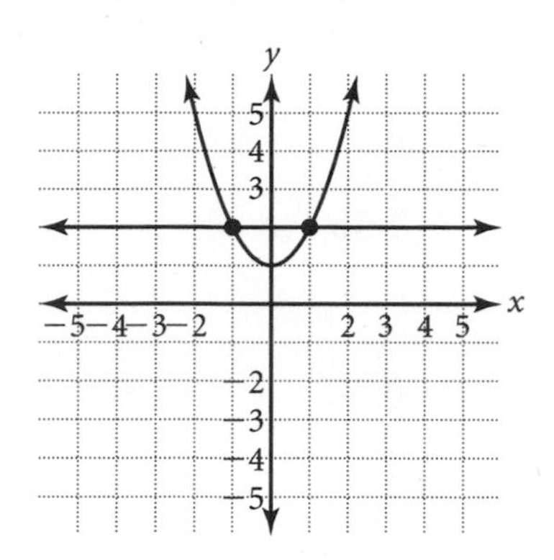

Lesson 10.4
Level B

1. 4 and $-\frac{3}{2}$
2. 6 and $\frac{4}{3}$
3. $x = \frac{9}{2}$ or $x = -\frac{7}{3}$
4. $x = \frac{11}{2}$ or $x = -\frac{11}{2}$
5. $x = 5$ or $x = 5$
6. $x = \frac{3}{2}$ or $x = -13$
7. $x = 12$ or $x = \frac{10}{3}$
8. $x = 15$ or $x = \frac{3}{2}$
9. $x = 2$ or $x = -4$
10. $x = 0.74$ or $x = -6.74$
11. $x = 8.61$ or $x = 1.39$
12. $x = 14.21$ or $x = -0.21$
13. $x = 3.23$ or $x = -0.23$
14. $x = 3$ or $x = 2$
15. $d = 3$ or $d = 0$
16. $r = 8.24$ or $r = -0.24$
17. $p = 4$ and $p = -\frac{11}{2}$
18. $q = 14$ or $q = -1$
19. $w = \frac{3}{2}$ or $w = -\frac{1}{2}$
20. $k = 27.99$ or $k = 2.01$
21. (2, 6) and (−3, 6)
22. (1, 5) and $\left(-\frac{5}{2}, 5\right)$
23. (2, −7) and (−3, 3)
24. (9, 14) and (1, −2)

Lesson 10.4
Level C

1. $-\frac{4}{3}$ and $\frac{9}{4}$
2. 15 and −6
3. $x = 1.61$ or $x = -3.11$
4. $x = 3.54$ or $x = 4.95$
5. $x = 9$ or $x = 4$
6. $x = 18.36$ or $x = -1.36$
7. 3, $-\frac{5}{2}$
8. $\frac{1 \pm \sqrt{7}}{2}$
9. 3 and 5

10. 50 feet and 125 feet

11. 1.25 feet and 10.5 feet

12. 1.5 inches

Lesson 10.5
Level A

1. $a = 1, b = -5, c = 14$
2. $a = 3, b = -4, c = 1$
3. $a = 2, b = 3, c = -10$
4. $a = 1, b = 0, c = 16$
5. $a = 3, b = -15, c = 0$
6. $a = 7, b = -1, c = 3$
7. 28; 2 solutions; no
8. -12; no solutions; no
9. 0; 1 solution; yes
10. 1; 2 solutions; yes
11. -7; no solutions; no
12. 144; 2 solutions; yes
13. 49; 2 solutions; yes
14. 4; 2 solutions; yes
15. $x = 1$ or $x = -2$
16. $x = 6$ or $x = -2$
17. $x = 2$ or $x = -5$
18. $x = 8$ or $x = 0$
19. $x = 2$ or $x = \frac{3}{2}$
20. $x = 2.26$ or $x = 0.74$
21. $x = 6.61$ or $x = -0.61$
22. $x = 1.5$ or $x = -2$
23. $x = 0.54$ or $x = -5.54$
24. $x = 3.41$ or $x = 0.59$
25. $x = 4.58$ or $x = -4.58$
26. $x = 4$ or $x = -7$
27. $d = 3.30$ or $d = -0.30$
28. $s = 2.62$ or $s = 0.38$
29. $w = 15$ or $w = 3$
30. $r = 4$ or $r = -\frac{2}{3}$

Lesson 10.5
Level B

1. -12; no solution; no
2. 212; 2 solutions; no
3. 4; 2 solutions; yes
4. 0; 1 solution; yes
5. -484; no solution; no
6. 225; 2 solutions; yes
7. $x = 2$ or $x = -14$
8. $x = 1.12$ or $x = -3.12$
9. $x = 2$ or $x = -\frac{1}{3}$
10. $x = 2.05$ or $x = 2.05$
11. $x = 0.44$ or $x = -3.77$
12. $x = 2.13$ or $x = -0.13$
13. $x = 2.67$ or $x = -2.5$
14. $x = 3.71$ or $x = -1.21$

15. $x = 3.23$ or $x = -0.23$

16. $x = 1.69$ or $x = -1.19$

17. $2(x - 4)(x + 3)$

18. $3(x - 2)(x + 5)$

19. $(x - 4)(2x - 3)$

20. $(2x + 5)(2x + 7)$

21. $x = 5$ or $x = 1.33$

22. $s = 3.75$ or $s = -3.75$

23. $y = 8.22$ or $y = -1.22$

24. $t = 1.93$ or $t = -1.55$

25. $p = 7$ or $p = -0.75$

26. $c = \frac{2}{3}$ or $c = \frac{1}{15}$

27. 2.5 inches, 5.5 inches

Lesson 10.5
Level C

1. $x = 5.87$ or $x = -2.13$

2. $x = 3$ or $x = -18$

3. $P = 16$

4. $P = 5$

5. $P = \pm 6$

6. $P = -4$

7. $P < \frac{4}{3}$

8. $P < 0$ or $0 < P < \frac{3}{4}$

9. 5.25 units and 7.5 units

10. a. No-the equation cannot be factored (the determinant is less than 0).
    b. 64

11. 5.17 seconds

12. 1.06 seconds

Lesson 10.6
Level A

1. $-3 < x < 3$

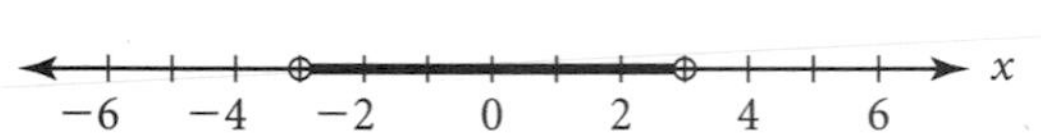

2. $x \leq -4$ or $x \geq 4$

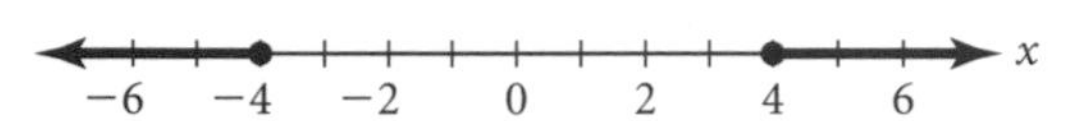

3. $1 \leq x \leq 3$

4. $x < -5$ or $x > 2$

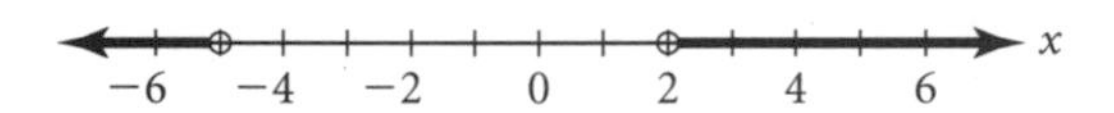

5.

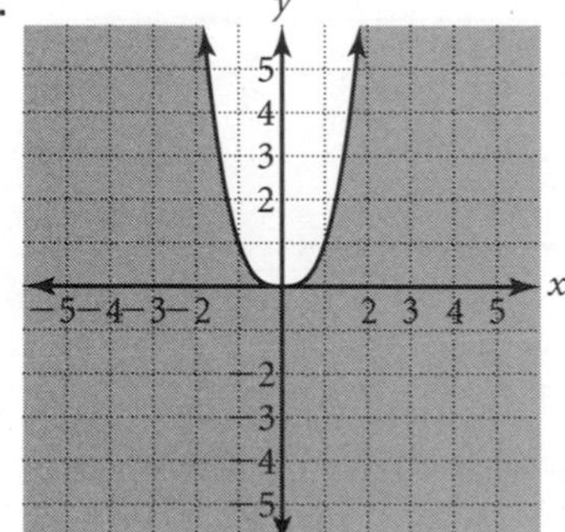

6.

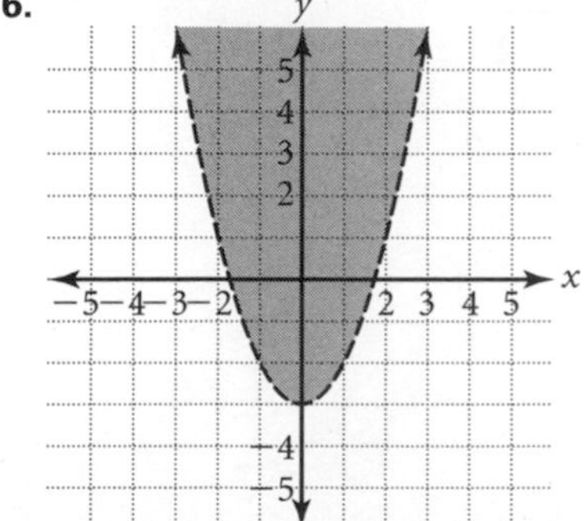

Answers

7.

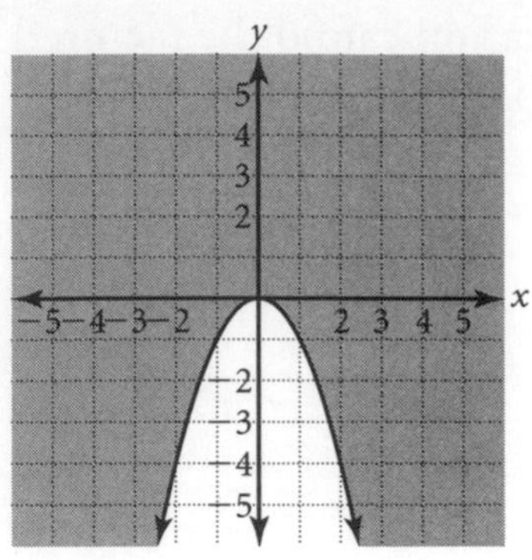

8.

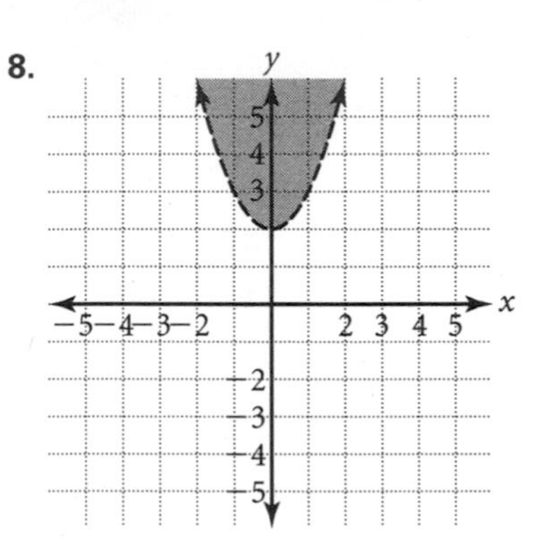

Lesson 10.6
Level B

1. $3 < x < 6$

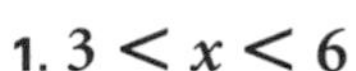

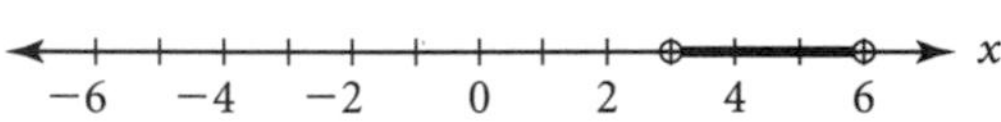

2. all real numbers

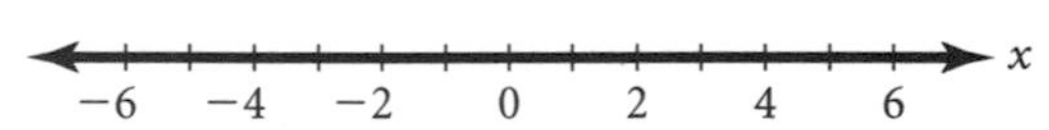

3. $x \geq 5$ or $x \leq -4$

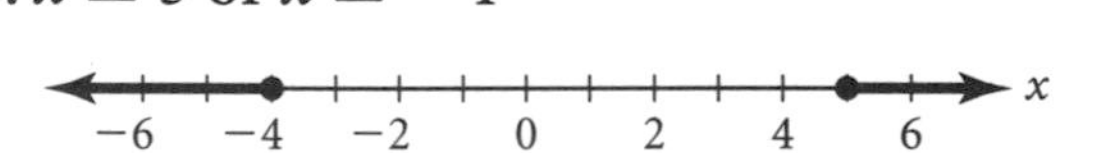

4. $x > \frac{3}{2}$ or $x < -\frac{5}{2}$

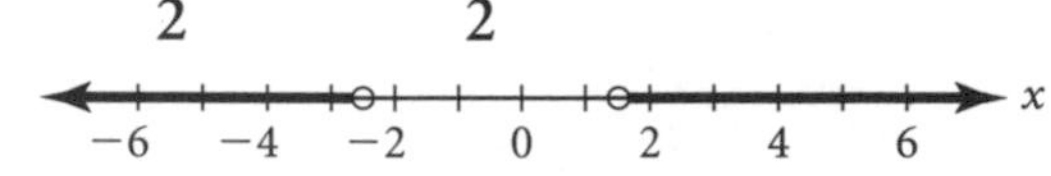

5.

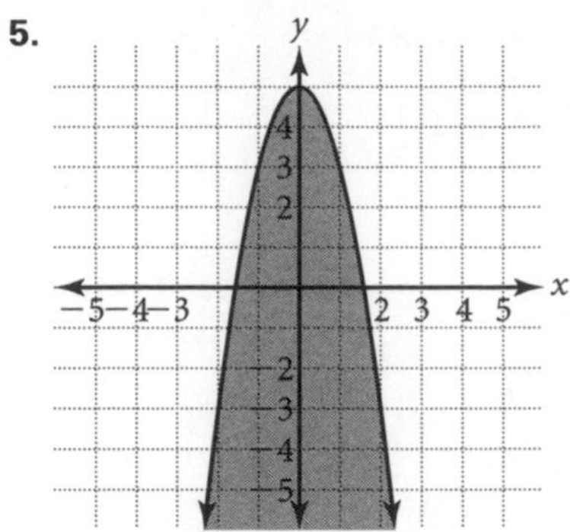

6.

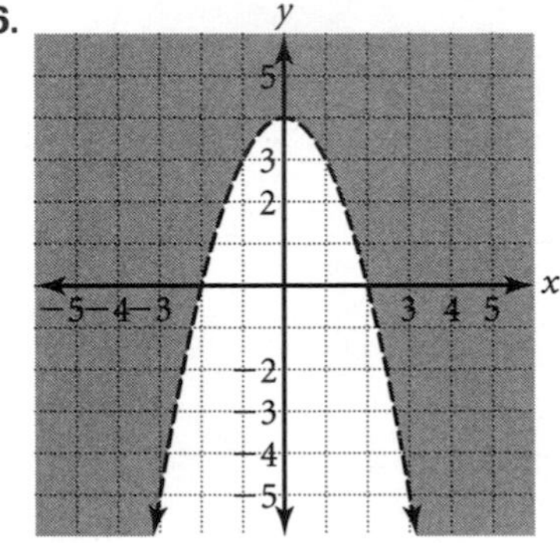

7.

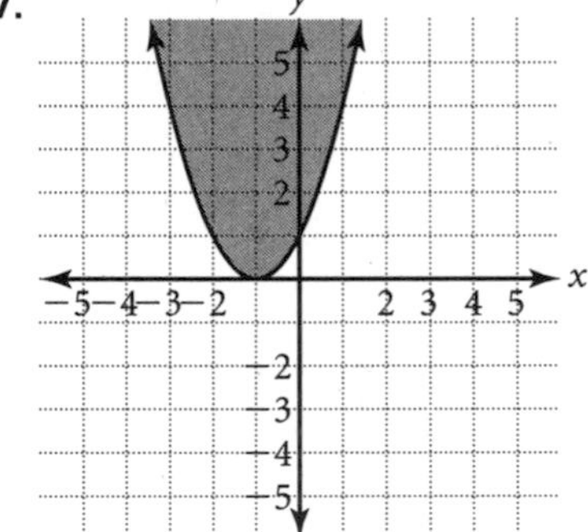

8.

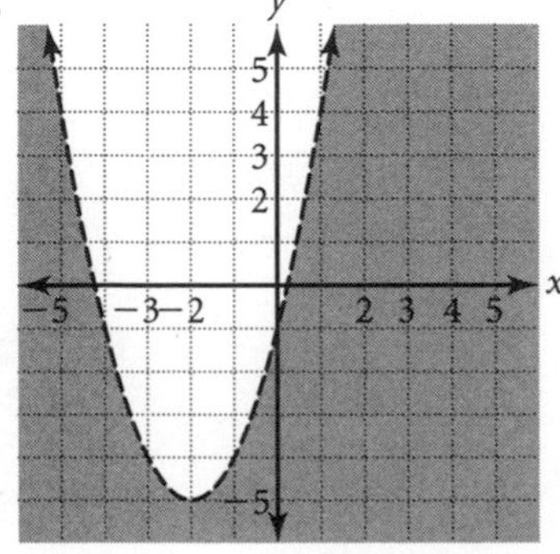

Answers

Lesson 10.6
Level C

1. $x \leq \frac{7}{2}$ or $x \geq \frac{13}{2}$

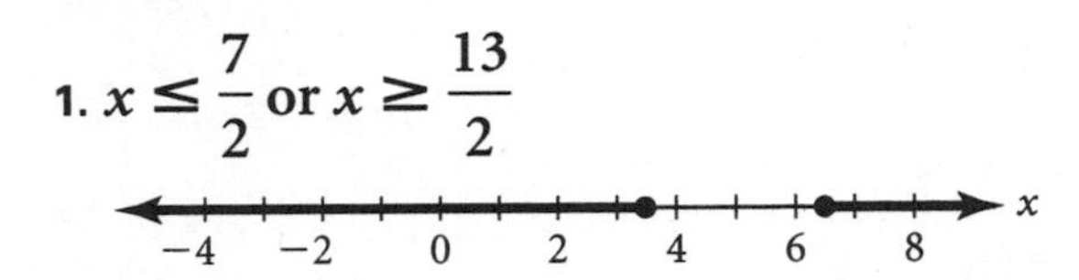

2. $-\frac{1}{3} < x < \frac{4}{3}$

3. $(2, 2)$; $(-2, 2)$

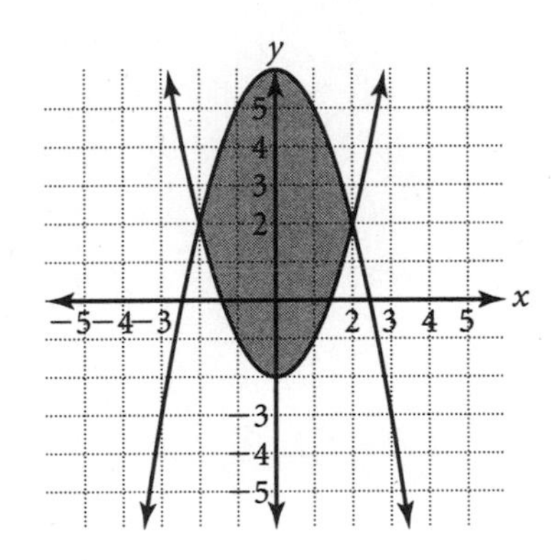

4. $(-2, 1)$; $\left(\frac{1}{2}, -\frac{11}{4}\right)$

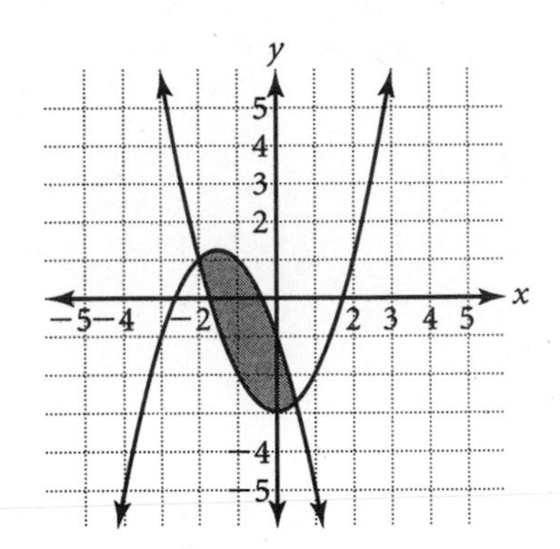

5. a. $t < 1$ or $t < 5$
   b. $1 < t < 5$
   c. 3 seconds
   d. 144 feet

Answers

Lesson 11.1
Level A

1. no
2. no
3. yes
4. yes
5. yes
6. no
7. no
8. no
9. yes
10. yes
11. no
12. yes
13. $x = 1$
14. $y = 6$
15. $x = -48$
16. $y = 16$
17. $x = 24$
18. $y = \frac{1}{2}$
19. $x = 12$
20. $x = \frac{9}{25}$
21. $y = 30$
22. $x = 2$
23. $y = -3$
24. $x = 3$
25. $y = 7$
26. $x = -4$

Lesson 11.1
Level B

1. d
2. c or d
3. 336
4. 3
5. $\frac{5}{3}$
6. 1
7. 20
8. -1
9. -5
10. 10.388
11. 13.3
12. $\frac{4}{3}$
13. $x = 18.15$
14. $y = 1.2$
15. $x = 3.375$
16. $y = \frac{4}{3}$

Lesson 11.1
Level C

1. $x = 25$
2. $y = 8$
3. $x = -9$
4. $y = 10$
5. $y = 25$
6. $x = 16$
7. $x = -48$

Answers

8. $y = -10$

9. $y = 7$

10. I and III; II and IV

11. $y = x$ and $y = -x$

12. $(1, 1)$

13. 15 Newtons

Lesson 11.2
Level A

1. $x = 0$

2. $m = 0, \frac{4}{3}$

3. $y = 4$

4. $z = -\frac{1}{3}$

5. $x = 5, -3$

6. $m = -\frac{5}{3}$

7. $x = 0$

8. $x = 0, \frac{-2}{3}$

9. $-\frac{1}{3}$; undefined

10. $-3\frac{1}{3}$; undefined

11. 3; $1\frac{1}{2}$

12. -3; $-2\frac{5}{8}$

13. -1; $\frac{1}{8}$

14. undefined; $\frac{1}{3}$

15. undefined; -2

16. $-\frac{4}{5}$; $-\frac{1}{8}$

17. $x = 3$

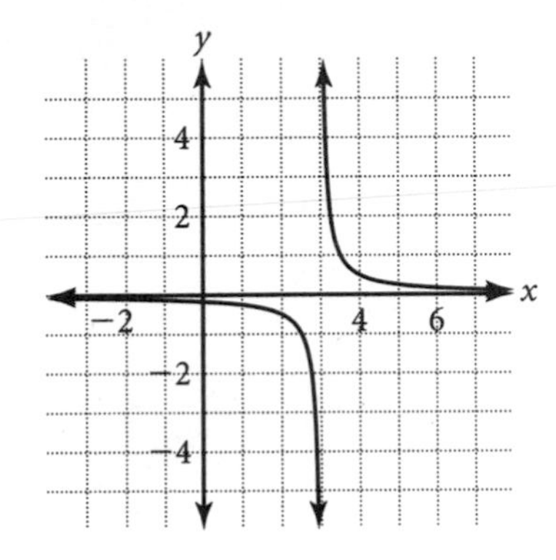

18. $x = -2$

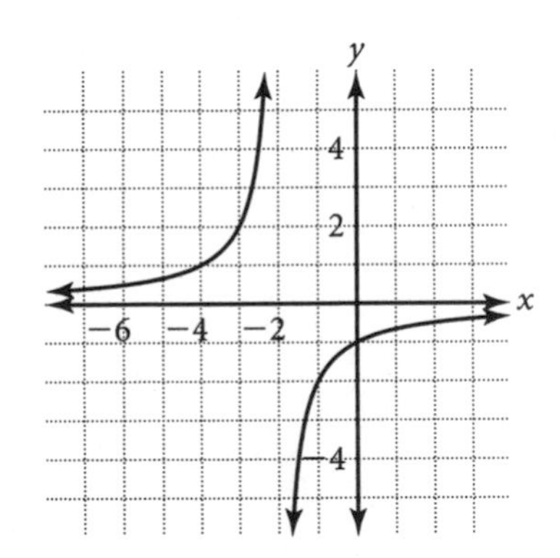

19. $x = 2$

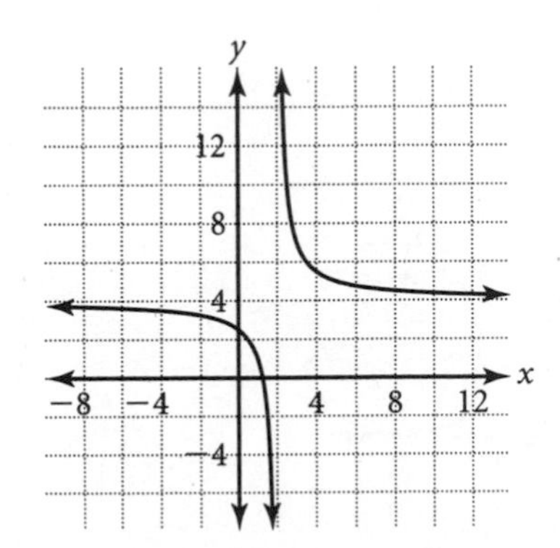

Answers

Lesson 11.2
Level B

1. If P and Q are polynomials and $Q \neq 0$, then an expression in the form $\frac{P}{Q}$ is a rational expression.
2. $m, n = 0$
3. $y = -3$
4. $z = 1, -8$
5. $x = -8, 6$
6. $p, q, r = 0$
7. $x = \frac{2}{3}$
8. $q = 4, -6$
9. $y = 4, 14$
10. $x \neq 5$
11. $x \neq 5$
12. $x \neq 0, \frac{3}{2}$
13. $x \neq \frac{1}{2}$
14. $x \neq \frac{3}{2}$
15. $x \neq 0, 3$
16. $x = 4$

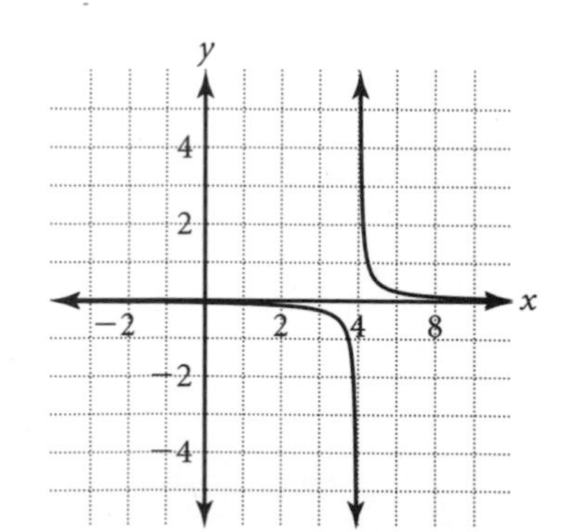

17. $x = 0$

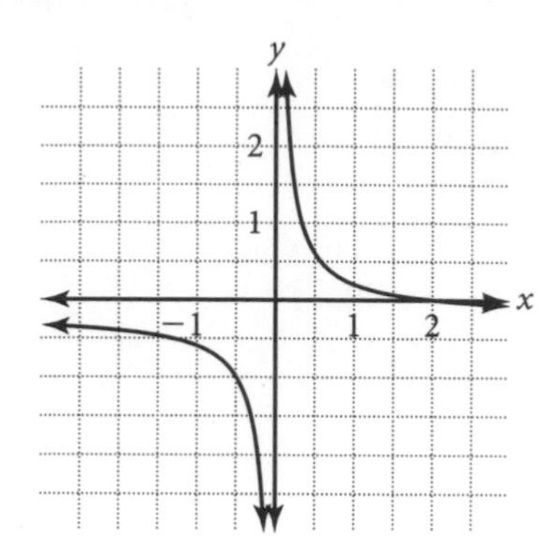

18. $x = 0$

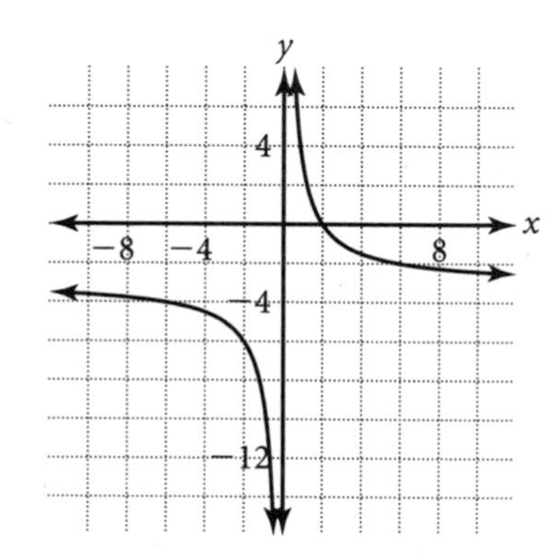

Lesson 11.2
Level C

1. A rational function is set equal to $f(x)$.
2. $x, y = 0$
3. $z = 5$
4. $z = 2, -6$
5. $x = -3, 6$
6. $m, n = 0$
7. $k = \frac{3}{2}$
8. $x \neq -6$
9. $x \neq 5, -5$
10. $x \neq 0, 2$
11. $x \neq 1$

Answers

12. up 2 and vertical stretch of 3; $x = 0$

13. right 3 and down 1; $x = 3$

14. $x = 3$

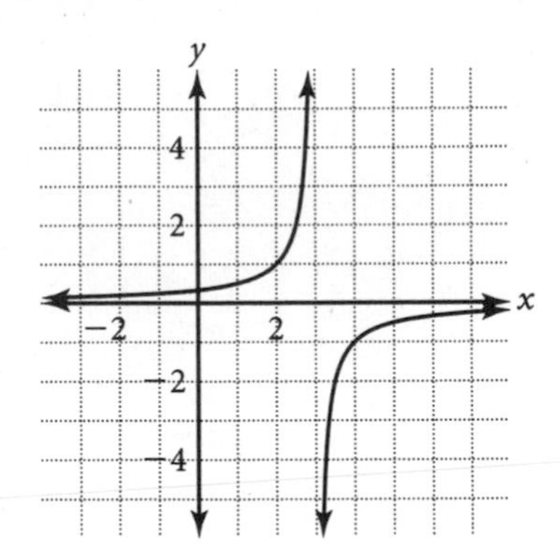

15. $x = -3$

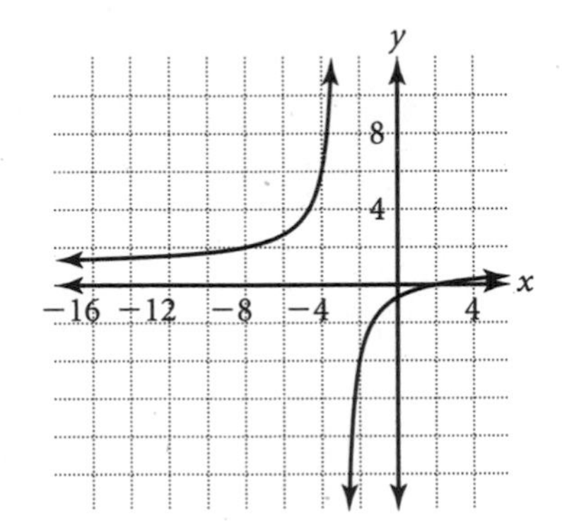

16. $x = 0$

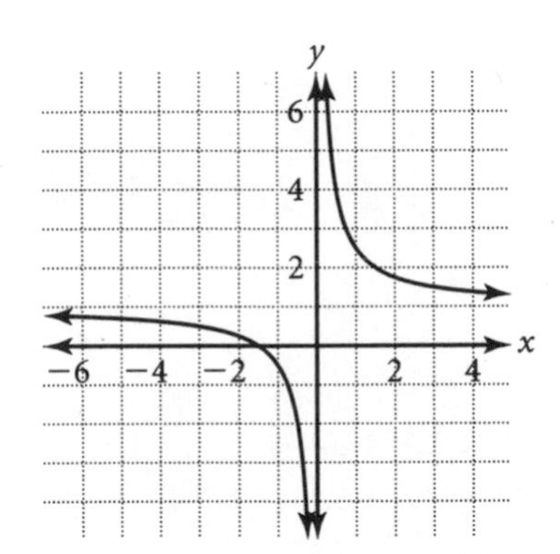

Lesson 11.3
Level A

1. 4
2. $x + 2$
3. 3
4. $3, x + 3$
5. 4
6. $x + 4$
7. $x + y$
8. $r + 2$
9. $x = 2$
10. $r = 0$
11. $q = 3$
12. $x = 3$
13. $p = 4$
14. $x = 0, 1, 2$
15. $m = -1$
16. $b = -2, 1$
17. $\frac{5(x-1)}{12(x-2)}; x \neq 2$
18. $7; t \neq -\frac{5}{2}$
19. $m + 3$
20. $\frac{4+x}{x}; x \neq 0$
21. $\frac{1}{c+2}; c \neq \pm 2$

Answers

22. $-\frac{1}{b+5}$; $b \neq -5, -1$

23. $\frac{4}{m-5}$; $m \neq 5, 6$

24. $-\frac{1}{n+5}$; $n \neq -5, 3$

18. $\frac{1}{9}$; $a \neq 9$

19. $4, a \neq 4$

20. $-\frac{1}{3}$; $b \neq \frac{4}{3}$

21. $2(x + 4)$

Lesson 11.3
Level B

1. $r + 2$
2. $15x$
3. $z + 2$
4. $x - 3$
5. $z - 3$
6. 3
7. $m = -2, 1$
8. $l = 3$
9. $x = \pm 1$
10. $x = 10$
11. $g = -1$
12. $x = 5$
13. $\frac{2}{x}$; $x \neq 0$
14. $\frac{2}{x}$; $x \neq 0$
15. $\frac{6y + 5}{12}$
16. $\frac{1}{x-7}$; $x \neq 7, -2$
17. $-\frac{1}{5}$; $b \neq \frac{1}{2}$

Lesson 11.3
Level C

1. $s - 5$
2. $15xy$
3. $z - 4$
4. $x - 3$
5. $z = -4, 2$
6. $k = 3$
7. $j = -4, -\frac{3}{2}$
8. $m = 4$
9. $h = -1$
10. $x = 7$
11. $-\frac{1}{4}, y \neq \frac{1}{2}$
12. $\frac{x + 2}{3}$
13. $\frac{1}{x - 13}$; $x \neq -3, 13$
14. $\frac{x+5}{x+6}$; $x \neq -6, -5$
15. $\frac{c-7}{c+8}$; $c \neq -8$
16. $\frac{(x+1)(x^2+1)}{x-1}$; $x \neq 1$

Answers

17. $\frac{x^2(x+5)}{(x-1)}$; $x \neq -5, 1, 0$

18. $\frac{x^n+6}{x^n+8}$; $x^n \neq -8, -1$

19. $15x^2, x \neq 0$; $\frac{3(x+1)}{2}, x \neq 5$; Both have variable restrictions that are not evident from their simplified forms.

20. Answers will vary. Sample answer: $\frac{1}{x(2x-1)(4x+1)}$

21. $p = q + 2$

Lesson 11.4
Level A

1. $\frac{8}{x}$; $x \neq 0$

2. $\frac{2}{z}$; $z \neq 0$

3. $\frac{5}{2x}$; $x \neq 0$

4. $2m$; $m \neq 0$

5. $\frac{9}{x}$; $x \neq 0$

6. $\frac{12}{3m}$; $m \neq 0$

7. $\frac{4}{3+n}$; $n \neq -2, -3$

8. $\frac{3}{4(x+2)}$; $x \neq \pm 2$

9. $\frac{3}{2y}$; $y \neq 0$

10. $\frac{-1}{4n}$; $n \neq 0$

11. $\frac{n^2+2n-8}{n^2+5n+6}$; $n \neq -3, -2$

12. $\frac{x+4}{2x}$; $x \neq 0, 4$

13. $\frac{15}{x+2}$; $x \neq -2$

14. $\frac{5}{p+3}$; $p \neq -3$

15. $\frac{3p}{(p-1)^2}$; $p \neq 1, -3$

16. $\frac{6}{5q}$; $q \neq 0$

17. $\frac{22}{w-2}$; $w \neq 2$

18. $\frac{7}{q-2}$; $q \neq 2$

19. $\frac{y^2}{(y+1)(y+4)}$; $y \neq 0, -1, -2, -4$

20. 60 joules

21. $\frac{x+7}{x-4}$ joules; $x \neq \pm 4, -6$

Lesson 11.4
Level B

1. $\frac{5a-3}{(a+6)(a-5)}$; $a \neq 5, -6$

2. $\frac{3t^2+5t+12}{6(t^2-9)}$; $t \neq \pm 3$

3. $\frac{1}{4}$; $m \neq 5, -6$

4. $\frac{2}{3cde}$; $c, d, e \neq 0$

Answers

5. $\frac{9a + 8}{(a + 7)(a - 4)}; a \neq 0, -7$

6. $\frac{4r - 9}{12(r + 3)}; r \neq -3$

7. $\frac{14(x + 5)}{x - 6}; x \neq \pm 6$

8. $\frac{1}{x + 6}; x \neq \pm 7, -6$

9. $\frac{32}{y^2 - 16}; y \neq \pm 4$

10. $\frac{v^3 - 11v^2 + 11}{121 - v^2}; v \neq \pm 11$

11. $2a^2bc^2; c, a \neq 0$

12. $\frac{4}{9q}; q, r, s \neq 0$

13. $\frac{6(x + 4)}{(x + 5)}; x \neq \pm 5$

14. $\frac{5v^2 - 40v - 1}{100 - v^2}; v \neq \pm 10$

15. $\frac{wv}{3}; v, w \neq 0$

16. $\frac{4k}{3}; k \neq 0$

17. $\frac{4}{3}$

18. 2

Lesson 11.4
Level C

1. $\frac{7 - 12t}{6t^2}; t \neq 0$

2. $\frac{3 - 12r}{8r^2}; r \neq 0$

3. $\frac{1 - g}{g + 2}; g \neq -2$

4. $\frac{(t - 1)(t + 1)}{(t - 3)(t + 4)}; t \neq 0, -1, 3, \pm 4$

5. $3p^2q^2; r, p \neq 0$

6. $6y; y \neq 0$

7. $\frac{22}{9x}; x \neq 0$

8. $\frac{9(x + 4)}{x - 2}; x \neq \pm 2$

9. $\frac{1}{3(x - 2)}; x \neq 2, \pm 6$

10. $\frac{10 - 11z}{6z^2}; z \neq 0$

11. $\frac{3t - 10}{6(t + 4)}; t \neq -4$

12. $\frac{1}{4}; n \neq 3, -7$

13. $15x; x \neq -1, 4$

14. $3(x + 5); x \neq 5$

15. $\frac{2}{x(x + 2)}; x \neq -5, -2, 0, -8$

16. $\frac{2x + 3}{2x}; x \neq 0, \pm 5$

17. $\frac{6x + 4}{x(x + 1)}$

Answers

18. $\dfrac{-6x^3 + 30x^2 - 4x}{(1 - x)(6x - 1)}$

19. $\dfrac{16x^2 + x - 2}{x(x + 1)(3x - 1)}$

20. Answers will vary. Sample answer: $\dfrac{1}{x} + \dfrac{1}{3}; \dfrac{1}{2x} + \dfrac{6}{4x}$

Lesson 11.5
Level A

1. $x = 4$
2. $x = 24$
3. $y = \dfrac{1}{2}$
4. $x = 96$
5. $x = -3$
6. $x = -3$
7. $a = 4$
8. $x = 1$
9. $n = 4$
10. $x = \dfrac{1}{4}$
11. $x = 1$
12. $y = -9$
13. $x = -4$
14. $x = -2$
15. $x = -2$
16. $x = \dfrac{5}{7}$
17. $x = 8$
18. $x = -32$
19. $x = 2$
20. $x = 9$
21. $x = -4$
22. $x = 2$

Lesson 11.5
Level B

1. $x = \dfrac{7}{4}$
2. $d = 5$
3. $x = \dfrac{8}{5}$
4. $k = \pm\sqrt{3}$
5. no solution
6. no solution
7. $k = -3$
8. $q = 18$
9. $m = \dfrac{11}{3}$
10. $x = \dfrac{-3}{20}$
11. $n = -6$
12. no solution
13. $x = 2$
14. no solution
15. $x = -6\dfrac{1}{3}$
16. $x = 2$
17. $x = 10$
18. $x = -1, -\dfrac{1}{2}$

Answers

Lesson 11.5
Level C

1. $d = -8$
2. $x = \frac{30}{7}$
3. $x = \frac{29}{13}$
4. $b = -\frac{1}{15}$
5. $y = -1$
6. $k = \frac{8}{45}$
7. $x = 1$
8. $x = -\frac{5}{2}$
9. $x = \frac{5}{2}$
10. Luis: $2300
    Sergio: $1250
11. 4 hours
12. 28 mi/h

Lesson 11.6
Level A

1. reason
2. hypothesis
3. conclusion
4. counter example
5. converse
6. Inductive reasoning
7. Additive Property of Equality
8. Additive Inverses
9. Additive Identity
10. Division Property of Equality
11. Substitution Property of Equality
12. Multiplicative Identity
13. Additive Property of Equality
14. Additive Inverse
15. Substitution Property of Equality
16. Division Property of Equality
17. Substitution Property of Equality
18. Substitution Property of Equality

Lesson 11.6
Level B

1. b
2. d
3. g
4. f
5. e
6. a
7. h
8. c

Answers

9. $-4x + 3 = 19$ Given
   $-4x + 3 - 3 = 19 - 3$ Additive Property of Equality
   $-4x + 0 = 16$ Substitution Property of Equality
   $-4x = 16$ Additive Identity
   $-4x \div 4 = 16 \div 4$ Division Property of Equality
   $-x = -4$ Substitution Property of Equality
   $x = -4$ Multiplicative Identity

10. $m = (cx - b) + cy$ Given
    $m = (cx + (-b)) + cy$ Definition of Subtraction
    $m = cx + ((-b) + cy)$ Associative Property of Equality
    $m = cx + (cy + (-b))$ Commutative Property of Equality
    $m = (cx + cy) + (-b)$ Associative Property of Equality
    $m = (cx + cy) - b$ Definition of Subtraction

Lesson 11.6
Level C

1. When you interchange the hypothesis and conclusion of a conditional statement.
2. The p portion of a conditional statement.
3. The q portion of a conditional statement.
4. An "If-p-then-q" statement.
5. A proven conditional statement.
6. The process of reasoning that a general principle is true because special cases that you have seen are true.
7. The process of reasoning that a special case of a proven general principle is true.
8. A counter example is needed to prove that the integers are not closed under division. Sample answer: $-6 \div 4 = -1.5$ is not an integer. Therefore the integers are not closed under division.
9. Two unique even integers are $2n$ and $2m$ when n and m are integers. Their sum is $2n + 2m = 2(m + n)$. Since m and n are integers, their sum is as well because the integers are closed under addition. By the definition of even numbers, any integer multiplied by 2 is even. Therefore $2(m + n)$ is even and it is shown that the sum of two even integers is an even integer.

Answers

Lesson 12.1
Lesson A

1. 9
2. -7
3. ± 16
4. 30
5. $11\sqrt{5}$
6. $-5\sqrt{2}$
7. $-9\sqrt{3}$
8. $2\sqrt{11}$
9. $2\sqrt{2}$
10. $4\sqrt{2}$
11. $5\sqrt{3}$
12. $6\sqrt{2}$
13. 7
14. 3
15. 11
16. 9
17. 4
18. 9
19. 6
20. 12
21. $2\sqrt{3}$
22. $2\sqrt{6}$
23. $5\sqrt{3}$
24. $7\sqrt{2}$
25. 6
26. 5
27. 4
28. 9

Lesson 12.1
Lesson B

1. 8
2. -3.16
3. ± 4.90
4. 15
5. -0.35
6. 1.00
7. t^6v^4
8. $f^2|g^3\sqrt{fg}|$
9. $|\frac{z^4}{a^3}|$
10. $\frac{x^2\sqrt{x}}{y^8}$
11. $2a^2|b^3|\sqrt{2ab}$
12. $4p^6f^8$
13. $4\sqrt{7}$
14. $3\sqrt{4} + 5$
15. $-5\sqrt{3} + 4$
16. $-7\sqrt{11} - 10$
17. $\sqrt{2} - 8\sqrt{3}$
18. $\sqrt{3} - 2\sqrt{5} + \sqrt{7}$
19. $2\sqrt{3} - 14$
20. $-8\sqrt{5} - 24$
21. $-14\sqrt{2} - \sqrt{10}$
22. -42
23. $-4\sqrt{7} + 56$
24. $18\sqrt{6} - 24$
25. $-4\sqrt{5} - 16$

Answers

26. $-2\sqrt{3} + 5$
27. 47
28. $-8\sqrt{5} + 21$
29. $-6\sqrt{7} + 16$
30. -7

Lesson 12.1
Lesson C

1. 14.14
2. ± 9.95
3. -0.41
4. $-\frac{5}{2}$
5. $9\sqrt{7}$
6. $15\sqrt{3} - 6\sqrt{2}$
7. $\sqrt{2} + \sqrt{3}$
8. -2
9. $-\sqrt{2}$
10. 9
11. $11 + 6\sqrt{3}$
12. $-29 + 9\sqrt{15}$
13. $24 + 3\sqrt{2} - 8\sqrt{7} - \sqrt{14}$
14. $18\sqrt{2} - 60\sqrt{3} - 10\sqrt{5} + \sqrt{30}$
15. $37 - 20\sqrt{3}$
16. $-215 + 60\sqrt{7}$
17. $-16 + 18\sqrt{2}$
18. $-98\sqrt{3} + 42\sqrt{15}$
19. 15 feet
20. 11.18034 meters
21. False
22. True
23. True
24. False

Lesson 12.2
Lesson A

1. $x = 2$
2. $x = -55$
3. $x = 21$
4. $x = 10$
5. $x = 12$
6. $x = -40$
7. $x = 16$
8. $x = 3$
9. $x = 2$
10. $x = 3$
11. $x = 9$
12. $x = \frac{5}{7}$
13. $x = \frac{7}{9}$
14. $x = 1$
15. $x = \frac{32}{7}$
16. $x = \frac{50}{9}$
17. $x = \pm 2\sqrt{10}$
18. $x = \pm 6\sqrt{2}$
19. $x = \pm 5\sqrt{3}$

Answers

20. $x = \pm 3\sqrt{7}$
21. $x = \pm 10\sqrt{5}$
22. $x = \pm 4\sqrt{11}$
23. $x = \pm 25\sqrt{2}$
24. $x = \pm 9\sqrt{13}$
25. $x = \pm\frac{\sqrt{11}}{3}$
26. $x = \pm\frac{\sqrt{7}}{2}$
27. $x = \pm 4$
28. $x = \pm 3$
29. $x = \pm 4$
30. $x = \pm\frac{\sqrt{21}}{5}$
31. first quadrant

Lesson 12.2
Lesson B

1. $x = 2$
2. $x = 5$
3. $x = 4$
4. $x = -4$ *or* $x = -1$
5. $x = 1$
6. $x = -1$
7. $x = -3$ or $x = 7$
8. no solution
9. $x = 4$
10. $x = 3$
11. $x = 1$
12. $x = 2$
13. $x = \frac{5}{4}$
14. no solution
15. $x = \frac{2}{5}$
16. no solution
17. $x = -1$
18. no solution
19. $x = 1$
20. $x = -\frac{3}{2}$
21. $x = \pm 8$
22. $x = \pm 2\sqrt{3}$
23. $x = \pm\frac{\sqrt{2}}{3}$
24. $x = \pm\frac{5}{4}$
25. $x = 2$
26. $x = -5$
27. $x = 25$
28. $x = -4$
29. $x = \frac{3}{2}$
30. $x = -\frac{5}{3}$
31. domain $x \leq 7$; range: all real numbers

Answers

Lesson 12.2
Lesson C

1. $x = \pm 2$
2. Cannot solve: It is not possible to take the absolute value of a number and get a negative result.
3. $x = 16$
4. Cannot solve: It is not possible to square a number and get a negative result.
5. $x = \pm 4$
6. Cannot solve: It is not possible to take the square root of a number and get a negative result.
7. $x = -2$ or $x = 1$
8. no solution
9. $x = 7$
10. no solution
11. $x = -4$ or $x = -1$
12. $x = -1$
13. $x = \frac{31}{12}$
14. $x = -\frac{1}{6}$
15. 0.63 seconds
16. 99.3 centimeters
17. 43.8 kilometers
18. 65 feet
19. 4 seconds

Lesson 12.3
Lesson A

1. No
2. Yes
3. Yes
4. No
5. No
6. No
7. Yes
8. No
9. Yes
10. Yes
11. 15
12. 10
13. 30
14. 4.8
15. 51.0
16. 24
17. $4\sqrt{13}$ units
18. 35 units
19. $7\sqrt{13}$ units
20. 127 feet
21. 112 feet
22. 22 inches

Lesson 12.3
Lesson B

1. Yes
2. No
3. Yes

4. Yes
5. No
6. Yes
7. No
8. No
9. Yes
10. Yes
11. 32.8
12. 61
13. 42.1
14. 96.8
15. 65.1
16. 10.6
17. 55 units
18. $13\sqrt{10}$ units
19. 70 units
20. 58 centimeters
21. 10.1 inches
22. 19.80 feet

Lesson 12.3
Lesson C

1. No
2. Yes
3. Yes
4. Yes
5. 65.6
6. 19
7. 36
8. 58.3
9. 113.1
10. 124.8
11. 9 inches
12. 12.7 miles
13. 60 inches
14. 2.83 feet
15. 8 feet
16. 40 feet
17. $151.60

Lesson 12.4
Level A

1. 2.83
2. 12.08
3. 5.66
4. 20.52
5. 10.20
6. 20
7. 2.24
8. 11.40
9. 8.25
10. 33.62
11. 9.49
12. 1.41
13. 13.93
14. 7.21
15. 8.94
16. 3.61
17. 11.40
18. 10

Answers

19. 2
20. 9
21. (1, 3.5)
22. (8, 1.5)
23. (1.5, 7)
24. (−2, 10)
25. (8, −5)
26. (−3, −9)
27. $\triangle PQR$ is an isosceles triangle.
28. $\triangle PQR$ is a scalene right triangle.
29. $\triangle PQR$ is an isosceles right triangle.
30. $\triangle PQR$ is a scalene triangle.
31. $\triangle PQR$ is a scalene right triangle.
32. $\triangle PQR$ is an isosceles triangle.
33. $\triangle PQR$ is a scalene right triangle.
34. $\triangle PQR$ is an isosceles triangle.
35. $\triangle PQR$ is an isosceles triangle.
36. $\triangle PQR$ is a scalene triangle.

Lesson 12.4
Level B

1. 4.47
2. 5
3. 5
4. 9.43
5. 6.71
6. 3.16
7. (6.5, 6)
8. (4, 5)
9. (−9.5, −4.5)
10. (9, 15.5)
11. (−1.5, 1)
12. (0, 0)
13. (2, 3)
14. (−2.5, 3)
15. (−1.5, −1)
16. (4, −3)
17. (8, 3.5)
18. (−38, −40)
19. (11.5, −7.5)
20. Yes, by the converse of the Pythagorean Theorem.

Lesson 12.4
Level C

1. (14.5, 17)
2. (5.5, −2)
3. (−3.5, −2.5)
4. (−2.5, −1.5)
5. $M(-1, 3)$
6. $P(7, 5)$ and $Q(-3, 9)$
7. $P(2, -8)$ and $Q(0, 6)$
8. $Q(-15, 10)$
9. $P(-3, -4)$
10. $P(6, -8)$ and $M(1.5, 1)$
11. $Q(10, 4)$ and $M(2, 1.5)$
12. $P(-9, 2)$
13. $Q(3, -11)$ and $M(-3, -2)$
14. $P(5, -11)$ and $M(3, -9)$

15. $M(1.5, -3)$

16. $P(12, -9)$ and $Q(-14, 15)$

17. $P(0, 3)$

18. $P(5, -2)$ and $M(1, -1)$

19. 32 miles

20. 9.5 miles

21. No, the lengths of the sides do not satisfy the Pythagorean Theorem.

22. true

Lesson 12.5
Lesson A

1. $x^2 + y^2 = 4$
2. $x^2 + y^2 = 49$
3. $x^2 + y^2 = 144$
4. $x^2 + y^2 = 625$
5. $x^2 + y^2 = 10.24$
6. $x^2 + y^2 = 32.49$
7. $x^2 + y^2 = 213.16$
8. $x^2 + y^2 = 1303.21$
9. $(x - 3)^2 + (y - 5)^2 = 4$
10. $(x - 7)^2 + (y - 9)^2 = 25$
11. $(x - 9)^2 + (y - 1)^2 = 9$
12. $(x - 10)^2 + (y - 15)^2 = 16$
13. $(x + 1)^2 + (y - 3)^2 = 49$
14. $(x - 4)^2 + (y + 6)^2 = 81$
15. $(x + 8)^2 + (y - 9)^2 = 1$
16. $(x - 7)^2 + (y + 1)^2 = 256$
17. $(x + 2)^2 + (y + 5)^2 = 25$
18. $(x + 7)^2 + (y + 2)^2 = 16$
19. center: (2, 3), radius: 5
20. center: (7, 9), radius: 3
21. center: (1, 8), radius: 2
22. center: (5, −1), radius: 11
23. center: (−6, 2), radius: 9
24. center:(−49, −64), radius: 1
25. midpoint of the hypotenuse, $C(2, 2)$, midpoint of vertical leg, $D(4, 2)$, slope of midsegment $\overline{CD} = 0$, slope of base $\overline{OB} = 0$. The slope is 0 in each case, therefore the lines are parallel. The length of midsegment $\overline{CD} = 2$ and the length of base $\overline{OB} = 4$. The length of the midsegment is one-half the length of the third side.

Lesson 12.5
Lesson B

1. $x^2 + y^2 = 5$
2. $x^2 + y^2 = 17$
3. $x^2 + y^2 = 57$
4. $x^2 + y^2 = 101$
5. $x^2 + y^2 = 12$
6. $x^2 + y^2 = 175$
7. $x^2 + y^2 = 32$
8. $x^2 + y^2 = 1331$
9. $(x + 12)^2 + (y + 13)^2 = 121$
10. $(x - 6)^2 + (y + 7)^2 = 20$
11. $x^2 + (y + 5)^2 = 4$
12. $(x - 9)^2 + y^2 = 150$
13. center: (11, 14), radius: 14
14. center: (9, −9), radius: 3

15. center: $(-3, 6)$, radius: 24
16. center: $(-2, -12)$, radius: 16
17. center: $(4.7, -6)$, radius: 1
18. center: $(-2.5, -1.6)$, radius: 2
19. $(x - 1)^2 + (y - 1)^2 = 4$
20. $(x - 3)^2 + (y - 4)^2 = 25$
21. midpoint of the leg $\overline{OA}$, $C(3, 3)$, midpoint of the leg $\overline{AB}$, $D(9, 3)$, slope of midsegment $\overline{CD} = 0$, slope of the leg $\overline{OB} = 0$; The slope is 0 in each case, therefore the lines are parallel. Length of midsegment $\overline{CD} = 6$ and the length of base $\overline{OB} = 12$. The length of the midsegment is one-half the length of the third side.

Lesson 12.5
Lesson C

1. center: $(-16, -32)$, radius: 36
2. center: $(-17, 9)$, radius: 8
3. center: $(21, 12)$, radius: 10
4. center: $(-10, -30)$, radius: 20
5. center: $(3, -7)$, radius: $\frac{3}{4}$
6. center: $(12, 6)$, radius: $\frac{15}{27}$
7. center: $(-4, -1)$, radius: 3
8. center: $(-3, 7)$, radius: 8
9. $x^2 + y^2 = 9$
10. $x^2 + y^2 = 36$
11. $(x + 1)^2 + (y - 3)^2 = 16$
12. $(x - 2)^2 + (y + 5)^2 = 25$
13. $(x - 4)^2 + (y - 8)^2 = 18$
14. $(x - 6)^2 + (y - 3)^2 = 32$
15. midpoint of the leg $\overline{OA}$, $D(2, 2)$, midpoint of the leg $\overline{OB}$, $C(0, 2)$, slope of midsegment $\overline{CD} = 0$, slope of the leg $\overline{AB} = 0$; The slope is 0 in each case, therefore the lines are parallel. Length of midsegment $\overline{CD} = 2$ and the length of base $\overline{AB} = 4$. The length of the midsegment is one-half the length of the third side.

Lesson 12.6
Lesson A

1. 0.2679
2. 0.7265
3. 1.1918
4. 0.8391
5. 27°
6. 10°
7. 45°
8. 51°
9. $\tan A = \frac{8}{15}$, $\tan B = \frac{15}{8}$,

   $A \approx 28°, B \approx 62°$
10. $\tan D = \frac{5}{12}$, $\tan E = \frac{12}{5}$,

   $D \approx 23°, \text{E} \approx 67°$
11. $\tan D = \frac{3}{4}$, $\tan E = \frac{4}{3}$,

   $D \approx 37°, E \approx 53°$
12. $\tan A = \frac{15}{112}$, $\tan B = \frac{112}{15}$,

   $A \approx 8°, B \approx 82°$

Answers

13. $\tan A = \frac{11}{60}, \tan B = \frac{60}{11},$

$A \approx 10°, B \approx 80°$

14. $\tan D = \frac{27}{364}, \tan E = \frac{364}{27},$

$D \approx 4°, E \approx 86°$

Lesson 12.6
Lesson B

1. 0.0981
2. 2.0413
3. 22.0217
4. 0.6056
5. 37°
6. 26°
7. 32°
8. 79°
9. $\tan A = \frac{10}{24}, \tan B = \frac{24}{10},$

   $A \approx 23°, B \approx 67°$
10. $\tan D = \frac{9}{40}, \tan E = \frac{40}{9},$

    $D \approx 13°, E \approx 77°$
11. $\tan D = \frac{25}{312}, \tan E = \frac{312}{25},$

    $D \approx 5°, E \approx 85°$
12. $\tan A = \frac{32}{60}, \tan B = \frac{60}{32},$

    $A \approx 28°, B \approx 62°$

13. $\tan A = \frac{39}{760}, \tan B = \frac{760}{39},$

$A \approx 3°, B \approx 87°$

14. $\tan D = \frac{33}{544}, \tan E = \frac{544}{33},$

$D \approx 3°, E \approx 87°$

Lesson 12.6
Lesson C

1. 28.6363
2. 1.4826
3. 2.8716
4. 3.4032
5. 67°
6. 50°
7. 29°
8. 84°
9. 45°
10. 2.31 inches
11. 0.61 feet
12. 21°
13. 3.58 yards
14. 5.30 meters
15. 6°
16. 12.43 inches
17. 16.64 yards
18. 106.28 feet
19. it also increases
20. 14.5 feet
21. 31.4°

Answers

Lesson 12.7
Lesson A

1. 0.8192
2. 0.7314
3. 0.9336
4. 0.2924
5. 0.5592
6. 0.5150
7. 71.8°
8. 13.3°
9. 29.3°
10. 54.1°
11. 31.8°
12. 62.0°
13. 58.0°
14. 77.9°
15. $\frac{a}{c}$
16. $\frac{a}{c}$
17. 53.1°
18. 3.9
19. 6
20. 27.0°
21. 58.7°
22. 5.0

Lesson 12.7
Lesson B

1. 0.4848
2. 0.8829
3. 0.9848
4. 0.1219
5. 0.9816
6. 0.9986
7. 37.4°
8. 11.3°
9. 19.0°
10. 0.5°
11. 24.6°
12. 58.4°
13. 71.7°
14. 89.9°
15. sin A or cos B
16. 60°
17. 11
18. 18.5
19. 45.6°
20. 46.0°
21. 10.1
22. 2165.1 yards
23. 3.1 feet

Lesson 12.7
Lesson C

1. 0.8192
2. 0.5736
3. 0.9511
4. 0.3090
5. 8.86
6. 33.39
7. 27.82°
8. 38.59°
9. 41.8°
10. 9.1
11. 33.1
12. 11.4°
13. If x represents $m\angle A$, $\sin x = \frac{a}{c}$, $\cos x = \frac{b}{a}$ and $\tan x = \frac{a}{b}$.

So, $\frac{\sin x}{\cos x} = \frac{\frac{a}{c}}{\frac{b}{c}} = \frac{a}{c} \cdot \frac{c}{b} = \frac{a}{b} = \tan x$.

Lesson 12.8
Lesson A

1. H and J
2. H and I
3. $\begin{bmatrix} 10 & -6 \\ 0 & -4 \end{bmatrix}$
4. $\begin{bmatrix} 4 & -2 \\ -10 & 10 \end{bmatrix}$
5. $\begin{bmatrix} 1 & 3 \\ 6 & -3 \end{bmatrix}$
6. $\begin{bmatrix} -7 & 15 \\ 4 & 1 \end{bmatrix}$
7. $[21\ 28\ -35]$
8. $[-6\ 18\ -12\ 0]$
9. $\begin{bmatrix} -12 & 18 \\ 6 & -24 \end{bmatrix}$
10. $\begin{bmatrix} 2 & 18 \\ 10 & -12 \end{bmatrix}$
11. $\begin{bmatrix} -15 & 25 \\ 10 & -20 \end{bmatrix}$
12. $\begin{bmatrix} -12 & 20 \\ -24 & 28 \end{bmatrix}$
13. $\begin{bmatrix} 1 & 0 & 0 & 0 \\ 0 & 1 & 0 & 0 \\ 0 & 0 & 1 & 0 \\ 0 & 0 & 0 & 1 \end{bmatrix}$
14. $\begin{bmatrix} 1 & 0 & 0 & 0 \\ 0 & 1 & 0 & 0 \\ 0 & 0 & 1 & 0 \\ 0 & 0 & 0 & 1 \end{bmatrix}$

Lesson 12.8
Lesson B

1. H and I
2. $\begin{bmatrix} -7 & 6 & -9 \\ 0 & -8 & 5 \\ -2 & 0 & -6 \end{bmatrix}$
3. $\begin{bmatrix} 3 & 2 & -7 \\ 6 & -4 & 13 \\ 0 & 2 & 4 \end{bmatrix}$
4. $\begin{bmatrix} -15 & 6 & 0 \\ 21 & -12 & -6 \\ 3 & 9 & -24 \end{bmatrix}$

5. $\begin{bmatrix} -35 & 5 & -20 \\ 25 & 30 & 0 \\ 10 & -15 & 45 \end{bmatrix}$

6. $\begin{bmatrix} -28 & 57 \\ 40 & -19 \end{bmatrix}$

7. $\begin{bmatrix} -4 & 36 & -24 \\ -24 & -4 & 6 \\ 8 & 38 & -27 \end{bmatrix}$

8. $\begin{bmatrix} 1 & 0 & 0 \\ 0 & 1 & 0 \\ 0 & 0 & 1 \end{bmatrix}$

9. $\begin{bmatrix} 1 & 0 & 0 & 0 \\ 0 & 1 & 0 & 0 \\ 0 & 0 & 1 & 0 \\ 0 & 0 & 0 & 1 \end{bmatrix}$

Lesson 12.8
Lesson C

1. $a = 9, b = 1, c = 4, d = -3$

2. $\begin{bmatrix} -2 & -5 & 1 \\ -3 & 10 & 3 \end{bmatrix}$

3. Does not exist because the dimensions are different.

4. $\begin{bmatrix} -95 & 85 \\ -43 & -118 \end{bmatrix}$

5. $\begin{bmatrix} 1499 \\ 1635 \\ 815 \end{bmatrix}, \begin{bmatrix} 1602 \\ 1549 \\ 785 \end{bmatrix}$

6. 7885 cases

Answers

Lesson 13.1
Level A

1. {1, 2, 3, 4, 5, 6}
2. {HH, HT, TH, TT}
3. {HHH, HHT, HTH, HTT, THH, THT, TTH, TTT}
4. {(1, 1), (1, 2), (1, 3), (1, 4), (1, 5), (1, 6), (2, 1), (2, 2), (2, 3), (2, 4), (2, 5), (2, 6), (3, 1), (3, 2), (3, 3), (3, 4), (3, 5), (3, 6), (4, 1), (4, 2), (4, 3), (4, 4), (4, 5), (4, 6), (5, 1), (5, 2), (5, 3), (5, 4), (5, 5), (5, 6), (6, 1), (6, 2), (6, 3), (6, 4), (6, 5) (6, 6)}
5. 3 favorable outcomes
6. 5 favorable outcomes
7. 1 favorable outcome
8. 3 favorable outcomes
9. $\frac{1}{12}$
10. $\frac{5}{36}$
11. $\frac{1}{8}$
12. $\frac{3}{8}$
13. 7 chances out of 10
14. 21 chances out of 30
15. 99 chances out of 100
16. 23 chances out of 40
17. $\frac{6}{25}$
18. $\frac{2}{25}$
19. $\frac{4}{25}$
20. $\frac{3}{25}$
21. $\frac{3}{25}$
22. $\frac{2}{25}$

Lesson 13.1
Level B

1. {HHHHH, HHHHT, HHHTH, HHHTT, HHTHH, HHTHT, HHTTH, HHTTT, HTHHH, HTHHT, HTHTH, HTHTT, HTTHH, HTTHT, HTTTH, HTTTT, THHHH, THHHT, THHTH, THHTT, THTHH, THTHT, THTTH, THTTT, TTHHH, TTHHT, TTHTH, TTHTT, TTTHH, TTTHT, TTTTH, TTTTT}
2. {1H, 1T, 2H, 2T, 3H, 3T, 4H, 4T, 5H, 5T, 6H, 6T}
3. 3 favorable outcomes
4. 5 favorable outcomes
5. 2 favorable outcomes
6. 1 favorable outcome
7. $\frac{2}{13}$
8. $\frac{3}{26}$
9. $\frac{9}{26}$
10. $\frac{3}{13}$
11. $\frac{7}{26}$

12. $\frac{5}{26}$

13. $\frac{2}{13}$

14. $\frac{1}{6}$

15. $\frac{1}{4}$

16. $\frac{1}{2}$

17. $\frac{1}{2}$

18. $\frac{3}{8}$

19. $\frac{11}{36}$

Lesson 13.1
Level C

1. {H1, H2, H3, H4, H5, H6, T1, T2, T3, T4, T5, T6}
2. {1HH, 1HT, 1TH, 1TT, 2HH, 2HT, 2TH, 2TT, 3HH, 3HT, 3TH, 3TT, 4HH, 4HT, 4TH, 4TT, 5HH, 5HT, 5TH, 5TT, 6HH, 6HT, 6TH, 6TT}
3. $\frac{7}{26}$
4. $\frac{9}{26}$
5. $\frac{5}{13}$
6. $\frac{9}{26}$
7. $\frac{3}{13}$
8. $\frac{5}{13}$
9. $\frac{7}{26}$
10. $\frac{7}{8}$
11. $\frac{3}{8}$
12. $\frac{5}{16}$
13. $\frac{5}{72}$
14. $\frac{1}{2}$
15. $\frac{125}{324}$
16. $\frac{24}{49}$
17. 1
18. $\frac{3}{7}$
19. $\frac{4}{49}$
20. 0
21. $\frac{6}{7}$
22. $\frac{2}{7}$

Answers

Lesson 13.2
Level A

1. 16
2. 26
3. 8
4. 13
5. 1
6. 2
7. 16
8. 6
9. {1, 3, 5, 7, 9}
10. {5, 10}
11. {5}
12. {1, 3, 5, 7, 9, 10}
13. {4, 8}
14. {6}
15. {4, 6, 8}
16. ∅
17. $\frac{4}{13}$
18. $\frac{1}{2}$
19. $\frac{1}{52}$
20. $\frac{1}{4}$
21. $\frac{8}{13}$
22. $\frac{3}{26}$
23. 16
24. 9
25. 23
26. 12
27. 26

Lesson 13.2
Level B

1. {6, 12, 18}
2. {9, 18}
3. {18}
4. {6, 9, 12, 18}
5. {2, 4, 6, 8, 10, 12, 14, 16, 18, 20}
6. {5, 10, 15, 20}
7. {2, 4, 5, 6, 8, 10, 12, 14, 15, 16, 18, 20}
8. {10, 20}
9. $\frac{2}{13}$
10. $\frac{17}{26}$
11. $\frac{8}{13}$
12. $\frac{3}{26}$
13. $\frac{4}{13}$
14. 4
15. 60
16. 9
17. 44
18. 50

19. 21

20. 37

Lesson 13.2
Level C

1. {3, 6, 9, 12, 15, 18, 21, 24}
2. {4, 8, 12, 16, 20, 24}
3. {5, 10, 15, 20, 25}
4. {10, 20}
5. {12, 24}
6. {15}
7. {5, 10, 15, 20, 25}
8. {10, 20}
9. {100, 110, 120, 130, 140, 150}
10. {100, 125, 150}
11. {100, 150}
12. {100, 110, 120, 125, 130, 140, 150}
13. $\frac{4}{13}$
14. $\frac{1}{26}$
15. $\frac{11}{13}$
16. 213
17. 500
18. 422
19. 135
20. 446

Lesson 13.3
Level A

1. 6 possible choices
2. 4 possible choices
3. 12 possible choices
4. 15 possible choices
5. 30 possible choices
6. 64 possible choices
7. 24 possible choices
8. 8 possible choices
9. 35 possible choices
10. 50 possible choices
11. 12 possible choices
12. 27 possible choices
13. 255 possible choices
14. 135 possible choices
15. 84 choices
16. 99 choices

Lesson 13.3
Level B

1. 54 possible choices
2. 405 possible choices
3. 84 possible choices
4. 462 possible choices
5. 12 possible choices
6. 30 possible choices
7. 20 possible choices

8. 48 possible choices
9. 63 possible choices
10. 40 possible choices
11. 1900 possible choices
12. 5225 possible choices
13. 784 possible choices

Lesson 13.3
Level C

1. 300 possible choices
2. 24 possible choices
3. 338 possible choices
4. 1296 possible choices
5. 676 possible choices
6. 144 possible choices
7. 78 possible choices
8. 1377 possible choices
9. 6561 possible choices
10. 7,000,000 possible choices
11. 7,000,000,000 possible choices
12. 630 possible choices

Lesson 13.4
Level A

1. $\frac{1}{4}$
2. $\frac{1}{9}$
3. $\frac{1}{4}$
4. $\frac{4}{9}$
5. $\frac{2}{9}$
6. $\frac{1}{4}$
7. 0.81%
8. 4.6%
9. 5.95%
10. $\frac{1}{169}$
11. $\frac{1}{16}$
12. $\frac{1}{4}$
13. $\frac{1}{676}$
14. $\frac{168}{169}$
15. $\frac{15}{16}$
16. $\frac{3}{4}$
17. $\frac{675}{676}$

Answers

Lesson 13.4
Level B

1. $\frac{1}{4}$
2. $\frac{5}{9}$
3. $\frac{1}{9}$
4. $\frac{25}{36}$
5. 0.56%
6. 50.7%
7. 0.32%
8. 83.3%
9. 93.1%
10. $\frac{9}{169}$
11. $\frac{81}{169}$
12. $\frac{4}{169}$
13. $\frac{160}{169}$
14. $\frac{88}{169}$
15. $\frac{165}{169}$

Lesson 13.4
Level C

1. 35.51%
2. 64.6%
3. 48.97%
4. $\frac{3}{20}$
5. $\frac{1}{10}$
6. $\frac{7}{20}$
7. $\frac{2}{5}$
8. $\frac{1}{4}$
9. $\frac{1}{4}$
10. $\frac{1}{4}$
11. $\frac{1}{2}$
12. $\frac{16}{49}$
13. $\frac{9}{49}$
14. $\frac{12}{49}$

Answers

Lesson 13.5
Level A

1. Answers may vary.
2. Answers may vary.
3. Answers may vary.
4. Answers may vary.
5. Answers may vary. Sample answer: Numbers 1 through 30 from a hat for June
6. Answers may vary. Sample answer: Tossing a marker onto a 6×4 rectangle of square.
7. Answers may vary. Sample answer: sum of two number cubes
8. Answers may vary. Sample answer: circular spinner divided into quarters
9. Answers may vary. Sample answer: random integers from 1 to 25
10. Answers may vary. Sample answer: one number cube

Lesson 13.5
Level B

1. Answers may vary.
2. Answers may vary.
3. Answers may vary.
4. Answers may vary.
5. Answers may vary. Sample answer: integers 1–5 = snow; 6–10 = no snow; generate twice.
6. Answers may vary. Sample answer: 1, 2, 3 = snow; 4, 5, 6 = no snow; roll twice.
7. Answers may vary. Sample answer: a slip for snow; a slip for no snow; draw twice.

Lesson 13.5
Level C

1. Answers may vary.
2. Answers may vary.
3. Answers may vary.
4. Answers may vary.
5. Answers may vary.
6. Answers may vary.

Answers

Lesson 14.1 Level A

1. no; Each x-value of 5 is paired with two y-values.
2. yes
3. no; Each x-value of 0 is paired with two y-values.
4. yes
5. 1, 13, -3
6. 1, 5, 3
7. 2, 50, 18
8. 2, -4, 4
9. 2, 4, 4
10. -3, -75, -27
11. $y = x$
12. $y = x^2$
13. $y = |x|$
14. $y = x^3$
15. $y = \frac{1}{x}$
16. $y = \sqrt{x}$

Lesson 14.1 Level B

1. yes
2. yes
3. no; Each x-value of 1 is paired with two y-values.
4. no; Each x-value of -2 is paired with two y-values.
5. yes; Each x goes to only one y.
6. no; The x-value 2 goes to the y-values 2 and -2, the x-value 4 goes to the y-values 4 and -4.
7. no; The x-value -2 goes to the y-values 4 and -2, the x-value 1 goes to the y-values 4 and -3.
8. -5
9. -4
10. $-4\frac{3}{4}$
11. -1
12. -4
13. 20
14. 40
15. 18
16. $\frac{-1}{3}$
17. 5
18. $f(x) = \frac{1}{x}$
19. $f(x) = \sqrt{x}$
20. $f(x) = x^2$

Lesson 14.1 Level C

1. Sample answer: $\{(0, 5), (1, 5), (1, 3), (2, 1)\}$
2. Sample answer: $f(x) = x + 2$
3. Sample answer:

1 3 5 → 2 5 7

Answers

4. Sample answer:

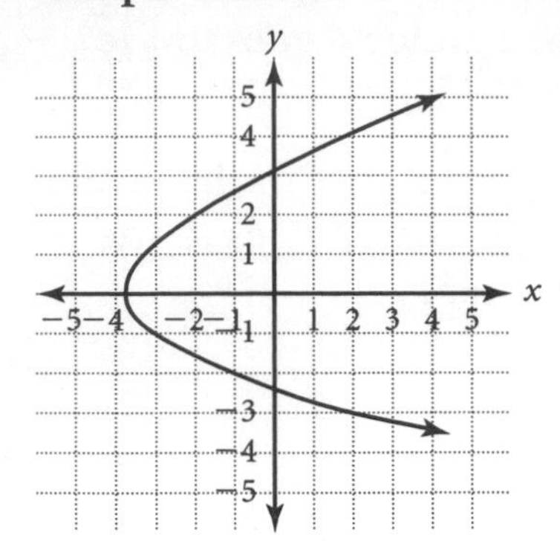

5. 33

6. $2(2b - 1)^2 + 1$

7. 33

8. 3

9. $2(a - 1)^2 + 1$

10. 19,603

11. $y = \frac{1}{x}$

12. $y = \sqrt{x}$

13. $y = x^2$

14. $y = |x|$

15. $y = x^2$

16. $y = \sqrt{x}$

17. $y = a^x, (a > 1)$

Lesson 14.2
Level A

1. up 2 units
2. down 6 units
3. right 3 units
4. left 2 units
5. left 2 units
6. up 5 units
7. down 8 units
8. up 6 units
9. right 5 units
10. up 6 units
11. (5, 4)
12. (3, 9)
13. (3, −2)
14. (3, 4)
15. $y = x^2$; down 4 units
16. $y = \sqrt{x}$; left 4 units
17. $y = \frac{1}{x}$; right 2 units

Lesson 14.2
Level B

1. left 1 unit, down 3 units
2. left 3 units, up 4 units
3. right 1 unit, up 3 units
4. right 2 units, up 5 units
5. left 1 unit, down 5 units
6. right 3 units, up 6 units
7. down 5 units
8. right 1 unit, up 2 units
9. left 2 units, down 5 units
10. left 1 unit, down 3 units
11. (5, 5)
12. (2, 2)
13. (8, 4)

14. (6, 3)

15. (−1, 9)

16. $y = x^2$; left 3 units, up 1 unit

17. $y = \sqrt{x}$; left 4 units, up 1 unit

18. $y = |x|$; left 2 units, down 3 units

Lesson 14.2
Level C

1. a. $y = x^2$, b. left 5 units, up 1 unit,

c.

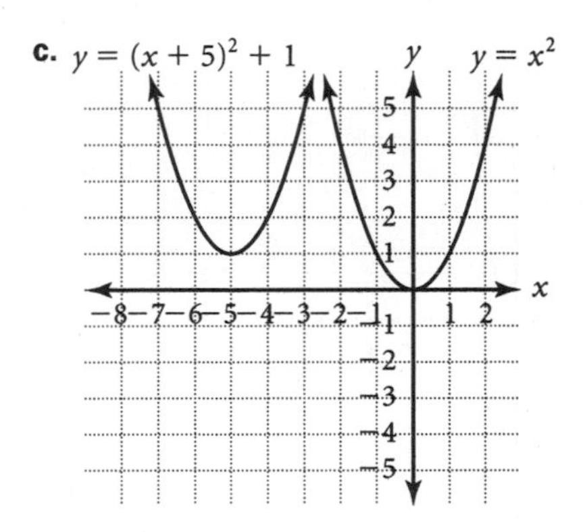

2. a. $y = |x|$, b. right 4 units, down 2 units,

c.

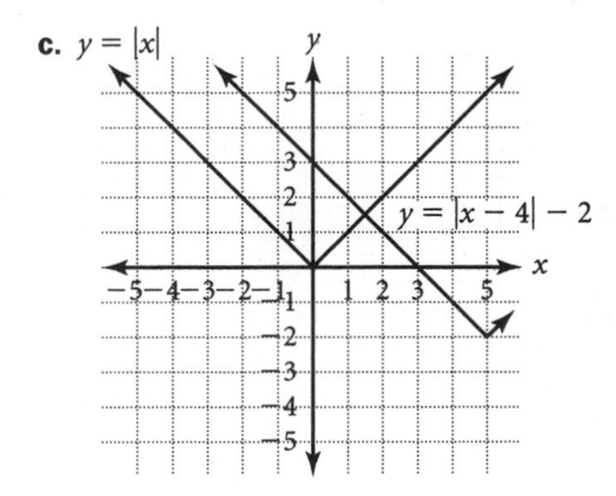

3. a. $y = \sqrt{x}$, b. left 3 units, down 4 units,

c.

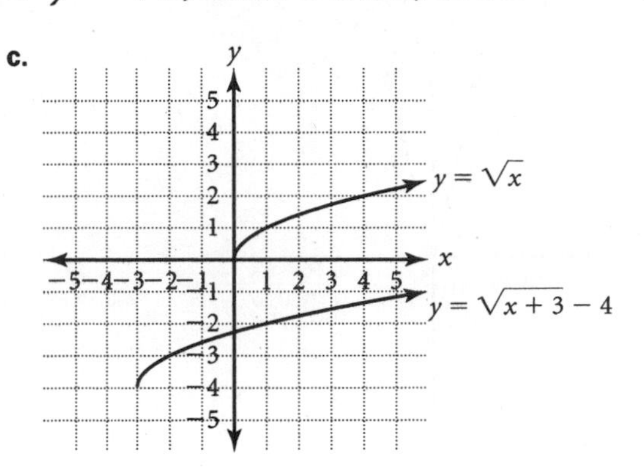

4. The x-value is decreased by 1.

5. The x-value is increased by 1.

6. The y-value is decreased by 4.

7. The x-value is increased by 4.

8. (4, −4)

9. (2, −7)

10. $(4, y - 5)$

11. $(4, -7 + d + b)$

Lesson 14.3
Level A

1. vertical stretch

2. horizontal compression

3. horizontal compression

4. horizontal stretch

5. vertical stretch

6. vertical compression

7. vertical compression

8. vertical compression

9. horizontal stretch

10. vertical stretch

Answers

11. $y = |x|$

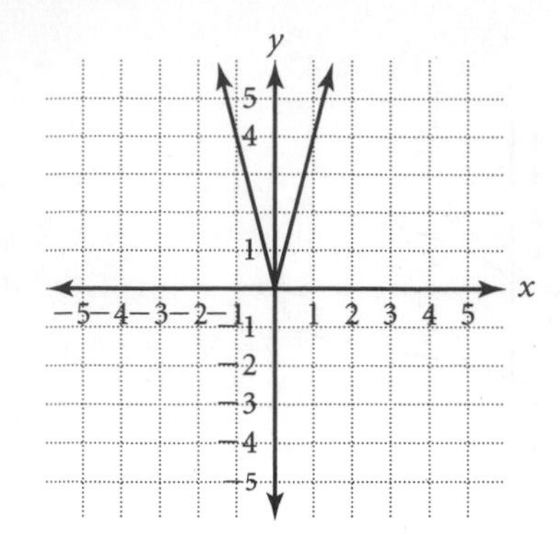

12. $y = \sqrt{x}$

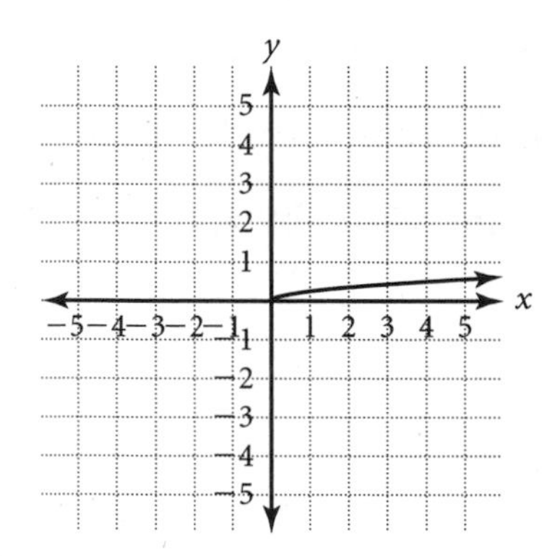

13. $y = x^2$

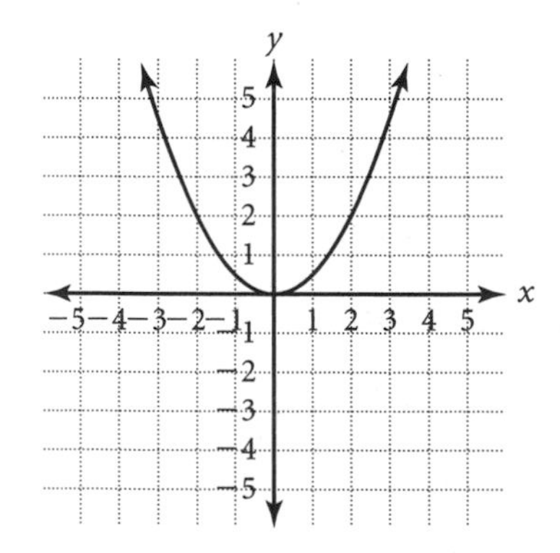

14. $y = |x|$

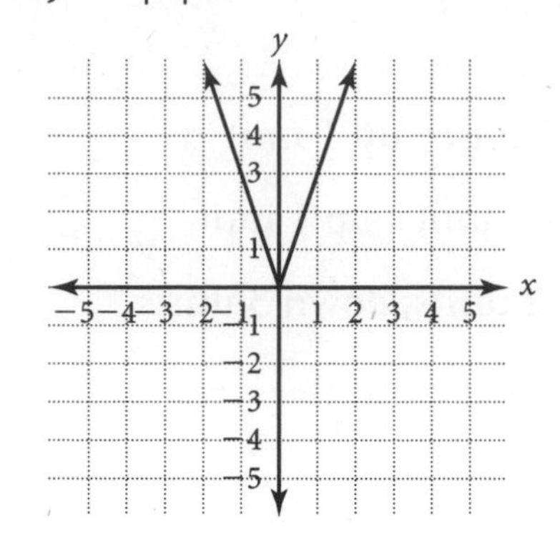

15. $y = x^2$

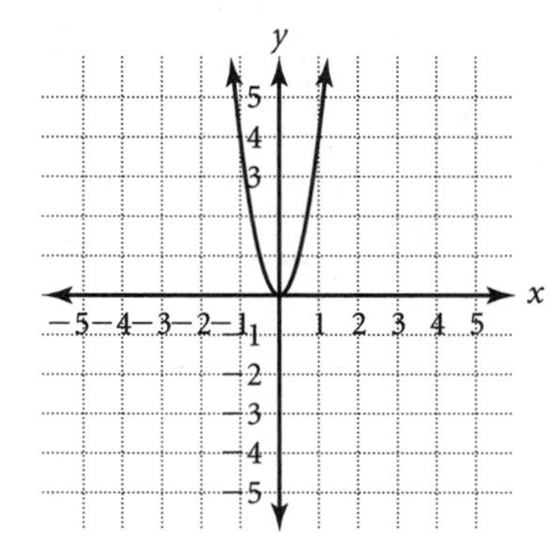

16. $y = \frac{1}{x}$

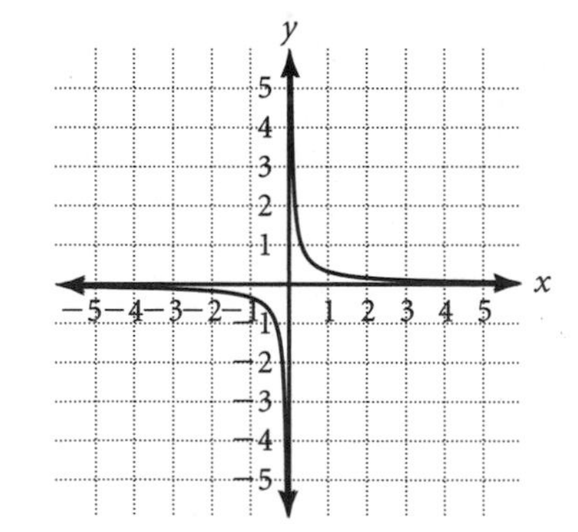

Lesson 14.3
Level B

1. vertical stretch
2. vertical compression
3. neither
4. vertical stretch

Answers

5. vertical compression
6. horizontal compression
7. horizontal compression
8. vertical stretch
9. neither
10. neither
11. $y = |x|$

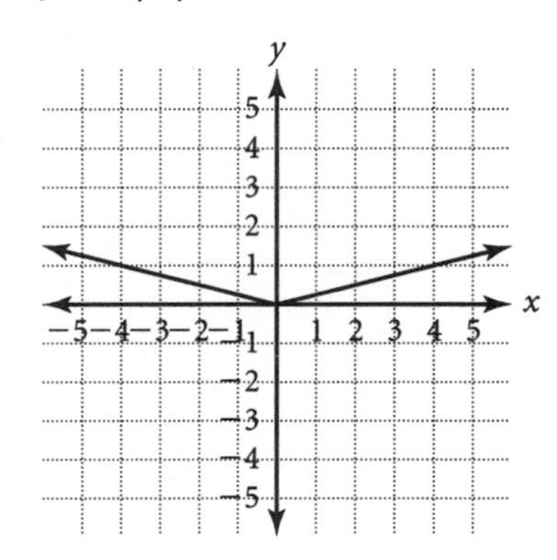

12. $y = \sqrt{x}$

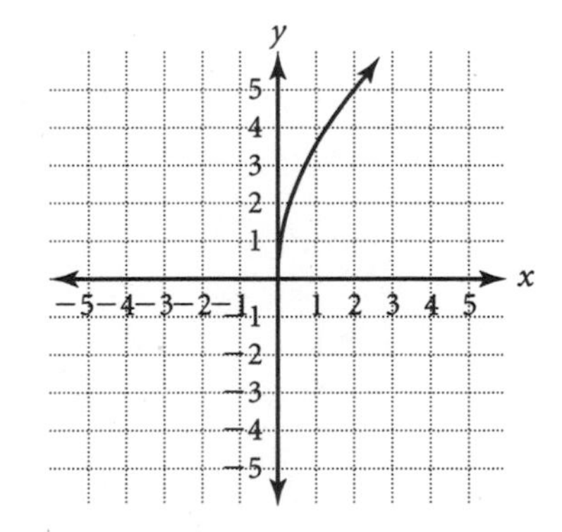

13. $y = x^2$

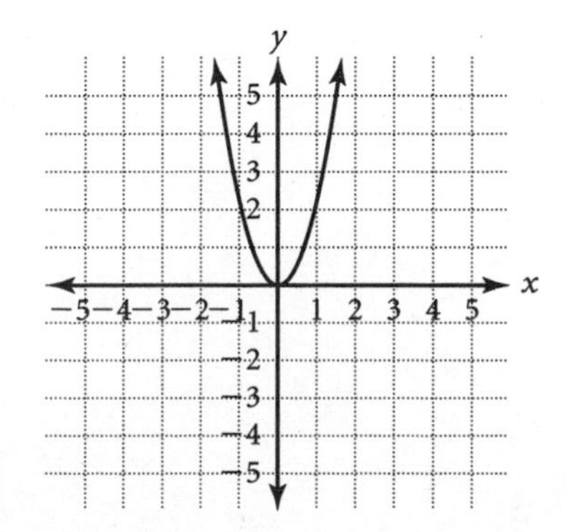

14. $y = |x|$

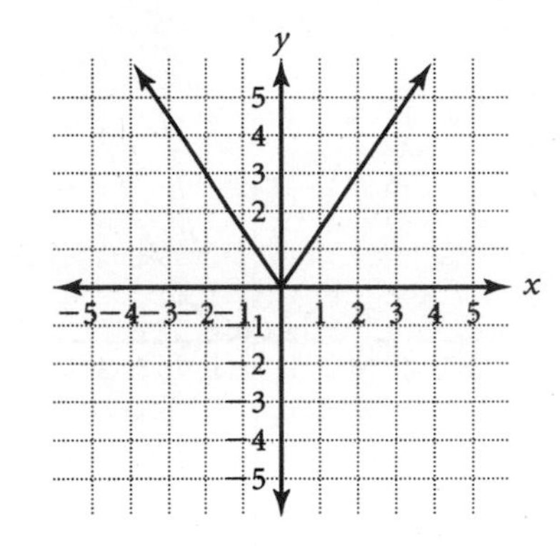

15. $y = x^2$

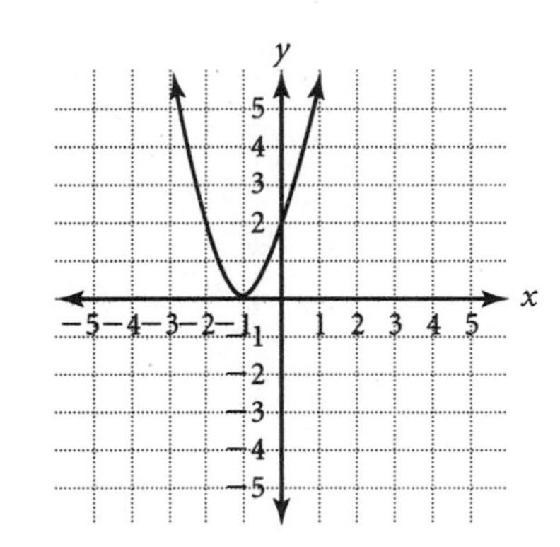

16. $y = \dfrac{1}{x}$

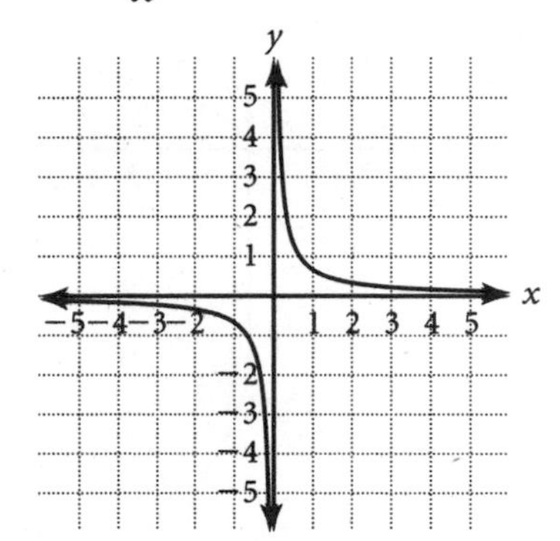

Answers

Lesson 14.3
Level C

1. $y = x^2$; $\left(x, \frac{1}{3}y\right)$;

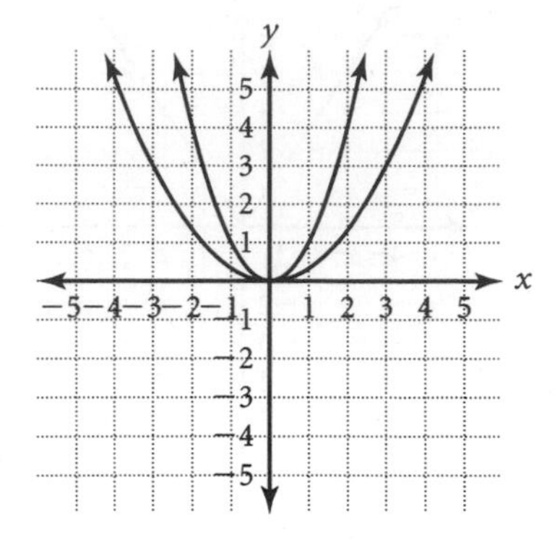

2. $y = |x|$; $\left(\frac{1}{2}x, y\right)$;

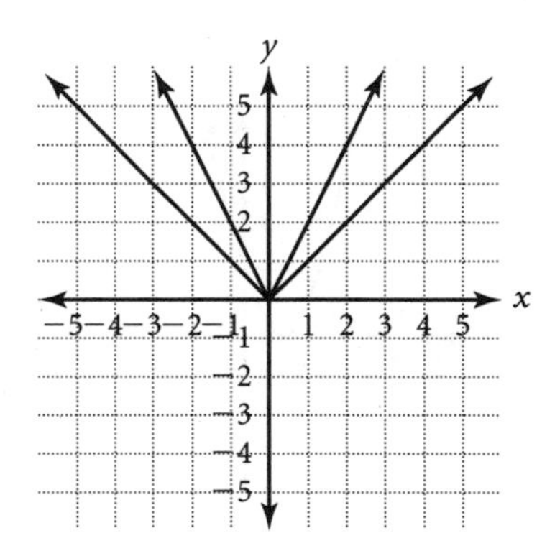

3. $y = \sqrt{x}$; $(x, 3y)$;

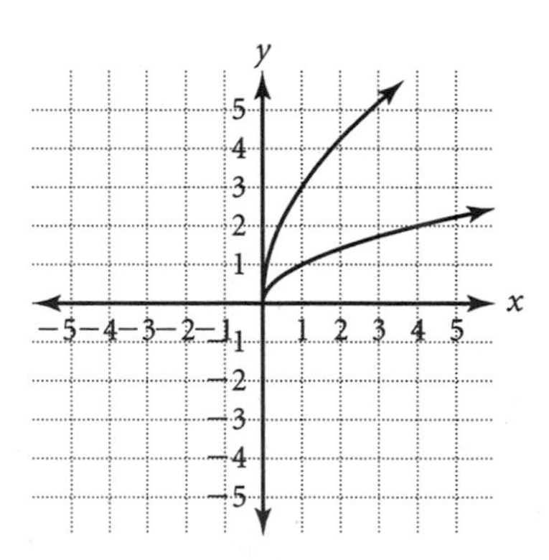

4. $y = 3x^2$
5. $y = \sqrt{\frac{1}{2}x}$
6. $y = \left|\frac{1}{3}x\right|$
7. The y-value is compressed by $\frac{1}{2}$.
8. The y-value is stretched by 2.
9. The x-value is stretched by 2.
10. The x-value is compressed by $\frac{1}{2}$.

Lesson 14.4
Level A

1. either
2. x-axis
3. y-axis
4. y-axis
5. x-axis
6. y-axis
7. either
8. y-axis
9. either
10. x-axis

11. reflected across x-axis

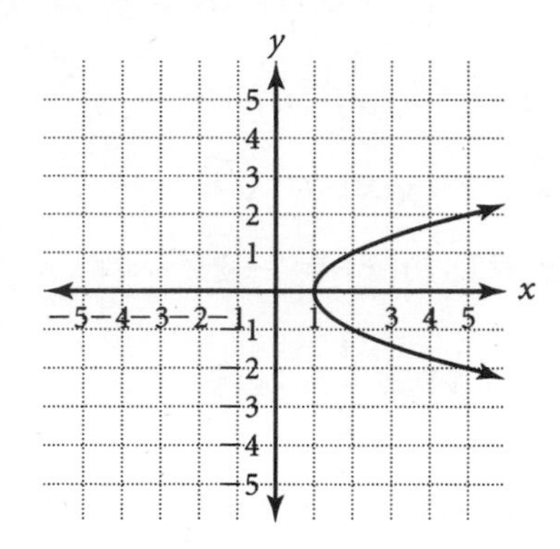

12. reflected across x-axis

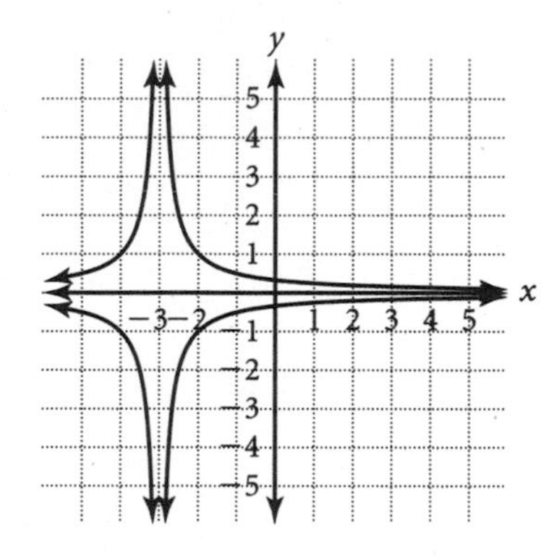

13. reflected across y-axis

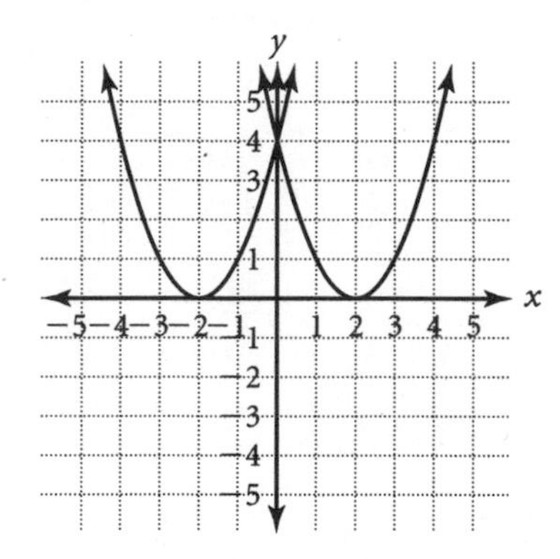

14. reflected across x-axis

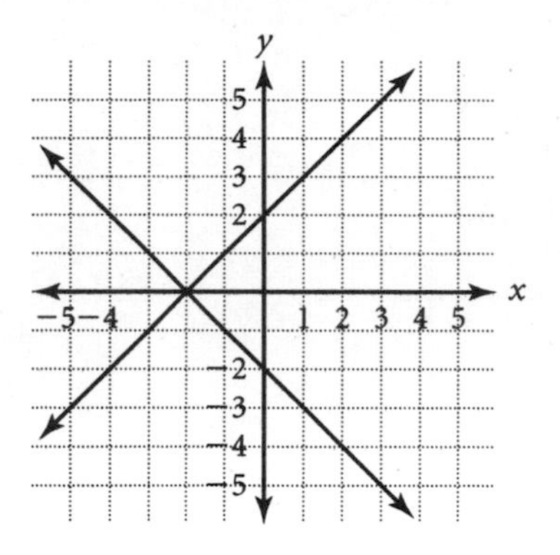

Lesson 14.4
Level B

1. left 1 unit, reflected over x-axis

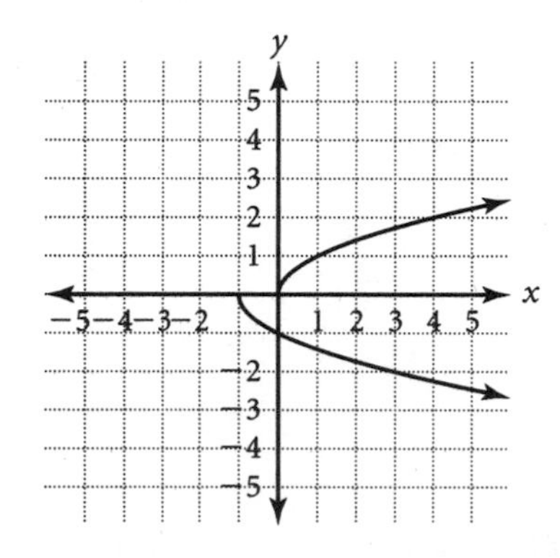

2. right 2 units, reflected across the y-axis

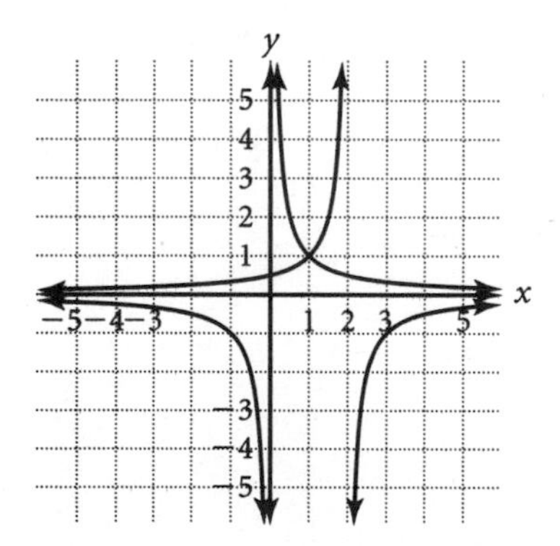

3. left 2 units, reflected over the x-axis

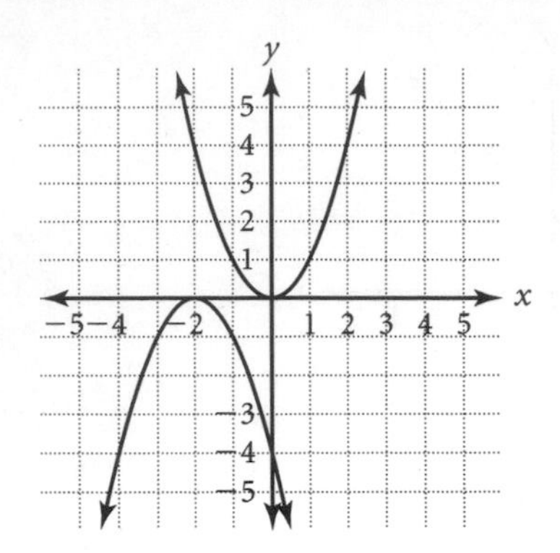

4. right 3 units, reflected over the x-axis

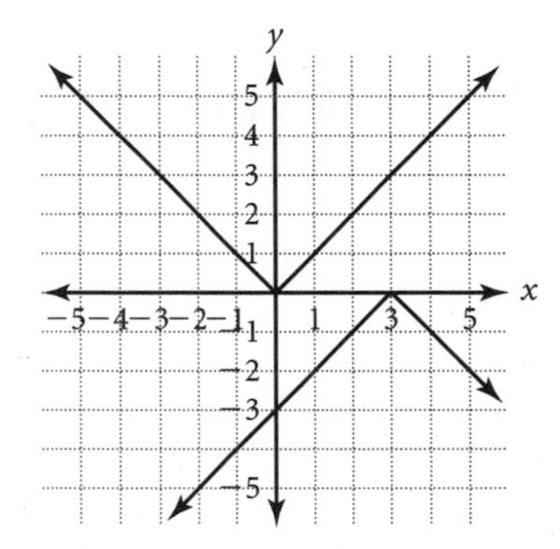

5. vertical reflection

6. horizontal reflection

7. $y = -2x - 1; y = -2x + 1$

8. $y = \sqrt{-x - 1}; y = -\sqrt{x - 1}$

9. $y = (-x + 5)^2; y = -(x + 5)^2$

10. $y = |-2x|; y = -|2x|$

11. $y = \dfrac{1}{-x - 1}; y = \dfrac{-1}{x - 1}$

12. $y = \dfrac{3}{-x}; y = \dfrac{-3}{x}$

Lesson 14.4
Level C

1. $f(x) = -(x + 2)^2; f(x) = (-x + 2)^2$

2. $g(x) = -\sqrt{x} + 2; g(x) = \sqrt{-x} - 2$

3. $m(x) = -2x + 1; m(x) = -2x - 1$

4. $n(x) = \dfrac{-1}{2}x^2; n(x) = \dfrac{1}{2}(-x)^2$

5. $r(x) = -x - 5; r(x) = -x + 5$

6. $d(x) = \dfrac{-1}{2x}; d(x) = \dfrac{1}{-2x}$

7. $y = -x^2$

8. $y = \sqrt{-x}$

9. $y = -|x|$

10. $\left(\dfrac{1}{4}, -4\right)$

11. $\left(-\dfrac{1}{4}, 4\right)$

12. $\left(-\dfrac{1}{4}, -4\right)$

13. $\left(-\dfrac{1}{4}, -4\right)$

Answers

Lesson 14.5
Level A

1.

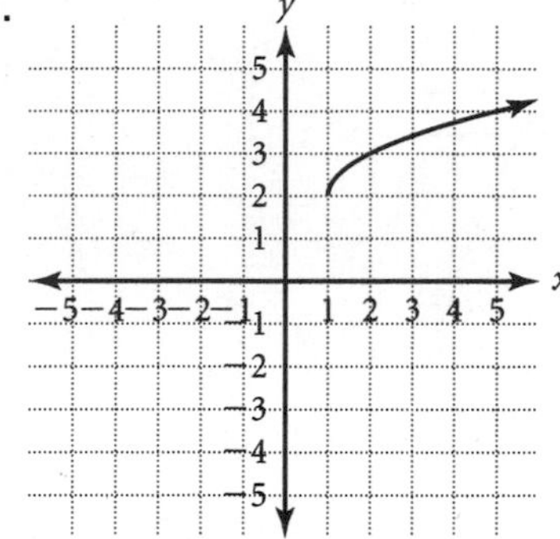

2.

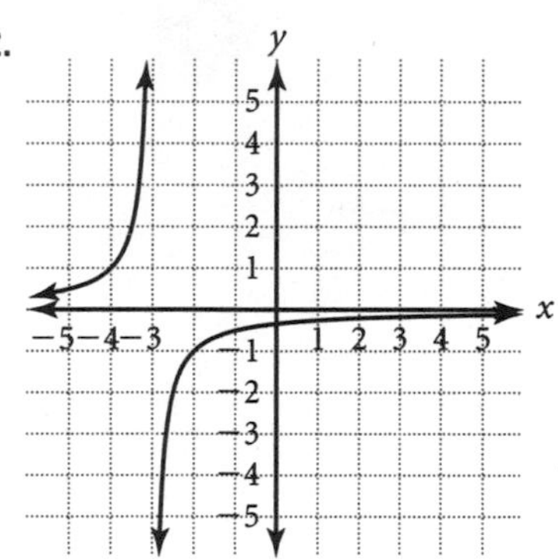

3.

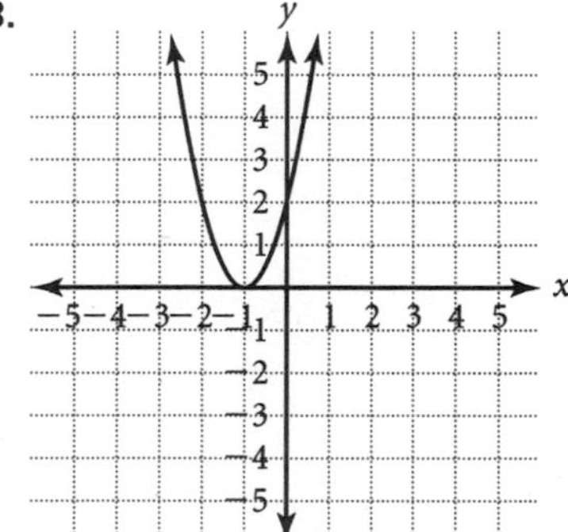

4.

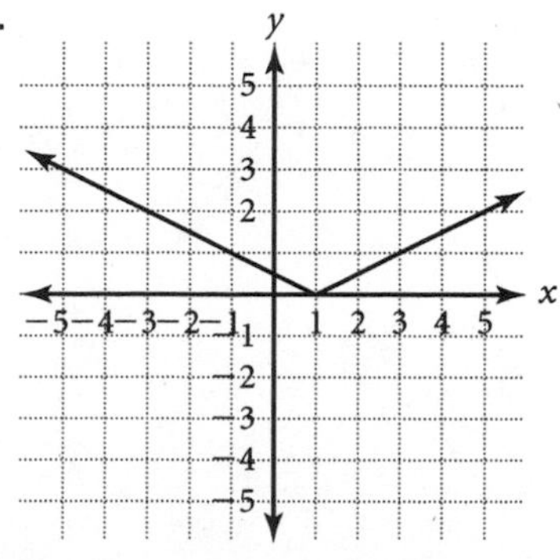

5.

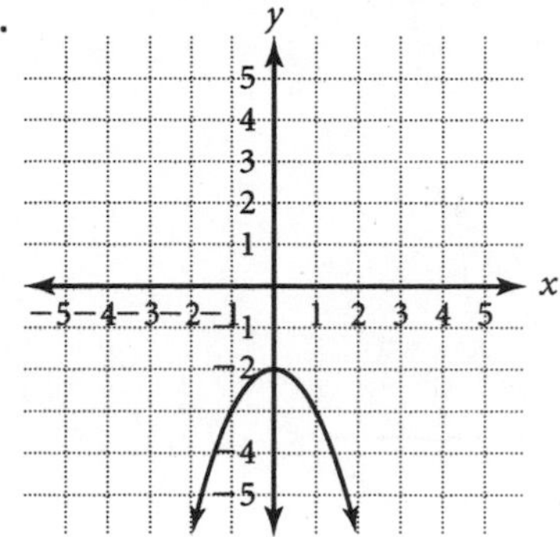

6.

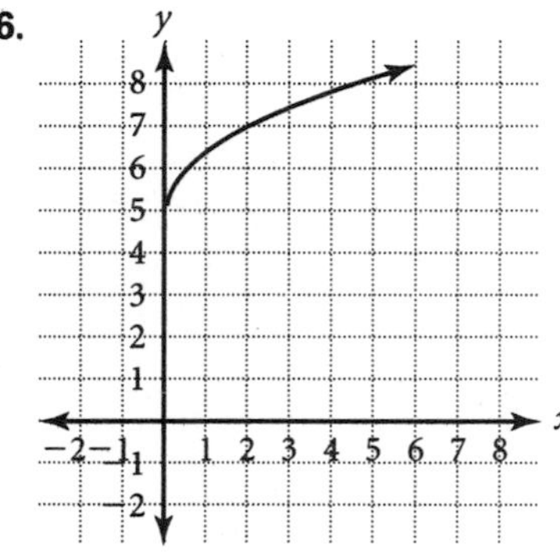

7. $(3, 2)$

8. $(-12, -2)$

9. $r = \dfrac{2000}{(h - 10)}; h > 10$

10. $r = \dfrac{1}{h};$

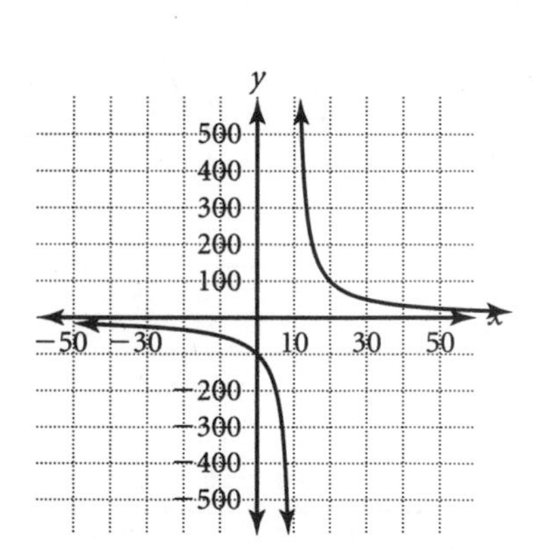

Answers

Lesson 14.5
Level B

1.

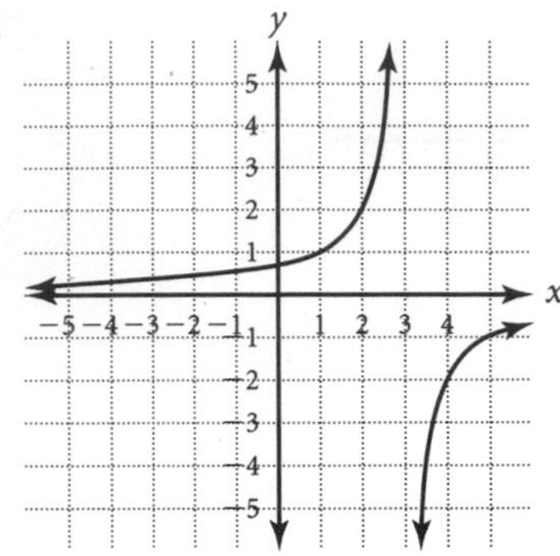

2.

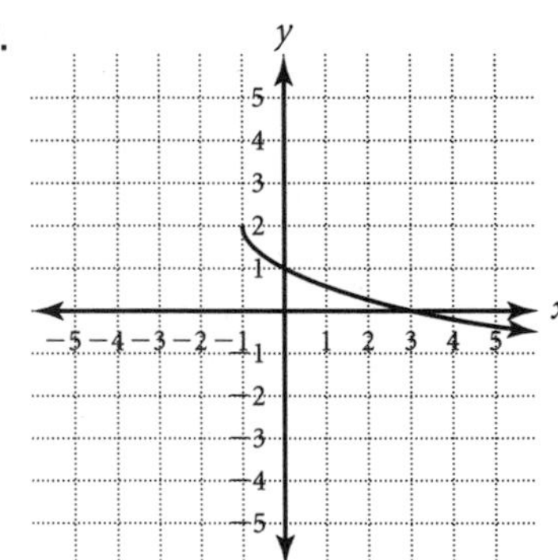

3.

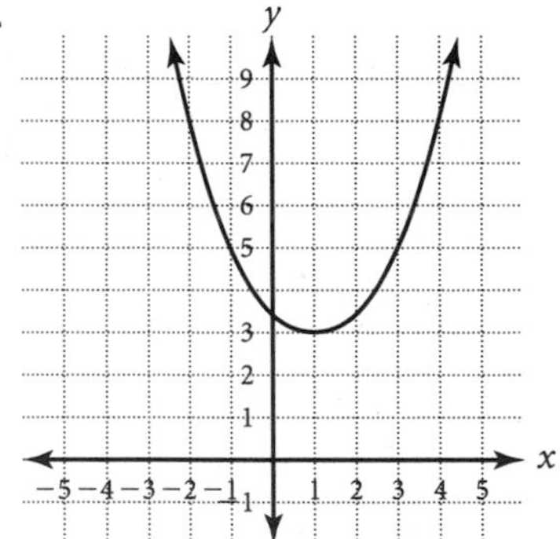

4.

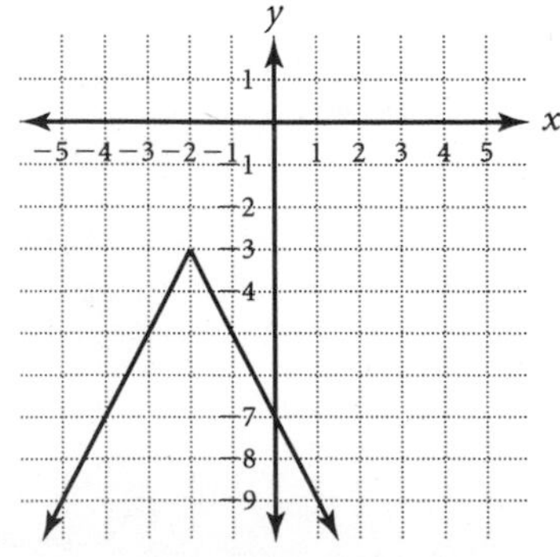

5.

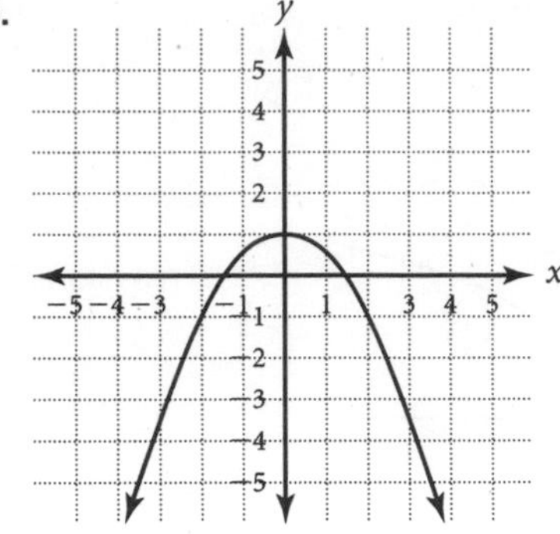

6.

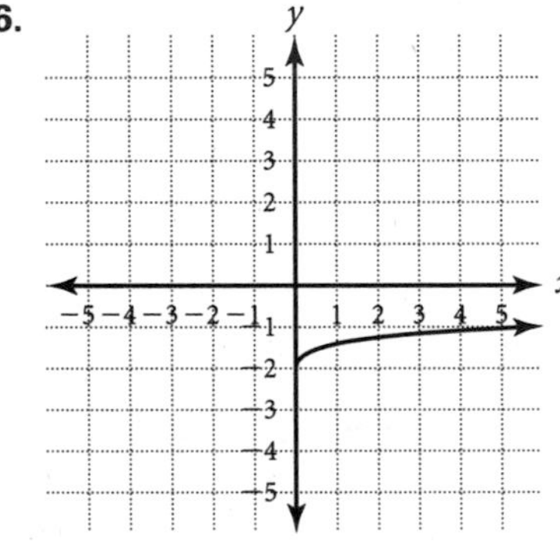

7. $(7, 6)$

8. $(-2, -6)$

9. $(-2, 10)$

Answers

Lesson 14.5
Level C

1.

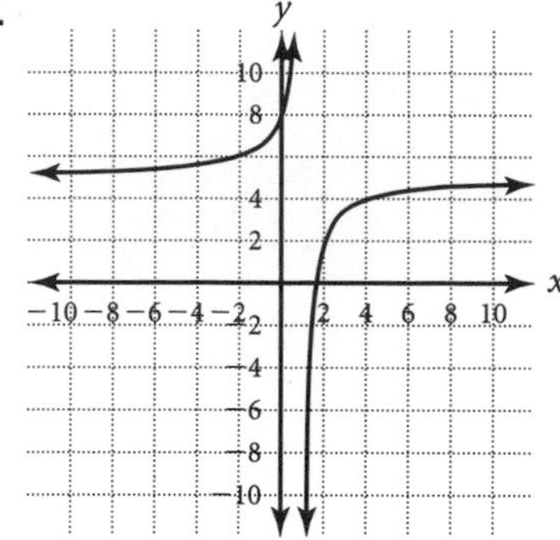

2.

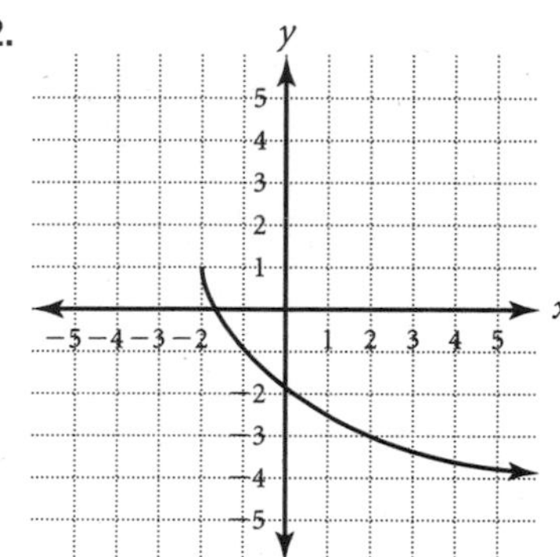

3.

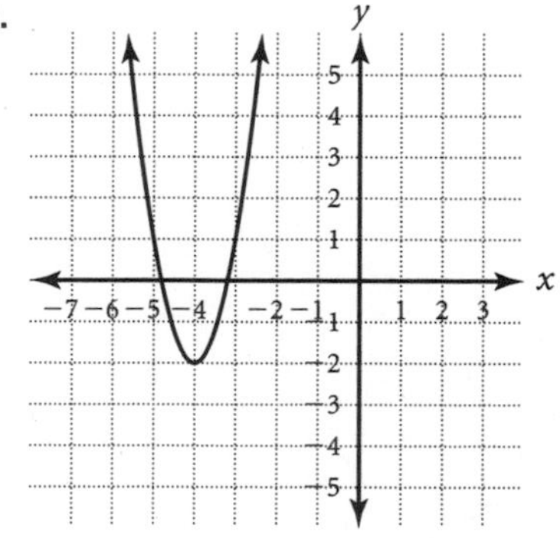

4.

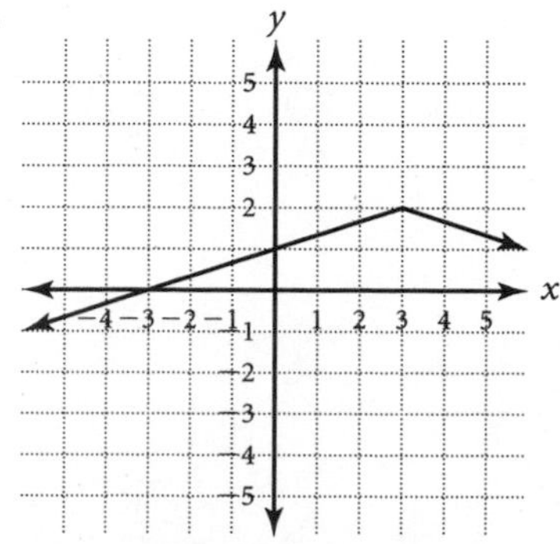

5.

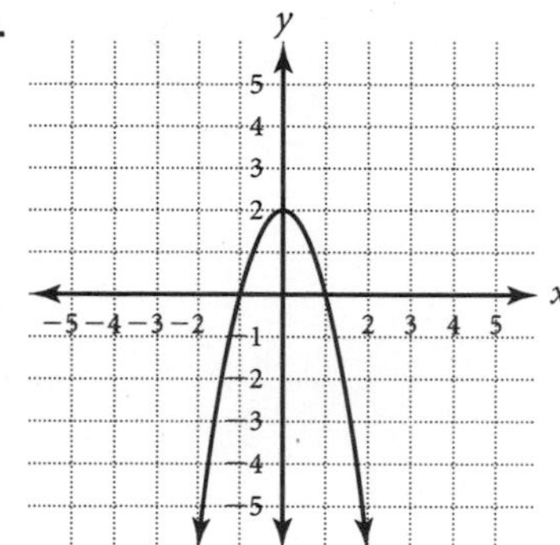

6.

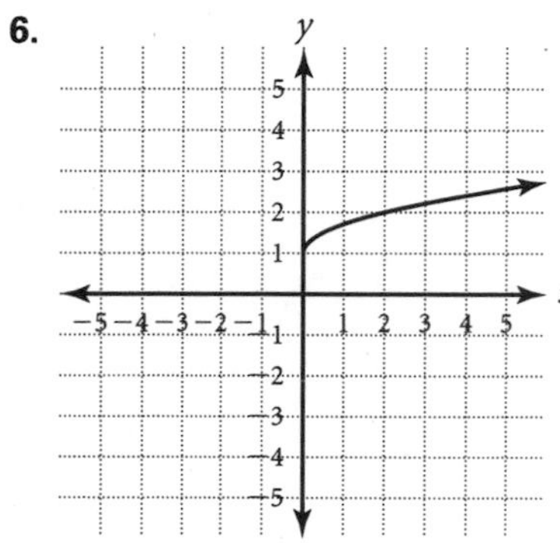

7. $(4, 3)$

8. $(-4, -4)$

9. $y = 2x^2 + 2$